AF509020

Kinematics and Robot Design II (KaRD2019) and III (KaRD2020)

Kinematics and Robot Design II (KaRD2019) and III (KaRD2020)

Editor

Raffaele Di Gregorio

MDPI • Basel • Beijing • Wuhan • Barcelona • Belgrade • Manchester • Tokyo • Cluj • Tianjin

Editor
Raffaele Di Gregorio
University of Ferrara (UNIFE)
Italy

Editorial Office
MDPI
St. Alban-Anlage 66
4052 Basel, Switzerland

This is a reprint of articles from the Special Issue published online in the open access journal *Robotics* (ISSN 2218-6581) (available at: http://www.mdpi.com).

For citation purposes, cite each article independently as indicated on the article page online and as indicated below:

LastName, A.A.; LastName, B.B.; LastName, C.C. Article Title. *Journal Name* **Year**, *Volume Number*, Page Range.

ISBN 978-3-0365-2928-8 (Hbk)
ISBN 978-3-0365-2929-5 (PDF)

Contents

About the Editor

Raffaele Di Gregorio is currently a Full Professor of machine mechanics at the Engineering Department of the University of Ferrara, Italy. He received an M.Sc. degree in nuclear engineering, an M.Sc. degree in mechanical engineering and an M.Sc. degree in automotive engineering from the Polytechnic University of Turin in 1982, 1985 and 1988, respectively, and a Ph.D. degree in applied mechanics from the University of Bologna, Italy, in 1992.

In 1983, he received a grant from the Italian Automotive Technical Association (ATA) for one year as a research fellow at the FIAT Research Center, Orbassano (Italy).

Between 1984 and 1992, he served as officer of the Technical Corp in the Italian Army: first, at the Military School of Turin and, successively, at the STAVECO (Military Vehicle Factory) of Bologna. In 1993, he taught at ITIS O. Belluzzi of Bologna, Italy. Since 1994, he has been part of the Engineering Department of the University of Ferrara, Italy.

His research interests include kinematics and dynamics of mechanisms and machines, biomechanics, robotics, vibration mechanics and vehicle mechanics.

He has been an author and co-author of about 80 technical papers published in refereed international journals and many papers published in conference proceedings.

He has been an "observer/member" of the IFToMM Technical Committee "Computational Kinematics" since 2007 and "member" of the IFToMM Permanent Commission "Standardization of Terminology" since 2009. Moreover, he is an ASME member and served as a "general member" of the ASME Mechanisms and Robotics Committee between 2007 and 2012; since 2013, he has continued to serve this committee as a past member. In these roles, he has actively participated in the organization of International Conferences on Mechanisms and/or Robotics and served as the chair of the "ASME—Mechanisms & Robotics' Honors and Awards Sub-Committee" from 2014 to 2016.

In addition to being an editorial board member of many international journals, he serves as associate editor of "ACTAPRESS International Journal of Robotics and Automation" and "Mechanism and Machine Theory" and served as associate editor of the "ASME Journal of Mechanisms and Robotics" from July 2014 to October 2020. Furthermore, he has been a "Guest Editor" of the Special Issue series "Kinematics and Robot Design" published by the journal Robotics since 2018 and served as the "Editor in Chief" of the "Journal of Mechanical Engineering and Automation" since 2012.

An updated list of his publication together with more details can be found on the following websites:

UNIFE: http://docente.unife.it/raffaele.digregorio-en

ResearchGate: http://www.researchgate.net/profile/Raffaele_Di_Gregorio

Scopus Author ID: 7003465861

ORCID: https://orcid.org/0000-0003-3925-3016

ResearchID: http://www.researcherid.com/rid/F-9316-2014

Google Scholar: http://scholar.google.it/citations?user=n_Ml9FEAAAAJ&hl=it

A list of his professional links can be found at

http://www.linkedin.com/profile/view?id=128075898&trk=tab_pro

Preface to "Kinematics and Robot Design II (KaRD2019) and III (KaRD2020)"

The Special Issue series on "Kinematics and Robot Design" (KaRD series) is hosted by the open access journal Robotics. KARD series started in 2018 and is now an open environment where researchers can present their studies and discuss all topics focused on the many aspects that involve kinematics in the design of robotic/automatic systems by using also supplementary multi-media materials uploaded during submission. Even though KaRD series publishes one Special Issue per year, all received papers are peer-reviewed as soon as they are submitted, and, if accepted, they are immediately published in Robotics and appear on the website of the KaRD issue. The open access nature of this series allows authors to easily share their papers and accompanying supplementary materials with the reference scientific community on many platforms where they can receive comments from other researchers. Furthermore, upon submission, authors can publish their preprint online at https://www.preprints.org/ for receiving comments to consider together with reviewer remarks when preparing revised versions of their papers. In short, the KaRD series is an "agora" where researchers meet with one another and efficiently exchange their experiences. These characteristics distinguish the KaRD series from numerous serial conferences/publications related to mechanisms and robotics.

KaRD2019 (https://www.mdpi.com/journal/robotics/special_issues/KRD2019) and KaRD2020 (https://www.mdpi.com/journal/robotics/special_issues/KaRD2020) are the second and third issues of the KaRD series. Beginning with KaRD2020, the activity of the Guest Editor has been supervised/supported by a scientific committee, as it is the case in all well-established serial international conferences/publications. The committee comprises the following members whose service is gratefully acknowledged: Massimo Callegari (Polytechnic University of Marche, Italy); Juan Antonio Carretero (University of New Brunswick, Canada), Yan Chen (Tianjin University; China), Daniel Condurache ("Gheorghe Asachi" Technical University of Iași, Romania); Xilun Ding (Beijing University of Aeronautics & Astronautics, China); Mary Frecker (Penn State—College of Engineering, USA); Clement Gosselin (Laval University, Canada); Just Herder (TU Deft, Netherlands); Larry Howell (Brigham Young University, USA); Xianwen Kong (Heriot-Watt University, UK); Pierre Larochelle (South Dakota School of Mines & Technology, USA); Giovanni Legnani (University of Brescia, Italy); Haitao Liu (Tianjin University, China); Daniel Martins (Universidade Federal de Santa Catarina, Brazil); Andreas Mueller (Johannes Kepler Universität, Austria); Andrew Murray (University of Dayton, USA), Leila Notash (Queen's University, Canada); Matteo Palpacelli (Polytechnic University of Marche, Italy); Alba Perez (Remy Robotics, Barcelona, Spain); Victor Petuya (University of the Basque Country, Spain), José Maria Rico Martinez (Universidad de Guanajuato, Mexico); Nina Robson (California State University, Fullerton, USA); Jon M. Selig (London South Bank University, UK); Bruno Siciliano (University of Naples Federico II, Italy); Tao Sun (Tianjin University, China), Yukio Takeda (Tokyo Institute of Technology, Japan); Federico Thomas (Institute of Industrial Robotics, Spain); Volkert Van Der Wijk (TU Deft, Netherlands).

KaRD2019, together with KaRD2020, received 22 papers and, after the peer-review process, accepted only 17 papers. This volume collects 17 accepted papers and is organized as follows. The first six papers [1–6] investigate or review the kinematics of parallel/serial manipulators from a theoretical point of view. Then, the successive four papers [7–10] address medical robotics issues; two papers [11, 12] that follow deal with performance analyses. Finally, paper [13] deals with education in robotics, and the last four papers [14–17] deal with robot design.

[1] Ruggiu, M.; Müller, A. Investigation of Cyclicity of Kinematic Resolution Methods for Serial and Parallel Planar Manipulators. Robotics 2021, 10(1), 9. doi: 10.3390/robotics10010009.

[2] Shen, H.; Xu, Q.; Li, J.; Yang, T. The Effect of the Optimization Selection of Position Analysis Route on the Forward Position Solutions of Parallel Mechanisms. Robotics 2020, 9(4), 93. doi: 10.3390/robotics9040093.

[3] Di Gregorio, R. A Review of the Literature on the Lower-Mobility Parallel Manipulators of 3-UPU or 3-URU Type. Robotics 2020, 9(1), 5. doi: 10.3390/robotics9010005.

[4] Ruggiu, M.; Kong, X. Reconfiguration Analysis of a 3-DOF Parallel Mechanism. Robotics 2019, 8(3), 66. doi: 10.3390/robotics8030066.

[5] Schappler, M.; Tappe, S.; Ortmaier, T. Modeling Parallel Robot Kinematics for 3T2R and 3T3R Tasks Using Reciprocal Sets of Euler Angles. Robotics 2019, 8(3), 68. doi: 10.3390/robotics8030068.

[6] Huber, G.; Wollherr, D. Efficient Closed-Form Task Space Manipulability for a 7-DOF Serial Robot. Robotics 2019, 8(4), 98. doi: 10.3390/robotics8040098.

[7] Oka, T.; Solis, J.; Lindborg, A.; Matsuura, D.; Sugahara, Y.; Takeda, Y. Kineto-Elasto-Static Design of Underactuated Chopstick-Type Gripper Mechanism for Meal-Assistance Robot. Robotics 2020, 9(3), 50. doi: 10.3390/robotics9030050.

[8] Yamine, J.; Prini, A.; Nicora, M.; Dinon, T.; Giberti, H.; Malosio, M. A Planar Parallel Device for Neurorehabilitation. Robotics 2020, 9(4), 104. doi: 10.3390/robotics9040104.

[9] Filippeschi, A.; Griffa, P.; Avizzano, C. Kinematic Optimization for the Design of a Collaborative Robot End-Effector for Tele-Echography. Robotics 2021, 10(1), 8. doi: 10.3390/robotics10010008.

[10] Castelli, K.; Carnevale, M.; Giberti, H. Development of an Automatic Robotic Procedure for Machining of Skull Prosthesis. Robotics 2020, 9(4), 108. doi: 10.3390/robotics9040108.

[11] Bussola, R.; Legnani, G.; Callegari, M.; Palmieri, G.; Palpacelli, M. Simulation Assessment of the Performance of a Redundant SCARA. Robotics 2019, 8(2), 45. doi: 10.3390/robotics8020045.

[12] Schulz, S. Performance Evaluation of a Sensor Concept for Solving the Direct Kinematics Problem of General Planar 3-RPR Parallel Mechanisms by Using Solely the Linear Actuators' Orientations. Robotics 2019, 8(3), 72. doi: 10.3390/robotics8030072.

[13] Castelli, K.; Giberti, H. Additive Manufacturing as an Essential Element in the Teaching of Robotics. Robotics 2019, 8(3), 73. doi: 10.3390/robotics8030073.

[14] Di Gregorio, R. A Novel 3-URU Architecture with Actuators on the Base: Kinematics and Singularity Analysis. Robotics 2020, 9(3), 60. doi: 10.3390/robotics9030060.

[15] Palpacelli, M.; Carbonari, L.; Palmieri, G.; D'Anca, F.; Landini, E.; Giorgi, G. Functional Design of a 6-DOF Platform for Micro-Positioning. Robotics 2020, 9(4), 99. doi: 10.3390/robotics9040099.

[16] Achilli, G.; Valigi, M.; Salvietti, G.; Malvezzi, M. Design of Soft Grippers with Modular Actuated Embedded Constraints. Robotics 2020, 9(4), 105. doi: 10.3390/robotics9040105.

[17] Castelli, K.; Zaki, A.; Dmytriyev, Y.; Carnevale, M.; Giberti, H. A Feasibility Study of a Robotic Approach for the Gluing Process in the Footwear Industry. Robotics 2021, 10(1), 6. doi: 10.3390/robotics10010006.

Raffaele Di Gregorio
Editor

 *robotics*

Article

Investigation of Cyclicity of Kinematic Resolution Methods for Serial and Parallel Planar Manipulators

Maurizio Ruggiu [1,*,†] **and Andreas Müller** [2,†]

1   Department of Mechanical, Chemical and Materials Engineering, University of Cagliari, Via Marengo, 2, 09123 Cagliari, Italy
2   Institut for Robotics, Johannes Kepler University, Altenberger Straße, 69-4040 Linz, Austria; a.mueller@jku.at
*   Correspondence: maurizio.ruggiu@dimcm.unica.it
†   These authors contributed equally to this work.

Abstract: Kinematic redundancy of manipulators is a well-understood topic, and various methods were developed for the redundancy resolution in order to solve the inverse kinematics problem, at least for serial manipulators. An important question, with high practical relevance, is whether the inverse kinematics solution is cyclic, i.e., whether the redundancy solution leads to a closed path in joint space as a solution of a closed path in task space. This paper investigates the cyclicity property of two widely used redundancy resolution methods, namely the projected gradient method (PGM) and the augmented Jacobian method (AJM), by means of examples. Both methods determine solutions that minimize an objective function, and from an application point of view, the sensitivity of the methods on the initial configuration is crucial. Numerical results are reported for redundant serial robotic arms and for redundant parallel kinematic manipulators. While the AJM is known to be cyclic, it turns out that also the PGM exhibits cyclicity. However, only the PGM converges to the local optimum of the objective function when starting from an initial configuration of the cyclic trajectory.

Keywords: kinematic redundancy; cyclicity; augmented Jacobian method; projected gradient method

Citation: Ruggiu, M.; Müller, A. Investigation of Cyclicity of Kinematic Resolution Methods for Serial and Parallel Planar Manipulators. *Robotics* **2021**, *10*, 9. https://doi.org/10.3390/robotics10010009

Received: 20 October 2020
Accepted: 30 December 2020
Published: 3 January 2021

Publisher's Note: MDPI stays neutral with regard to jurisdictional claims in published maps and institutional affiliations.

1. Introduction

Kinematically redundant serial kinematic manipulators (SKM) are gaining importance for dedicated solution in flexible automation due their advantageous properties. Only a few kinematically redundant parallel kinematic manipulators (PKM) have been proposed, and they have not yet seen a similar application in industry. Kinematic redundancy offers various advantages, and it can in particular be exploited to improve dexterity, to avoid singularities, to circumvent obstacles, and eventually to increase the workspace. These advantages are accompanied with an increased complexity of the control and with increased costs, however. The complexity is due to the non-uniqueness of the solution to the geometric inverse kinematics problem (IKP), which consists of finding a vector $\mathbf{q} \in \mathbb{V}^n$ of generalized coordinates (joint coordinates in SKM) for a given vector $\mathbf{p} \in \mathbb{R}^m$ of end-effector (EE) coordinates satisfying the forward kinematics relation

$$\mathbf{p} = f(\mathbf{q}) \tag{1}$$

where f is the forward kinematics mapping. The corresponding velocity IKP, in a given pose $\mathbf{q}$, is to find a vector of generalized coordinates $\dot{\mathbf{q}}$ for given EE-velocity $\dot{\mathbf{p}}$ satisfying the velocity forward kinematics

$$\dot{\mathbf{p}} = \mathbf{J}(\mathbf{q})\dot{\mathbf{q}} \tag{2}$$

where $\mathbf{J}(\mathbf{q}) \in \mathbb{R}^{m \times n}$ is the forward kinematics Jacobian.

A manipulator is kinematically redundant if $r = n - m > 0$, where r is called the degree of redundancy, n is the degree of freedom (DOF), and m denotes the dimension of the task space. For a redundant manipulator, the Jacobian is not square, and the inverse kinematics has r-dimensional solution set.

A number of solution techniques for solving the geometric IKP problem for redundant manipulators have been introduced in the last three decades, almost exclusively focusing on SKM [1]. Most of them determine a solution of the geometric inverse kinematics problem from a solution of the velocity inverse kinematics problem with the aim to provide an online solution scheme that can be executed on the robot controller with a well-defined computation time. As a consequence, all these method yield a local solution rather than the global solution of the geometric inverse kinematics problem. The two most widely used local solution schemes are the following:

1. *Null space methods:* A particular solution $\dot{\mathbf{q}}$ of (2) is determined by adding a vector in the null-space of the Jacobian. The latter is usually the gradient of a scalar objective function that is to be maximized (or minimized).
2. *Task augmentation methods:* Redundancy is eliminated by adding r auxiliary tasks, in order to make the overall system non-redundant.

Most industrial use cases involve repetitive tasks. That is, the EE performs a cyclic motion. An obvious requirement for safe application of redundant robots is that the IK solution is well-determined in the sense that for given pose $\mathbf{p}$ of the EE the IK solution scheme always yields the same pose $\mathbf{q}$ of the robot. This property is called cyclicity, which means that for a closed EE-path the IK solution method gives a closed path in joint space. It is well-known that null-space methods cannot be guaranteed to be cyclic [2]. This is an important drawback of the null-space methods since the motion of the chain is unpredictable during the motion. The cyclicity problem was first analyzed in a differential geometric framework in [3] and then explored in [4,5] for the pseudoinverse solution. In [6] the authors were able to design a feedback control law to produce convergence of the joint motion towards a cyclic trajectory assuming the null space vector as a linear map of the EE trajectory.

While task augmentation methods are cyclic [6], they suffer from algorithmic singularities [7]. It must be mentioned that this topic was extensively used and studied for SKM (aiming to avoid collision and singularities). To the authors knowledge these same techniques have been applied very rarely to PKMs as in [8]. For the sake of the completeness, we must point out that recently these classical schemes have been applied to the hyper-redundant manipulators [9,10] as well as algorithms based on the machine learning have been applied to the redundancy resolution [11].

In this paper the projected gradient method (PGM) and the augmented Jacobian method (AJM) are compared by means of examples when applied to SKM and PKM. The methods are designed to minimize an objective function. In the following, and w.l.o.g., the objective function will be the sum of the squared joint variables. The comparison regards (1) the cyclicity of the joint trajectories, and (2) the dependency of the (cyclic) joint trajectory with respect to changes in the initial configuration. The sensitivity is an aspect that has not yet been addressed in the literature as well as the cyclicity of the solutions for PKMs. It will be shown that only the projected gradient method is capable of minimizing the objective function independently of the initial configurations. Results are reported for both SKM and PKM.

This is the first publication that addresses the cyclicity of PGM. The novel contribution is a numerical investigation and comparison of the cyclicity of PGM and AJM when applied to SKM as well as to PKM.

The paper is organized as follows. In Section 2 the two methods for redundancy resolution at the velocity level are presented. Numerical results are shown in Section 3. The paper concludes with a discussion in Section 4.

2. Redundancy Resolution Methods

Redundancy resolution of the differential kinematics of a manipulator is expressed as:

$$\dot{\mathbf{q}} = \mathbf{J}^{\#}\dot{\mathbf{p}}. \tag{3}$$

$\mathbf{J}^{\#}$ is the Moore–Penrose pseudoinverse of the manipulator Jacobian $\mathbf{J} \in \mathbb{R}^{m\times n}$. In order to avoid integration drift, an algorithm solution (CLIK: closed-loop inverse kinematics) of Equation (3) is widely adopted and the geometric inverse kinematic problem expressed as:

$$\dot{\mathbf{q}} = \mathbf{J}^{\#}(\dot{\mathbf{p}}_d + \mathbf{G}\mathbf{e}), \tag{4}$$

where $\mathbf{e} = \mathbf{p}_d - \mathbf{p}$ represents the error between the desired and actual EE location and $\mathbf{G}$ is a positive–definite diagonal gain matrix. Numerical integration of the Equation (4) given the initial configuration $\mathbf{q}_0$ followed by the solution of the forward position kinematic equations closes the feedback loop.

1. PGM

   The PGM exploits the fact that a general solution of the differential kinematics can be substituted to Equation (4) when a desired joint rate vector $\mathbf{v}$ is projected into the null space of $\mathbf{J}$:

$$\dot{\mathbf{q}} = \mathbf{J}^{\#}(\dot{\mathbf{p}}_d + \mathbf{G}\mathbf{e}) - \mathbf{P}\mathbf{v}, \tag{5}$$

   with $\mathbf{P} = (\mathbf{I} - \mathbf{J}^{\#}\mathbf{J})$ defined as the projector matrix. The added term generates self–motion of the kinematic chain without affecting the EE velocity. Vector $\mathbf{v}$ can be chosen in order to make a scalar objective function $h(\mathbf{q})$ stationary by using the gradient projection method, such that $\mathbf{v} = (\frac{\partial h(\mathbf{q})}{\partial \mathbf{q}})^T$, with $\mathbf{v} \in \mathbb{R}^n$. $h(\mathbf{q})$ may be any analytical differentiable function expressed in terms of the joint variables $\mathbf{q}$ only.

2. AJM

   A different approach is followed in the AJM. An additional constraint task is imposed to the original task of the EE. Following [7,12], the objective function $h(\mathbf{q})$ is projected onto the null space of $\mathbf{J}$ and imposed to be zero. Formally we can write:

$$\mathbf{g}(\mathbf{q}) = \mathbf{Z}^T\mathbf{v} = 0. \tag{6}$$

   $\mathbf{Z} \in \mathbb{R}^{n\times r}$ is an orthonormal basis for the null space of $\mathbf{J}$ and $\mathbf{v}$ is the gradient of $h(\mathbf{q})$ with respect to the joint variables as in the previous method. Therefore, Equation (6) yields r independent constraints keeping $h(\mathbf{q})$ at the extreme at each time of the trajectory starting from the initial configuration $\mathbf{q}_0$.

   The added Jacobian $\mathbf{J}_a \in \mathbb{R}^{r\times n}$ can simply obtained as

$$\mathbf{J}_a = \frac{\partial \mathbf{g}(\mathbf{q})}{\partial \mathbf{q}}, \tag{7}$$

   that leads to the CLIK kinematic resolution expressed as:

$$\dot{\mathbf{q}} = \mathbf{J}_{aug}^{-1}\left(\begin{bmatrix}\dot{\mathbf{p}}_d \\ 0\end{bmatrix} + \begin{bmatrix}\mathbf{G} & 0 \\ 0 & 0\end{bmatrix}\begin{bmatrix}\mathbf{e} \\ 0\end{bmatrix}\right), \quad \mathbf{J}_{aug} = \begin{bmatrix}\mathbf{J} \\ \mathbf{J}_a\end{bmatrix}. \tag{8}$$

The two methods are, therefore, designed to reach the same goal in order to have a direct comparison between them.

As pointed out in [6] for SKM, the cyclicity of the joints trajectories for a closed EE path can be reached exactly, asymptotically or it cannot even be reached. Exact cyclicity is obtained whenever $\mathbf{v} = \mathbf{L}(\mathbf{q})\dot{\mathbf{p}}$ at the steady–state, i.e., $\mathbf{e} = 0$. It has been shown in [6] that the augmented Jacobian can be expressed in this form thus proving its cyclicity. Furthermore, the cyclicity property may be verified on–line by calculating that the Lie brackets formed by any columns of the control matrix, $\mathbf{J}_{aug}^{-1}$, are linear combinations of the columns (involutive property) [13]. This calculation was indeed performed in this paper.

On the contrary, PGM cannot guarantee to lead to exact or even asymptotic cyclicity *a priori*. Indeed, it has been proved that control scheme in Equation (4) can only be asymptotic cyclic with $\mathbf{v} = \mathbf{K}(\mathbf{q}_{cyclic} - \mathbf{q})$, with $\mathbf{q}_{cyclic}$ representing the joint trajectory to converge to in order to reach repeatability.

3. Numerical Simulations

PGM and AJM are implemented for the serial manipulators 4R and the parallel 2RRP planar manipulators and for a 6R industrial UR10e robot. The planar manipulators have a degree of redundancy $r = 2$, while the 6DOF arm has $r = 3$. The end effectors of the manipulators can track a circle at a given constant speed.

For all the manipulators, the objective function to be minimized was chosen to be:

$$h(\mathbf{q}) = \sum_{i=1}^{n} q_i^2, \tag{9}$$

with $q_i \equiv \theta_i$ for the SKMs and $\mathbf{q} = [\psi_1, \rho_1, \psi_2, \rho_2]$ for the PKM. The simple nature of $h(\mathbf{q})$ strongly simplifies the computations. In the PGM method, indeed, only the analytical calculation of the gradient of the objective function is required. Instead, AJM requires the analytical calculation of the gradient of $\mathbf{g}(\mathbf{q})$ in Equation (6) to obtain the added rows of the Jacobian. First, an analytical calculation of $\mathbf{Z}$ is needed. This can be avoided using an alternative procedure proposed in [14] to calculate only $\mathbf{Z}$ numerically.

The geometrical data and the data of the EE trajectory are given in Tables 1–3.

Table 1. 4R. *a*: links length; (x_0, y_0), *R*: coordinates of the centre and radius of the EE trajectory.

$a = 0.5$ m	$x_0 = 1.0$ m	$y_0 = 1.0$ m	$R = 0.5$ m

Table 2. 2RRP. *a*: links length; *H*: base width, (x_0, y_0), *R*: coordinates of the centre and radius of the EE trajectory.

$a = 0.2$ m	$H = 1$ m	$x_0 = 0.5$ m	$y_0 = 0.5$ m	$R = 0.1$ m

Table 3. 6R. (x_0, y_0, z_0), *R*: coordinates of the centre and radius of the EE trajectory. EE trajectory on the plane $y = -0.256$ m. $d_1, a_2, a_3, d_4, d_5, d_6$: *Denavit-Hartenberg* geometrical parameters.

$x_0 = 0.572$ m	$y_0 = -0.256$ m	$z_0 = 0.496$ m	$R = 0.153$ m	$d_1 = 0.127$ m
$a_2 = 0.612$ m	$a_3 = 0.572$ m	$d_4 = 0.164$ m	$d_5 = 0.116$ m	$d_6 = 0.092$ m

3.1. Serial 4R

The Jacobian of the manipulator in Figure 1a is trivial and it is not shown in the paper. Instead, Equation (10) shows $\mathbf{g}(\mathbf{q})$ used in the AJM:

$$\mathbf{g}(\mathbf{q}) = \frac{2}{(s_{234} + s_{23} + s_2)} \left(\begin{bmatrix} (\theta_3 - \theta_2)(s_{234} + s_{23}) + (\theta_1 - \theta_2)(s_{34} + s_3) + \theta_3 s_2 \\ (\theta_4 - \theta_2)s_{234} + (\theta_1 - \theta_2)(s_{34} + s_4) + \theta_4(s_{23} + s_2) \end{bmatrix} \right) \tag{10}$$

$\mathbf{g}(\mathbf{q})$ is then differentiated with respect to the joint angles to obtain $\mathbf{J}_a \in \mathbb{R}^{2 \times 4}$.

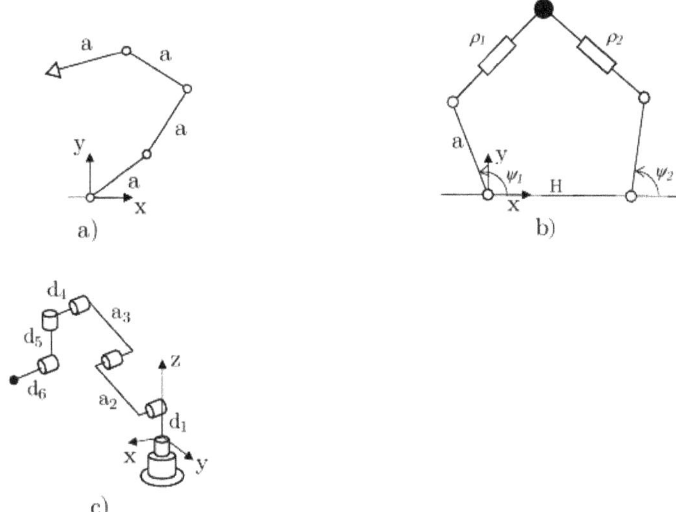

Figure 1. The test cases: (**a**) Planar 4R, (**b**) Planar 2RRP, (**c**) UR10 industrial 6R robot.

In Figure 2 we show $h(\mathbf{q})$ from both the methods for the planar 4R. The initial configuration $\mathbf{q}_0$ was calculated such to be a solution of the inverse position problem and that $\mathbf{g}(\mathbf{q}_0) = \mathbf{0}$.

From Figure 2 we see that $h(\mathbf{q})$ from the methods are in very good agreement. Both can keep the objective function stationary. Accordingly, also the joint trajectories of the two methods, shown in Figure 3, only exhibit small differences.

From Figure 3 we also see that both methods produce cyclic joints trajectories. In the case of the PGM we can verify that only by inspection while for the AJM we performed the numerical calculations of the Lie brackets associated to the augmented Jacobian. To have cyclicity we need to have that for any two columns $\mathbf{c}_i$ and $\mathbf{c}_j$ of $\mathbf{J}_{aug}^{-1}$, their Lie bracket $\mathcal{L}_k$ is a linear combination of the columns of the *control matrix* $\mathbf{J}_{aug}^{-1}$. In this case, for each step of calculation we obtain $\mathcal{L}_k$ of the order of 10^{-15} proving the cyclicity. Figure 4 shows the objective function whenever $\mathbf{g}(\mathbf{q}_0^*) \neq \mathbf{0}$ as $\mathbf{q}_0^* = \mathbf{q}_0 + \boldsymbol{\epsilon}_v$ with $\boldsymbol{\epsilon}_v = \epsilon[1, -1, 1, -1]$. As expected, the results show that AJM can only keep the constraint as it is, while PGM quickly reaches the minimum.

The inability of AJM to converge to the minimum is stressed in Figure 5 where $h(\mathbf{q})$ is plotted with different initial conditions.

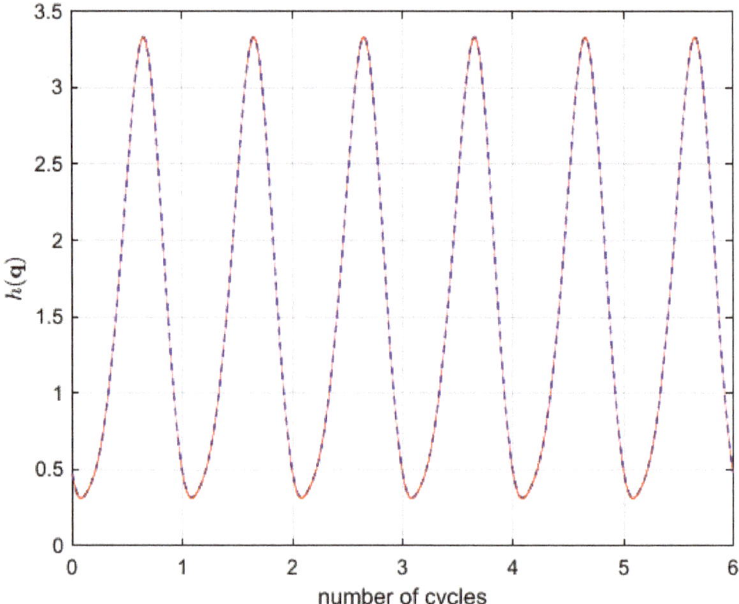

Figure 2. $h(\mathbf{q})$ in 4$\underline{\text{R}}$. PGM: − (red line), AJM: −− (blue line).

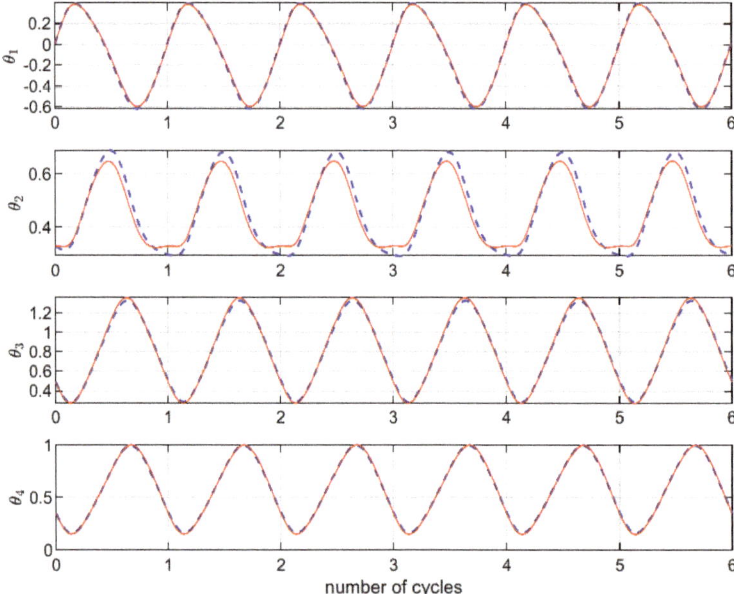

Figure 3. θ_i in 4$\underline{\text{R}}$. PGM: − (red line), AJM: −− (blue line).

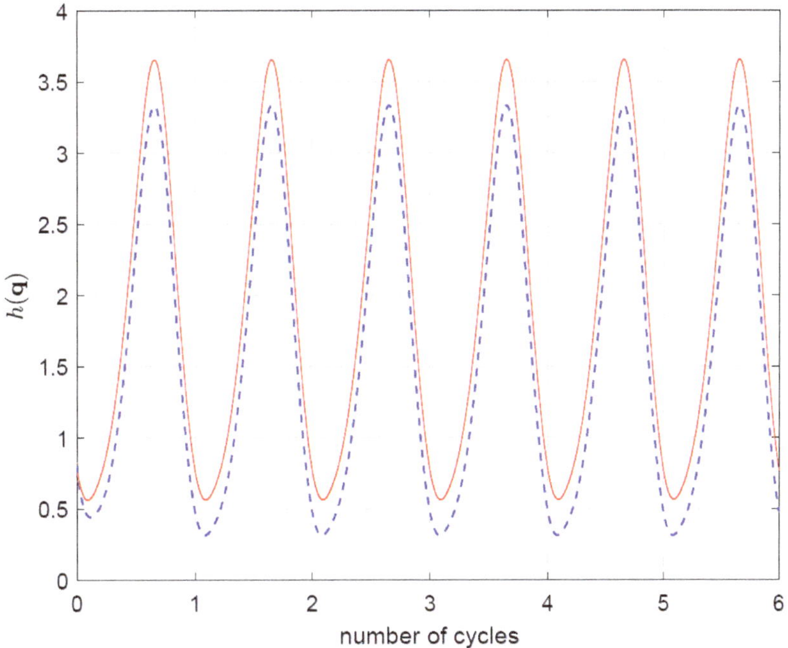

Figure 4. $h(\mathbf{q})$ in 4$\underline{\mathrm{R}}$ with $\epsilon = 0.3$. PGM: $-$ (red line), AJM: $--$ (blue line).

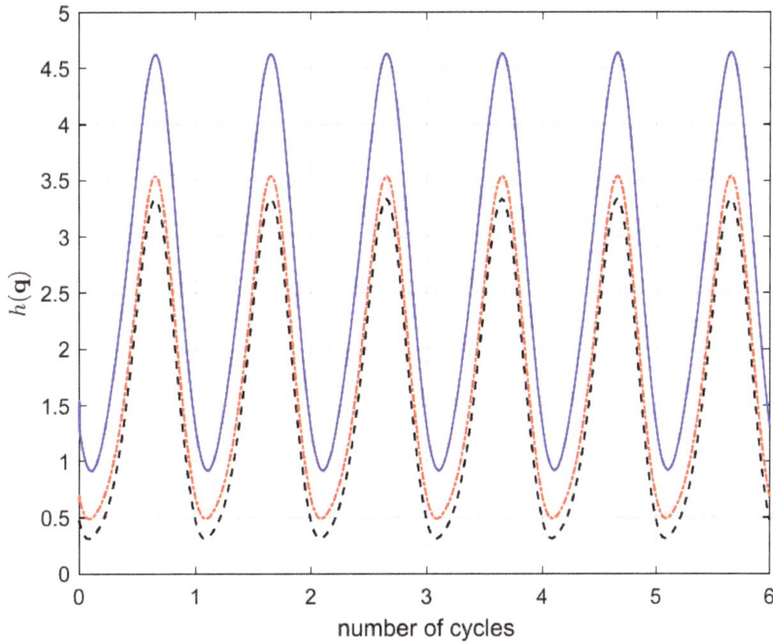

Figure 5. AJM: $h(\mathbf{q})$ in 4$\underline{\mathrm{R}}$. $\epsilon = 0$: $--$ (black line), $\epsilon = 0.25$: $-.-$ (red line), $\epsilon = 0.5$: $-$ (blue line).

We then tested the robustness of PGM as the initial conditions varied. In Figure 6, $h(\mathbf{q})$ is computed by PGM when starting from different initial configurations. The method allows reaching the minimum even when $\epsilon = 1.5$. In this case the method converges in $N_c = 1.2570$ cycles chosen $\Delta h = (h_\epsilon - h_0) = 1/1000$ as threshold. Similarly, the joints trajectories coincide quickly. This is a striking characteristic of this method that proves to be robust with respect to the initial configuration of the motion that can be inaccurate for some reasons. In this case the method can reach the minimum curve even when the gap from the correct initial configuration is about $90°$.

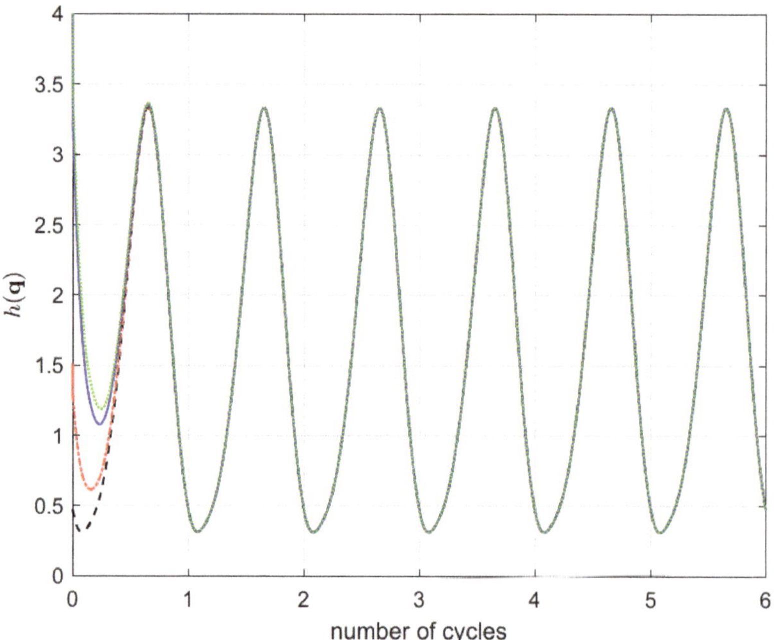

Figure 6. PGM: $h(\mathbf{q})$ in 4<u>R</u>. $\epsilon = 0$: $--$ (black line), $\epsilon = 0.5$: $-.-$ (red line), $\epsilon = 1.0$: $-$ (blue line), $\epsilon = 1.5$: $\cdots$ (green line).

3.2. Parallel 2<u>R</u>RP

The kinematic relation of the PKM is [15]:

$$\mathbf{J}_p \dot{\mathbf{p}} = \mathbf{J}_q \dot{\mathbf{q}}. \tag{11}$$

The forward kinematics Jacobian in (2) is thus $\mathbf{J} = \mathbf{J}_p^{-1}\mathbf{J}_q$. For the planar PKM in Figure 1b), the Jacobians $\mathbf{J}_p$ and $\mathbf{J}_q$ are

$$\mathbf{J}_p = \begin{bmatrix} x - ac_{\psi_1} & y - as_{\psi_1} \\ (x - H) - ac_{\psi_2} & y - ac_{\psi_2} \end{bmatrix},$$

$$\mathbf{J}_q = \begin{bmatrix} a(yc_{\psi_1} - xs_{\psi_1}) & \rho_1 & 0 & 0 \\ 0 & 0 & a(yc_{\psi_2} - (x-H)s_{\psi_2}) & \rho_2 \end{bmatrix}. \tag{12}$$

The term $\mathbf{g}(\mathbf{q})$ used in the AJM is

$$\mathbf{g}(\mathbf{q}) = 2\left(\begin{bmatrix} \rho_1\left(1 - \frac{\psi_1}{a(yc_{\psi_1} - xs_{\psi_1})}\right) \\ \rho_2\left(1 - \frac{\psi_2}{a(yc_{\psi_2} - (x-H)s_{\psi_2})}\right) \end{bmatrix} \right). \tag{13}$$

As with the planar serial case, Figure 7 shows $h(\mathbf{q})$ from both the methods for the PKM. The initial configuration $\mathbf{q}_0$ was calculated such to be a solution of the inverse position problem and that $\mathbf{g}(\mathbf{q}_0) = \mathbf{0}$.

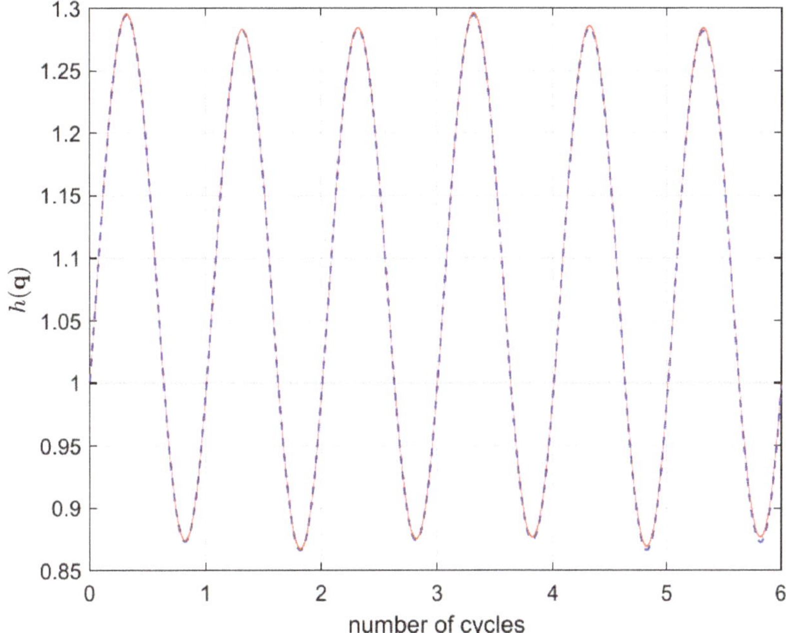

Figure 7. $h(\mathbf{q})$ in 2RRP. PGM: $-$ (red line), AJM: $--$ (blue line).

Also in this case the methods coincide.

We repeat the tests considering $\mathbf{g}(\mathbf{q}_0^*) \neq \mathbf{0}$. However, in this case because of the different types of joints we use $\mathbf{q}_0^* = \mathbf{q}_0 + \boldsymbol{\epsilon}_v$ with $\boldsymbol{\epsilon}_v = \epsilon\left[1, 1\ m, 1, 1\ m\right]$. The methods work consistently for the PKM as well. Only PGM can quickly reach the minimum of the objective function as shown in Figure 8.

Therefore, AJM cannot reach the minimum as pointed out in Figure 9.

The robustness of PGM when varying the initial conditions for the PKM is essentially the same of that for the serial 4$\underline{R}$. The results of the test are shown in Figure 10. The convergence rate in terms of number of cycles was computed of $N_c = 0.9975$.

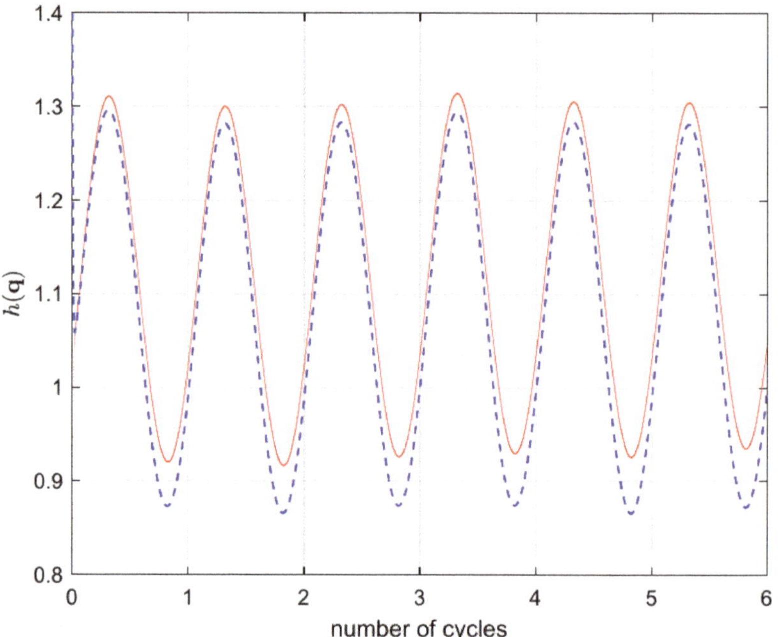

Figure 8. $h(\mathbf{q})$ in 2<u>R</u>R<u>P</u> with $\epsilon = 0.15$. PGM: − (red line), AJM: −− (blue line).

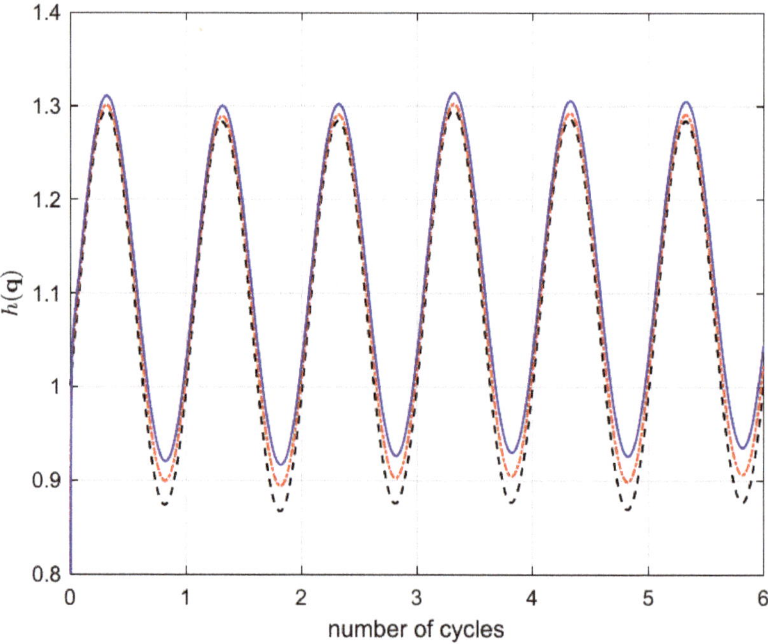

Figure 9. AJM: $h(\mathbf{q})$ in 2<u>R</u>R<u>P</u>. $\epsilon = 0$: −− (black line), $\epsilon = 0.1$: −.− (red line), $\epsilon = 0.15$: − (blue line).

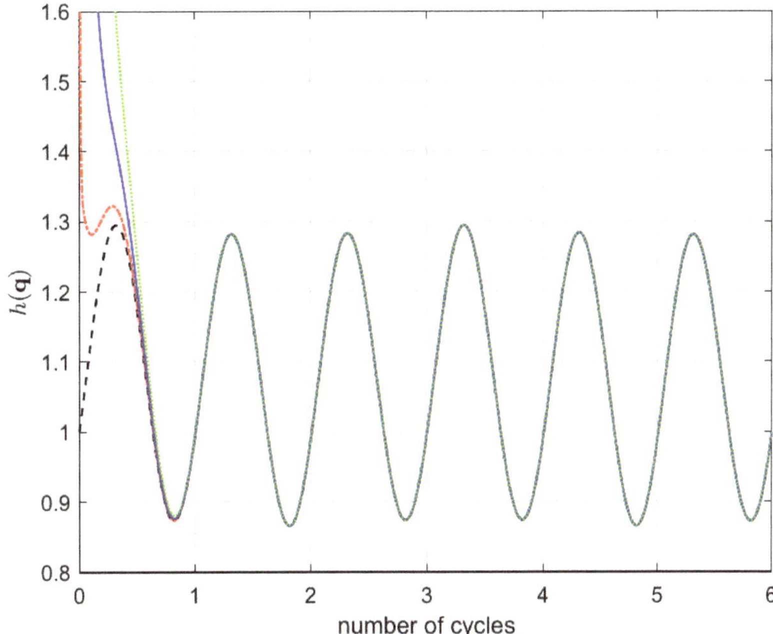

Figure 10. PGM: $h(\mathbf{q})$ in 2R̲R̲P̲. $\epsilon = 0$: $--$ (black line), $\epsilon = 0.5$: $-.-$ (red line), $\epsilon = 1.0$: $-$ (blue line), $\epsilon = 1.5$: $\cdots$ (green line).

3.3. Serial 6R̲

Finally the AJM and PGM methods were tested for a 6DOF industrial robotic arm UR10e. In this case, due to their sizes, it is not possible to show explicitly the terms involved into calculations. Figure 11 shows that also in this case the methods coincide when $\mathbf{g}(\mathbf{q}_0) = \mathbf{0}$, i.e., the objective function is at the minimum at the initial configuration.

Also, in the case of $\mathbf{q}_0^* = \mathbf{q}_0 + \boldsymbol{\epsilon}_v$ with $\boldsymbol{\epsilon}_v = \epsilon[1, -1, 1, -1, 1, -1]$, the methods work as in the other cases as shown in Figure 12.

The response of the AJM in terms of variation of the initial conditions is shown in Figure 13.

Sensitivity of the PGM method to the initial configuration is shown in Figure 14. The behavior is similar to the other cases.

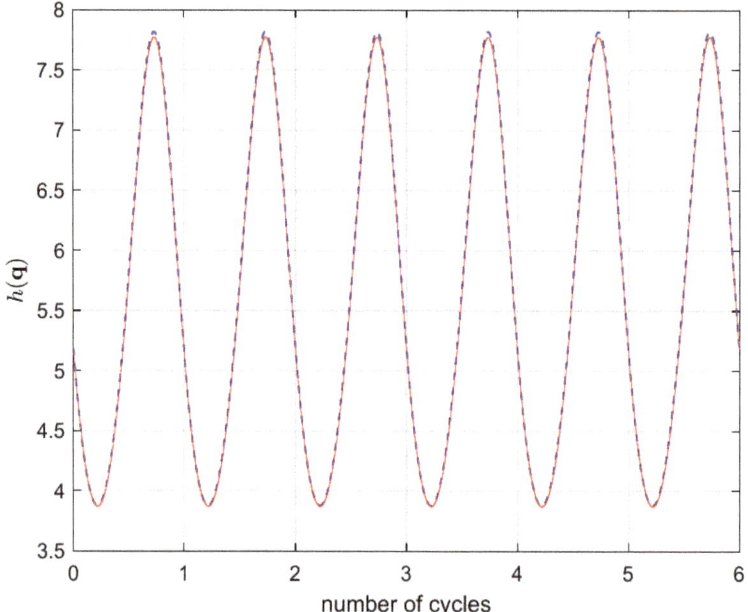

Figure 11. $h(\mathbf{q})$ in 6$\underline{\text{R}}$. PGM: − (red line), AJM: −− (blue line).

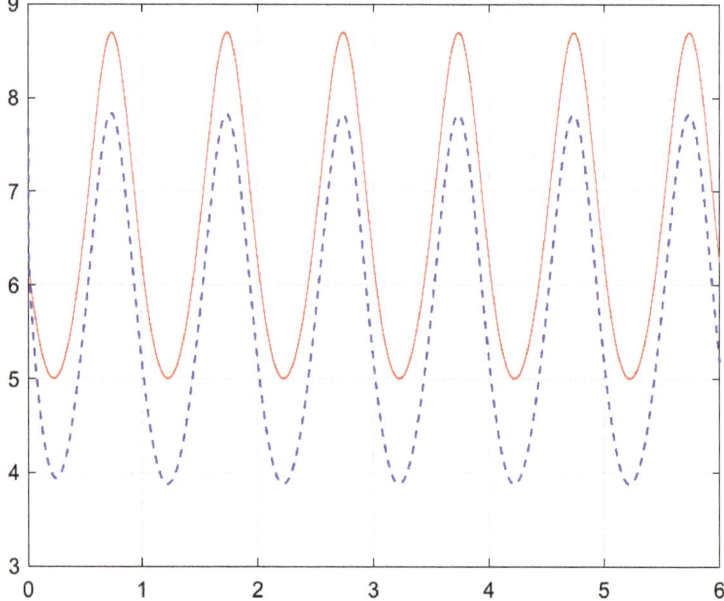

Figure 12. $h(\mathbf{q})$ in 6$\underline{\text{R}}$ with $\epsilon = 0.5$. PGM: − (red line), AJM: −− (blue line).

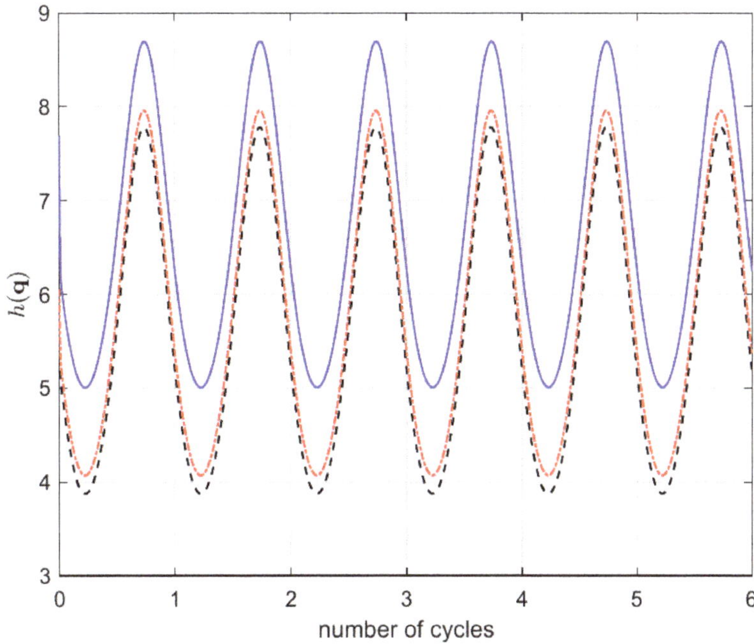

Figure 13. AJM: $h(\mathbf{q})$ in 6$\underline{\mathrm{R}}$. $\epsilon = 0$: $--$ (black line), $\epsilon = 0.25$: $-.-$ (red line), $\epsilon = 0.5$: $-$ (blue line).

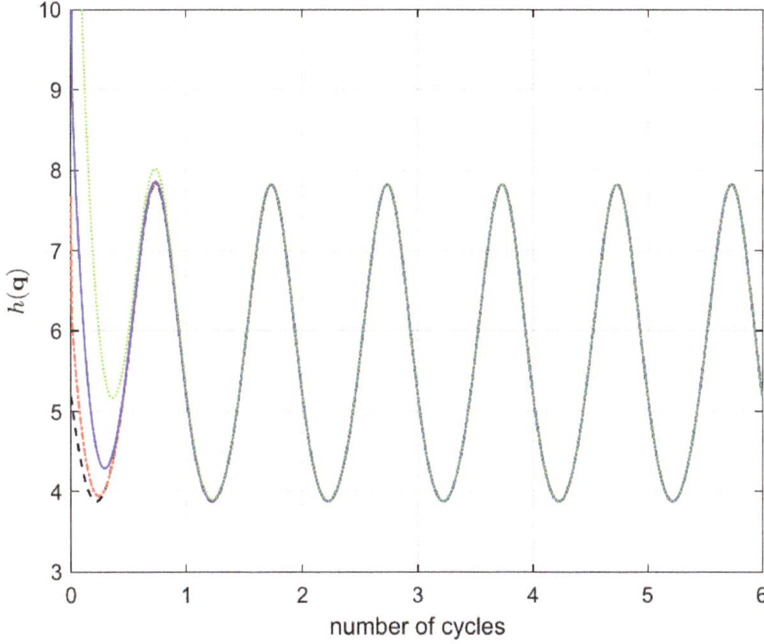

Figure 14. PGM: $h(\mathbf{q})$ in 6$\underline{\mathrm{R}}$. $\epsilon = 0$: $--$ (black line), $\epsilon = 0.5$: $-.-$ (red line), $\epsilon = 1.0$: $-$ (blue line), $\epsilon = 1.5$: $\cdots$ (green line).

The convergence measure in terms of number of cycles is $N_c = 2.1145$. In order to check the robustness of the methods, two further test were performed. First, the circular trajectory was tracked with a variable EE velocity obtained as $v = \dot{\theta}R$ with

$$\dot{\theta} = \frac{2\pi}{T}(1 - \cos 2\pi\tau), \quad \tau = \frac{t}{T}. \tag{14}$$

Then, a test with a trajectory represented by *Bernoulli's Lemniscate* (∞-like curve) tracked with an EE velocity $v = \dot{\theta}r$. In this case, the angular velocity follows Equation (14) while the radial velocity is kept constant. The authors did not find discrepancies with respect to the case of the circle trajectory tracked at the constant velocity for all the manipulators. As an example, Figure 15 shows the robustness of the PGM for the PKM when the EE is tracking the ∞-like curve at the v velocity.

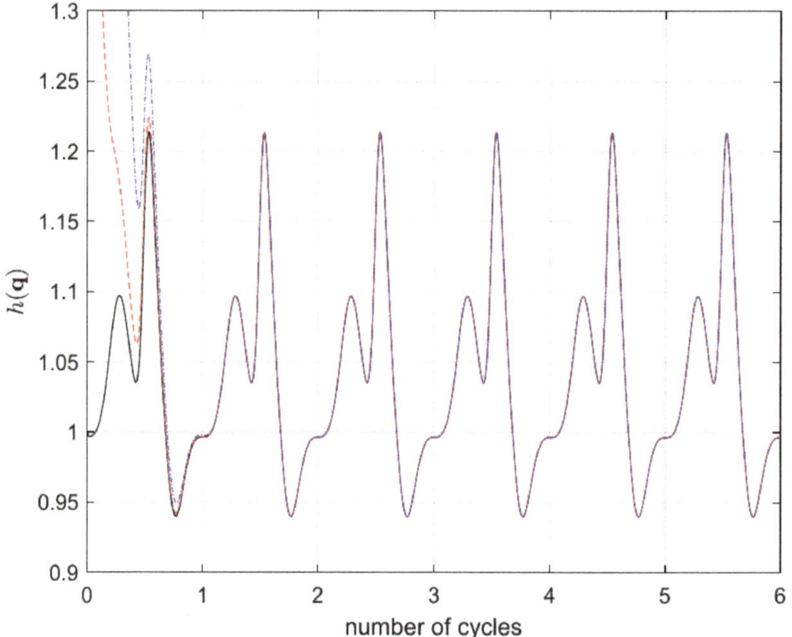

Figure 15. PGM with v and ∞-trajectory: $h(\mathbf{q})$ in 2RRP. $\epsilon = 0$: − (black line), $\epsilon = 0.75$: −− (red line), $\epsilon = 1.5$: −. (blue line).

4. Conclusions

This paper is the combination of two research directions pursued independently: investigation of the trajectories cyclicity and of the gradient-based algortihms for redundancy resolution. In the past, the augmented Jacobian method was shown to lead to cyclic trajectories while the projected gradient method for the inverse kinematics was only used to maximize an objective function. It was never attempted to use the projected gradient method to generate cyclic inverse kinematics solution. In this paper, the cyclicity property of the gradient-based and of the augmented Jacobian method for the redundancy resolution is investigated by means of numerical examples. This paper is the first presented in the literature in which a such comparison is performed.

Briefly, the main results of this paper are: (*a*) The PGM leads to cyclic trajectories at which the objective function attains a minimum, and converges to such cyclic trajectory even if the initial configuration is far from a point where the objective is minimized. The tests on the robustness with respect to the initial configurations showed similar behavior for all the manipulators investigated. On the contrary, as expected, the AJM can only keep the objective function at the value at the initial configuration. (*b*) The results are identical for serial and parallel manipulators since both kinematically characterized by a forward/inverse kinematic mapping. Both methods are cyclic either for the SKMs or for the PKM investigated. The cyclicity property of the AJM was confirmed by numerical calculation of the relevant Lie brackets, while the asymptotic cyclicity of the PGM was confirmed by numerical calculation. These results are novel and it remains to further investigate the properties of the PGM. Moreover, the analyses carried out in this paper are numeric and further research needs to address this topic from an analytical point of view.

Author Contributions: Conceptualization, M.R. and A.M.; software, M.R. and A.M.; investigation, M.R. and A.M.; writing—original draft preparation, M.R. and A.M.; writing—review and editing, M.R. and A.M.; Both authors have equally contributed to this paper. All authors have read and agreed to the published version of the manuscript.

Funding: The second author acknowledges support from the LCM K2 Center for Symbiotic Mechatronics within the framework of the Austrian COMET-K2 program.

Institutional Review Board Statement: Not applicable.

Informed Consent Statement: Not applicable.

Data Availability Statement: The data presented in this study are available on request from the corresponding author.

Conflicts of Interest: The authors declare no conflict of interest.

References

1. Nenchev, D.N. Redundancy Resolution through Local Optimization: A Review. *J. Robot. Syst.* **1989**, *6*, 769–798. [CrossRef]
2. Siciliano, B. Kinematic Control of Redundant Robot Manipulators: A Tutorial. *J. Lntelligent Robot. Syst.* **1990**, *3*, 201–212. [CrossRef]
3. Shamir, T.; Yomdin, Y. Repeatability of Redundant Manipulators: Mathematical Solution of the Problem. *IEEE Trans. Autom. Control* **1988**, *33*, 1004–1009. [CrossRef]
4. Klein, C.A.; Kee, K.B. The nature of drift in pseudo-inverse control of kinematically redundant manipulators. *IEEE Trans. Robot. Automat.* **1989**, *RA-5*, 231–234. [CrossRef]
5. Luo, S.; Ahmad, S. Predicting the Drift Motion for Kinematically Redundant Robots. *IEEE Trans. Syst. Man Cybern.* **1992**, *22*, 717–728. [CrossRef]
6. De Luca, A.; Lanari, L.; Oriolo, G. Control of Redundat Robots on Cyclic Trajectories. In Proceedings of the IEEE International Conference on Robotics and Automation, Nice, France, 12–14 May 1992; pp. 500–506.
7. Baillieul, J. Kinematic programming alternatives for redundant manipulators. In Proceedings of the IEEE International Conference on Robot and Automation, St. Louis, MO, USA, 25–28 March 1985; pp. 772–778.
8. Simas, H.; Di Gregorio, R. A Technique Based on Adaptive Extended Jacobians for Improving the Robustness of the Inverse Numerical Kinematics of Redundant Robots. *J. Mech. Robot.* **2019**, *11*, 020913. [CrossRef]
9. Chembuly, V.V.M.J.S.; Voruganti, H.K. An Efficient Approach for Inverse Kinematics and Redundancy Resolution Scheme of Hyper-Redundant Manipulators. *AIP Conf. Proc.* **2018**, *1943*, 020019.
10. Yahya, S.; Moghavvemi, M.; Mohamed, H.A.F. Geometrical Approach of Planar Hyper-Redundant Manipulators: Inverse Kinematics, Path Planning and Workspace. *Simul. Modell. Pract. Theory* **2011**, *19*, 406–422. [CrossRef]
11. Perrusquía, A.; Yu, W.; Li, X. Multi-agent reinforcement learning for redundant robot control in task-space. *Int. J. Mach. Learn. Cybern.* **2020**. [CrossRef]
12. Chang, P.H. A closed-form solution for inverse kinematics of robot manipulators. *IEEE J. Robot. Automat.* **1987**, *3*, 393–403. [CrossRef]
13. Boothby, W.M. *An Introduction to Differentiable Manifolds and Riemannian Geometry*; Academic Press: Cambridge, MA, USA, 2003.
14. Klein, C.A.; Chu-Jenq, C.; Ahmed, S. A New Formulation of the Extended Jacobian Method and its Use in Mapping Algorithmic Singularities for Kinematically Redundant Manipulators. *IEEE Trans. Robot. Autom.* **1995**, *11*, 50–55. [CrossRef]
15. Gosselin, C.; Angeles, J. Singularity Analysis of Closed-Loop Kinematic Chains. *IEEE Trans. Robot. Autom.* **1990**, *6*, 281–290. [CrossRef]

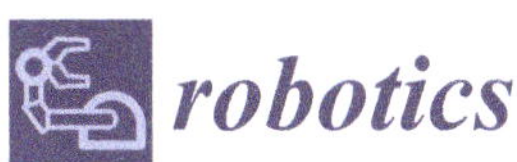 *robotics*

Article

The Effect of the Optimization Selection of Position Analysis Route on the Forward Position Solutions of Parallel Mechanisms

Huiping Shen *, Qing Xu, Ju Li and Ting-li Yang

School of Mechanical Engineering, Changzhou University, Changzhou 213164, China; xq950220@163.com (Q.X.); wangju0209@163.com (J.L.); yangtl@126.com (T.-l.Y.)
* Correspondence: shp65@126.com

Received: 13 October 2020; Accepted: 11 November 2020; Published: 13 November 2020

Abstract: The forward position solution (FPS) of any complex parallel mechanism (PM) can be solved through solving in sequence all of the independent loops contained in the PM. Therefore, when solving the positions of a PM, all independent loops, especially the first loop, must be correctly selected. The optimization selection criterion of the position analysis route (PAR) proposed for the FPS is presented in this paper, which can not only make kinematics modeling and solving efficient but also make it easy to get its symbolic position solutions. Two three-translation PMs are used as the examples to illustrate the optimization selection of their PARs and obtain their symbolic position solutions.

Keywords: parallel mechanism; forward kinematics; coupling degree

1. Introduction

The forward position solution (FPS) of a PM is one of the most important and basic problems in the parallel mechanisms (PMs) research community.

At present, most researchers use the loop vector method [1–3] to establish the input–output position equations of parallel mechanism (PM) and then use numerical or algebraic methods to find its solutions. The numerical method is used directly to solve the position equations, whose advantage is that the real number solutions can be obtained, and the disadvantage is that the iteration is easy to diverge and large calculations are needed. The algebraic methods [4–7], by eliminating the unknowns in the position equations, finally express it as a one-variable higher-order equation, from which all possible solutions can be found. However, algebraic methods need advanced mathematical elimination methods.

It is noticed that the establishment of the above-mentioned position equations of a PM based on the loop vector method does not consider the effect of topological characteristics [8,9] of the PM on its kinematics. For the loop vector method, on the one hand, the loop that some researchers use is a mostly "nature loop" but not actual independent loop. On the other hand, all loops are treated as having equal "ranking status". Therefore, there exist more variables contained in the position equations. Especially when the algebraic method is used, the mathematical elimination processes take more time and are complicated.

For PMs with the same topology of branch chains, there is no need for selecting the optimal position analysis route (PAR). However for PMs with branch chains of different topology, the correct selection of the PAR is necessary, which will affect the efficiency of the FPS and even the solution forms of the forward position (i.e., numerical, closed-form, symbolic). If the PAR is not selected properly, the process of FPS of PMs may be more complicated and difficult, and even their symbolic solutions cannot be obtained. On the contrary, the process of the FPS is convenient, and their symbolic position solutions can be obtained as much as possible.

It can be seen from Ref. [9,10] that the FPS of a PM can be performed by handling its sub-kinematic chains included in the PM in an orderly manner, while an SKC can be solved in accordance with the order of the contained independent loops. Then, for PMs with branch chains of different topologies, the problem of how to sequentially select the first, the 2nd, ..., v-th independent loops, especially the first loop, will become the key to solving FPS. This problem will directly affect whether the kinematics modeling and FPS can be carried out smoothly.

Through previous work [10,11], the authors find that (1) based on the "principle of least constraint degree" [8,9], there may be multiple topological structure decomposition schemes, which are not all the best PAR. (2) If the PAR is not selected properly, it may make the FPS complicated or impossible. Otherwise, the FPS is simple and convenient, and the symbolic position solutions can be obtained as much as possible. These observations lead the investigation of the paper.

The remainder of the paper is organized as follows. In Section 2, optimization criteria and procedures of optimization selection of PAR are presented. In Section 3, a group of basic formulas of the topological characteristic index used for topological property analysis [12] is described. Two three-translation (3T) PMs with branch chains of different topologies are used as examples to illustrate the optimization selection of PAR and their effect on the efficiency of the FPS and on the symbolic position solutions in Section 4, respectively. Finally, conclusions are drawn in the last part.

2. The Optimization Selection Criteria for the Position Analysis Route (PAR)

2.1. Optimization Criteria for PAR

In the FPS of the PMs, the principle of the minimum constraint degree [8,9] $\Delta_{\min}$ and the minimum number of independent displacement equations (NIDE) $\xi_{\min}$ should be satisfied at the same time when selecting an optimal PAR. Once choosing the optimal PAR correctly, the FPS can be carried out efficiently, and its symbolic position solutions can be obtained as much as possible.

2.2. Procedures of Optimization Selection of PAR

i.   For all loops inside the sub-kinematic chains (SKC) [9] of a PM, the loop with the smallest constraint degree value ($\Delta \geq 0$) should be used as the first loop for the FPS, which can minimize the number of virtual variables [10] that should be assigned when performing FPS to the smallest degree.

ii. If there are several optional first loops with the same minimum constraint degree Δ, the loop with the smallest of the NIDE $\xi_{\min}$ should be selected as the first loop. In this way, the number of position equations required to solve the loop positions can be minimized, which is exactly equal to $\xi_{\min}$.

iii. If there are both planar $SKC_{(s)}$ and space $SKC_{(s)}$ in a PM, the FPS should be started from the planar $SKC_{(s)}$ first, and then the space $SKC_{(s)}$ should be analyzed. This is because the NIDE of the planar mechanism loop is always the smallest, i.e., $\xi = 3$.

The above-mentioned criteria and procedures of optimization selection of PAR will be applied to explain the optimization selection of PAR and their effect on the efficiency of the FPS and on the symbolic position solutions of the two three-translation (3T) PMs with branch chains of different topology.

3. Basic Formulas for Topological Characteristics Analysis

3.1. Analysis of the POC Set

The position and orientation characteristic (POC) set equations for the serial mechanism and parallel mechanism are expressed as respectively:

$$M_{bi} = \overset{m}{\underset{i=1}{\cup}} M_{Ji} \tag{1}$$

$$M_{Pa} = \bigcap_{i=1}^{n} M_{bi} \tag{2}$$

where

M_{Ji}—POC set generated by the ith joint.

M_{bi}—POC set generated by the end link of the ith chain.

M_{Pa}—POC set generated by the moving platform of PM.

3.2. Determining the DOF

The proposed general and full-cycle degrees of freedom (DOF) formula for PMs is given below:

$$F = \sum_{i=1}^{m} f_i - \sum_{i=1}^{v} \xi_{Lj} \tag{3}$$

$$\xi_{Lj} = \mathrm{dim.}\left\{ \left(\bigcap_{i=1}^{j} M_{b_i} \right) \cup M_{b_{(j+1)}} \right\} \tag{4}$$

where

F—DOF of PM.

f_i—DOF of the ith joint.

v—number of independent loops, and $v = m - n + 1$.

m, n—number of all joints and links of the whole PM, respectively.

ξ_{Lj}—number of independent displacement equations (NIDE) of the jth loop.

$\bigcap_{i=1}^{j} M_{b_i}$—POC set generated by the sub-PM formed by the former j branches.

$M_{b(j+1)}$—POC set generated by the end link of $(j + 1)$th sub-chains.

3.3. Determining the Coupling Degree

According to the mechanism composition principle based on single-opened-chains (SOC) units, any PM can be decomposed into groups of SKC, and an SKC with v independent loops can be decomposed into v SOC. The constraint degree, denoted as Δ_j, of the jth SOC is defined by:

$$\Delta_j = \sum_{i=1}^{m_j} f_i - I_j - \xi_{Lj} = \begin{cases} \Delta_j^- = -5, -4, -2, -1 \\ \Delta_j^0 = 0 \\ \Delta_j^+ = +1, +2, +3, \cdots \end{cases} \tag{5}$$

where

m_j—number of joints contained in the jth SOC$_j$.

I_j—the number of actuated joints in the jth SOC$_j$.

For an SKC, the following equation must be satisfied:

$$\sum_{j=1}^{v} \Delta_j = 0 \tag{6}$$

Then, the coupling degree of an SKC is expressed as:

$$k = \left| \Delta_j^- \right| = \Delta_j^+ = \frac{1}{2} \min \left\{ \sum_{j=1}^{v} |\Delta_j| \right\} \tag{7}$$

Here, min{•} operation means that the decomposition sequence with the smallest $\sum_{j=1}^{v}|\Delta_j|$ should be selected.

The constraint degree of the *jth* SOC indicates the constraint influences of the chain on the kinematic performance of the mechanism. Its physical meaning will be explained below.

(a) An SOC with a negative constraint degree, denoted as SOC$^-$, will apply $\left|\Delta_j^-\right|$ constraint equations to a mechanism, and the number of DOF of the mechanism will be decreased by DOF$_s$ of $\left|\Delta_j^-\right|$.

(b) An SOC with a positive constraint degree, denoted as SOC$^+$, will increase the number of DOF of the mechanism by Δ_j^+. Therefore, its forward kinematics solutions could not be solved immediately. Its assembly can be determined only on the condition that Δ_j^+ virtual variables are assigned. When the number of the virtual variables is equal to the number of $\left|\Delta_j^-\right|$ constraint equations applied in SOC$^-$, i.e., $k=\left|\Delta_j^-\right|=\Delta_j^+$, the motion of the mechanism is defined, and its forward kinematics can be obtained.

(c) An SOC with zero constraint degree, denoted as SOC^0, does not affect the *DOF*. Its forward kinematics solutions can be obtained immediately without assigning virtual variables.

Therefore, the coupling degree k describes the complexity level of the topological structure of a PM, and it also represents the complexity level of its kinematic and dynamic analysis. The lower the coupling degree k, the easier the treatment of its forward kinematic and dynamic analysis [9,10].

The detailed explanations for these concepts and notations of topological characteristic index used for topological property analysis can be found in Ref. [9,12].

4. Case Studies

4.1. Three-Translation PM (3T-CU)

The three-translation PM proposed by authors [13], denoted as 3T-CU, as shown in Figure 1a, consists of a base platform 0, a moving platform 1, and three different branch chains. Among them, the first branch is $R_{11}//R_{21}//C_{31}$, which is connected to the base platform 0 through the revolute joint R_{11} and connected to the moving platform 1 through the cylindrical joint C_{31}. The second branch is R_{12}-U_{22}-U_{32}, which is connected to the base platform 0 through the revolute joint R_{12} and connected to the moving platform 1 through the moving joint U_{32}. The third branch is a hybrid branch chain, which includes a parallelogram, and its two ends are respectively connected to the base platform 0 and the moving platform 1 through revolute joints R_{13} and R_{33}. Here, R_{11}, R_{12}, and R_{13} on the base platform 0 could be the actuated joints.

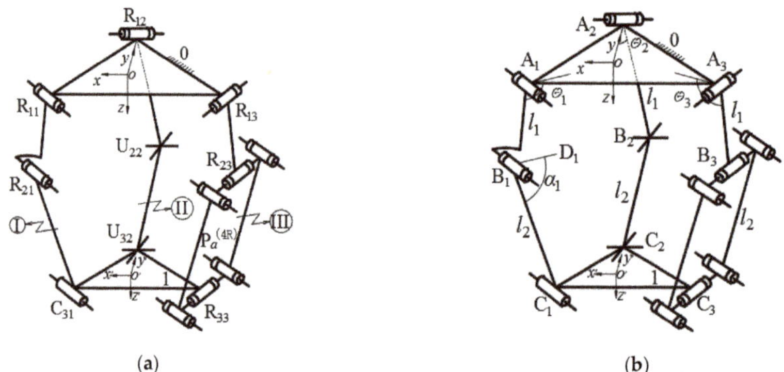

(a)   (b)

Figure 1. Three-translation parallel mechanisms (PM) (3T-CU). (a) Topology structure; (b) Kinematic modeling of the 3T PM.

4.1.1. Topology Analysis

(i) Analysis of the POC Set

Obviously, the topological architecture of chain I, II, and III of the PM can be denoted as, respectively:

$$SOC_1:\{R_{11}//R_{21}//C_{31}\},\ SOC_2:\{R_{12}\text{-}U_{22}\text{-}U_{32}\},\ SOC_3:\{R_{13}//R_{23}(P_a{}^{(4R)})//R_{33}\}.$$

Here, symbol "//" stands for "parallel", while symbol "-" stands for no special geometrical relation, which applies to the whole context of the paper.

The POC sets of the end of the three SOC$_S$ are determined according to Equation (1) as follows, respectively.

$$M_{SOC_1} = \begin{bmatrix} t^3 \\ r^1(\|R_{11}) \end{bmatrix},\ M_{SOC_2} = \begin{bmatrix} t^3 \\ r^2 \end{bmatrix},\ M_{SOC_3} = \begin{bmatrix} t^3 \\ r^1(\|R_{13}) \end{bmatrix}.$$

The POC set of the moving platform of this PM is determined from Equation (2) by

$$M_{P_a} = M_{SOC_1} \cap M_{SOC_2} \cap M_{SOC_3} = \begin{bmatrix} t^3 \\ r^1(\|R_{11}) \end{bmatrix} \cap \begin{bmatrix} t^3 \\ r^2 \end{bmatrix} \cap \begin{bmatrix} t^3 \\ r^1(\|R_{13}) \end{bmatrix} = \begin{bmatrix} t^3 \\ r^0 \end{bmatrix}.$$

Hence, the moving platform 1 of the PM has a three-translation motion output.

(ii) Determining the DOF

It is easy for this PM to have two selection methods for PAR, as follows,

Case ①: If the first loop of the PM (that is, sub-PM) consists of the first and third branch chains, namely $R_{11}//R_{21}//C_{31}\text{-}R_{33}//(P_a{}^{(4R)})R_{23}//R_{13}$, the number of independent displacement equations (NIDE) is obtained by Equation (4)

$$\xi_{L_1} = \dim\left\{\begin{bmatrix} t^3 \\ r^1(\|R_{11}) \end{bmatrix} \cup \begin{bmatrix} t^3 \\ r^1(\|R_{13}) \end{bmatrix}\right\} = \dim\left\{\begin{bmatrix} t^3 \\ r^2\|(\diamond(R_{11},R_{13})) \end{bmatrix}\right\} = 5.$$

From Equation (3), the degree of freedom (DOF) of the first loop (1st sub-PM) is:

$$F_{(1-2)} = \sum f - \xi_{L_1} = (4+4) - 5 = 3.$$

The second loop is composed of the above-mentioned sub-PM and the second branch chain. From Equation (4), the NIDE is calculated by

$$\xi_{L_2} = \dim\left\{\begin{bmatrix} t^3 \\ r^1(\|R_{11}) \end{bmatrix} \cap \begin{bmatrix} t^3 \\ r^1(\|R_{13}) \end{bmatrix} \cup \begin{bmatrix} t^3 \\ r^2 \end{bmatrix}\right\} = \dim\left\{\begin{bmatrix} t^3 \\ r^2 \end{bmatrix}\right\} = 5.$$

Case ②: If the first loop (sub-PM) of the PM consists of the second and the third branch chains, namely $R_{12}\text{-}U_{22}\text{-}U_{32}\text{-}R_{33}//(P_a{}^{(4R)})R_{23}//R_{13}$, the NIDE is obtained by Equation (4):

$$\xi_{L_1} = \dim\left\{\begin{bmatrix} t^3 \\ r^2 \end{bmatrix} \cup \begin{bmatrix} t^3 \\ r^1(\|R_{13}) \end{bmatrix}\right\} = \dim\left\{\begin{bmatrix} t^3 \\ r^2 \end{bmatrix}\right\} = 5.$$

From Equation (3), the degree of freedom of the first loop (1st sub-PM) is:

$$F_{(1-2)} = \sum f - \xi_{L_1} = (5+4) - 5 = 4.$$

The second loop is composed of the above-mentioned 1st sub-PM and the first branch chain. From Equation (4), the number of independent displacement equation is:

$$\xi_{L_2} = \dim\left\{\begin{bmatrix} t^3 \\ r^2 \end{bmatrix} \cap \left[\begin{matrix} t^3 \\ r^1(\|R_{13}) \end{matrix}\right] \cup \left[\begin{matrix} t^3 \\ r^1(\|R_{11}) \end{matrix}\right]\right\} = \dim\left\{\begin{matrix} t^3 \\ r^2(\|\diamond(R_{11}, R_{13}) \end{matrix}\right\} = 5.$$

Thus, the DOF of the PM is calculated from Equation (3) as:

$$F = \sum_{i=1}^{m} f_i - \sum_{i=1}^{v} \xi_{L_j} = (8+5) - (5+5) = 3.$$

Therefore, when the revolute joints R_{11}, R_{12}, and R_{13} on the base platform 0 are selected as the actuated joints, the moving platform 1 can realize 3T motion outputs.

(iii) Determining the coupling degree

Form Equation (5), the constraint degree of the two loops are respectively given by:

$$\text{For Case : } \Delta_1 = \sum_{i=1}^{m_1} f_i - I_1 - \xi_{L_1} = 8 - 2 - 5 = +1, \Delta_2 = \sum_{i=1}^{m_2} f_i - I_2 - \xi_{L_2} = 5 - 1 - 5 = -1.$$

From Equation (6), the PM contains one SKC. Further, from Equation (7), the coupling degree of the SKC is given by

$$\kappa = \frac{1}{2}\min\left\{\sum_{j=1}^{v} |\Delta_j|\right\} = \frac{1}{2}(|+1|+|-1|) = 1.$$

Therefore, the coupling degree of the PM is $\kappa = 1$, which means that one virtual variable needs to be assigned when solving its positions.

$$\text{For Case } ② \ \Delta_1 = \sum_{i=1}^{m_1} f_i - I_1 - \xi_{L_1} = 9 - 2 - 5 = +2,. \ \Delta_2 = \sum_{i=1}^{m_2} f_i - I_2 - \xi_{L_2} = 4 - 1 - 5 = -2.$$

From Equation (6), the PM contains one SKC. Furthermore, from Equation (7), the coupling degree of the SKC is given by

$$\kappa = \frac{1}{2}\min\left\{\sum_{j=1}^{v} |\Delta_j|\right\} = \frac{1}{2}(|+2|+|-2|) = 2$$

This moment, the coupling degree of the PM is $\kappa = 2$, which means that two virtual variables need to be assigned when solving its positions, and it undoubtedly makes the FPS much more complicated than the case ① with only one virtual variable.

(iv) Optimization selection for PAR

So far, it can be found that according to the cases ① and ②, the NIDE of the two cases obtained are the same, i.e., $\xi_{L1} = \xi_{L2} = 5$, but their constraint degree Δ (or coupling degree κ) is different, i.e., κ of the case ① is one, while κ of the case ② is two. Therefore, according to the optimization criteria for the PAR, the case ① that has the smallest constraint degree value ($\Delta_{min} = 1$) and the minimum NIDE ($\xi_{min} = 5$) should be used to solve the FPS. The details are described below.

4.1.2. Position Analysis

(i) The coordinate system and parameterization

The kinematic modelling of the PM is shown in Figure 1b. The base platform 0 is an equilateral triangle with a circle radius R, and select the geometric center O as the origin of the base coordinate system. The x- and y-axis are perpendicular and parallel to the line OA_2. Let moving platform 1 be an equilateral triangle with a circle radius r, and select the O' point on the moving platform as the origin of the moving coordinate system. The x'- and y'-axis are perpendicular and parallel to the line $O'C_2$. The z and z' axis are determined by the Cartesian coordinate rule.

Let the angle θ_i between vectors A_iB_i and A_iO be the input angle, and the length of the line A_iB_i and B_iC_i ($i = 1$–3) is equal to l_1 and l_2, respectively.

(ii) Direct kinematics

To perform the FPS, i.e., it is to compute the position $O'(x, y, z)$ of the moving platform with the known actuated joints θ_1, θ_2, and θ_3.

The coordinates of points A_i and B_i ($i = 1$–3) are easily known as:

$$A_1 = (R\cos 30°, -R\sin 30°, 0)^{\mathrm{T}}, A_2 = (0, R, 0)^{\mathrm{T}}, A_3 = (-R\cos 30°, -R\sin 30°, 0)^{\mathrm{T}},$$
$$B_1 = ((R - l_1\cos\theta_1)\cos 30°, -(R - l_1\cos\theta_1)\sin 30°, l_1\sin\theta_1)^{\mathrm{T}},$$
$$B_2 = (0, R - l_1\cos\theta_2, l_1\sin\theta_2)^{\mathrm{T}}, B_3 = (-(R - l_1\cos\theta_3)\cos 30°, -(R - l_1\cos\theta_3)\sin 30°, l_1\sin\theta_3)^{\mathrm{T}}.$$

① Solving the first loop (A_1-B_1-C_1-C_3-B_3-A_3) with a positive constraint ($\Delta_1 = 1$)

Since the coupling degree of the PM is 1, it is necessary to assign one virtual variable when performing the FPS. Let the angle α_1 between B_1C_1 and B_1D_1 be the virtual variable, where $B_1D_1//A_1O$. It is easy to know that the coordinates of point C_1 are below:

$$C_1 = (x_{B_1} - l_2\cos 30°\cos\alpha_1 \,,\, y_{B_1} + l_2\sin 30°\cos\alpha_1 \,,\, z_{B_1} + l_2\sin\alpha_1)^{\mathrm{T}}. \tag{8}$$

Thus, the three points $C_i(i = 1$–3) in the base coordinate system are calculated as:

$$C_1 = (r\cos 30° + x, \, -r\sin 30° + y, \, z)^{\mathrm{T}} \tag{9}$$

$$C_2 = (x, r + y, z)^{\mathrm{T}} \tag{10}$$

$$C_3 = (-r\cos 30° + x, \, -r\sin 30° + y, \, z)^{\mathrm{T}}. \tag{11}$$

From Equations (9) and (11), we can get:

$$C_3 = (x_{B_1} - l_2\cos 30°\cos\alpha_1 - 2r\cos 30° \,,\, y_{B_1} + l_2\sin 30°\cos\alpha_1 \,,\, z_{B_1} + l_2\sin\alpha_1)^{\mathrm{T}}. \tag{12}$$

Due to the length constraint defined by $B_3C_3 = l_2$, we can get:

$$\alpha_1 = 2\arctan\left(\frac{-G_2 \pm \sqrt{G_2{}^2 + G_1{}^2 - G_3{}^2}}{G_3 - G_1}\right) \tag{13}$$

where

$$p = x_{B_1} - 2r\cos 30° - x_{B_3}, q = y_{B_1} - y_{B_3}, g = z_{B_1} - z_{B_3}, G_1 = 2l_2(q\sin 30° - p\cos 30°),$$
$$G_2 = 2gl_2, G_3 = p^2 + q^2 + g^2.$$

Substituting Equation (13) into Equations (8) and (12), the points C_1 and C_3 can be obtained.

② Solving the second loop (*Loop*$_2$: A_2-B_2-C_2) with negative constraint ($\Delta_2 = -1$)

After the positions of all joints on the first loop are obtained, it is easier to solve that on the second loop. From Equations (9) and (10), the coordinates of point C_2 are:

$$C_2 = (x_{C_1} - r\cos 30° \,, y_{C_1} + r + r\sin 30° \,, z_{C_1})^{\mathrm{T}}. \tag{14}$$

Therefore, the origin coordinates of the moving platform can be easily obtained.

It can be seen that from FPS of the PM there is a total of five position equations, i.e., (1) three geometric constraint equations: $B_1C_1 = l_2$, $B_3C_3 = l_2$, $y_{C_3} = y_{C_1}$; (2) two topological constraint equations introduced by the POC feature (3T): $z_{C_2} = z_{C_1}, z_{C_3} = z_{C_1}$, which are exactly equal to the NIDE $\xi_L = 5$, and we ensure that the positions of the PM can be solved.

The inverse kinematics of this PM is omitted here due to its easiness.

(iii) Verification

Set the parameters of the PM as $R = 90$, $r = 55$, $l_1 = 40$, $l_2 = 40$(unit: mm), and the input angles of the three actuated joints are $\theta_1 = 30°$, $\theta_2 = 60°$, and $\theta_3 = 60°$.

From Equations (8)–(14), two sets of real solutions are obtained, i.e., (1) $x = -33.9339$, $y = 19.5917$, $z = 13.9672$, and (2) $x = 23.5901$, $y = -13.6197$, and $z = 49.6216$. It has been verified that both the FPS and inverse solutions derived above are correct.

The authors also analyze the position of the 3T-CU PM according to case ②, but only numerical solutions were obtained, and calculations are more complicated.

So far, for route selection cases ① and ②, the NIDE are $\xi_{L_1} = \xi_{L_2} = 5$, but the constraint degree of case ① is $\Delta_1 = +1$, $\Delta_2 = -1$ and the constraint degree of case ② is $\Delta_1 = +2$, $\Delta_2 = -2$. Therefore, case ① is the optimization selection for PAR. It not only guarantees the efficient kinematic modeling but also obtains the symbolic solutions.

4.2. Three-Translation PM (Delta-CU)

Figure 2a shows another three-translation PM [14,15], denoted as Delta-CU PM, designed by the authors, which consists of base platform 0, moving platform 1, and three branch chains. Among them, the first and third ones are hybrid branches containing a parallelogram structure (same as the traditional Delta mechanism). The first and third branches are connected to the base platform 0 through the revolute joints R_{11} and R_{13}, and they are connected to the moving platform 1 through the revolute joints R_{31} and R_{33}, respectively. The second branch is R_{12}-U_{22}-U_{32}, and its two ends are connected to base platform 0 and moving platform 1 through R_{12} and U_{32}, respectively. The PM is called Delta-CU, i.e., 2-R//R//P_a//R+R-U-U.

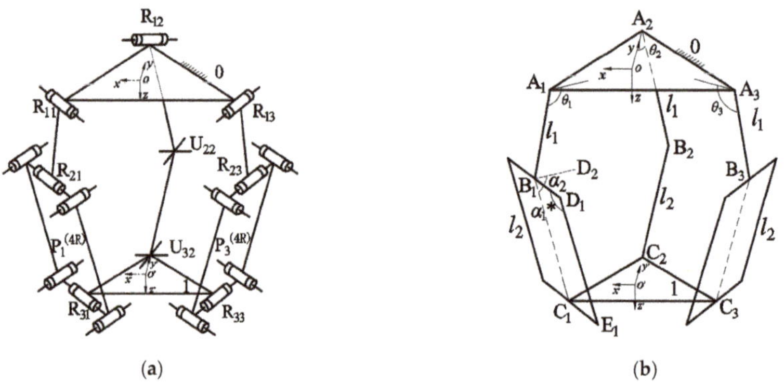

(a) (b)

Figure 2. Three-translation PM (Delta-CU). (**a**) Topology structure; (**b**) Kinematic modeling of the 3T PM.

Compared with the Delta mechanism with the same three branches (all of which are hybrid branches with a parallelogram structure), this Delta-CU mechanism decreases two revolute joints and three links, and hence the structure is simpler. Since the topological structure of the three branch chains are not all different, it involves the problem of optimization selection of PAR, and different PARs will lead to a big difference in the difficulty of the FPS of the PM.

4.2.1. Topological Analysis

(i) Analysis of the POC Set

The first and third mixed branches are denoted as $SOC_i\{R_{1i}//R_{2i}//Pa^{(4R)}\text{-}R_{3i}\}(i = 1, 3)$, while the second branch is denoted as $SOC_2\{R_{12}\text{-}U_{22}\text{-}U_{42}\}$.

The POC sets of the end of the three SOC$_S$ are determined according to Equation (1) as follows, respectively:

$$M_{SOC_1} = \begin{bmatrix} t^3 \\ r^1(\|R_{11}) \end{bmatrix}, M_{SOC_2} = \begin{bmatrix} t^3 \\ r^2 \end{bmatrix}, M_{SOC_3} = \begin{bmatrix} t^3 \\ r^1(\|R_{13}) \end{bmatrix}$$

The POC set of the moving platform of this PM is determined from Equation (2) by:

$$M_{P_a} = M_{SOC_1} \cap M_{SOC_2} \cap M_{SOC_3} = \begin{bmatrix} t^3 \\ r^1(\|R_{11}) \end{bmatrix} \cap \begin{bmatrix} t^3 \\ r^2 \end{bmatrix} \cap \begin{bmatrix} t^3 \\ r^1(\|R_{13}) \end{bmatrix} = \begin{bmatrix} t^3 \\ r^0 \end{bmatrix}$$

Hence, the moving platform 1 of the PM has three-translation motion output.

(ii) Determining the DOF

The PM has three branches; therefore, it contains two independent loops. Since the three branches belong to two different topological structures, that is, the first and third branches are hybrid branches with four degrees of freedom (DOF) (one 4R parallelogram mechanism is equivalent to one prismatic joint). The DOF of the second branch is 5. Therefore, there are at least two schemes in the topology decomposition of the PM as follows.

Case ①: The first loop (sub-PM) is composed of the first and third hybrid branch chains with the least degree of freedom (DOF = 4), while the second loop is composed of the sub-PM and the second branch (R_{12}-U_{22}-U_{32}) with 5-DOF.

Case ②: The first loop (sub-PM) is composed of the first branch chain with 4-DOF and the second branch chain with 5-DOF, while the second loop is composed of the sub-PM and the third branch chain with 4-DOF.

For Case ①: The NIDE of the two loops can be obtained by Equation (4):

$$\xi_{L_1} = \dim\left\{ \begin{bmatrix} t^3 \\ r^1(\|R_{11}) \end{bmatrix} \cup \begin{bmatrix} t^3 \\ r^1(\|R_{13}) \end{bmatrix} \right\} = \dim\left\{ \begin{bmatrix} t^3 \\ r^2 \end{bmatrix} \right\} = 5, \xi_{L_2} = \dim\left\{ \begin{bmatrix} t^3 \\ r^1(\|R_{11}) \end{bmatrix} \cap \begin{bmatrix} t^3 \\ r^1(\|R_{13}) \end{bmatrix} \cup \begin{bmatrix} t^3 \\ r^2 \end{bmatrix} \right\} = \dim\left\{ \begin{bmatrix} t^3 \\ r^2 \end{bmatrix} \right\} = 5.$$

From Equation (3), the DOF of the PM is:

$$F = \sum_{i=1}^{m} f_i - \sum_{i=1}^{v} \xi_{L_j} = (8+5) - (5+5) = 3.$$

Therefore, when the revolute joints R_{11}, R_{12}, and R_{13} on the base platform 0 are selected as the actuated joints, the moving platform 1 can realize 3T motion outputs.

For Case ②: The NIDE of the two loops can be obtained by Equation (4):

$$\xi_{L_1} = \dim\left\{ \begin{bmatrix} t^3 \\ r^1 \end{bmatrix} \cup \begin{bmatrix} t^3 \\ r^2 \end{bmatrix} \right\} = \dim\left\{ \begin{bmatrix} t^3 \\ r^3 \end{bmatrix} \right\} = 6, \xi_{L_2} = \dim\left\{ \begin{bmatrix} t^3 \\ r^1 \end{bmatrix} \cap \begin{bmatrix} t^3 \\ r^2 \end{bmatrix} \cup \begin{bmatrix} t^3 \\ r^1 \end{bmatrix} \right\} = \dim\left\{ \begin{bmatrix} t^3 \\ r^1 \end{bmatrix} \right\} = 4.$$

From Equation (3), the DOF of the PM is:

$$F = \sum_{i=1}^{m} f_i - \sum_{i=1}^{v} \xi_{L_j} = (4+5+4) - (5+5) = 3.$$

The DOF of the PM is still three. Therefore, when R_{11}, R_{12}, and R_{13} on the base platform are actuated joints, the moving platform 1 can realize 3T motion outputs.

(iii) Determining the coupling degree

Case ①: From Equation (5), the constraint degrees of the two loops are respectively given by:

$$\Delta_1 = \sum_{i=1}^{m_1} f_i - I_1 - \xi_{L_1} = (4+4) - 2 - 5 = +1, \Delta_2 = \sum_{i=1}^{m_2} f_i - I_2 - \xi_{L_2} = 5 - 1 - 5 = -1.$$

From Equation (6), the PM contains one SKC. Furthermore, from Equation (7), the coupling degree of the SKC is given by:

$$k = \frac{1}{2}\min\left\{\sum_{j=1}^{v}|\Delta_j|\right\} = \frac{1}{2}(|+1|+|-1|) = 1.$$

The above equation shows the following: ① This PM contains only one SKC, and its coupling degree is one. When performing FPS, it is necessary to assign one virtual variable on the first loop with the constraint degree of positive one, and it is necessary to establish a constraint position equation containing the virtual variable on the second loop with a constraint degree of minus one. ② Since the three actuated joints are in one SKC, according to the input–output (I-O) motion decoupling determination principle [9], it can be determined that the PM does not have an input–output (I-O) motion decoupling property.

For Case ②: From Equation (5), the constraint degrees of the two loops are respectively given by:

$$\Delta_1 = \sum_{i=1}^{m_1} f_i - I_1 - \xi_{L_1} = (4+5) - 2 - 6 = +1, \Delta_2 = \sum_{i=1}^{m_2} f_i - I_2 - \xi_{L_2} = 4 - 1 - 4 = -1.$$

From Equation (6), the PM contains one SKC. Furthermore, from Equation (7), the coupling degree of the SKC is calculated as:

$$k = \frac{1}{2}\min\left\{\sum_{j=1}^{v}|\Delta_j|\right\} = \frac{1}{2}(|+1|+|-1|) = 1.$$

Of course, the topology decomposition scheme ② of the first loop of the PM can also be composed of the second and third branches, while the second loop is composed of the sub-PM and the first branch, and the result is identical.

(iv) Optimization selection for PAR

Comparing the topological decomposition Cases ① and ②, it can be seen that if the position analysis is performed according to Case ②, the constraint degree of the two loops are $\Delta_1 = 1$ and $\Delta_2 = -1$, and the NIDEs are $\xi_{L1} = 6$, $\xi_{L2} = 4$, respectively, which means that when performing the FPS, six position constraint equations must be found in the first loop. Obviously, the difficulty will increase, and it will even not be solved. On the contrary, if the position analysis is carried out according to Case ①, the constraint degrees of the two loops are $\Delta_1 = 1$, $\Delta_2 = -1$, and the NIDEs are $\xi_{L1} = 5$ and $\xi_{L2} = 5$, respectively, which means that when performing the FPS, only five position constraint equations are needed to be found in the first loop, which is obviously easier.

Therefore, when performing the FPS of the Delta-CU PM, the PAR should be selected as Case ①.

4.2.2. Position Analysis

(i) The coordinate system and parameterization

The kinematic modeling of the Delta-CU PM is shown in Figure 2b. The base platform 0 is an equilateral triangle with a circle radius R, and select the geometric center O as the origin of the base coordinate system. The x- and y-axis are perpendicular and parallel to the line OA_2. Let moving platform 1 be an equilateral triangle with a circle radius r, and select the O′ point on the moving platform as the origin of the moving coordinate system. The x'- and y'-axis are perpendicular and parallel to the line $O'C_2$. The z- and z'-axis are all determined by the Cartesian coordinate rule. Let the angle θ_i between vectors A_iB_i and A_iO be the input angle, and the lengths of the lines A_iB_i and B_iC_i ($i = 1$–3) are equal to l_1 and l_2, respectively. Let the coordinates of the origin of the moving coordinate system be O′ (x, y, z).

(ii) Direct kinematics

To perform the FPS, compute the position O′ $= (x, y, z^*)$ of the platform with the known actuated joints θ_2, θ_2, and θ_3.

Since the coupling degree of the PM is $\kappa = 1$, one virtual variable needs to be assigned. Furthermore, the first loop passes through the moving platform 1, and there are two methods for selecting virtual variables, namely method A (i.e., the virtual variable starts from one side of the loop) and method B (i.e., the virtual variable starts from the moving platform) [16]. Since method B has the advantages of a short calculation path, fewer calculations, and available symbolic solutions [16], method B is now used for the calculation of the PM.

Since the PM is a 3T PM and the coupling degree is $\kappa = 1$, take one of the position parameters (x, y, z^*) of the platform 1, for example, z^*, as a virtual variable.

① Solving the first loop with the positive constraint ($\Delta_1 = 1$)

The coordinates of C_i point ($i = 1$–3) on the moving platform are given, respectively.

$$C_1 = (r\cos 30° + x, -r\sin 30° + y, z*)^{\mathrm{T}}, C_2 = (x, r + y, z*)^{\mathrm{T}}, C_3 = (-r\cos 30° + x, -r\sin 30° + y, z*)^{\mathrm{T}}.$$

Due to the length constraint defined by $B_iC_i = l_2$ ($i = 1, 3$), we can get:

$$(x + m)^2 + (y + n)^2 + (z * -z_{B_1})^2 = l_2{}^2 \tag{15}$$

$$(x + u)^2 + (y + v)^2 + (z * -z_{B_3})^2 = l_2{}^2 \tag{16}$$

where

$$m = r\cos 30° - x_{B_1}, n = -\left(r\sin 30° + y_{B_1}\right), u = -\left(r\cos 30° + x_{B_3}\right), v = -\left(r\sin 30° + y_{B_3}\right).$$

② Solve the second loop with the negative constraint ($\Delta_2 = -1$)

Due to the length constraint defined by $B_2C_2 = l_2$, we can get:

$$(x - x_{B_2})^2 + (y + t)^2 + (z * -z_{B_2})^2 = l_2{}^2 \tag{17}$$

where $t = r - y_{B_2}$.

From Equations (15)–(17), we can get the symbolic solutions as follows:

$$\begin{cases} x = Nz * +D \\ y = Hz * +K \\ z* = \dfrac{-C_2 \pm \sqrt{C_2{}^2 - 4C_1C_3}}{2C_1} \end{cases} \tag{18}$$

where

$$M = \frac{m^2 + n^2 + z_{B_1}{}^2 - x_{B_2}{}^2 - t^2 - z_{B_2}{}^2}{2}, T = m^2 + n^2 + z_{B_1}{}^2 - u^2 - v^2 - z_{B_3}{}^2 - \frac{2M(m-u)}{m+x_{B_2}}, P = \frac{2(m-u)(t-n)}{m+x_{B_2}} + 2(n - v),$$

$$Q = \frac{2(m-u)\left(z_{B_1} - z_{B_2}\right)}{m+x_{B_2}} + 2\left(z_{B_3} - z_{B_1}\right), H = -\frac{Q}{P}, K = -\frac{T}{P},$$

$$N = -\frac{(n-t)H + \left(z_{B_2} - z_{B_1}\right)}{x_{B_2} + m}, D = -\frac{(n-t)K + M}{x_{B_2} + m},$$

$$C_1 = N^2 + H^2 + 1, C_2 = 2\left(ND + HK + mN + nH - z_{B_1}\right), C_3 = D^2 + K^2 + 2mD + 2nK + m^2 + n^2 + z_{B_1}{}^2 - l_2{}^2.$$

It can be seen that from the FPS of the PM there is also a total of five position equations in the first loop, i.e., ① two position constraint equations $z_{C_1} = z_{C_3} = z$ introduced by the topological constraint that the moving platform 1 always performs three-dimensional translation, ② two length constraint conditions $B_1C_1 = l_2$ and $B_3C_3 = l_2$, and ③ the position of C_2 is also constrained by the length constraint equation $B_2C_2 = l_2$ inside the second loop. In this way, there are five position constraint equations, which are equal to the NIDE $\zeta_L = 5$. Therefore, the position of the first loop is guaranteed to be solvable, since these position equations are nonlinear ones and simper, symbolic solutions can be easily obtained.

The inverse kinematics of this PM and numerical verification are omitted here due to its easiness.

5. Conclusions

The optimization selection for PAR can affect the effectiveness of FPS and the solution forms. For this reason, the optimization selection criteria for PAR are proposed, i.e., when solving the position, the loop should be selected according to the criteria of "minimum constraint degree value (Δ_{min}) and minimum number of independent displacement equations (ξ_{min})" in order to effectively perform FPS and obtain the symbolic solutions to the greatest possible extent. Otherwise, the FPS may be difficult, complicated, or symbolic solutions cannot be obtained. Two examples are used to illustrate the procedures of how to select the PAR.

In the recent years, the authors have performed a large number of FPS of the PMs [17,18], which proves that the procedures of the selection criteria of PAR for FPS proposed in this paper are universal.

For PMs with branch chains of different topology, the correct selection of the PAR is necessary and important. This work provides new inspiration and the road map for the FPS of these complex PMs.

Author Contributions: Conceptualization, H.S. and T.-l.Y.; methodology, Q.X. and J.L.; validation, Q.X.; writing—original draft preparation, H.S.; writing—review and editing, Q.X. and J.L.; funding acquisition, H.S. All authors have read and agreed to the published version of the manuscript.

Funding: This research is sponsored by the NSFC (No. 51975062, 51375062).

Conflicts of Interest: The authors declare no conflict of interest.

References

1. Kinzel, G.L.; Chang, C. The analysis of planar linkage using a modular approach. *Mech. Mach. Theory* **1984**, *19*, 165–172. [CrossRef]
2. Kong, X.; Gosselin, C.M. Generation and forward displacement analysis of RPR-PR-RPR analytic planar parallel manipulators. *J. Mech. Des.* **2001**, *8*, 2195–2304.
3. Pennock, G.R.; Hasan, A. A Polynomial Equation for a Coupler Curve of the Double Butterfly Linkage. *J. Mech. Des.* **2002**, *124*, 39–46. [CrossRef]
4. Husain, M.; Waldron, K.J. Direct Position Kinematics of the 3-1-1-1 Stewart Platforms. *J. Mech. Des.* **1994**, *116*, 1102–1107. [CrossRef]
5. Wampler, C.W. Forward displacement analysis of general six-in-parallel sps (Stewart) platform manipulator using soma coordinates. *Mech. Mach. Theory* **1996**, *31*, 311–337. [CrossRef]
6. Husty, M.L. Algorithm for Solving the direct kinematic of Stewart-Gough-Type platforms. *Mech. Mach. Theory* **1996**, *31*, 365–380. [CrossRef]
7. Rojas, N.; Thomas, F. On closed-form Solutions to the position analysis of Baranov trusses. *Mech. Mach. Theory* **2012**, *50*, 179–196. [CrossRef]
8. Yang, T.; Liu, A.; Shen, H.; Hang, L.; Luo, Y.; Jin, Q. *Theory and Appication of of Robot Mechanism Topology*; Science Press: Beijing, China, 2012.
9. Yang, T.; Liu, A.; Shen, H.; Hang, L.; Luo, Y.; Jin, Q. *Topology Design of Robot Mechanisms*; Springer: Singapore, 2018.
10. Shen, H. Research on Forward Position Solutions for 6-SPS Parallel Mechanisms Based on Topology Structure Analysis. *Chin. J. Mech. Eng.* **2013**, *49*, 70–80. [CrossRef]
11. Shen, H.; Chablat, D.; Zen, B.; Li, J.; Wu, G.; Yang, T. A Translational Three-Degrees-of-Freedom Parallel Mechanism with Partial Motion Decoupling and Analytic Direct Kinematics. *J. Mech. Robot.* **2020**, *12*, 1–7. [CrossRef]
12. Shen, H.; Yang, T.; Li, J.; Zhang, D.; Deng, J.; Liu, A. Evaluation of Topological Properties of Parallel Manipulators Based on the Topological Characteristic Indexes. *Robotica* **2020**, *38*, 1381–1399. [CrossRef]
13. Shen, H.; Xu, Q.; Li, J.; Yang, T. The Effect of the Optimal Route Selection on the Forward Position Solutions of Parallel Mechanisms. In *Advances in Mechanism and Machine Science*; Springer: Cham, Switzerland, 2020; pp. 450–458.

14. Li, J.; Shen, H.; Meng, Q.; Deng, J. A delta-CU—Kinematic analysis and dimension design. In Proceedings of the 10th International Conference on Intelligent Robotics and Applications, Wuhan, China, 16–18 August 2017; Springer: Cham, Switzerland, 2017; pp. 371–382.
15. Shen, H.; Wang, Y.; Wu, G.; Meng, Q. Stiffness Analysis of a Semi-symmetrical Three-Translation Delta-CU Parallel Robot. In Proceedings of the 6th IFToMM International Symposium on Robotics and Mechatronics, ISRM 2019, Taipei, Taiwan, 28 October–3 November 2019.
16. Shen, H.; Xu, Q.; Li, J.; Wu, G.; Yang, T.-L. The Effect of Selection of Virtual Variable on the Direct Kinematics of Parallel Mechanisms. In *New Trends in Mechanism and Machine Science*; Springer: Cham, Switzerland, 2020.
17. Shen, H.; Ji, H.; Xu, Z.; Yang, T. Design, Kinematic Symbolic Solution and Performance Evaluation of a New Three Translation Mechanism. *Trans. Chin. Soc. Agric. Mach.* **2020**, *51*, 397–407.
18. Shen, H.; Zhu, Z.; Meng, Q.; Wu, G.; Deng, J. Kinematics and Stiffness Modeling Analysis of a Spatial 2T1R Parallel Mechanism with Zero Coupling Degree. *Trans. Chin. Soc. Agric. Mach.* **2020**, *51*, 411–420.

Publisher's Note: MDPI stays neutral with regard to jurisdictional claims in published maps and institutional affiliations.

Review

A Review of the Literature on the Lower-Mobility Parallel Manipulators of 3-UPU or 3-URU Type

Raffaele Di Gregorio

Department of Engineering, University of Ferrara, 44122 Ferrara, Italy; raffaele.digregorio@unife.it;
Tel.: +39-0532-974-828

Received: 10 December 2019; Accepted: 10 January 2020; Published: 13 January 2020

Abstract: Various 3-UPU architectures feature two rigid bodies connected to one another through three kinematic chains (limbs) of universal–prismatic–universal (UPU) type. They were first proposed in the last decade of the 20th century and have animated discussions among researchers for more-or-less two decades. Such discussions brought to light many features of lower-mobility parallel manipulators (PMs) that were unknown until then. The discussions also showed that such architectures may be sized into translational PMs, parallel wrists, or even reconfigurable (metamorphic) PMs. Even though commercial robots with these architectures have not yet been built, the interest in them remains. Consequently, a review of the literature on these architectures, highlighting their contribution to the progress of lower-mobility PM design, is still of interest for the scientific community. This paper aims at presenting a critical review of the results that have been obtained up until now.

Keywords: parallel manipulators; lower mobility; reconfigurable mechanism; singularity locus; constraint singularities; structural singularity

---

1. Introduction

The most common parallel manipulators with three degrees of freedom (DOF) are constituted by two rigid bodies, the end effector (platform) and the frame (base), joined by three kinematic chains (limbs) with the same topology and connectivity[1] equal to five. Among these parallel manipulators (PMs), 3-UPU architectures[2] (Figure 1) are those with three limbs of UPU type. Such architectures are special cases of the 3-UTU ones, where T denotes a generic single-DOF kinematic pair that makes the axes of the two intermediate R-pairs of the limb translate with respect to one another. A T pair, over a P pair, could be, for instance, an R-pair with axis parallel to the two that have to translate with respect to one another (see Figure 2). Even though most of the literature refers to 3-U$\underline{P}$U, the presented results hold for any 3-UTU.

In 1996, Tsai [2] proposed a translational 3-U$\underline{P}$U. The Tsai 3-U$\underline{P}$U (Figure 1c) had the three R-pair axes fixed in the base (in the platform) in a coplanar arrangement. In [2], Tsai solved, in explicit form, the direct (DPA) and inverse (IPA) position analyses of this translational PM (TPM), which gave two DPA solutions and one IPA solution. Such results made him conclude that the translational 3-U$\underline{P}$U was simple to manufacture and to control.

Following Tsai's proposal, Park built a prototype, named SNU 3-U$\underline{P}$U [3–5] (Figure 1d), which had the three R-pair axes fixed in the base (in the platform) in a coplanar arrangement and with a common intersection point. In 2001 [3], he presented the prototype and highlighted that the SNU 3-U$\underline{P}$U

---

[1] According to [1], here, we use the term "limb connectivity" to denote the DOF number the platform would have if it were connected to the base only through that limb.

[2] Hereafter, U, S, R, and P stand for universal joint, spherical pair, revolute pair, and prismatic pair, respectively. Also, the underscore denotes an actuated kinematic pair.

exhibited an unforeseen extra mobility at the home position, which made the platform orientation change. Such strange behavior started animated discussions [3–9] that, in 2002, identified the constraint singularities [3,6] as the main cause of the strange behavior and as a negative feature that occurs in most of the lower-mobility PMs.

In the meantime, Parenti-Castelli et al. [10–16] studied more general families of TPMs that included both the Tsai 3-UPU and the SNU 3-UPU. In 1998, Di Gregorio and Parenti-Castelli [10] showed that a PM of 3-RRPRR type becomes a TPM if it is manufactured and assembled out of particular singular configurations (later called constraint singularities [6]), so that, in each RRPRR limb,

(i)   the axes of the two intermediate R-pairs are parallel to one another, and

(ii)   the axes of the two ending R-pairs are parallel to each other.

Then, from 1999 to 2000, together with Bubani, Di Gregorio and Parenti-Castelli studied in depth the whole family of translational 3-UPUs with the three R-pair axes fixed in the base (in the platform) in a coplanar arrangement. In [11], Di Gregorio and Parenti-Castelli presented the general expressions of the singularity loci of any translational 3-UPU and demonstrated that, in the above-mentioned case of coplanar axes, the rotation (constraint) singularity locus is constituted by a right circular cylinder (Figure 3), which could degenerate [11,12] (Figure 4) for particular platform (base) geometries, and a plane. In the SNU 3-UPU (Figure 1d), the cylinder equation, reported in [11], becomes the equation of a line perpendicular to the base plane and passing through the home position, which explains the strange behavior found by Park. These results were exploited by Parenti-Castelli et al. to build a prototype [13,14] of the Tsai 3-UPU that worked correctly. Later, the same results together with an in-depth analysis of the joint-clearance effects allowed for Bhutani and Dwarakanath [17,18] to build a "high-precision" Tsai 3-UPU that could be used as a measuring machine.

In 2000, Karouia and Hervè [19] identified the geometric conditions that make a 3-UPU architecture become a parallel wrist[3] (PW). In particular, by using group theory, they demonstrated that, out of singular configurations, a 3-UPU architecture is a PW (Figure 1b), if

(a)   the platform and the base are manufactured so that the three R-pair axes embedded in them have a common intersection point;

(b)   each UPU limb (Figure 5) is manufactured and assembled so that the axes of the two intermediate R-pairs are parallel to one another; and

(c)   the 3-UPU is assembled so that the axes of the six R-pairs adjacent to the base or to the platform share a common intersection point (such point becomes the spherical motion center).

Karouia and Hervè [19] also advised that such conditions do not exclude the existence of singular configurations where the platform locally acquires an additional translational DOF. Later, Di Gregorio [20,21], by analyzing statics and kinematics of 3-UPU wrists, provided both the geometric (Figure 6) and the analytic conditions that identify the translation (constraint) singularities of these wrists. In [22], Ashith-Shyam and Ghosal presented a 3-UPU wrist prototype for sun tracking; and, in [23,24], Huda and Takeda presented the prototype of a 3-URU wrist for a machine tool and the adopted design methodology.

The consideration that translational 3-UPUs have rotation singularities and 3-UPU wrists have translation singularities pushed Zlatanov et al. [6] to introduce the concept of "constraint singularity" and to highlight that these singularities may occur in any lower-mobility PMs. A constraint singularity is a configuration where a lower-mobility PM may change its operating mode. They may occur when the limbs' connectivity is higher than the DOF number of the PM. Also, Zlatanov et al. [25] illustrated this concept through the DYMO 3-URU prototype (Figure 7), which exploited its constraint singularities

---

3   Parallel wrists (PWs) are PMs in which the relative motion between platform and base can only be a spherical motion with a fixed center.

to change its operating mode: It was able to become a TPM, a PW, or a 3-DOF planar PM. DYMO showed for the first time that a 3-UTU can be a reconfigurable machine.

In particular, a 3-UTU can switch from TPM to PW and vice versa if it satisfies conditions (a) and (b) from Karouia and Hervè [19] as stated above; whereas, it can switch from TPM or PW to 3-DOF planar PM and vice versa, if in addition to satisfying conditions (a) and (b), the three R-pair axes fixed in the base (in the platform) are coplanar. The central issue for actually getting a reconfigurable 3-UTU is how to manage the passage through a singular configuration? In [26], Carbonari et al. bypassed the problem by proposing a reconfigurable 3-URU (Figure 8) that could switch from TPM to PW and vice versa at a given non-singular configuration through an ad-hoc-conceived device, which modifies the geometry of the U-joints adjacent to the base. In the same line, Sarabandi et al. [27] presented a particular 3-UPU geometry that can switch from TPM to PW and vice versa by simply turning the platform assembly upside down. This 3-UPU could do the same switch by passing through a singular configuration without disassembling the platform, but the authors did not propose any strategy to go through the singularity.

Eventually, structural singularities of 3-UPU architectures were used to ideate a Shoenflies motion generator [28] of 4-UPU type [29–32], and a rolling mechanism [33].

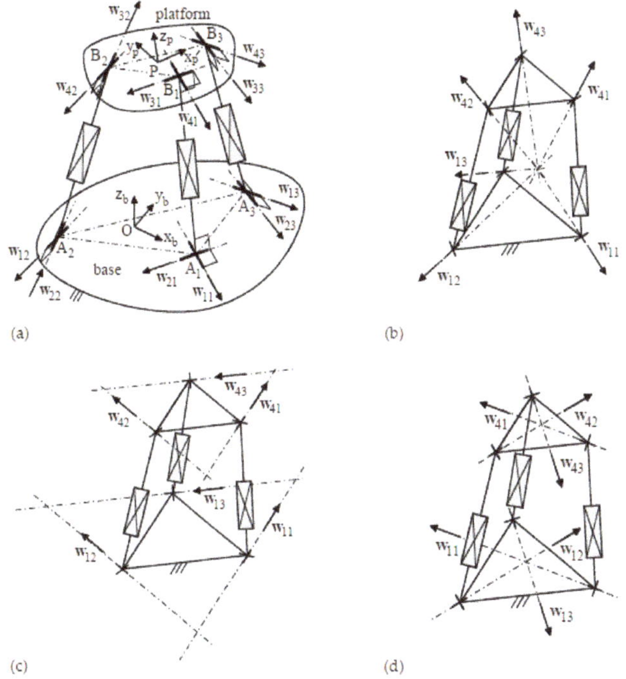

Figure 1. 3-UPU architectures: (**a**) general geometry and notations, (**b**) 3-UPU wrist, (**c**) Tsai 3-UPU, (**d**) SNU 3-UPU.

This review paper aims at summarizing the relevant analytic and geometric aspects of the above-mentioned results in a unique framework that should be useful for designers and researchers. The paper is organized as follows. Section 2 gives some background concepts and presents the adopted notations together with some general comments. Sections 3 and 4 analyze translational and wrist architectures, respectively. Then, Section 5 analyzes reconfigurable and structurally singular architectures, and Section 6 draws the conclusions.

2. Background, Notations, and General Comments

The instantaneous input–output relationship of a PM is a linear and homogeneous system that relates the platform twist and the actuated-joint rates. In this relationship, the two coefficient matrices (Jacobians) that multiply the platform twist, and the actuated-joint rates depend only on the PM configuration. Singularities are PM configurations that makes either or both these Jacobians rank-deficient. The PM configuration is not controllable at a singularity. In particular, singularities of the Jacobian that multiplies the platform twist (named parallel singularities) occur inside the reachable workspace, and make the platform gain one or more instantaneous DOFs locally (i.e., they are a particular type of uncertainty configurations [1]). From a statics' point of view [21], at a parallel singularity, the platform is not able to carry external loads, even small ones, without overloading at least one link (i.e., without breaking down at least one link). Consequently, parallel singularities must be identified during design and avoided during operation by locating the useful workspace in free-from-singularity regions of the operational space.

That is why the possibility of building a particular type of 3-UPU is related to the identification of those geometries that provide wide free-from-singularity regions where the platform can perform only one type of motion (spatial translation (TPMs) or spherical motion (PWs) or planar motion). The following part of this paper illustrates the main results reported in the literature by analyzing the input–output instantaneous relationships of these architectures.

Figure 1a shows a general 3-UPU architecture together with the adopted notations. With reference to Figure 1a,

- $Ox_by_bz_b$ and $Px_py_pz_p$ are two Cartesian references fixed to the base and the platform, respectively;
- A_i (B_i) for i = 1, 2, 3 are the centers of the U joints adjacent to the base (platform);
- in each UPU limb, the four R-pairs are numbered with an index, j, that increases by moving from the base toward the platform;
- w_{ji}, for j = 1, ... , 4, is the j-th R-pair axis' unit vector of the i-th UPU limb, i = 1, 2, 3;
- w_{2i} and w_{3i} are perpendicular to the axis of the i-th limb (i.e., the line through A_i and B_i), for i = 1, 2, 3.

Moreover, the following parameters/vectors are defined:

- θ_{ji}, for j = 1, ... ,4, is the angular joint variable, counterclockwise with respect to w_{ji}, of the j-th R-pair of the i-th UPU limb, i = 1, 2, 3;
- $d_i = |B_i - A_i|$ is the linear joint variable of the P-pair (hereafter named "limb length") of the i-th UPU limb, i = 1, 2, 3;
- $p = (P - O)$; $b_i = (B_i - O) = p + b_{0i}$ with $b_{0i} = (B_i - P)$, for i = 1, 2, 3;
- $a_i = (A_i - O)$, for i = 1, 2, 3; $c_i = (b_{0i} - a_i)$ for i = 1, 2, 3; $g_i = (b_i - a_i)/d_i$ for i = 1, 2, 3;
- $r_i = w_{1i} \times w_{2i}$ for i = 1, 2, 3; $h_i = w_{3i} \times w_{4i}$ for i = 1, 2, 3; $n_i = [(b_i - a_i) \cdot r_i] h_i$ for i = 1, 2, 3.

With these notations, the following instantaneous relationships can be written:

$$\dot{b}_i = \dot{d}_i g_i + d_i\left(\dot{\theta}_{1i} w_{1i} + \dot{\theta}_{2i} w_{2i}\right) \times g_i \quad i = 1, 2, 3, \tag{1a}$$

$$\dot{b}_i = \dot{p} + \omega \times (b_i - p) \quad i = 1, 2, 3, \tag{1b}$$

$$\omega = \sum_{j=1,4} \dot{\theta}_{ji} w_{ji} \quad i = 1, 2, 3, \tag{1c}$$

where ω is the angular velocity of the platform, and $\dot{x}$ denotes the time derivative of x. Equations (1a), (1b), and (1c) are formally the same that appeared in [34] for the 3-nSPU manipulator and, with the same algebraic manipulations reported in [34], they yield the following instantaneous input–output relationship:

$$\begin{bmatrix} \mathbf{1} \\ \mathbf{0} \end{bmatrix} \dot{\mathbf{d}} = \begin{bmatrix} \mathbf{G} & \mathbf{K} \\ \mathbf{S} & \mathbf{J} \end{bmatrix} \begin{pmatrix} \dot{\mathbf{p}} \\ \boldsymbol{\omega} \end{pmatrix}, \tag{2}$$

where **1** and **0** are the 3×3 identity and null matrices, respectively; $\dot{\mathbf{d}} = (\dot{d}_1, \dot{d}_2, \dot{d}_3)^{\mathrm{T}}$ is the vector collecting the P-pairs' joint rates, which are the instantaneous inputs, and

$$\mathbf{G}^{\mathrm{T}} = (\mathbf{g}_1, \mathbf{g}_2, \mathbf{g}_3), \ \mathbf{K}^{\mathrm{T}} = (\mathbf{k}_1, \mathbf{k}_2, \mathbf{k}_3), \ \mathbf{S}^{\mathrm{T}} = (\mathbf{s}_1, \mathbf{s}_2, \mathbf{s}_3), \ \mathbf{J}^{\mathrm{T}} = (\mathbf{j}_1, \mathbf{j}_2, \mathbf{j}_3) \tag{3}$$

with

$$\mathbf{k}_i = (\mathbf{b}_i - \mathbf{p}) \times \mathbf{g}_i, \ \mathbf{s}_i = \mathbf{h}_i \times \mathbf{r}_i - [\mathbf{g}_i \cdot (\mathbf{h}_i \times \mathbf{r}_i)] \, \mathbf{g}_i, \ \mathbf{j}_i = (\mathbf{b}_i - \mathbf{p}) \times \mathbf{s}_i - [(\mathbf{b}_i - \mathbf{a}_i) \cdot \mathbf{r}_i] \, \mathbf{h}_i, \qquad i = 1, 2, 3. \tag{4}$$

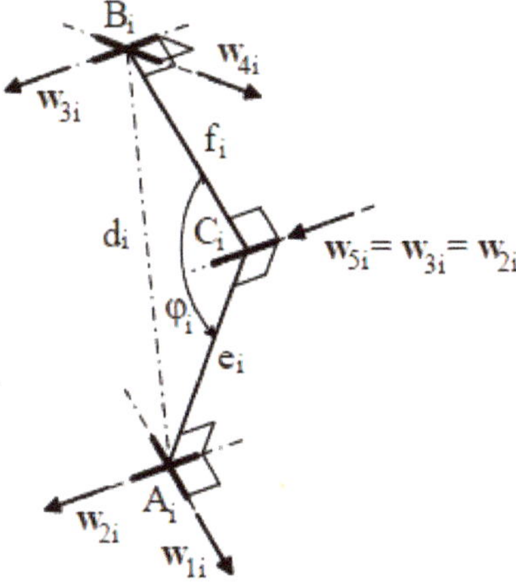

Figure 2. The i-th limb of a 3-U<u>R</u>U.

System (2) holds for any 3-UPU no matter if it is a TPM or a PW. It becomes specific when the above-mentioned geometric conditions that make the 3-UPU a TPM or a PW are inserted into it. The last three equations of system (2) do not involve the input-joint rates and are general no matter which input variables (i.e., the chosen actuated joints) are used; when their coefficient matrix (i.e., the 3×6 matrix [**S J**]) is rank-deficient, a constraint singularity occurs.

The first three equations of system (2) relate the input-joint rates to the platform twist; in these equations, when the coefficient matrix that multiplies the platform twist (i.e., the 3×6 matrix [**G K**]) is rank deficient, the platform can perform elementary motions without changing its operating mode, even though the actuated joints are locked. These three equations vary together with the associated singularity conditions if the input variables are changed. Nevertheless, if the actuated joints just control the limb lengths, the changes involve only the left-hand sides of these equations, leaving the 3×6 matrix [**G K**] unchanged together with the associated singularities. For instance, with reference to Figure 2, if the actuated P-pair of the i-th limb is replaced by an actuated R-pair with an axis parallel to $\mathbf{w}_{2i}$ and $\mathbf{w}_{3i}$, the following relationships hold

$$d_i^2 = e_i^2 + f_i^2 - 2e_i f_i \cos \varphi_i \qquad\qquad i = 1, 2, 3 \tag{5}$$

whose 1st time derivatives are

$$\dot{d}_i = \frac{e_i f_i \sin \varphi_i}{\sqrt{e_i^2 + f_i^2 - 2e_i f_i \cos \varphi_i}} \dot{\varphi}_i \qquad\qquad i = 1, 2, 3. \qquad\qquad (6)$$

Thus, the simple substitution of the right-hand side of Equation (6) for $\dot{d}_i$, $i = 1, 2, 3$ into the left-hand side of system (2) transforms the instantaneous input–output relationship of a 3-U$\underline{P}$U into the one of a 3-U$\underline{R}$U.

3. Translational 3-U$\underline{T}$U

The geometric conditions, (i) and (ii), for getting a TPM, with the adopted notations, become (i) $\mathbf{w}_{2i} = \pm\mathbf{w}_{3i}$ and (ii) $\mathbf{w}_{1i} = \pm\mathbf{w}_{4i}$ for $i = 1, 2, 3$, which yield $\mathbf{h}_i = \pm\mathbf{r}_i$, $\mathbf{s}_i = 0$ and $\mathbf{j}_i = \pm[(\mathbf{b}_i - \mathbf{a}_i)\cdot\mathbf{h}_i]\,\mathbf{h}_i$. Consequently, the instantaneous input–output relationship (2) becomes

$$\begin{bmatrix} 1 \\ 0 \end{bmatrix} \dot{\mathbf{d}} = \begin{bmatrix} \mathbf{G} & \mathbf{K} \\ \mathbf{0} & \mathbf{J} \end{bmatrix} \begin{pmatrix} \dot{\mathbf{p}} \\ \boldsymbol{\omega} \end{pmatrix}. \qquad\qquad (7)$$

3.1. Rotation (Constraint) Singularities

By canceling the coefficients $\pm[(\mathbf{b}_i - \mathbf{a}_i)\cdot\mathbf{h}_i]^4$ the last three equations of system (7) become [12]

$$\mathbf{h}_i\cdot\boldsymbol{\omega} = 0 \qquad i = 1, 2, 3. \qquad\qquad (8)$$

Equation (8) admits a non-null solution for $\boldsymbol{\omega}$ (i.e., a rotation (constraint) singularity occurs) if and only if

$$\mathbf{h}_1\cdot(\mathbf{h}_2 \times \mathbf{h}_3) = 0. \qquad\qquad (9)$$

From a geometric point of view, Equation (9) is satisfied when the three vectors $\mathbf{h}_i$, for $i = 1, 2, 3$, are coplanar (i.e., when all the intersections among the planes parallel to the U-joints' cross links are parallel lines). Consequently, if the unit vectors $\mathbf{w}_{1i}$ ($\mathbf{w}_{4i}$), for $i = 1, 2, 3$, are all parallel, this geometric condition is always satisfied[5] and a structural rotation (constraint) singularity occurs [29–31].

From an analytic point of view [11], Equation (9) is an algebraic equation, whose unknowns are the coordinates of a platform point[6] measured in $Ox_b y_b z_b$, which represents a surface (rotation (constraint) singularity locus) in $Ox_b y_b z_b$ (the operational space) whose points locate the singular configurations where the platform can rotate. The deduction of this algebraic equation is as follows[7]:

$$\mathbf{w}_{3i} = \frac{\mathbf{w}_{4i} \times (\mathbf{b}_i - \mathbf{a}_i)}{|\mathbf{w}_{4i} \times (\mathbf{b}_i - \mathbf{a}_i)|} = \frac{\mathbf{w}_{1i} \times [\mathbf{p} + (\mathbf{b}_{0i} - \mathbf{a}_i)]}{|\mathbf{w}_{1i} \times [\mathbf{p} + (\mathbf{b}_{0i} - \mathbf{a}_i)]|} = \frac{\mathbf{w}_{1i} \times \mathbf{p} + \mathbf{c}_i}{|\mathbf{w}_{1i} \times \mathbf{p} + \mathbf{c}_i|} \quad i = 1, 2, 3 \qquad (10)$$

which yields

$$\mathbf{h}_i = \mathbf{w}_{3i} \times \mathbf{w}_{4i} = \frac{(\mathbf{w}_{1i} \times \mathbf{p} + \mathbf{c}_i) \times \mathbf{w}_{1i}}{|\mathbf{w}_{1i} \times \mathbf{p} + \mathbf{c}_i|} = \frac{\mathbf{p} - (\mathbf{w}_{1i} \cdot \mathbf{p})\mathbf{w}_{1i} + \mathbf{c}_i \times \mathbf{w}_{1i}}{|\mathbf{w}_{1i} \times \mathbf{p} + \mathbf{c}_i|} \quad i = 1, 2, 3 \qquad (11)$$

---

[4] Since this coefficient has no effect on the value of $\boldsymbol{\omega}$ when it is different from zero, the value of $\boldsymbol{\omega}$ as this coefficient goes to zero is unchanged. Therefore, the zeroing of this coefficient does not affect the angular velocity of the platform and does not identify a rotation (constraint) singularity.

[5] Indeed, in this case, all the intersections among the cross-links' planes are lines parallel to the unit vectors $\mathbf{w}_{1i}$ ($\mathbf{w}_{4i}$).

[6] In TPMs, the coordinates of a platform point are sufficient to identify the platform pose in the operational space since the platform translates with respect to the base.

[7] Note that, in a TPM, the above-defined vectors $\mathbf{c}_i$ (= $\mathbf{b}_{0i} - \mathbf{a}_i$), $i = 1, 2, 3$, are constant vectors since the platform translates.

and

$$\mathbf{h}_1 \cdot (\mathbf{h}_2 \times \mathbf{h}_3) = \frac{[\mathbf{p}-(\mathbf{w}_{11}\cdot\mathbf{p})\mathbf{w}_{11}+\mathbf{c}_1\times\mathbf{w}_{11}]\cdot\{[\mathbf{p}-(\mathbf{w}_{12}\cdot\mathbf{p})\mathbf{w}_{12}+\mathbf{c}_2\times\mathbf{w}_{12}]\times[\mathbf{p}-(\mathbf{w}_{13}\cdot\mathbf{p})\mathbf{w}_{13}+\mathbf{c}_3\times\mathbf{w}_{13}]\}}{|\mathbf{w}_{11}\times\mathbf{p}+\mathbf{c}_1||\mathbf{w}_{12}\times\mathbf{p}+\mathbf{c}_2||\mathbf{w}_{13}\times\mathbf{p}+\mathbf{c}_3|}. \tag{12}$$

Since the denominator of expression (12) is constituted by the product of vector magnitudes, it does not provide zeros of Equation (9); hence, in Equation (9), it can be eliminated to give the following algebraic equation of the singularity locus:

$$[\mathbf{p}-(\mathbf{w}_{11}\cdot\mathbf{p})\mathbf{w}_{11}+\mathbf{c}_1\times\mathbf{w}_{11}]\cdot\{[\mathbf{p}-(\mathbf{w}_{12}\cdot\mathbf{p})\mathbf{w}_{12}+\mathbf{c}_2\times\mathbf{w}_{12}]\times[\mathbf{p}-(\mathbf{w}_{13}\cdot\mathbf{p})\mathbf{w}_{13}+\mathbf{c}_3\times\mathbf{w}_{13}]\}=0 \tag{13a}$$

whose expansion yields

$$\begin{aligned}
&(\mathbf{w}_{12}\cdot\mathbf{p})(\mathbf{w}_{13}\cdot\mathbf{p})[(\mathbf{w}_{12}\times\mathbf{w}_{13})\cdot\mathbf{p}]-(\mathbf{w}_{12}\cdot\mathbf{p})\{[\mathbf{w}_{12}\times(\mathbf{c}_3\times\mathbf{w}_{13})]\cdot\mathbf{p}\}-(\mathbf{w}_{13}\cdot\mathbf{p})\{[(\mathbf{c}_2\times\mathbf{w}_{12})\times\mathbf{w}_{13}]\cdot\mathbf{p}\}+\\
&+\{[(\mathbf{c}_2\times\mathbf{w}_{12})\times(\mathbf{c}_3\times\mathbf{w}_{13})]\cdot\mathbf{p}\}+(\mathbf{w}_{11}\cdot\mathbf{p})\{[(\mathbf{w}_{13}\cdot\mathbf{p})\mathbf{w}_{13}-(\mathbf{w}_{12}\cdot\mathbf{p})\mathbf{w}_{12}-(\mathbf{c}_3\times\mathbf{w}_{13})+\mathbf{c}_2\times\mathbf{w}_{12}]\times\mathbf{w}_{11}\}\cdot\mathbf{p}+\\
&-(\mathbf{w}_{11}\cdot\mathbf{p})(\mathbf{w}_{12}\cdot\mathbf{p})(\mathbf{w}_{13}\cdot\mathbf{p})[(\mathbf{w}_{12}\times\mathbf{w}_{13})\cdot\mathbf{w}_{11}]+(\mathbf{w}_{11}\cdot\mathbf{p})(\mathbf{w}_{12}\cdot\mathbf{p})[\mathbf{w}_{12}\times(\mathbf{c}_3\times\mathbf{w}_{13})]\cdot\mathbf{w}_{11}+\\
&+(\mathbf{w}_{11}\cdot\mathbf{p})(\mathbf{w}_{13}\cdot\mathbf{p})[(\mathbf{c}_2\times\mathbf{w}_{12})\times\mathbf{w}_{13}]\cdot\mathbf{w}_{11}-(\mathbf{w}_{11}\cdot\mathbf{p})[(\mathbf{c}_2\times\mathbf{w}_{12})\times(\mathbf{c}_3\times\mathbf{w}_{13})]\cdot\mathbf{w}_{11}-\{[(\mathbf{w}_{13}\cdot\mathbf{p})\mathbf{w}_{13}+\\
&-(\mathbf{w}_{12}\cdot\mathbf{p})\mathbf{w}_{12}-(\mathbf{c}_3\times\mathbf{w}_{13})+\mathbf{c}_2\times\mathbf{w}_{12}]\times(\mathbf{c}_1\times\mathbf{w}_{11})\}\cdot\mathbf{p}+(\mathbf{w}_{12}\cdot\mathbf{p})(\mathbf{w}_{13}\cdot\mathbf{p})[(\mathbf{w}_{12}\times\mathbf{w}_{13})\cdot(\mathbf{c}_1\times\mathbf{w}_{11})]+\\
&-(\mathbf{w}_{12}\cdot\mathbf{p})\{[\mathbf{w}_{12}\times(\mathbf{c}_3\times\mathbf{w}_{13})]\cdot(\mathbf{c}_1\times\mathbf{w}_{11})\}-(\mathbf{w}_{13}\cdot\mathbf{p})\{[(\mathbf{c}_2\times\mathbf{w}_{12})\times\mathbf{w}_{13}]\cdot(\mathbf{c}_1\times\mathbf{w}_{11})\}+\\
&+[(\mathbf{c}_2\times\mathbf{w}_{12})\times(\mathbf{c}_3\times\mathbf{w}_{13})]\cdot(\mathbf{c}_1\times\mathbf{w}_{11})=0.
\end{aligned} \tag{13b}$$

Equation (13) is cubic in the coordinates of point P (see Figure 1a). Since the coefficients that appear in Equation (13) depend on the shape of the platform and base, the rotation (constraint) singularity locus depends only on the platform and base geometries.

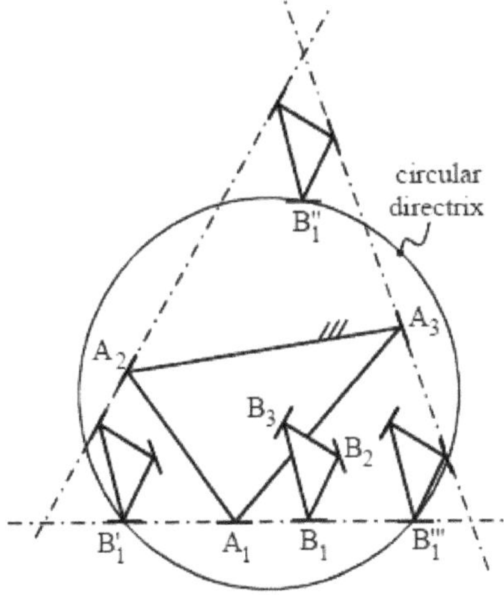

Figure 3. Singularity cylinder: graphic determination of the circular directrix.

In the SNU 3-UPU (Figure 1d), the base (platform) triangle $A_1A_2A_3$ ($B_1B_2B_3$) is an equilateral triangle, and the unit vectors $\mathbf{w}_{1i}$ ($\mathbf{w}_{4i}$), $i = 1, 2, 3$, lie on three R-pair axes that have the center of this triangle as a common intersection. By choosing this center as origin O (P), and the triangle plane as the x_by_b (x_py_p) coordinate plane for $Ox_by_bz_b$ ($Px_py_pz_p$), it is easy to realize that, in Equation (13a), the vectors $\mathbf{c}_i \times \mathbf{w}_{1i}$, for $i = 1, 2, 3$, are all null vectors since $\mathbf{c}_i$ is parallel to $\mathbf{w}_{1i}$. Consequently, when P lies on the line through O perpendicular to the base triangle $A_1A_2A_3$ (i.e., the SNU 3-UPU is at its home position), the dot products ($\mathbf{w}_{1i} \cdot \mathbf{p}$), for $i = 1, 2, 3$, are equal to zero and the left-hand side of Equation (13a), which becomes $\mathbf{p} \cdot (\mathbf{p} \times \mathbf{p})$, is identically equal to zero, that is, the platform can rotate.

In the Tsai 3-UPU (Figure 1c), the base (platform) triangle $A_1A_2A_3$ ($B_1B_2B_3$) is an equilateral triangle, but the i-th unit vector $\mathbf{w}_{1i}$ ($\mathbf{w}_{4i}$), $i = 1, 2, 3$, is parallel to the base-triangle (platform-triangle) side opposite to the vertex A_i (B_i), which the corresponding R-pair axis passes through. For this geometry, Equation (13) yields, as a singularity locus, a cubic surface (see [14]), that is the product of the base-triangle plane by a right circular cylinder whose generatrix is a line perpendicular to the base-triangle plane. The analytic expression of this cylinder is reported in [14].

The set of all the 3-UPUs with the axes of the three R-pairs, adjacent to the base (platform), that lie on the plane of the base (platform) triangle contains both SNU and Tsai 3-UPUs. This set was studied in [12]. The introduction of the geometric conditions that identify this set into Equation (13) shows that [12] the rotation (constraint) singularity locus of all these 3-UPUs is always the product of the base-triangle plane by a right circular cylinder whose generatrix is a line perpendicular to the base-triangle plane. By choosing A_1 (B_1) as origin O (P), and the base-triangle (platform-triangle) plane as $x_b y_b$ ($x_p y_p$) coordinate plane for $Ox_b y_b z_b$ ($Px_p y_p z_p$), the analytic expression of this cylinder is reported in [12] together with the simple geometric construction shown in Figure 3, which allows to draw immediately the singularity cylinder. The construction of Figure 3 relies on the fact that three points are sufficient to identify a circle, and that three singularities, B_1', B_1'' and B_1''' in Figure 3, are easy to find. The same construction highlights (Figure 4) that, when any two R-pair axes (together with the corresponding unit vectors w_{1i}) are parallel, the singularity cylinder degenerates into a singularity plane (see [12] for details).

3.2. Translation Singularities

Out of rotation singularities, the platform angular velocity, ω, is a null vector. Thus, the first three equations of system (7) become [12]:

$$g_i \cdot \dot{p} = \dot{d}_i \quad i = 1, 2, 3. \tag{14}$$

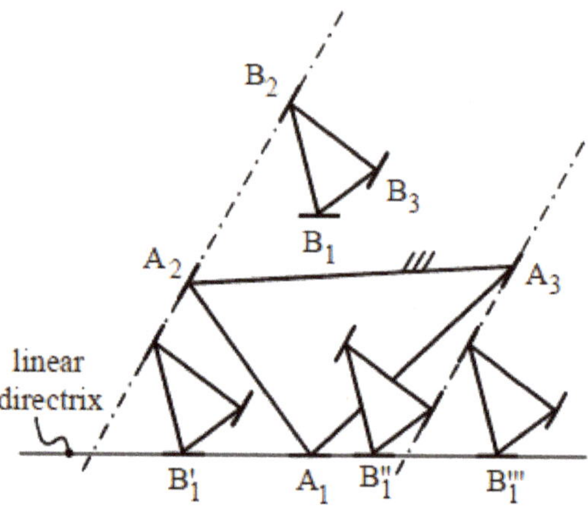

Figure 4. Case with two parallel R-pair axes: The circular directrix of the singularity cylinder degenerates into a linear directrix (i.e., the cylinder degenerates into a plane).

When the actuators are locked, Equations (14) admit a non-null solution for $\dot{p}$ (i.e., a translation singularity occurs) if and only if

$$g_1 \cdot (g_2 \times g_3) = 0. \tag{15}$$

From a geometric point of view, Equation (15) is satisfied when the three unit vectors g_i, for $i = 1, 2, 3$, are parallel to a plane (i.e., when all the limb axes are parallel to a unique plane). If the base and platform triangles are equal, this condition is always satisfied and a structural translation singularity occurs [2,12].

From an analytic point of view [11], Equation (15) is an algebraic equation, whose unknowns are the coordinates of a platform point measured in $Ox_by_bz_b$, which represents a surface (translation singularity locus) in $Ox_by_bz_b$ whose points locate the singular configurations where the platform translation is not controllable by the actuators. The deduction of this algebraic equation is as follows:

$$\mathbf{g}_1 \cdot (\mathbf{g}_2 \times \mathbf{g}_3) = \frac{(\mathbf{p} + \mathbf{b}_{01} - \mathbf{a}_1) \cdot [(\mathbf{p} + \mathbf{b}_{02} - \mathbf{a}_2) \times (\mathbf{p} + \mathbf{b}_{03} - \mathbf{a}_3)]}{d_1 d_2 d_3}. \tag{16}$$

Since the denominator of expression (16) is constituted by the product of the limb lengths, it does not provides zeros of Equation (15); hence, in Equation (15), it can be eliminated to give the following algebraic equation of the singularity locus

$$(\mathbf{p} + \mathbf{b}_{01} - \mathbf{a}_1) \cdot [(\mathbf{p} + \mathbf{b}_{02} - \mathbf{a}_2) \times (\mathbf{p} + \mathbf{b}_{03} - \mathbf{a}_3)] = 0 \tag{17a}$$

whose expansion is

$$\mathbf{p} \cdot \{[(\mathbf{b}_{03} - \mathbf{a}_3) \times (\mathbf{b}_{01} - \mathbf{a}_1)] + [(\mathbf{b}_{01} - \mathbf{a}_1) \times (\mathbf{b}_{02} - \mathbf{a}_2)] + [(\mathbf{b}_{02} - \mathbf{a}_2) \times (\mathbf{b}_{03} - \mathbf{a}_3)]\} + (\mathbf{b}_{01} - \mathbf{a}_1) \cdot [(\mathbf{b}_{02} - \mathbf{a}_2) \times (\mathbf{b}_{03} - \mathbf{a}_3)] = 0. \tag{17b}$$

Equation (17b) is linear in the coordinates of P. Therefore, the translation singularity locus is always a plane [11], which is perpendicular to the vector in curly brackets that dot multiplies $\mathbf{p}$ in Equation (17b). By choosing A_1 (B_1) as origin O (P), and the base-triangle (platform-triangle) plane as the x_by_b (x_py_p) coordinate plane for $Ox_by_bz_b$ ($Px_py_pz_p$), it is easy to realize that, if the base and platform triangles are equal, the vectors $(\mathbf{b}_{0i} - \mathbf{a}_i)$, for i = 1, 2, 3, are all null vectors. Consequently, the left-hand side of Equation (17b) is identically null (i.e., a structural singularity occurs). Also, in the case of the 3-UPUs with the axes of the three R-pairs, adjacent to the base (platform), that lie on the plane of the base (platform) triangle, it is easy to realize that the singularity plane is the base-triangle plane [12]. Indeed, in this case, the vectors $(\mathbf{b}_{0i} - \mathbf{a}_i)$, for i = 1, 2, 3, are all parallel to the base plane, which implies that, in Equation (17b), the mixed product $(\mathbf{b}_{01} - \mathbf{a}_1) \cdot [(\mathbf{b}_{02} - \mathbf{a}_2) \times (\mathbf{b}_{03} - \mathbf{a}_3)]$ is equal to zero and the vector in curly brackets is perpendicular to the base-triangle plane.

4. 3-U$\underline{\text{T}}$U Wrist

The geometric condition (a) for getting a PW, allows for the choice of the point O (P), see Figure 1a,b, coincident with the common intersection point of the three axes of the R-pairs adjacent to the base (platform). Such a choice makes $\mathbf{a}_i$ ($\mathbf{b}_i$) parallel to $\mathbf{w}_{1i}$ ($\mathbf{w}_{4i}$), for i = 1, 2, 3. In addition, condition (b) yields $\mathbf{w}_{2i} = \pm\mathbf{w}_{3i}$; whereas, condition (c) implies that O coincides with P (i.e., $\mathbf{p} = 0$). Consequently, $\mathbf{k}_i = \mathbf{b}_i \times \mathbf{g}_i$, $\mathbf{s}_i = \mathbf{h}_i \times \mathbf{r}_i$ since $\mathbf{g}_i$ is perpendicular to $\mathbf{h}_i \times \mathbf{r}_i$, and $\mathbf{j}_i = 0$, for i = 1, 2, 3 (see Equation (4)). These formulas allow for the conclusion that $\mathbf{k}_i$ and $\mathbf{s}_i$ are both parallel to $\mathbf{w}_{2i}$ and $\mathbf{w}_{3i}$ (see Figure 5), which are unit vectors perpendicular to the plane of the triangle A_iB_iP.

Therefore, the instantaneous input–output relationship (2) for the 3-UPU wrist becomes

$$\begin{bmatrix} 1 \\ 0 \end{bmatrix} \dot{\mathbf{d}} = \begin{bmatrix} \mathbf{G} & \mathbf{K} \\ \mathbf{S} & 0 \end{bmatrix} \begin{pmatrix} \dot{\mathbf{p}} \\ \boldsymbol{\omega} \end{pmatrix}. \tag{18}$$

4.1. Translation (Constraint) Singularities

The last three equations of system (18) become

$$\mathbf{s}_i \cdot \dot{\mathbf{p}} = 0 \quad i = 1, 2, 3. \tag{19}$$

Equations (19) admit a non-null solution for $\dot{\mathbf{p}}$ (i.e., a translation (constraint) singularity occurs) if and only if

$$\mathbf{s}_1 \cdot (\mathbf{s}_2 \times \mathbf{s}_3) = 0, \tag{20a}$$

which, since s_i is parallel to w_{2i} for i = 1, 2, 3, can be simplified as follows [20,21]:

$$\mathbf{w}_{21} \cdot (\mathbf{w}_{22} \times \mathbf{w}_{23}) = 0. \tag{20b}$$

Equation (20b) is satisfied when the three vectors $\mathbf{w}_{2i}$, for i = 1, 2, 3, are all parallel to a unique plane, that is, when the planes of the three triangles A_iB_iP, for i = 1, 2, 3, have a line as a common intersection (Figure 6) [20]. Also, it is worth noting that each unit vector $\mathbf{w}_{2i}$ is indeterminate when the triangle A_iB_iP is flattened (i.e., the points A_i, B_i, and P are aligned); in this case (see Figure 5), a simple inspection of the flattened limb reveals that the i-th limb can freely rotate around its axis.

A general analytic expression of Equation (20b) has been deduced in [21] where, by using the Rodrigues parameters [35] to parameterize the platform orientation, a fourth-degree polynomial equation in the three Rodrigues parameters has been obtained.

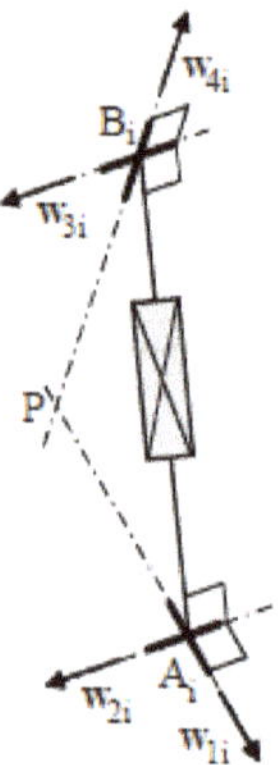

Figure 5. The i-th limb of a 3-UPU wrist.

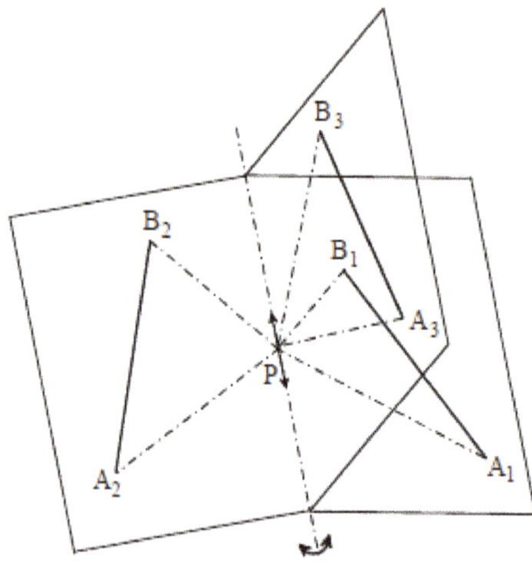

Figure 6. Singular configuration of a 3-UPU wrist.

4.2. Rotation Singularities

Out of the translation singularities, $\dot{\mathbf{p}}$ is a null vector. Thus, the first three equations of system (18) become [20]:

$$\mathbf{k}_i \cdot \boldsymbol{\omega} = \dot{d}_i \qquad i = 1,\, 2,\, 3. \tag{21}$$

When the actuators are locked, Equation (21) admit a non-null solution for $\boldsymbol{\omega}$ (i.e., a rotation singularity occurs) if and only if

$$\mathbf{k}_1 \cdot (\mathbf{k}_2 \times \mathbf{k}_3) = 0, \tag{22a}$$

which, since $\mathbf{k}_i$ is parallel to $\mathbf{w}_{2i}$ for i = 1, 2, 3, can be simplified as follows [20,21]:

$$\mathbf{w}_{21} \cdot (\mathbf{w}_{22} \times \mathbf{w}_{23}) = 0. \tag{22b}$$

Equation (22b) coincides with Equation (20b). Thus, the locus of the rotation singularities coincides with that of the translation (constraint) singularities in a 3-UPU wrist.

5. Reconfigurable and Structurally Singular 3-6UTUs

The presence of constraint singularities in 3-UTUs allows for the building of reconfigurable PMs, that is, machines that can change their operating mode. In [25], Zlatanov et al. presented DYMO (Figure 7a), a 3-URU that is able to become a TPM (Figure 7b), a PW (Figure 7c), or a 3-DOF planar PM (Figure 7d). DYMO (Figure 7a) satisfies geometric conditions (a) and (b), and has the three R-pair axes fixed in the base (in the platform) in a coplanar arrangement. So, the constraint singularity that occurs when the intersection of the three R-pair axes fixed in the platform coincides with the intersection of the three R-pair axes fixed in the base is present in all the three operating modes and can be exploited to change the operating mode.

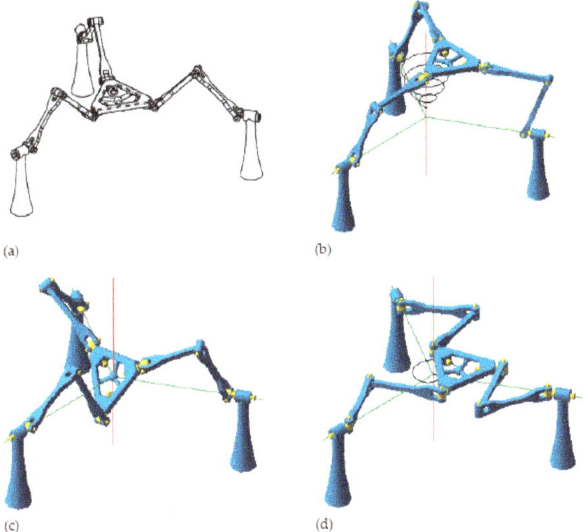

Figure 7. DYMO 3-URU: (**a**) 3D model, (**b**) translational parallel manipulator (TPM) operating mode, (**c**) parallel wrist (PW) operating mode, (**d**) 3-DOF planar PM operating mode. Figures downloaded from http://www.parallemic.org/Reviews/Review008.html and reproduced with the permission of the authors.

Unfortunately, the platform pose is out of control at a constraint singularity, and this simple method for reconfiguring the machine cannot be implemented. Carbonari et al. [26] bypassed the problem by proposing a reconfigurable 3-URU that could switch from TPM to PW and vice versa at a given non-singular configuration. Their 3-URU (Figure 8) satisfies conditions (a) and (b), and adopts an ad-hoc-conceived device, which modifies the geometry of the U-joints adjacent to the base. In the same line, Sarabandi et al. [27] presented a particular 3-UPU geometry that satisfies conditions (a) and (b) and can switch from TPM to PW and vice versa by simply turning the platform assembly upside down.

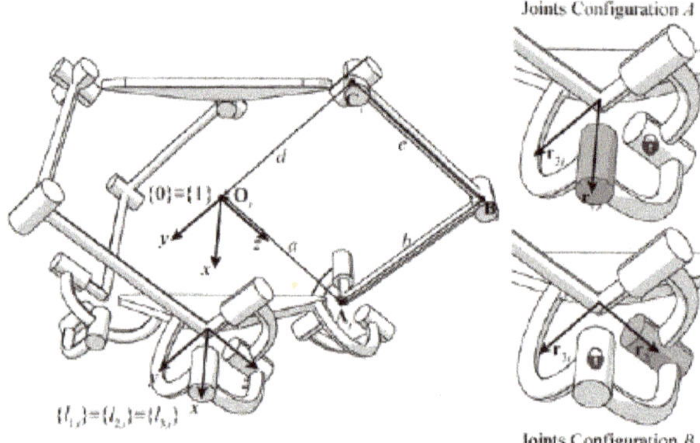

Figure 8. Carbonari et al. reconfigurable 3-URU [26]: Joint configurations A and B refer to the TPM and PW modes, respectively, the padlock denotes the locked R-pair, the darker R-pair is the actuated pair. Figure downloaded from [36] https://www.mdpi.com/2218-6581/7/3/42 and reproduced with the permission of the authors.

The rotation (constraint) structural singularity (see Section 3.1) that occurs in translational 3-UPUs when (Figure 1) the unit vectors $\mathbf{w}_{1i}$ ($\mathbf{w}_{4i}$), for i = 1, 2, 3, are all parallel makes the platform able to rotate around axes parallel to these unit vectors. Such additional finite DOF allows for the introduction of one more UPU limb to control the platform rotation. The resulting 4-U$\underline{P}$U is a Shoenflies motion generator [28]. It was presented in [29] and studied in [30–32].

Eventually, in a translational 3-UPU, if, over the above-mentioned structural singularity, the structural translation singularity (see Section 3.2) that occurs when the platform and base triangles are equal is introduced, the resulting 3-UPU acquires two additional finite DOFs: one rotation and one translation. Such a geometry has been used in [33] to conceive a 5-DOF rolling mechanism able to move on the floor.

6. Conclusions

A critical review of the extensive literature on 3-UTU architectures has been presented. The presented review allows for the following conclusions: The study of these architectures contributed to the mechanism theory by revealing the presence of "constraint singularities" in most of the lower-mobility PMs. All the design tools have been developed for both translational 3-UPUs and 3-UPU wrists. Even though commercial robots with these architectures are not present on the market, prototypes that work correctly have been built, and the structural singularities of these architectures have been exploited. The majority of the published works on the translational 3-UTUs refer to the set of 3-UTUs with coplanar axes of the three R-pairs adjacent to the base (platform).

Possible future research on these architectures should investigate translational 3-UTUs with base and platform geometries in which the R-pair axes are not coplanar and strategies to pass through a constraint singularity.

Funding: This work has been developed at the Laboratory of Mechatronics and Virtual Prototyping (LaMaViP) of Ferrara Technopole, supported by FAR2019 UNIFE funds.

Conflicts of Interest: The author declares no conflict of interest.

References

1. Hunt, K.H. *Kinematic Geometry of Mechanisms*; Clarendon Press: Oxford, UK, 1990.
2. Tsai, L.W. Kinematics of a Three-dof Platform with Three Extensible Limbs. In *Recent Advances in Robot Kinematics*; Lenarcic, J., Parenti-Castelli, V., Eds.; Kluwer Academic Publishers: Dordrecht, The Netherlands, 1996; pp. 401–410, ISBN 978-94-010-7269-4.
3. Bonev, I.; Zlatanov, D. The Mystery of the Singular SNU Translational Parallel Robot. ParalleMIC-The Parallel Mechanisms Information Center, (http://www.parallemic.org) 12 June 2001. Available online: https://www.parallemic.org/Reviews/Review004.html (accessed on 18 October 2019).
4. Han, C.; Kim, J.; Kim, J.; Park, F.C. Kinematic sensitivity analysis of the 3-UPU parallel mechanism. *Mech. Mach. Theory* **2002**, *37*, 787–798. [CrossRef]
5. Walter, D.R.; Husty, M.L.; Pfurner, M. A complete kinematic analysis of the SNU 3-UPU parallel robot. In *Interactions of Classical and Numerical Algebraic Geometry*; Bates, D.J., Besana, G.M., Di Rocco, S., Wampler, C.W., Eds.; Contemporary Mathematics; AMS: Providence, RI, USA, 2009; Volume 496, pp. 331–346, ISBN 978-0-8218-4746-6.
6. Zlatanov, D.; Bonev, I.A.; Gosselin, C.M. Constraint singularities of parallel mechanisms. In Proceedings of the IEEE International Conference on Robotics and Automation, Washington, DC, USA, 11–15 May 2002; pp. 496–502. [CrossRef]
7. Joshi, S.A.; Tsai, L.W. Jacobian analysis of limited-DOF parallel manipulators. *ASME J. Mech. Des.* **2002**, *124*, 254–258. [CrossRef]
8. Wolf, A.; Shoham, M.; Park, F.C. Investigation of singularities and self-motions of the 3-UPU robot. In *Advances in Robot Kinematics*; Lenaric, J., Thomas, F., Eds.; Kluwer: Norwell, MA, USA, 2002; pp. 165–174, ISBN 978-90-481-6054-9.
9. Liu, G.; Lou, Y.; Li, Z. Singularities of Parallel Manipulators: A Geometric Treatment. *IEEE Trans. Rob. Autom.* **2003**, *19*, 579–594. [CrossRef]
10. Di Gregorio, R.; Parenti-Castelli, V. A Translational 3-DOF Parallel Manipulator. In *Advances in Robot Kinematics: Analysis and Control*; Lenarcic, J., Husty, M.L., Eds.; Kluwer: Norwell, MA, USA, 1998; pp. 49–58, ISBN 978-90-481-5066-3.
11. Di Gregorio, R.; Parenti-Castelli, V. Influence of the geometric parameters of the 3–UPU parallel mechanism on the singularity loci. In Proceedings of the International Workshop on Parallel Kinematic Machines (PKM'99), Milan, Italy, 30 November 1999; pp. 79–86, ISBN 88-900426-0-5.
12. Di Gregorio, R.; Parenti-Castelli, V. Mobility analysis of the 3–UPU parallel mechanism assembled for a pure translational motion. *ASME J. Mech. Des.* **2002**, *124*, 259–264. [CrossRef]
13. Parenti-Castelli, V.; Bubani, F. Singularity loci and dimensional design of a translational 3-dof fully-parallel manipulator. In *Advances in Multibody Systems and Mechatronics*; Kecskeméthy, A., Schneider, M., Woernle, C., Eds.; Technische Universität Graz. Institut für Mechanik und Getriebelehre: Duisburg, Germany, 1999; pp. 319–331, ISBN 9783950110807.
14. Parenti-Castelli, V.; Di Gregorio, R.; Bubani, F. Workspace and Optimal Design of a Pure Translational Parallel Manipulator. *Meccanica* **2000**, *35*, 203–214. [CrossRef]
15. Di Gregorio, R.; Parenti-Castelli, V. Benefits of twisting the legs in the 3-UPU Tsai mechanism. In Proceedings of the Year 2000 Parallel Kinematic Machines International Conference, Ann Arbor, MI, USA, 13–15 September 2000; pp. 201–211.
16. Di Gregorio, R.; Parenti-Castelli, V. A New Approach for the evaluation of kinematic and static performances of a family of 3-UPU translational manipulators. In *Romansy 16: Robot Design, Dynamics, and Control*; Zielińska, T., Zieliński, C., Eds.; Springer: Vienna, Austria, 2006; pp. 47–54, ISBN 978-3-211-36064-4.
17. Bhutani, G.; Dwarakanath, T.A. Practical feasibility of a high-precision 3-UPU parallel mechanism. *Robotica* **2014**, *32*, 341–355. [CrossRef]
18. Bhutani, G.; Dwarakanath, T.A. Novel design solution to high precision 3 axes translational parallel mechanism. *Mech. Mach. Theory* **2014**, *75*, 118–130. [CrossRef]
19. Karouia, M.; Hervé, J.M. A three-dof tripod for generating spherical rotation. In *Advances in Robot Kinematics*; Lenarcic, J., Stanisic, M.M., Eds.; Springer: Dordrecht, The Netherlands, 2000; pp. 396–402, ISBN 978-94-010-5803-2.
20. Di Gregorio, R. Kinematics of the 3-UPU wrist. *Mech. Mach. Theory* **2003**, *38*, 253–263. [CrossRef]

21. Di Gregorio, R. Statics and singularity loci of the 3-UPU wrist. *IEEE Trans. Robot.* **2004**, *20*, 630–635. [CrossRef]

22. Ashith Shyam, R.B.; Ghosal, A. Path planning of a 3-UPU wrist manipulator for sun tracking in central receiver tower systems. *Mech. Mach. Theory* **2018**, *119*, 130–141. [CrossRef]

23. Huda, S.; Takeda, Y. Kinematic analysis and synthesis of a 3-URU pure rotational parallel mechanism with respect to singularity and workspace. *J. Adv. Mech. Des. Syst. Manuf.* **2007**, *1*, 81–92. [CrossRef]

24. Huda, S.; Takeda, Y. Kinematic Design of 3-URU Pure Rotational Parallel Mechanism with Consideration of Uncompensatable Error. *J. Adv. Mech. Des. Syst. Manuf.* **2008**, *2*, 874–886. [CrossRef]

25. Zlatanov, D.; Bonev, I.A.; Gosselin, C.M. Constraint Singularities as C-Space Singularities. In *Advances in Robot Kinematics: Theory and Applications*; Lenarčič, J., Thomas, F., Eds.; Springer: Dordrecht, The Netherlands, 2002; pp. 183–192, ISBN 978-90-481-6054-9.

26. Carbonari, L.; Corinaldi, D.; Palpacelli, M.; Palmieri, G.; Callegari, M. A Novel Reconfigurable 3-URU Parallel Platform. In *Advances in Service and Industrial Robotics*; Ferraresi, C., Quaglia, G., Eds.; Springer: Dordrecht, The Netherlands, 2018; pp. 63–73, ISBN 978-3-319-61275-1.

27. Sarabandi, S.; Grosch, P.; Porta, J.M.; Thomas, F. A Reconfigurable Asymmetric 3-UPU Parallel Robot. In Proceedings of the 2018 International Conference on Reconfigurable Mechanisms and Robots (ReMAR2018), Deft, The Netherlands, 20–22 June 2018; pp. 2–9.

28. Hervé, J.M. The lie group of rigid body displacements, a fundamental tool for mechanism design. *Mech. Mach. Theory* **1999**, *34*, 719–730. [CrossRef]

29. Zhao, T.S.; Dai, J.S.; Huang, Z. Geometric Analysis of Overconstrained Parallel Manipulators with Three and Four Degrees of Freedom. *JSME Int. J. Ser. C* **2002**, *45*, 730–740. [CrossRef]

30. Zhao, T.S.; Li, Y.W.; Chen, J.; Wang, J.C. Novel Four-DOF Parallel Manipulator Mechanism and Its Kinematics. In Proceedings of the 2006 IEEE Conference on Robotics, Automation and Mechatronics (RAM 2006), Bangkok, Thailand, 1–3 June 2006. [CrossRef]

31. Solazzi, M.; Gabardi, M.; Frisoli, A.; Bergamasco, M. Kinematics Analysis and Singularity Loci of a 4-UPU Parallel Manipulator. In *Advances in Robot Kinematics*; Lenarcic, J., Khatib, O., Eds.; Springer: Cham, Switzerland, 2014; pp. 467–474, ISBN 978-3-319-06698-1.

32. Coste, M.; Demdah, K.M. Extra Modes of Operation and Self-Motions in Manipulators Designed for Schoenflies Motion. *ASME J. Mech. Rob.* **2015**, *7*, 041020. [CrossRef]

33. Miao, Z.; Yao, Y.; Kong, X. A rolling 3-UPU parallel mechanism. *Front. Mech. Eng.* **2013**, *8*, 340–349. [CrossRef]

34. Grosch, P.; Di Gregorio, R.; Thomas, F. Generation of under-actuated manipulators with non-holonomic joints from ordinary manipulators. *ASME J. Mech. Rob.* **2010**, *2*, 011005. [CrossRef]

35. Roberson, R.E.; Schwertassek, R. *Dynamics of Multibody Systems*; Springer: New York, NY, USA, 1988; p. 77, ISBN 978-3-642-86466-7.

36. Palpacelli, M.; Carbonari, L.; Palmieri, G.; Callegari, M. Design of a Lockable Spherical Joint for a Reconfigurable 3-URU Parallel Platform. *Robotics* **2018**, *7*, 42. [CrossRef]

Article

Reconfiguration Analysis of a 3-DOF Parallel Mechanism

Maurizio Ruggiu [1,*] and Xianwen Kong [2]

[1] Department of Mechanical, Chemical and Materials Engineering, University of Cagliari, Piazza d'Armi, 09123 Cagliari, Italy

[2] School of Engineering and Physical Sciences, Heriot-Watt University, Edinburgh EH14 4AS, UK

* Correspondence: maurizio.ruggiu@dimcm.unica.it

Received: 25 June 2019; Accepted: 1 August 2019; Published: 2 August 2019

Abstract: This paper deals with the reconfiguration analysis of a 3-DOF (degrees-of-freedom) parallel manipulator (PM) which belongs to the cylindrical parallel mechanisms family. The PM is composed of a base and a moving platform shaped as equilateral triangles connected by three serial kinematic chains (legs). Two legs are composed of two universal (U) joints connected by a prismatic (P) joint. The third leg is composed of a revolute (R) joint connected to the base, a prismatic joint and universal joint in sequence. A set of constraint equations of the 1-RPU$-$2-UPU PM is derived and solved in terms of the Euler parameter quaternion (*a.k.a.* Euler-Rodrigues quaternion) representing the orientation of the moving platform and of the Cartesian coordinates of the reference point on the moving platform. It is found that the PM may undergo either the 3-DOF PPR or the 3-DOF planar operation mode only when the base and the moving platform are identical. The transition configuration between the operation modes is also identified.

Keywords: mobility; multi-mode parallel manipulator; quaternion; Euler parameters

---

1. Introduction

During the past two decades, a great effort has been made in the research on multi-mode mechanisms (also known as kinematotropic mechanisms, variable-degrees of freedom (DOF) mechanisms, mechanisms with bifurcation or multifurcation and disassembly-free reconfigurable PMs), which are a class of reconfigurable parallel mechanisms (PMs). In multi-mode PMs, fewer actuators are required for the moving platform to perform two or more operation modes and less time is needed in reconfiguring the PM because the process does not need to disassemble the mechanism [1–7]. The main issue when dealing with a reconfigurable mechanism is its complete kinematic analysis (the reconfiguration analysis) [4,5], consisting of finding all the operation modes (motion patterns) and the transition configurations from one operation mode to an other, which, as it has been noted, represent constraint singular configurations of the PM. There are numerous examples of reconfiguration analyses in the recent literature [8–13]. The methods commonly used to solve the analysis are based on algebraic geometry and its numerical implementations [14–19]. For example, in References [13,20,21] the Study coordinates are used to represent the motion of the moving platform of a PM and for the kinematic analysis while in References [8,9,22] the position and orientation of the moving platform are represented by using the Cartesian coordinates of a point on the moving platform and the Euler parameter quaternion, respectively. The kinematic interpretation of all the 15 possible cases of the values with a vanished component taken by the Euler parameter quaternion to represent orientation of a rigid body has been presented in Reference [8], firstly. This classification is used in this paper for the reconfiguration analysis of the 3-DOF PM under study.

The paper aims to investigate the operation modes and the transition configurations of the 1-RPU-2-UPU cylindrical parallel mechanism. It is worth noticing that this work will extend the special case of the same PM analyzed by the same authors [23] to the general case. In Reference [23] the mobility analysis was conducted via an analytical procedure based on the screw theory and the kinematic analysis was carried out by taking into account additional constraints to the UPU legs. These constraints forced the UPU legs and the RPU leg to lay on parallel planes in the planar operation mode. In this work the mentioned constrains were released leading to a general case with a wider range of mechanisms with the same function for further optimization. Besides, the method used here proved to be simpler and more convenient than that based on the screw theory providing a more complete analysis of the operation modes of the PM under study. Indeed, the method adopted has allowed to find every possible (even theoretical) operation mode of the PM that the authors did not find in Reference [23] and to clarify which are the operation modes when the platforms' circumscribed circles have different radii (non-identical case).

The paper is organized as follows: Section 2 briefly recalls the mathematical definition of the Euler parameter quaternion; Section 3 describes the PM architecture under study; Section 4 deals with the reconfiguration analysis of the PM with base and moving platform shaped as identical equilateral triangles; Section 5 revises the reconfiguration analysis by considering the base and moving platform as different equilateral triangles (non-identical case), while Section 6 shows an alternative architecture of the same family of PMs. Finally, the conclusions are drawn.

2. Mathematical Preamble

A rotation ϕ about an arbitrary axis $\mathbf{u}$ can be expressed by four parameters e_i, $(i = 0, \ldots, 3)$, in the Euler parameter quaternion [24]:

$$q = e_0 + e_1\mathbf{i} + e_2\mathbf{j} + e_3\mathbf{k} = \cos(\frac{\phi}{2}) + \sin(\frac{\phi}{2})\mathbf{u}. \tag{1}$$

In Equation (1) $\{\mathbf{i}, \mathbf{j}, \mathbf{k}\}$ is the basis in the Euclidean space $\mathbb{V}^3$. The Euler parameters are isomorphic to the *unit quaternion* such that:

$$e_0^2 + e_1^2 + e_2^2 + e_3^2 = 1. \tag{2}$$

In general, a vector $\mathbf{p}'$ obtained by a rotation ϕ about $\mathbf{u}$ of a vector $\mathbf{p}$ can be expressed as:

$$\mathbf{p}' = q\mathbf{p}q^*, \tag{3}$$

where $q^* = e_0 - e_1\mathbf{i} - e_2\mathbf{j} - e_3\mathbf{k}$ is the conjugate of q. A rotation q_1 followed by a rotation q_2 may be represented by $q = q_2q_1$. The product of quaternions follows the multiplication rules:

$$\begin{aligned}
\mathbf{i}^2 = \mathbf{j}^2 = \mathbf{k}^2 = \mathbf{i}\mathbf{j}\mathbf{k} &= -1, \\
\mathbf{i}\mathbf{j} = \mathbf{k} &= -\mathbf{j}\mathbf{i}, \\
\mathbf{j}\mathbf{k} = \mathbf{i} &= -\mathbf{k}\mathbf{j}, \\
\mathbf{k}\mathbf{i} = \mathbf{j} &= -\mathbf{i}\mathbf{k}.
\end{aligned} \tag{4}$$

3. Description of the 1-RPU−2-UPU PM

Figure 1 shows the 1-RPU−2-UPU PM under study. The PM is composed by the fixed (base) and the moving platforms connected by three legs. One leg is a serial kinematic chain with R, P and U joints in sequence starting from the base. The other two legs are identical and they are serial kinematic chains with U, P and U joints in sequence starting from the base. The vertical axis of the mounting arrangement intersects the axis of the R joint at B_1 at the base. The axes of the U joints intersect at B_2, B_3 on the base and P_2, P_3 on the moving platform. Points $B_1B_2B_3$ and $P_1P_2P_3$ form two identical equilateral triangles with the radius of their circumscribed circles equal to r. In all legs, the axes of two R joints connected by the P joint are parallel to each other. The direction of each P joint is perpendicular to the axes of its two adjacent R joints.

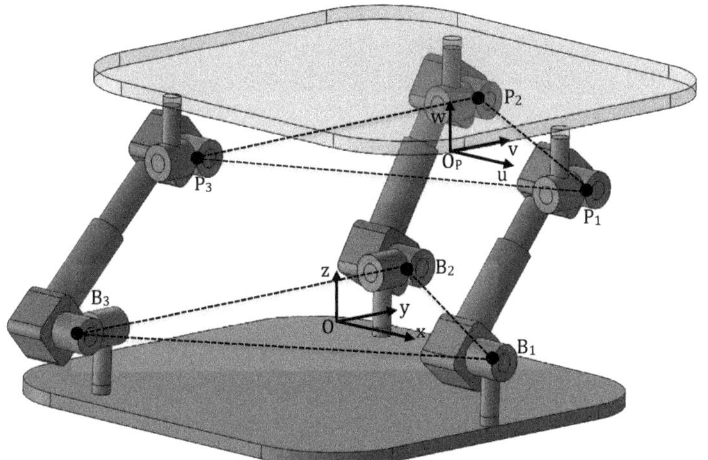

Figure 1. The 1-RPU−2-UPU PM geometry.

Let $\{O\_xyz\}$ and $\{O_P\_uvw\}$ denote the coordinate frames fixed on the base and on the moving platform, respectively. The x-axis is along the axis of R joint in the RPU leg and normal to the axes of R joints on the base in the U joints of the UPU legs. The u-axis is normal to the axes of R joints on the moving platform in the U joints of all legs. The x- and u-axes pass through the joint's centers B_1 and P_1, respectively. The y- and v-axes are, respectively, located on the plane defined by the axes of the R joints on the base and that defined by the axes of the R joints on the moving platform. The points O and O_P are located at the centroides of the triangles $B_1B_2B_3$ and $P_1P_2P_3$, respectively. The third axes of the reference systems are normal to the platforms. α, β, γ are the the unit vectors along with x-, y-, z-axes while ξ, v, ζ are the unit vectors along with u-, v-, w-axes expressed in $\{O\_xyz\}$. The location of $\{O_P\_uvw\}$ in the fixed reference system is given by the position of its center $\mathbf{p} = \begin{pmatrix} x_P & y_P & z_P \end{pmatrix}^T$ and the orientation denoted by the Euler parameter quaternion q. The unit vectors α along the x-axis and β along the y-axis can be written in quaternion form as $\mathbf{i}$ and $\mathbf{j}$, respectively, such that:

$$\xi = qiq^* = \begin{pmatrix} e_0^2 + e_1^2 - e_2^2 - e_3^2 & 2(e_1e_2 + e_0e_3) & 2(e_1e_3 - e_0e_2) \end{pmatrix}^T,$$

$$v = qjq^* = \begin{pmatrix} 2(e_1e_2 - e_0e_3) & e_0^2 - e_1^2 + e_2^2 - e_3^2 & 2(e_2e_3 + e_0e_1) \end{pmatrix}^T. \tag{5}$$

The positions of B_i are:

$$\mathbf{r}_{B_1} = r\mathbf{i};$$
$$\mathbf{r}_{B_2} = r/2(-\mathbf{i} + \sqrt{3}\mathbf{j}); \tag{6}$$
$$\mathbf{r}_{B_3} = r/2(-\mathbf{i} - \sqrt{3}\mathbf{j}).$$

The positions of P_i are:

$$\mathbf{r}_{P_1} = \mathbf{p} + r\xi;$$
$$\mathbf{r}_{P_2} = \mathbf{p} + r/2(-\xi + \sqrt{3}v); \tag{7}$$
$$\mathbf{r}_{P_3} = \mathbf{p} + r/2(-\xi - \sqrt{3}v).$$

4. Reconfiguration Analysis

The set of the constraints equations of the legs to the moving platform motion are:

1. Leg 1: RPU. This leg provides two constraint conditions:

    -   The R joint-axis at the base is perpendicular to R joint-axis attached to the moving platform:

$$(\mathbf{r}_{B_1}) \cdot (\mathbf{r}_{P_3} - \mathbf{r}_{P_2}) = 0. \tag{8}$$

    -   The R joint-axis connected to the moving platform belongs to the plane $x = r$:

$$(\mathbf{i}) \cdot (\mathbf{r}_{P_1} - \mathbf{r}_{B_1}) = 0. \tag{9}$$

2. Legs 2 and 3: UPU. Each of these legs provide one constraint condition. The constraint condition is the same for both the legs.

    -   The R joints-axes attached to the base and the R joints-axes attached to the moving platform are coplanar:

$$(\mathbf{r}_{B_3} - \mathbf{r}_{B_2}) \times (\mathbf{r}_{P_3} - \mathbf{r}_{P_2}) \cdot (\mathbf{r}_{P_k} - \mathbf{r}_{B_k}) = 0. \tag{10}$$

    where $k = 2$ or $k = 3$, depending on which leg is considered.

After simple manipulations, the foregoing constraint equations lead to:

$$\begin{aligned}
h_1 &= e_0 e_3 - e_1 e_2 = 0 \\
h_2 &= x_P - 2r(e_2^2 + e_3^2) = 0 \\
h_3 &= x_P(e_0 e_1 + e_2 e_3) + z_P(e_0 e_3 - e_1 e_2) + r e_2 e_3 = 0
\end{aligned} \tag{11}$$

Equation (11) are the kinematic equations sought.

4.1. Operation Modes

Now, we look for all the sets of positive dimension solutions of Equation (11). In other words, we search any possible combination of Euler parameters such to satisfy the Equation (11). The result will give us all the possible operation modes of the PM. Because of the simple form of h_i, $i = 1, 2, 3$, the sets of positive dimension solutions can be obtained by inspection by searching the values of the Euler parameters able to vanish the left-hand side of Equation (11) and ensuring Equation (2). The results obtained were, then, verified by carrying out the primary decomposition of the ideal $H = \langle h_1, h_2, h_3 \rangle$ associated with the constraint Equation (2).

Mode I: 3-DOF PPR operation mode

$$\begin{cases} e_1 = 0 \\ e_3 = 0 \end{cases} : q = e_0 + e_2 \mathbf{j}, \quad x_P = 2r e_2^2.$$

It represents a rotation by $2 atan2(e_2, e_0)$ about the y-axis with the position of O_P given by $\mathbf{p} = \begin{pmatrix} 2r e_2^2 & y_P & z_P \end{pmatrix}^T$. These two results are in agreement as it can be seen by a simple geometrical proof.

This is a 3-DOF motion of the moving platform. Indeed, there are 3 constraint equations more than Equation (2) that constrain the free motion of the moving platform represented by 7 parameters: $\{e_0, e_1, e_2, e_3, x_P, y_P, z_P\}$. Figure 2 shows the PM undergoing this motion. The transformation matrix is:

$$\mathbf{T}_a = \begin{pmatrix} e_0^2 - e_2^2 & 0 & 2 e_0 e_2 & 2r e_2^2 \\ 0 & 1 & 0 & y_P \\ -2 e_0 e_2 & 0 & e_0^2 - e_2^2 & z_P \\ 0 & 0 & 0 & 1 \end{pmatrix},$$

with $e_0^2 + e_2^2 = 1$.

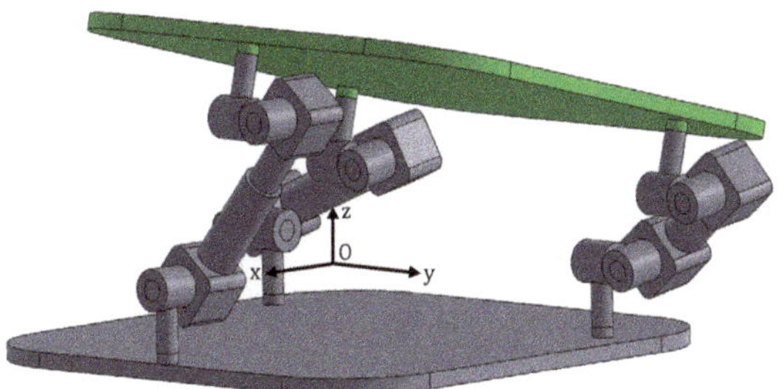

Figure 2. The 1-RPU$-$2-UPU PM undergoing the 3-DOF PPR operation mode (*Mode I*).

The operation mode I is defined PPR in [23].

Mode II: 3-DOF planar operation mode

$$\begin{cases} e_2 = 0 \\ e_3 = 0 \end{cases} : \; q = e_0 + e_1 \mathbf{i}, \; x_P = 0.$$

It represents a rotation by $2atan2(e_1, e_0)$ about the x-axis. Point O_P can only move on the plane $x = 0$. As for mode I, also this motion is a 3-DOF motion. Figure 3 shows the PM undergoing this motion.

Figure 3. The 1-RPU$-$2-UPU PM undergoing the 3-DOF planar operation mode (*Mode II*).

The transformation matrix is:

$$\mathbf{T}_b = \begin{pmatrix} 1 & 0 & 0 & 0 \\ 0 & e_0^2 - e_1^2 & -2e_0 e_1 & y_P \\ 0 & 2e_0 e_1 & e_0^2 - e_1^2 & z_P \\ 0 & 0 & 0 & 1 \end{pmatrix},$$

with $e_0^2 + e_1^2 = 1$. The operation mode II is defined planar motion in [23].

There are other Euler parameters combinations that satisfy Equations (2) and (11) . Only one of them leads to an independent motion mode of the PM, namely Mode III, that, on the other side, can be obtained only in theory by disconnecting and reassembling the PM. The other combinations of Euler parameters lead to solutions being subsets of other solutions that cannot be considered independent motion modes of the PM. We call them Solution IV and Solution V. For the sake of completeness, all the results are given in Table 1.

Table 1. Other solutions from the kinematic equations.

Mode III: $e_0 = e_1 = e_2 = 0$, $x_P = 2r$: $q = e_3\mathbf{k}$.		Half-turn rotation about the z-axis.
Solution IV: $e_0 = e_1 = e_3 = 0$, $x_P = 2r$: $q = e_2\mathbf{j}$.		Half-turn rotation about the y-axis.
Solution V: $e_0 = e_2 = 0$, $x_P = 2re_3^2$: $q = e_1\mathbf{i} + e_3\mathbf{k} = (e_1 - e_3\mathbf{j})\mathbf{i}$.		Half-turn rotation about the x-axis followed by a rotation by $2atan2(-e_3, e_1)$ about the y-axis.

It can be noted that Solution IV is nothing but the limit of mode I whenever the rotation of the moving platform reaches π: $e_2 \to 1$. Solution V is a composition of mode II (which reaches the half-turn rotation) and mode I. It cannot be obtained without bodies interferences.

4.2. Transition Configurations

Transition configurations are the configurations that allow the PM to switch from one operation mode to an other. In the transition configuration the constraint equations of the associated operation modes have to be guaranteed at the same time.

- Transition configuration between Mode I and Mode II: $(Mode\ I) \wedge (Mode\ II)$
  The constraint equations are:

$$\begin{cases} e_1 = 0 \\ e_2 = 0 \\ e_3 = 0 \end{cases} : q = e_0, \quad x_P = 0.$$

This configuration represents any translation on the $x = 0$ plane.

Its transformation matrix is: $\mathbf{T}_{I\wedge II} = \begin{pmatrix} \mathbf{1} & \mathbf{p} \\ \mathbf{0} & 1 \end{pmatrix}$, where $\mathbf{1}$ denotes the unity matrix and

$$\mathbf{p} = \begin{pmatrix} 0 \\ y_P \\ z_P \end{pmatrix}.$$

Figure 1 shows the PM at the transition configuration.
- Other transition configurations
  There can be no transitions between Mode I and Mode III or Mode II and Mode III that can be physically reached. It can be noticed that Solution IV is the transition configuration between Mode I and Mode II when the rotation about $x-$axis is π.

5. Reconfiguration Analysis: Non-Identical Case

In this section, the reconfiguration analysis presented in Section 4 is revised when considering the base and the moving platforms as two equilateral triangles with the radius of their circumscribed circles of r_b and r_p, respectively. In this case, Equation (11) become:

$$e_0 e_3 - e_1 e_2 = 0$$
$$x_P - (r_b - r_p)(e_0^2 + e_1^2) - (r_b + r_p)(e_2^2 + e_3^2) = 0 \tag{12}$$
$$(r_b - r_p)e_0 e_1 + (r_b + r_p)e_2 e_3 + 2x_P(e_0 e_1 + e_2 e_3) + z_P(e_0 e_3 - e_1 e_2) = 0$$

Only the positive dimension solutions of Equation (12) that lead to independent motion modes are considered in the following analysis.

The 3-DOF PPR operation mode (Mode I) is still possible in this case. Indeed, the moving platform can rotate about the y-axis and $x_P = 2r_p e_2^2 + (r_b - r_p)$. The 3-DOF planar operation mode (Mode II) is no longer possible in this case. Indeed, it is trivial to show that Equation (12) may be guaranteed if and only if $(r_b - r_p) = 0$. Mode III is not affected by the dimensions of the base and moving platform and it is only possible in theory.

Transition configurations are not considered in this case as the PM may undergo an unique operational mode (PPR) with physics and geometry always guaranteed.

6. Other PMs Architectures

In general, any PM with the same set of constraint equations of that used for the 1-RPU$-$2-UPU leads to the operation modes analyzed [25]. For example, in Figures 4 and 5 the 1-URU-2-RRU PM is shown in the 3-DOF PPR and in the 3-DOF planar operation mode, respectively. As can be seen there are two legs with two constraint equations, one leg with one. Further, the intermediate P joint is substituted by an R joint to form a leg with $3-$R planar kinematic chain.

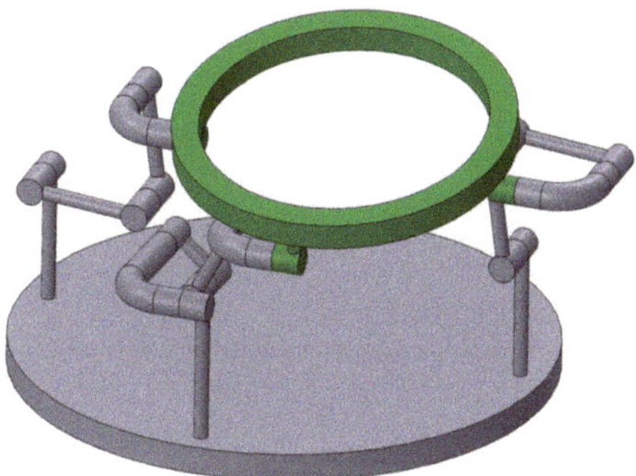

Figure 4. The 1-URU$-$2-RRU PM undergoing the 3-DOF PPR operation mode.

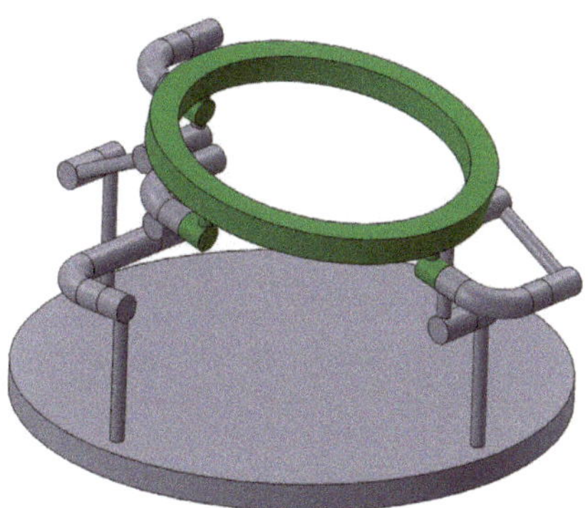

Figure 5. The 1-URU$-$2-RRU PM undergoing the 3-DOF planar operation mode.

7. Conclusions

The paper has presented a systematic reconfiguration analysis of a 1-RPU$-$2-UPU and other PMs with the same set of constraint equations as the 1-URU$-$2-RRU PM. The use of the Euler parameter quaternion and of the Cartesian coordinates of a point on the moving platform has proved to be convenient and simple for the analysis of this type of PMs. Indeed, the analysis leads to the operation modes of the PM in a straightforward manner with no need of any algebraic geometry method that was used only to verify the results obtained. In the case of two identical equilateral triangles as base and moving platform of the PM, the analysis shows two possible operation modes: the 3-DOF PPR and the 3-DOF planar modes. The former is a rotation about the y-axis whilst the latter is a rotation about the x-axis and a translation on the $x = 0$ plane of the reference point on the moving platform. An other theoretical operation mode was found since it requires to disconnect and reassemble the PM. As expected, the transition between the two operation modes occurs when the platform has no rotation and the reference point of the moving platform lies onto $x = 0$ plane.

When the base and the moving platform are equilateral triangles of different sizes, the only operation mode is the 3-DOF PPR. Therefore, in this case, the PM no longer has multiple operation modes.

Future work will focus on the optimization of this multi-mode PM and developing a prototype able to switch between the different modes by eliminating the constraint singularity. The main idea is to use lockable Pi (planar parallelogram) and R joints as proposed in Reference [26].

Author Contributions: Conceptualization, X.K. and M.R.; formal analysis, M.R. and X.K.; writing—original draft preparation, M.R.; writing—review and editing, X.K.

Funding: This research received no external funding.

Conflicts of Interest: The authors declare no conflict of interest.

References

1. Ding, X.; Kong, X.; Dai, J.S. *Advances in Reconfigurable Mechanisms and Robots II*; Springer International Publishing: Cham, Switzerland, 2016.
2. Kong, X.; Gosselin, C.; Richard, P.L. Type synthesis of parallel mechanisms with multiple operation modes. *J. Mech. Des.* **2007**, *129*, 595–601. [CrossRef]
3. Kong, X. Type synthesis of 3-DOF parallel manipulators with both a planar operation mode and a spatial translational operation mode. *J. Mech. Robot.* **2013**, *5*, 041015-1–041015-8. [CrossRef]
4. Fanghella, P.; Galletti, C.; Gianotti, E. Parallel robots that change their group of motion. In *Advances in Robot Kinematics*; Lenarcic, J., Roth, B., Eds.; Springer: Dordrecht, The Netherlands, 2006; pp. 49–56.
5. Refaat, S.; Hervé, J.M.; Nahavandi, S.; Trinh, H. Two-mode overconstrained three-DOFs rotational-translational linear-motor-based parallel kinematics mechanism for machine tool applications. *Robotica* **2007**, *25*, 461–466. [CrossRef]
6. Li, Q.; Hervé, J.M. Parallel mechanisms with bifurcation of Schoenflies motion. *IEEE Trans. Robot.* **2009**, *25*, 158–164.
7. Gogu, G. Maximally regular T2R1-type parallel manipulators with bifurcated spatial motion. *J. Mech. Robot.* **2011**, *3*, 011010-1–011010-8. [CrossRef]
8. Kong, X. Reconfiguration analysis of a 3-DOF parallel mechanism using Euler parameter quaternions and algebraic geometry method. *Mech. Mach. Theory* **2014**, *74*, 188–201. [CrossRef]
9. Kong, X. Reconfiguration analysis of a 4-DOF 3-RER parallel manipulator with equilateral triangular base and moving platform. *Mech. Mach. Theory* **2016**, *98*, 180–189. [CrossRef]
10. Kong, X. Reconfiguration Analysis of Multimode Single-Loop Spatial Mechanisms Using Dual Quaternions. *ASME J. Mech. Robot.* **2017**, *9*, 051002-1–051002-8. [CrossRef]
11. Carbonari, L.; Callegari, M.; Palmieri, G.; Palpacelli, M.C. A new class of reconfigurable parallel kinematic machines. *Mech. Mach. Theory* **2014**, *79*, 173–183. [CrossRef]
12. Carbonari, L.; Callegari, M.; Palmieri, G.; Palpacelli, M.C. Analysis of kinematics and reconfigurability of a spherical parallel manipulator. *IEEE Trans.Robot.* **2014**, *30*, 1541–1547. [CrossRef]
13. Nurahmi, L.; Schadlbauer, J.; Caro, S.; Husty, M.; Wenger, P. Kinematic analysis of the 3-RPS cube parallel manipulator. *J. Mech. Robot.* **2015**, *7*, 011008-1–011008-11. [CrossRef]
14. Cox, D.A.; Little, J.B.; O'Shea, D. *Ideals, Varieties, and Algorithms*; Springer: New York, NY, USA, 2007.
15. Sommese, A.J.; Wampler, C.W. *The Numerical Solution of Systems of Polynomials Arising in Engineering and Science*; World Scientific Press: Singapore, 2005.

16. Decker, W.; Pfister, G. *A First Course in Computational Algebraic Geometry*; Cambridge University Press: New York, NY, USA, 2013.

17. Kong, X. A variable-DOF single-loop 7R spatial mechanism with five motion modes. *Mech. Mach. Theory* **2018**, *120*, 239–249. [CrossRef]

18. Walter, D.R.; Husty, M.L. Kinematic analysis of the TSAI 3-UPU parallel manipulator using algebraic methods. In Proceedings of the 13th IFToMM World Congress in Mechanism and Machine Science, Guanajuato, Mexico, 19–25 June 2011; pp. 1–10.

19. Schadlbauer, J.; Walter, D.R.; Husty, M. The 3-RPS parallel manipulator from an algebraic viewpoint. *Mech. Mach. Theory* **2014**, *75*, 161–176. [CrossRef]

20. Walter, D.R.; Husty, M.L.; Pfurner, M. *Chapter A: Complete Kinematic Analysis of the SNU3-UPU Parallel Manipulator, Contemporary Mathematics*; American Mathematical Society: Providence, RI, USA, 2009; Volume 496, pp. 331–346.

21. Nurahmi, L.; Caro, S.; Wenger, P.; Schadlbauer, J.; Husty, M. Reconfiguration analysis of a 4-RUU parallel manipulator. *Mech. Mach. Theory* **2016**, *96*, 269–289. [CrossRef]

22. Kong, X. Reconfiguration Analysis of a Variable Degrees-of-Freedom Parallel Manipulator With Both 3-DOF Planar and 4-DOF 3T1R Operation Modes. In Proceedings of the ASME 40th Mechanisms and Robotics Conference, Charlotte, NC, USA, 21–24 August 2016.

23. Ruggiu, M.; Kong, X. Mobility and kinematic analysis of a parallel mechanism with both PPR and planar operation modes. *Mech. Mach. Theory* **2012**, *55*, 77–90. [CrossRef]

24. Spring, K.W. Euler parameters and the use of quaternion algebra in the manipulation of finite rotations: A review. *Mech. Mach. Theory* **1986**, *21*, 365–373. [CrossRef]

25. Kong, X.; Gosselin, C.M. *Type Synthesis of Parallel Mechanisms*; Springer: Berlin, Germany, 2007.

26. Chablat, D.; Kong, X.; Zhang, C. Kinematics, Workspace, and Singularity Analysis of a Parallel Robot with Five Operation Modes. *J. Mech. Robot.* **2018**, *10*, 035001-1–035001-12. [CrossRef]

 *robotics*

Article

Modeling Parallel Robot Kinematics for 3T2R and 3T3R Tasks Using Reciprocal Sets of Euler Angles

Moritz Schappler *, Svenja Tappe and Tobias Ortmaier

Institute of Mechatronic Systems, Leibniz University Hannover, 30167 Hannover, Germany
*   Correspondence: moritz.schappler@imes.uni-hannover.de

Received: 27 June 2019; Accepted: 1 August 2019; Published: 6 August 2019

Abstract: Industrial manipulators and parallel robots are often used for tasks, such as drilling or milling, that require three translational, but only two rotational degrees of freedom ("3T2R"). While kinematic models for specific mechanisms for these tasks exist, a general kinematic model for parallel robots is still missing. This paper presents the definition of the rotational component of kinematic constraints equations for parallel robots based on two reciprocal sets of Euler angles for the end-effector orientation and the orientation residual. The method allows completely removing the redundant coordinate in 3T2R tasks and to solve the inverse kinematics for general serial and parallel robots with the gradientdescent algorithm. The functional redundancy of robots with full mobility is exploited using nullspace projection.

Keywords: parallel robot; five-DoF task; 3T2R task; functional redundancy; task redundancy; redundancy resolution; reciprocal Euler angles; inverse kinematics

---

1. Introduction

Industrial tasks like welding, gluing, milling or drilling represent a major part of the applications of industrial robots, which generally have full mobility, i.e., the operational space of their end-effector has three translational and three rotational ("3T3R") degrees of freedom ("DoF"). Parallel robots like the Stewartplatform have especially been proposed for milling tasks regarding their high structural stiffness. The task space of the named applications can be defined by three translational DoF and only two rotations due to a symmetry around the tool axis ("3T2R"). This results in a functional or task redundancy, which is not exploited to full extend yet for *parallel* robots.

1.1. Inverse Kinematics and Resolution of Task Redundancy for Serial-Link Robots

Various general gradient-based methods exist to solve the inverse kinematics for *serial* robots; either by augmenting the joint space [1] or by reducing the task space [2–5]. The different approaches each define a residual vector and a gradient matrix considering the properties of 3T2R tasks, e.g., by adding a virtual joint axis [1], orthogonal decomposition of the task space [2], rotation of the residual into a task frame and removing the corresponding component [3], defining the tool axis by two points for constructing a nullspace [4] or by defining the absolute orientation and the orientation residual with two reciprocal sets of Euler angles [5]. The gradient matrices corresponding to the different residuals are used for an iterative Newton-Raphson algorithm [6,7] by exploiting the functional redundancy with a null space projection of additional performance criteria [6]. Without the definition of a proper gradient, a global optimization has to be performed outside of the inverse kinematics algorithm [8,9].

1.2. Overview of Parallel Robots Structures for 3T2R Tasks

Parallel robots in 3T2R tasks can be ordered in classes according to their kinematic structure into

I mechanisms with full platform mobility (3T3R) that are redundantly controlled to five DoF,

II mechanisms with 3T2R platform mobility enforced with a passive five-DoF constraining leg and five other legs with six DoF each,

III mechanisms with 3T2R platform mobility resulting from the mobility of five actuated legs with five or six DoF each,

IV mechanisms with 3T2R platform mobility and five legs with only five DoF each.

The classes I and II were introduced in [10], where class IV is analyzed regarding leg symmetry and singularities. Class III is mainly influenced by the systematic synthesis of [11] and several existing prototypes and is demarcated against class II by the absence of the passive constraining leg. Class IV can be seen as a subclass of III, but is differentiated in this paper due to its characteristics. Other classifications are provided e.g., by [12], where IV and II are termed "families".

Examples for the first class are hexapods (6U$\underline{\text{P}}$S) [13] or the Eclipse [14] machine tool (2$\underline{\text{PP}}$RS-$\underline{\text{PP}}$RS). (The joint structure of the parallel robots is denoted by the number of the legs and the order of universal ("U"), prismatic ("P"), spherical ("S"), helical ("H") and revolute ("R") joints in the leg chains. Different actuated legs are connected by hyphens ("-"), passive constraint legs are connected by a slash ("/"). Actuated joints are underlined.) Any other parallel robot with full mobility (see e.g., [11,15,16]) may be used as well.

The second class allows for more variety, since the six-DoF mechanism and the five-DoF constraining leg can manifest in different kinematic structures: The U$\underline{\text{P}}$S structure is used for the six-DoF part of the mechanism by [17] with a focus on kinematic analysis of 5U$\underline{\text{P}}$S/US, by [18] with a focus on kinetostatic modeling at the example of 5U$\underline{\text{P}}$S/RUU (see Figure 1a), by [19] with focus on trajectory control of 5U$\underline{\text{P}}$S/PRPU and by [20] for pose measurement with the passive leg of 5U$\underline{\text{P}}$S/PRPU. Other possible general base structures are $\underline{\text{R}}$US at the 5$\underline{\text{R}}$US/US example in [17], $\underline{\text{P}}$US, which has been investigated for the control of a redundantly actuated 6$\underline{\text{P}}$US/UPU regarding the force control of the redundant leg with inverse dynamics compensation [21] or force optimization [22].

The most-straightforward member of the third class is the 4U$\underline{\text{P}}$S-U$\underline{\text{P}}$U of Figure 1b, which is investigated in [23] for a simulation and feasibility study together with a survey on possible architectures for a technical realization of this class. Other possible structures are the 4U$\underline{\text{R}}$S-U$\underline{\text{R}}$U, which is analyzed kinematically in [24] and the 4$\underline{\text{P}}$SU-$\underline{\text{P}}$U*U, which has a special parallelogram structure in one leg (termed "U*") and is presented in [25]. A subclass of III consists of mechanisms [26–28], where the last joint axis of the legs is coaxial with the tool axis and is constructed as rotating ring. It contains the Metrom machine tool (4S$\underline{\text{P}}$RR-S$\underline{\text{P}}$R), depicted in Figure 1c, which is analyzed regarding inverse and forward kinematics in [26] or its variants, the redundant 4S$\underline{\text{P}}$RR-$\underline{\text{P}}$S$\underline{\text{P}}$R from [28] or the hybrid 4UR$\underline{\text{H}}$U-UR$\underline{\text{H}}$R with an additional linear actuator at the platform [27]. A structural synthesis based on linear transformations and evolutionary morphology [11] led e.g., to the Isoglide5 mechanisms (3$\underline{\text{P}}$RRRRR-2$\underline{\text{P}}$RRRR), which are analyzed and optimized regarding the isotropy of the Jacobian in [29].

The simplest member of class IV, the 5U$\underline{\text{P}}$U is shown in [30] with the help of screw theory to only have local mobility and no global mobility, since the twist systems of the leg chains have no intersection and the resulting twist system of the platform is empty. Members of class IV have been found by systematic structural synthesis with screw theory, which has been performed in [31] for symmetric 3T2R mechanisms. The resulting 5R$\underline{\text{P}}$UR of Figure 1d and 5$\underline{\text{P}}$RUR are analyzed in [10,32].

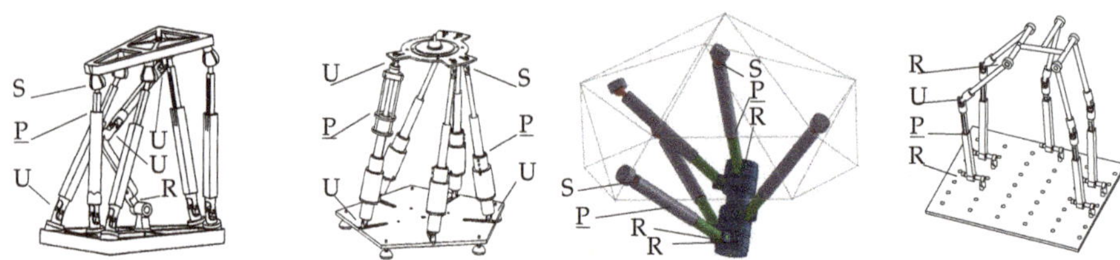

(a) 5U$\underline{\text{P}}$S/RUU (class II)   (b) 4U$\underline{\text{P}}$S-U$\underline{\text{P}}$U (class III)   (c) 4S$\underline{\text{P}}$RR-S$\underline{\text{P}}$R (Metrom, class III)   (d) 5R$\underline{\text{P}}$UR (class IV)

Figure 1. Typical mechanisms of the different classes. Adapted from [18] (**a**), [23] (**b**), [28] (**c**), [10] (**d**).

In a practical application with competing requirements on workspace, stiffness, costs and precision, each of the existing systems has its legitimization. Nevertheless, each of the classes has inherent disadvantages: In tasks like milling with high process forces and requirements on stiffness and precision, robots from class II with one constraining leg have the drawback that the passive leg takes the complete reaction wrench in the blocked rotational degree of freedom, which strongly affects the mechanisms stiffness in this direction [18]. The same is argued by [27] at the example of the Metrom machine tool, but can be extended to all the members of class III, where most leg chains have six DoF and usually only one leg chain has five DoF. This leg chain also has to take the reaction moments in the blocked DoF which affects the overall stiffness. Therefore, members of the classes I and IV can be expected to reach a higher stiffness. Mechanisms of the class IV may further suffer from an increased sensitivity of manufacturing tolerances, which may cause a high pretension of the bearings or even reduce the DoF, since the five DoF of all leg chains have to coincide exactly to allow the platform to also have five DoF. Additionally, only members of class I provide redundancy which allows using the additional DoF for performance optimizations, e.g., to avoid singularities and to compensate the smaller workspace caused by the sixth leg.

Therefore, the remainder of the paper focuses on mechanisms of the first class to allow an optimization of their performance criteria using the degree of task redundancy.

1.3. Inverse Kinematics of Parallel Robots for 3T2R Tasks

The parallel robots with five DoF presented above have a kinematic structure which allows for an analytic model of the inverse kinematics. All references define the end-effector orientation with two consecutive elementary rotations, i.e., define two Euler angles to represent the tool axis orientation in minimal coordinates [10,13,19–22,26,28,32,33], which is called "partial pose" in [13]. The inverse kinematics problem (IKP) is first solved for the first chain, which is called "leading chain" in this paper. Due to the geometry of the leading chain, this solution can be found algebraically. Then the IKP is solved for the other "following" chains with the given orientation from the leading chain and standard methods. For robots of class II, the constraining leg is selected as the leading leg chain and for class III the 3T2R leg is selected.

To the best knowledge of the authors, only one reference [13] for the IKP of functionally redundant parallel robots of the class I is known. The reason presumably is that a solution of the 3T3R IKP for these robots is possible with standard methods, as used in [14] for the 3T3R Eclipse. It is always possible to transfer the 3T2R IKP into 3T3R by adding an arbitrary value for the desired rotation around the tool axis. An optimization of additional performance criteria is possible by varying the redundant rotation angle [8,9]. This approach was chosen in [13] by first defining the IKP with the redundant rotation as a parameter and then performing an optimization of this parameter using analytical computation of the dexterity and interval analysis to ensure a minimum determinant of the inverse Jacobian. The drawback of this method is the need for a cascaded optimization which is more complex than the gradient-based approach presented in this paper.

1.4. Motivation and Summary of the State of the Art

The overview over the literature shows that no general, machine-independent method for the resolution of functional redundancy for 3T3R PKM in 3T2R tasks exists. The works either focus on a general structural synthesis of machines, e.g., via screw theory [31] or linear transformations [29] or the description and improvement of specific, manually selected machines. To choose the best machine for given requirements, a structural synthesis is only the first step. Additionally, a dimensional synthesis should be performed for all possible structures to select the most suitable mechanism. This combined structural and dimensional synthesis [34] is sketched in Figure 2. To be able to perform the dimensional synthesis for all structures, the inverse kinematics has to be implemented in a general form to calculate the performance criteria over a given trajectory for further optimization of the structures and their comparison. For the generation of task redundant parallel robots, the inverse kinematics has to include an optimization of the performance criteria and the restrictions such as joint limits to ensure a comparability of the results.

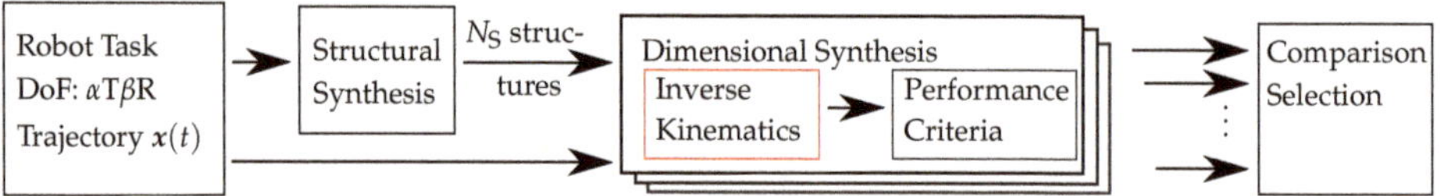

Figure 2. Overview of the procedure for combined structural and dimensional synthesis.

To address this, the contributions of this paper are

- a general kinematics model for parallel robots using the concept of reciprocal Euler angles [5],
- a complete elimination of the redundant operational space coordinate in this formulation for 3T2R tasks allowing a nullspace optimization in the gradient-based inverse kinematics,
- proofs, examples and simulations to show the performance for single serial kinematic leg chains and complete parallel robots.

The remainder of the paper is structured as follows: The description of the inverse kinematics problem and prior definitions are given in Section 2 and the concept of reciprocal Euler angles from [5] is adapted in Section 3 for parallel robots. This leads to the full kinematic constraints of parallel robots in 3T2R tasks, introduced in Section 4 and applied to the differential kinematics in Section 5. The theoretical analysis is followed by examples and simulations in Section 6. The appendix contains proofs and additional details on the mathematical formulation.

2. Inverse Kinematics Problem for Parallel Robots

Before addressing the specific model for parallel robots in 3T2R tasks in the next section, the standard kinematics model of parallel kinematic machines ("PKM") is repeated in the following, corresponding to the state of the art [11,15,35] and serving as a reference to highlight its shortcomings for 3T2R tasks. The regarded parallel robot consists of m legs, which each have the joint coordinates q_i. All joints are considered to be single-DoF and additionally to the active joints $q_{i,a}$ explicitly all passive joints at the base and at the platform $q_{i,p}$ are included in the coordinates q_i of leg i. The coordinates

$$x = \begin{bmatrix} x_t^T & x_r^T \end{bmatrix}^T \in \mathbb{R}^6 \tag{1}$$

of the end-effector platform describe the position and orientation of the end-effector frame $\mathscr{F}_D$ with respect to the base frame $\mathscr{F}_0$. In the equations, this is marked with left subscript "(0)" for vectors and left superscript "0" for rotation matrices. The platform-related end-effector frame is the desired frame in the inverse kinematics problem and is therefore abbreviated with "D". The position

$$x_t = {}_{(0)}r_D \in \mathbb{R}^3 \tag{2}$$

is defined as the origin of the platform frame and the rotation matrix

$${}^0R_D(x_r) = \begin{bmatrix} n_D & o_D & a_D \end{bmatrix} \in SO(3) \tag{3}$$

of the platform frame is expressed with Euler angles

$$x_r = \begin{bmatrix} \beta_1 & \beta_2 & \beta_3 \end{bmatrix}^T =: \beta \in \mathbb{R}^3 \tag{4}$$

as a minimal representation of the orientation coordinates. The symbol "β" will be used to denote orientations relative to the base frame throughout this paper. The XYZ convention

$$R(\beta) = R_x(\beta_1)R_y(\beta_2)R_z(\beta_3) \in \mathrm{SO}(3) \tag{5}$$

is used for the Euler angles without loss of generality. The relation between joint coordinates q and platform coordinates x is established with the kinematic constraint equations, for which most commonly the vector loop

$$\Phi_{\mathrm{t},i}(q_i, x) = -_{(0)}r_{A_iB_i}(x) + _{(0)}r_{A_iB_i}(q_i) \tag{6}$$

between the position of the platform coupling point B_i relative to the base coupling point A_i is used for each leg chain i [15]. The second term $_{(0)}r_{A_iB_i}(q_i)$ corresponds to the forward kinematics of the serial leg chain. The vector

$$_{(0)}r_{A_iB_i}(x) = -_{(0)}r_{A_i} + x_{\mathrm{t}} + {}^0R_D(x_{\mathrm{r}})_{(D)}r_{B_i} \tag{7}$$

includes the term ${}^0R_D(x_{\mathrm{r}})$ that depends on the full orientation x_{r} of the end-effector. For the bigger part of existing parallel robots, the passive joint coordinates can be eliminated analytically from Equation (6), e.g., by using the Euclidean distance for U$\underline{\mathrm{P}}$S or R$\underline{\mathrm{P}}$R leg chains or via trigonometry for R$\underline{\mathrm{R}}$Rchains. This is termed "minimal kinematics set" in [15] and leads to the scalar constraint equation

$$\Phi_i = \Phi_i(q_{i,\mathrm{a}}, x) \tag{8}$$

for each leg i, which can be assembled to the vector of constraint equations

$$\Phi(q_{\mathrm{a}}, x) = \begin{bmatrix} \Phi_1 & \Phi_2 & \cdots & \Phi_m \end{bmatrix}^{\mathrm{T}} \tag{9}$$

for all m legs of the PKM. The differential kinematics of the PKM is calculated with the time derivative

$$\frac{\mathrm{d}}{\mathrm{d}t}\Phi(q_{\mathrm{a}}, x) = \Phi_{\partial q_{\mathrm{a}}}\dot{q}_{\mathrm{a}} + \Phi_{\partial x}\dot{x} = 0 \tag{10}$$

where the passive joint coordinates q_{p} do not occur, since they have been eliminated in a previous step. Following [11] to avoid the term "Jacobian" for the elements of (10), the inverse-kinematics matrix

$$\Phi_{\partial q_{\mathrm{a}}} = \frac{\partial\Phi}{\partial q_{\mathrm{a}}} = \begin{bmatrix} \Phi_{1,\partial q_{1,\mathrm{a}}} & 0 & 0 & 0 \\ 0 & \Phi_{2,\partial q_{2,\mathrm{a}}} & \ddots & 0 \\ 0 & \ddots & \ddots & 0 \\ 0 & 0 & 0 & \Phi_{m,\partial q_{m,\mathrm{a}}} \end{bmatrix} \tag{11}$$

of this model has diagonal form and the direct-kinematics matrix

$$\Phi_{\partial x} = \frac{\partial\Phi}{\partial x} = \begin{bmatrix} (\partial\Phi_1/\partial x)^{\mathrm{T}} & (\partial\Phi_2/\partial x)^{\mathrm{T}} & \cdots & (\partial\Phi_m/\partial x)^{\mathrm{T}} \end{bmatrix}^{\mathrm{T}} \tag{12}$$

is fully populated. This definition of the constraints has the following drawbacks:

1. For parallel robots with arbitrary leg chains like those generated by a structural synthesis [11,31,34], it is generally not possible to analytically eliminate the passive joint coordinates.
2. If more than three joint coordinates per leg influence the coupling point position B_i, the three kinematic constraints per joint in (6) are not sufficient to generate enough equations for the matrix of (11) to become invertible.

3. The velocity-based theory of linear transformations used by [11] allows determining the mobility of arbitrary parallel robots. The linear transformation is generalized in [12] to accuracy and stiffness modeling by means of screw theory resulting in the "generalized Jacobian". Both concepts are similar to (10) and do not provide a direct appliance to solve the IKP, since this requires a formulation of the orientation at position level, not velocity level.
4. Exploiting the reduction of end-effector coordinates for 3T2R tasks is not possible, since all end-effector coordinates are included in (7).

An alternative kinematic model to encounter the combination of these points is presented in the next section, where the concept of reciprocal sets of Euler angles for the inverse kinematics problem for serial-link robots [5] is transferred to the leading leg of parallel robots.

3. Reciprocal Sets of Euler Angles for the Kinematics of a Serial Leg Chain

To take the rotational symmetry around the tool axis in 3T2R tasks into account, a new set of task space coordinates

$$\eta = \begin{bmatrix} \eta_t^T & \eta_r^T \end{bmatrix}^T \in \mathbb{R}^5 \tag{13}$$

has to be defined. The translational part

$$\eta_t = x_t = {}_{(0)}r_D \in \mathbb{R}^3 \tag{14}$$

remains unchanged relative to the operational space coordinates x. The rotational part

$$\eta_r = \begin{bmatrix} \beta_1 & \beta_2 \end{bmatrix}^T = \underbrace{\begin{bmatrix} 1 & 0 & 0 \\ 0 & 1 & 0 \end{bmatrix}}_{=P_{\eta_r}} x_r \in \mathbb{R}^2 \tag{15}$$

only contains the first two rotational coordinates of x. The last operational space coordinate β_3, the rotation around the z-axis a_D of $\mathscr{F}_D$, is excluded from the task space by the selection matrix P_{η_r}. To be able to set the rotational DoF around the tool axis in 3T2R tasks arbitrarily and use gradient-based inverse kinematics, β_3 has to be eliminated completely from the kinematics Equations (6). To simplify the following elaborations, the platform frame $\mathscr{F}_D$ is still identified as the desired frame of the inverse kinematics problem and the end-effector frame that results from the joint angles of leg i is now termed $\mathscr{F}_E$, which corresponds to the forward kinematics of the leg chain. For a formulation without the tool axis rotation, a different constraint definition

$$\Phi_{t,i}(q_i, x) = -{}_{(0)}r_D + {}_{(0)}r_E(q_i) = -x_t + {}_{(0)}r_E(q_i) \in \mathbb{R}^3 \tag{16}$$

containing the vector loop from the robot base frame $\mathscr{F}_0$ to the platform $\mathscr{F}_D$ and the leg chain end-effector $\mathscr{F}_E$ can be used, where in contrast to (6) only the translational part x_t of the end-effector coordinates appears and not the rotational part x_r. The vector loop is depicted in Figure 3 for a planar robot with opened (Figure 3a,b) and closed loops (Figure 3c). The triangle represents the end-effector platform and only one leg chain is drawn in the figure.

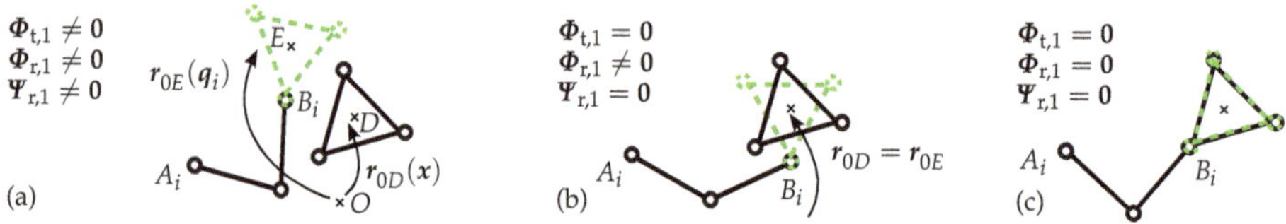

Figure 3. Different cases for the kinematic constraints of the leading chain for the 3RRR example: (**a**) no constraints complied; (**b**) position and tool axis rotation complied; (**c**) all constraints complied.

As a drawback of (16), all joint angles q_i of the leg i and not only the coordinates of the first joints counted from the base are now included in the vector

$$_{(0)}\boldsymbol{r}_E(\boldsymbol{q}_i) = {}_{(0)}\boldsymbol{r}_{A_i} + {}_{(0)}\boldsymbol{r}_{A_iB_i}(\boldsymbol{q}_i) + {}^0\boldsymbol{R}_{B_i}(\boldsymbol{q}_i)_{(B_i)}\boldsymbol{r}_E. \tag{17}$$

This implies that the platform is now part of the last link of the considered leg chain, as sketched in Figure 3 by the dashed triangle. To account for the increased number of included joints in (17), the full kinematic constraints

$$\boldsymbol{\Phi}_i = \begin{bmatrix} \boldsymbol{\Phi}_{t,i}^T & \boldsymbol{\Phi}_{r,i}^T \end{bmatrix}^T \in \mathbb{R}^6, \tag{18}$$

have to be considered, including the rotational part

$$\boldsymbol{\Phi}_{r,i}(\boldsymbol{q}_i, \boldsymbol{x}) = \begin{bmatrix} \alpha_1 & \alpha_2 & \alpha_3 \end{bmatrix}^T = \boldsymbol{\alpha}\left({}^D\boldsymbol{R}_E(\boldsymbol{x}_r, \boldsymbol{q}_i)\right) = \boldsymbol{\alpha}\left({}^0\boldsymbol{R}_D^T(\boldsymbol{x}_r){}^0\boldsymbol{R}_E(\boldsymbol{q}_i)\right), \tag{19}$$

which is needed to generate enough equations for an invertible matrix in the differential equations. The constraints again contain the deviation between the desired end-effector frame $\mathscr{F}_D$ expressed with $\boldsymbol{x}$ and the actual robots end-effector frame $\mathscr{F}_E$ expressed with $\boldsymbol{q}$. Figure 3b,c show cases, where the translational constraints are met, but the rotational constraints have different values. For 3T3R tasks, only Figure 3c represents a valid solution of the inverse kinematics. For 3T2R tasks, Figure 3b,c represent valid solutions.

The goal of eliminating the tool rotation β_3 from the equations is not achieved yet, since all three components of the platform orientation $\boldsymbol{x}_r$ affect the rotation matrix ${}^0\boldsymbol{R}_D$. This can be addressed by the selection of the Euler angles: Similar to the definition of the rotational operational space coordinates $\boldsymbol{x}_r$ in (4), the constraints $\boldsymbol{\Phi}_{r,i}$ are also expressed with a set of Euler angles $\boldsymbol{\alpha}$. In the following, "$\boldsymbol{\alpha}$" will always refer to the rotation error/residual and "$\boldsymbol{\beta}$" to an orientation relative to the base frame. The Euler angle convention of $\boldsymbol{\alpha}$ can be chosen independently of the choice for the orientation representation in $\boldsymbol{\beta}$. The intuitive approach of choosing

$$\boldsymbol{R}(\boldsymbol{\alpha}^*) := \boldsymbol{R}_x(\alpha_1^*)\boldsymbol{R}_y(\alpha_2^*)\boldsymbol{R}_z(\alpha_3^*) \in \mathrm{SO}(3) \tag{20}$$

the same way as $\boldsymbol{\beta}$ leads to a set of transformations depicted in Figure 4a, where the intermediate steps of the single elementary rotations are omitted since they have no technical meaning. The superscript asterisk in (20) demarcates this specific example and the following elaborations on the calculation of $\boldsymbol{\alpha}$.

To be able to remove the redundant coordinate β_3 from the rotational constraints of (19), it is necessary to change the expression of the orientation error $\boldsymbol{\alpha}$ to be reciprocal to the expression of the absolute orientation $\boldsymbol{\beta}$. By using the ZYX Euler angles with

$$\boldsymbol{R}(\boldsymbol{\alpha}) := \boldsymbol{R}_z(\alpha_1)\boldsymbol{R}_y(\alpha_2)\boldsymbol{R}_x(\alpha_3) \in \mathrm{SO}(3), \tag{21}$$

only the error component α_1 is affected by rotations around the tool axis, which is the z-axis of the intermediate frames $\mathscr{F}_{A1}$, $\mathscr{F}_{A2}$ and the platform frame $\mathscr{F}_D$ in Figure 4b, where the frame rotations with the reciprocal set of Euler angles are sketched.

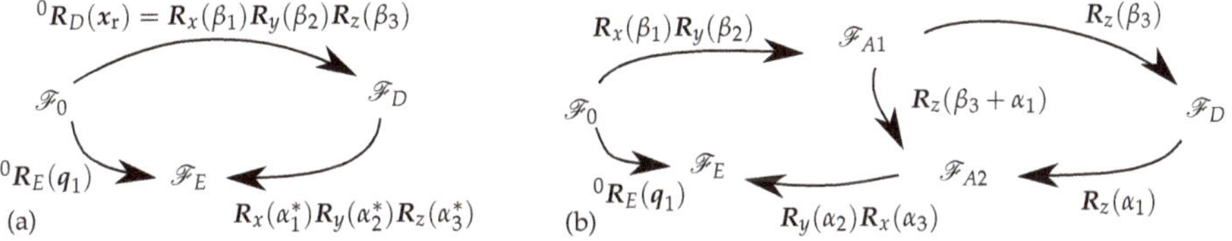

Figure 4. Overview of the different frames (a) for six-DoF tasks with standard Euler angle convention and (b) for five-DoF tasks with reciprocal Euler angle convention; taken from [5].

The mathematical proof is given in Appendix A.1 and in [5]. The new, reduced rotational part of the kinematic constraints

$$\Psi_{\mathrm{r},i}(q_i,\eta) = \begin{bmatrix} \alpha_2 & \alpha_3 \end{bmatrix}^{\mathrm{T}} = \overbrace{\begin{bmatrix} 0 & 1 & 0 \\ 0 & 0 & 1 \end{bmatrix}}^{=P_{\Psi_{\mathrm{r}}}} \Phi_{\mathrm{r},i}(q_i,x) \in \mathbb{R}^2 \tag{22}$$

does not contain the tool rotation any more. The full kinematic constraints

$$\Psi_i = \begin{bmatrix} \Phi_{\mathrm{t},i}^{\mathrm{T}} & \Psi_{\mathrm{r},i}^{\mathrm{T}} \end{bmatrix}^{\mathrm{T}} \in \mathbb{R}^5 \tag{23}$$

for the reduced coordinates can be used for the inverse kinematics of the leg chain i in 3T2R tasks, where $\Psi_i = 0$ leads to a valid position and orientation of the tool axis. The condition $\Phi_i = 0$ leads to a valid configuration of leg i of the parallel robot in 3T3R tasks. In the following, "Φ" is always used for 3T3R kinematic descriptions and "Ψ" for 3T2R. Note that by omitting the corresponding lines in the operational space coordinates x and the constraint equations Φ, it is also possible to use the 3T3R approach for systems with reduced mobility of 2T1R, 3T0R and 3T1R platform DoF.

4. Full Kinematic Constraints for Parallel Robots Using Reciprocal Sets of Euler Angles

The definition of the full kinematic constraints (16,19,18) of a single leg chain of the parallel robot from the previous chapter can be used to write the kinematic constraints in a general form. The full kinematic constraint equations can only be defined for 3T3R tasks without further adjustments as

$$\Phi = \begin{bmatrix} \Phi_1^{\mathrm{T}} & \Phi_2^{\mathrm{T}} & \cdots & \Phi_m^{\mathrm{T}} \end{bmatrix}^{\mathrm{T}}. \tag{24}$$

The constraints Ψ_i from (23) for the reduced coordinates η can only be defined for one leg chain: Figure 5a show an open-loop second leg chain for a given first leg chain from Figure 3.

By also closing the *3T2R* kinematic constraints Ψ_2 for the second loop, as depicted in Figure 5b, the tool axis stays arbitrary and the platform pose demanded from the two legs would be different and therefore would not be a valid solution for the complete mechanism, i.e., $\Phi_2 \neq 0$. Only if the second leg fulfills the *3T3R* kinematic constraints for all platform coordinates, as shown in Figure 5c, a valid configuration of the mechanism emerges. This approach has already been used for many specific robots systems, as introduced in Section 1. As a generalization, the first leg of the parallel robot is now termed the "leading leg chain" (index "1") and the other legs are termed as "following leg chains" (index "*j*").

The translational part of the constraints is not coupled by the platform orientation and therefore left unchanged relative to (16) with

$$\Phi_{\mathrm{t},j}(q_j,x) = -_{(0)}r_D + {}_{(0)}r_E(q_j) = -x_{\mathrm{t}} + {}_{(0)}r_E(q_j) \in \mathbb{R}^3 \tag{25}$$

for the following legs *j*.

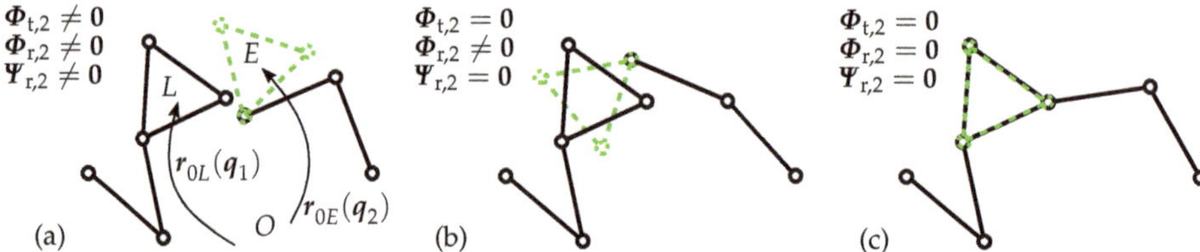

Figure 5. Different cases for the kinematic constraints of the following chain: (**a**) wrong position and orientation; (**b**) correct position and wrong orientation; (**c**) all kinematic constraints are complied.

The orientation for the platform is given with the rotation matrix

$$^{0}R_{L}(q_1) := {}^{0}R_{E}(q_1) \tag{26}$$

which gives the reference end-effector frame $\mathscr{F}_L$ resulting from the leading leg 1. The rotational part of the kinematic constraints

$$\Phi_{r,j}(q_j, q_1) = \alpha({}^{0}R_{L}^{\mathsf{T}}(q_1)\,{}^{0}R_{E}(q_j)) \tag{27}$$

for the following leg is given by the Euler angle representation of the deviation between the orientation of the platform frame $\mathscr{F}_L$ given by the leading ("L") leg and the frame $\mathscr{F}_E$ given by the respective following leg j. The choice of the Euler angle convention is arbitrary. The full kinematic constraints for the complete parallel robot with m legs for 3T2R tasks

$$\Psi = \begin{bmatrix} \Psi_1^{\mathsf{T}} & \Phi_2^{\mathsf{T}} & \cdots & \Phi_m^{\mathsf{T}} \end{bmatrix}^{\mathsf{T}} \tag{28}$$

are assembled from the 3T2R constraints Ψ_1 from (23) for the leading leg and the 3T3R constraints Φ_j, $2 \le j \le m$ from (25,27) for the following legs. The index "j" is used to distinguish the following legs of the 3T2R case and all legs "i" of the general case in Section 3. The full constraints of (28) lead to a 35-dimensional vector for the kinematic constraints for parallel robots with six legs in 3T2R tasks, which are aggregated as class I in Section 1.2. This formulation can be reduced by combining the mechanism-specific approach for the constraints from (6) or (8) with the principle of leading and following legs of this section. For the 6UPS structure this would result in a 10-dimensional constraint vector with five entries for the leading leg and only one entry for each of the five following legs.

5. Differential Kinematics for Parallel Robots

To be able to compute the differential kinematics of the constraints Φ (24) and Ψ (28) to

$$\frac{\mathrm{d}}{\mathrm{d}t}\Phi(q,x) = \Phi_{\partial q}\dot{q} + \Phi_{\partial x}\dot{x} = 0 \quad \text{and} \quad \frac{\mathrm{d}}{\mathrm{d}t}\Psi(q,\eta) = \Psi_{\partial q}\dot{q} + \Psi_{\partial\eta}\dot{\eta} = 0, \tag{29}$$

the full inverse-kinematics matrices

$$\Phi_{\partial q}(q,x) = \begin{bmatrix} \Phi_{1,\partial q_1} & 0 & 0 & 0 \\ 0 & \Phi_{2,\partial q_2} & \ddots & 0 \\ \vdots & \ddots & \ddots & 0 \\ 0 & 0 & 0 & \Phi_{m,\partial q_m} \end{bmatrix} \quad \text{and} \quad \Psi_{\partial q}(q,\eta) = \begin{bmatrix} \Psi_{1,\partial q_1} & 0 & 0 & 0 \\ \Phi_{2,\partial q_1} & \Phi_{2,\partial q_2} & \ddots & 0 \\ \vdots & \ddots & \ddots & 0 \\ \Phi_{m,\partial q_1} & 0 & 0 & \Phi_{m,\partial q_m} \end{bmatrix} \tag{30}$$

and the full direct-kinematics matrices

$$\Phi_{\partial x}(q,x) = \frac{\partial\Phi}{\partial x} = \begin{bmatrix} \Phi_{1,\partial x} \\ \Phi_{2,\partial x} \\ \vdots \\ \Phi_{m,\partial x} \end{bmatrix} \quad \text{and} \quad \Psi_{\partial\eta}(q,\eta) = \frac{\partial\Psi}{\partial\eta} = \begin{bmatrix} \Psi_{1,\partial\eta} \\ \Phi_{2,\partial\eta} \\ \vdots \\ \Phi_{m,\partial\eta} \end{bmatrix} \tag{31}$$

have to be calculated for the 3T3R and the 3T2R case respectively. The gradient $\Phi_{\partial q}$ has block diagonal form, indicating that the inverse kinematics problem can be solved for each leg independently. The structures of the gradient $\Psi_{\partial q}$ results from the coupling of the leading and following joints in the rotational constraints equation.

The gradient matrices $\Phi_{\partial q}$ and $\Phi_{\partial x}$ contain nested nonlinear functions related to the orientation error. They are calculated with the chain rule and a syntax for stacking matrix columns to avoid differentiating matrices or with respect to matrices, which was introduced in [5]. The product operator

$\overline{\Pi}$, the stacking operator $\overline{R}$ and the transpose operator P_{T} used for the implementation in the next section are explained in Appendix A.4.

5.1. Constraint Gradients for the Leading Leg of the 3T2R and All Legs of the 3T3R Case

The constraint definition $\mathbf{\Psi}_1$ for the leading leg of the 3T2R case (28) and $\mathbf{\Phi}_i$ with $i = 1, ..., m$ for all legs of the 3T3R case (24) are subject to the same model of (16), (19), (18). In the following, the 3T3R constraints are displayed. The form $\mathbf{\Psi}_1$ for the 3T2R case is obtained by removing the corresponding line of the rotational component according to (22) and replacing "$\mathbf{\Phi}$" by "$\mathbf{\Psi}$" in the following equations. For the analysis, the constraint gradient matrix w. r. t. the joint coordinates has to be divided out to

$$\mathbf{\Phi}_{1,\partial q_1} = \left[\mathbf{\Phi}_{t,1,\partial q_1}^{\mathrm{T}} \quad \mathbf{\Phi}_{r,1,\partial q_1}^{\mathrm{T}} \right]^{\mathrm{T}}, \tag{32}$$

where the translational component can be calculated with the geometric Jacobian of the leg chain, as derived in Appendix A.3. The rotational part is written down as a function composition of the three functions α (Euler angles), $\overline{\Pi}$ (matrix product) and ^0R_E (rotation matrix) as

$$\mathbf{\Phi}_{r,1,\partial q_1} = \frac{\partial}{\partial q_1} \alpha \left(^0R_D^{\mathrm{T}}(x)\,^0R_E(q_1) \right) = \frac{\partial}{\partial q_1} \alpha \left(\overline{\Pi} \left(^0\overline{R}_D^{\mathrm{T}}(x), ^0\overline{R}_E(q_1) \right) \right), \tag{33}$$

which is then expanded with the chain rule for differentiation and the stack operators to

$$\mathbf{\Phi}_{r,1,\partial q_1} = \underbrace{\frac{\partial \alpha}{\partial \overline{R}}}_{\mathrm{I} \in \mathbb{R}^{3 \times 9}} \underbrace{\frac{\partial \overline{\Pi} \left(^0\overline{R}_D^{\mathrm{T}}, ^0\overline{R}_E \right)}{\partial\,^0\overline{R}_E}}_{\mathrm{II} \in \mathbb{R}^{9 \times 9}} \underbrace{\frac{\partial\,^0\overline{R}_E(q_1)}{\partial q_1}}_{\mathrm{III} \in \mathbb{R}^{9 \times \dim(q_1)}} \in \mathbb{R}^{3 \times \dim(q_1)}. \tag{34}$$

The two first partial derivatives from (34) are sparse matrices and can be calculated efficiently as shown in Appendix A.4. The factor "I" contains (A24) with $\overline{R} = {}^D\overline{R}_E(x_r, q_1)$ and the factor "II" is (A27) with the contents of $^0\overline{R}_D^{\mathrm{T}}(x_r)$. The last partial derivative "III" can be derived with computer algebra systems from the analytic expression of the rotation matrix $^0R_E(q_1)$ or from the geometric Jacobian, see Appendix A.2. The leading legs constraint gradient matrix w. r t. the platform coordinates can be expanded in the same manner into

$$\mathbf{\Phi}_{1,\partial x} = \begin{bmatrix} \mathbf{\Phi}_{t,1,\partial x_t} & \mathbf{\Phi}_{t,1,\partial x_r} \\ \mathbf{\Phi}_{r,1,\partial x_t} & \mathbf{\Phi}_{r,1,\partial x_r} \end{bmatrix} = \begin{bmatrix} -1 & 0 \\ 0 & \mathbf{\Phi}_{r,1,\partial x_r} \end{bmatrix}, \tag{35}$$

where the definitions from (16) and (19) only leave the rotational part

$$\mathbf{\Phi}_{r,1,\partial x_r} = \frac{\partial}{\partial x_r} \alpha \left(\left(^0R_E^{\mathrm{T}}(q_1)\,^0R_D(x_r) \right)^{\mathrm{T}} \right) = \frac{\partial}{\partial x_r} \overline{\alpha} \left(P_{\mathrm{T}} \overline{\Pi} \left(^0\overline{R}_E^{\mathrm{T}}(q_1), ^0\overline{R}_D(x_r) \right) \right) \tag{36}$$

$$= \underbrace{\frac{\partial \overline{\alpha}}{\partial \overline{R}}}_{\mathrm{I} \in \mathbb{R}^{3 \times 9}} \underbrace{P_{\mathrm{T}}}_{\mathrm{II} \in \mathbb{R}^{9 \times 9}} \underbrace{\frac{\partial \overline{\Pi} \left(^0\overline{R}_E^{\mathrm{T}}, ^0\overline{R}_D \right)}{\partial\,^0\overline{R}_D}}_{\mathrm{III} \in \mathbb{R}^{9 \times 9}} \underbrace{\frac{\partial\,^0\overline{R}_D(x_r)}{\partial x_r}}_{\mathrm{IV} \in \mathbb{R}^{9 \times 3}} \in \mathbb{R}^{3 \times 3},$$

where the simplicity of the single expression "I"-"IV" is demonstrated in Appendix A.4. The factors are (A24) with $\overline{R} = {}^D\overline{R}_E(x_r, q_1)$ in "I", the permutation matrix for transposition from (A23) in "II", (A27), where the contents of $^0\overline{R}_E^{\mathrm{T}}$ are inserted in "III" and (A25) with the elements of x_r for β in "IV".

5.2. Constraint Gradients for the Following Leg in the 3T2R Case

As explained regarding (24), the constraints (16) and (19) and their gradients (34) and (36) are used for all legs in the 3T3R case and the leading leg in the 3T2R case. For the following legs in the 3T2R case, the gradients $\Phi_{j,\partial q_1}$, $\Phi_{j,\partial q_j}$ and $\Phi_{j,\partial x}$ with $j = 2, ..., m$ from the right part of (30) and (31) have to be calculated in a similar way. Due to the absence of the platform orientation in (27), (35) simplifies for the following leg to

$$\Phi_{j,\partial x} = \begin{bmatrix} -1 & 0 \\ 0 & 0 \end{bmatrix}.$$ (37)

The gradient w. r. t. the joint coordinates of the following leg contains again the translational part of the legs Jacobian regarding the end-effector platform position in $\Phi_{t,j,\partial q_j}$ and has the rotational part

$$
\Phi_{r,j,\partial q_j} = \frac{\partial}{\partial q_j} \alpha \left({}^0R_L^T(q_1) {}^0R_E(q_j) \right) = \frac{\partial}{\partial q_j} \alpha \left(\prod \left({}^0\overline{R}_L^T(q_1), {}^0\overline{R}_E(q_j) \right) \right)
$$ (38)

$$
= \underbrace{\frac{\partial \alpha}{\partial \overline{R}}}_{\text{I} \in \mathbb{R}^{3 \times 9}} \underbrace{\frac{\partial \overline{\prod} \left({}^0\overline{R}_D^T, {}^0\overline{R}_E \right)}{\partial {}^0\overline{R}_E}}_{\text{II} \in \mathbb{R}^{9 \times 9}} \underbrace{\frac{\partial {}^0\overline{R}_E(q_j)}{\partial q_j}}_{\text{III} \in \mathbb{R}^{9 \times \dim(q_j)}} \in \mathbb{R}^{3 \times \dim(q_j)},
$$

which is similar to the expression in (34). The factors of Equation (38) are (A24) with $\overline{R} = {}^L\overline{R}_E(q_1, q_j)$ in "I", (A27), where the elements of ${}^0R_L^T(q_1)$ have to be inserted in "II" and the partial derivative of the platform orientation calculated from leg j w. r. t. the legs joint coordinates in "III", similar to term "III" from (34). The gradient w. r. t. the joint coordinates of the leading leg

$$
\Phi_{r,j,\partial q_1} = \frac{\partial}{\partial q_1} \alpha \left(\left({}^0R_E^T(q_j) {}^0R_L(q_1) \right)^T \right) = \frac{\partial}{\partial q_1} \alpha \left(P_T \prod \left({}^0\overline{R}_E^T(q_j) {}^0\overline{R}_L(q_1) \right) \right)
$$ (39)

$$
= \underbrace{\frac{\partial \overline{\alpha}}{\partial \overline{R}}}_{\text{I} \in \mathbb{R}^{3 \times 9}} \underbrace{P_T}_{\text{II} \in \mathbb{R}^{9 \times 9}} \underbrace{\frac{\partial \overline{\prod} \left({}^0\overline{R}_E^T, {}^0\overline{R}_L \right)}{\partial {}^0\overline{R}_L}}_{\text{III} \in \mathbb{R}^{9 \times 9}} \underbrace{\frac{\partial {}^0\overline{R}_L(q_1)}{\partial q_1}}_{\text{IV} \in \mathbb{R}^{9 \times \dim(q_1)}} \in \mathbb{R}^{3 \times \dim(q_1)}.
$$

is similar to the expression in (36). The order of the residual expression (27) has to be switched by exploiting the associative property of matrix transposition to avoid differentiating a transposed matrix. The factors of equation (39) are (A24) with $\overline{R} = {}^L\overline{R}_E(q_1, q_j)$ in "I", the permutation matrix for transposition from (A23) in "II", (A27) with ${}^0R_E^T(q_j)$ in "III" and the term "III" from (34) in "IV".

5.3. Gradient-Based Solution of the Inverse Kinematics Problem with Redundancy Resolution

The presented kinematic constraints and their gradient matrices can be used to solve the inverse kinematics problem (IKP) of single leg chains and complete parallel robots. Since all active and passive joint angles are involved for the case of parallel robots, solving their IKP results in solving the IKP for all leg chains. As first introduced in [7] for Euler angle residuals in the IKP, the Taylor series expansion of $\Phi(q, x)$ leads to the definition of

$$\Phi(q^{k+1}, x) = \Phi(q^k, x) + \frac{\partial}{\partial q} \Phi(q, x) \Big|_{q^k} (q^{k+1} - q^k)$$ (40)

in an iterative algorithm at the step $k + 1$, which can be used to solve the IKP using a given initial value q^0 and the condition

$$\Phi(q^{k+1}, x) = 0.$$ (41)

Defining the solution of the IKP as the main task ("T"), the stepwise solution for the joint coordinates results to

$$\Delta q^k = \Delta q_T^k = q^{k+1} - q^k = -\Phi_{\partial q}^\dagger(q^k, x) \Phi(q^k, x).$$ (42)

Depending on the dimension, $(\cdot)^{\dagger}$ denotes the matrix inverse or the pseudo-inverse. Again, the Equations (40)–(42) can be written with "$\mathbf{\Psi}$" from (28) instead of "$\mathbf{\Phi}$" from (24) for the 3T2R case.

In the latter case, the corresponding gradient matrix $\mathbf{\Psi}_{\partial q}(q,\eta)$ from (30) allows defining of a nullspace in the case of $\dim(q_1) > \dim(\eta)$. This redundancy can be exploited by using the nullspace ("N") projection Δq_N from [6] additionally to the solution Δq_T of the IKP in (42) with the new increment

$$\Delta q^k = q^{k+1} - q^k = \Delta q_T^k + \Delta q_N^k = -\mathbf{\Psi}_{\partial q}^{\dagger}\mathbf{\Psi} + (1 - \mathbf{\Psi}_{\partial q}^{\dagger}\mathbf{\Psi}_{\partial q})h_{\partial q} \tag{43}$$

in the iterative algorithm. The optimization of additional performance criteria h requires their gradient $h_{\partial q}$ w.r.t the joint positions. One criterion is the summed W_1-weighted quadratic distance

$$h_1(q) = \frac{1}{2}(q - \bar{q})^{\mathrm{T}}W_1(q - \bar{q}), \quad h_{1,\partial q} = \frac{\partial h_1}{\partial q} = W_1(q - \bar{q}) \tag{44}$$

of the joint positions q from their respective reference position $\bar{q}$, e.g., used in [2,36]. Defining $\bar{q}$ to be in the middle of the joint limits and minimizing $h_1(q)$ reduces the risk of joints reaching their technical limits, but does not guarantee it, since exceeding the limit for one joint can be compensated by improving other joints. The W_2-weighted hyperbolic joint limit distance

$$h_2(q) = \frac{1}{n}\sum_{i=1}^{n} w_{2,i}\frac{(q_{i,\max} - q_{i,\min})}{8}\left(\frac{1}{(q_i - q_{i,\min})^2} + \frac{1}{(q_i - q_{i,\max})^2}\right) \tag{45}$$

from [8] (written element-wise for $n = \dim(q)$) circumvents this problem by generating infinitely high values when reaching the limits. In contrast to h_1, the criterion h_2 is only defined for joints within their limits with $q_{i,\min} < q_i < q_{i,\max}$, which is ensured by setting $w_{2,i} = 0$ for joints exceeding their limits and $w_{2,i} = 1$ otherwise. To combine the effect of drawing joint positions to their middle with h_1 of (44) and of strongly rejecting joints directly near their limits with h_2 of (45), their weighted sum

$$h_3(q) = K_{h_1}h_1(q) + K_{h_2}h_2(q) \tag{46}$$

is used in the simulation studies of Section 6. Other criteria not related to the joint limits are for example stiffness [9] or singularity avoidance via Frobeniusnorm condition number [8] or squared condition number [4]. The method can be used for serial-link robots as well by removing all entries for the following legs from the formulas, as presented in [5]. The platform pose x/η corresponds to the desired pose for the serial robots end-effector and the kinematic constraints $\mathbf{\Phi}/\mathbf{\Psi}$ correspond to the residual of the IKP.

In the practical implementation, it has proven to be useful to extend the basic principle of (43) to

$$\Delta q^k = K_{\mathrm{Lim}}(q^k)K_{\mathrm{Rel}}(q^k)(K_{\mathrm{T}}\Delta q_T^k + K_{\mathrm{N}}\Delta q_N^k), \tag{47}$$

where the constant damping coefficients K_{T} for Δq_T^k and K_{N} for Δq_N^k were introduced to avoid overshooting of the solution for the prize of slower convergence. The damping term K_{N} has to be chosen according to the optimization criterion. Further damping was introduced for the 3T2R case with task redundancy to reduce a Δq^k that would lead to overshoot over the joint limits with $K_{\mathrm{Lim}}(q^k)$. The value $K_{\mathrm{Lim}} = 1$ is set if no limits would be violated by the increment Δq^k. For the 3T3R case, $K_{\mathrm{Lim}} := 1$ is set permanently, since slowing down when approaching the limits does not change the direction of the increment and violating the limits is inevitable. The maximum step size for one iteration Δq^k was ensured with $K_{\mathrm{Rel}}(q^k)$ to stay below 5% of the joint limit range to prevent leaving the validity of the first-order approximation of (40). For smaller increments, $K_{\mathrm{Rel}} = 1$ holds. The damping terms are always applied to the full vector and not to single elements and therefore only change the norm and not the direction of Δq^k.

5.4. Differential Kinematics for the Parallel Robot and Its Applications

The reasoning so far only considered the inverse kinematics of the parallel robot. The kinematic definitions can also be used in the differential kinematics (29) to establish the connection between joint and platform velocity. This was already presented in general form in [15] and also corresponds to the theory of linear transformation which is the base of the works of Gogu on structural synthesis [11]. The derivation in this paper is based on the position *and* orientation, but comes to the same result as the already-existing velocity-based approach. Furthermore, using $\mathbf{\Psi}/\eta$ as elaborated before allows for the first time defining differential kinematics specifically for 3T2R tasks in a general form. The differential equation of (29) is expanded to

$$\frac{\mathrm{d}}{\mathrm{d}t}\mathbf{\Phi}(q_{\mathrm{a}}, q_{\mathrm{p}}, x) = \mathbf{\Phi}_{\partial q_{\mathrm{a}}}\dot{q}_{\mathrm{a}} + \mathbf{\Phi}_{\partial q_{\mathrm{p}}}\dot{q}_{\mathrm{p}} + \mathbf{\Phi}_{\partial x}\dot{x} = 0 \tag{48}$$

to distinguish active ("a") and passive ("p") joints. The latter also contain the coordinates of the platform-connecting joints. Reordering the equation leads to the full inverse differential kinematics

$$\mathbf{\Phi}_{\partial x}\dot{x} = -\begin{bmatrix}\mathbf{\Phi}_{\partial q_{\mathrm{a}}} & \mathbf{\Phi}_{\partial q_{\mathrm{p}}}\end{bmatrix}\begin{bmatrix}\dot{q}_{\mathrm{a}}\\\dot{q}_{\mathrm{p}}\end{bmatrix} = \mathbf{\Phi}_{\partial\mathrm{ap}}\begin{bmatrix}\dot{q}_{\mathrm{a}}\\\dot{q}_{\mathrm{p}}\end{bmatrix}, \qquad \begin{bmatrix}\dot{q}_{\mathrm{a}}\\\dot{q}_{\mathrm{p}}\end{bmatrix} = -\mathbf{\Phi}_{\partial\mathrm{ap}}^{-1}\mathbf{\Phi}_{\partial x}\dot{x}, \tag{49}$$

which has been addressed in the previous sections, and the full direct differential kinematics

$$\mathbf{\Phi}_{\partial q_{\mathrm{a}}}\dot{q}_{\mathrm{a}} = -\begin{bmatrix}\mathbf{\Phi}_{\partial x} & \mathbf{\Phi}_{\partial q_{\mathrm{p}}}\end{bmatrix}\begin{bmatrix}\dot{x}\\\dot{q}_{\mathrm{p}}\end{bmatrix} = \mathbf{\Phi}_{\partial xp}\begin{bmatrix}\dot{x}\\\dot{q}_{\mathrm{p}}\end{bmatrix}, \qquad \begin{bmatrix}\dot{x}\\\dot{q}_{\mathrm{p}}\end{bmatrix} = -\mathbf{\Phi}_{\partial xp}^{-1}\mathbf{\Phi}_{\partial q_{\mathrm{a}}}\dot{q}_{\mathrm{a}}. \tag{50}$$

By only selecting the first rows for $\dot{q}_{\mathrm{a}}$ in (49) and for $\dot{x}$ in (50), the well-known analytic Jacobian of the parallel robot (called "Euler angles Jacobian matrix" in [15] and "design Jacobian" in [11]), relating actuator velocities $\dot{q}_{\mathrm{a}}$ and platform velocities $\dot{x}$, can be obtained from both equations. For the case of task or kinematic redundancy, the pseudo-inverse can be used for $\mathbf{\Phi}_{\partial\mathrm{ap}}$ in (49) together with a nullspace optimization as shown in Section 5.3. This allows exploiting the functional redundancy in 3T2R tasks also for trajectories $\eta(t)$, $\dot{\eta}(t)$ and not only for single poses η. When solving the IKP for a trajectory, the IK on position level of Section 5.3 is only needed to correct linearization errors. The case of task redundancy does not affect (50), since the full platform velocity $\dot{x}$ is obtained from given actuator velocities $\dot{q}_{\mathrm{a}}$.

6. Results

To evaluate the inverse kinematics (IK) algorithm presented in the previous Section 5.3, first the solution of the IKP is shown for the trajectory of a serial-link industrial robot in Section 6.1 and for the trajectory of a parallel robot in Section 6.2. The results are generalized by the statistical analysis of random point-to-point movements of arbitrary serial link chains in Section 6.3.

6.1. Resolution of Functional Redundancy of a Serial-Link Six-DoF Robot in 3T2R tasks

The first evaluation of the IK algorithm from Section 5.3 is performed with simulations at the basic example of a six-DoF industrial robot with a 500 mm × 800 mm rectangular trajectory centered at 1200 mm/−200 mm/200 mm with a constant orientation of the z-axis pointing into the ground plane. The manipulator Fanuc M-710 iC/50 was taken from the example of [36] with the tabulated kinematics parameters and a sketch of the trajectory in Figure 6. Deviations in the parameters relative to [36] result from the use of the *modified* Denavit–Hartenberg notation from [35] for the joint transformations and the use of only positive axis alignment parameters α_i for consistency with the results of the structural synthesis from [34].

i	α_i	d_i	θ_i	r_i	$q_{i,\min}$	$q_{i,\max}$
	R_x	T_x	R_z	T_z		
1	0	0	q_1	0	$-180°$	$180°$
2	$90°$	150 mm	q_2	0	$30°$	$165°$
3	0	870 mm	q_3	0	$-132°$	$230°$
4	$90°$	170 mm	q_4	1016 mm	$-360°$	$360°$
5	$90°$	0	q_5	0	$-125°$	$125°$
6	$90°$	0	q_6	-175 mm	$-360°$	$360°$

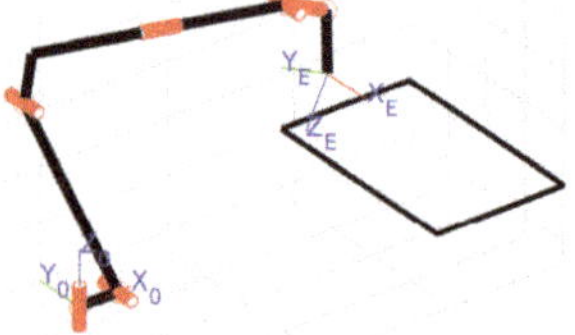

Figure 6. Left: Table with the kinematic parameters of the industrial manipulator Fanuc M-710 iC/50. **Right**: sketch of the robot scenario.

The IKP is solved with two settings: setting the tool axis rotation to different constant values β_3 with the 3T3R algorithm and solving the IKP only for the desired pointing direction with the 3T2R algorithm. The algorithm from (43) was used in the extended version of (47) for both cases with different settings caused by their nature. For the 3T3R case, $K_T = 0.7$ and $K_N = 0.7$ were set. The terms K_N and K_{Lim} have no effect, since no nullspace movement is possible. For the 3T2R case, with $K_{h_1} = 0$ and $K_{h_2} = 1$ only the hyperbolic limit rejection criterion from (45) was used. The first criterion was not used, since in the trajectory example the limits are not even temporarily exceeded by principle. All IKP algorithms had the same initial value from Figure 6.

The results of the inverse kinematics for different settings are given in Figure 7, where the representative joint coordinates q_1 and q_5, the redundant coordinate of the end-effector orientation β_3, as well as the optimization criterion (45) are depicted over time for the trajectory from Figure 6. The positions are normalized to the joint limits from -1 to 1. The first three lines in Figure 7 represent IKP solutions with a given constant end-effector orientation β_3 of $-150°$, $-15°$ and $45°$ and the 3T3R algorithm. The 3T2R algorithm without nullspace optimization is plotted with dotted lines for each first sample of the 3T3R cases as initial value with the same colors. Using these initial values for a 3T2R IK with optimization leads to strong nullspace movements at the beginning, quickly converging to a local minimum. Therefore, the 3T2R case with optimization, plotted as the green line with triangle markers, is shown only for the initial condition from Figure 6. It can be observed that the optimization of the criterion leads to the best solution of the IKP. The lines for the criterion for $\beta_3 = -150°$ and $\beta_3 = -15°$ partly exceed the limits of the plot, indicating that the limit is violated, which can also be seen at the plot for q_5. This exposes the need for keeping the solution always within the limits by the measures described. The 3T2R IK without optimization with dotted lines tends to lower changes in the joint positions than the 3T3R IK, since this corresponds to the solution of the matrix pseudo-inverse in (43).

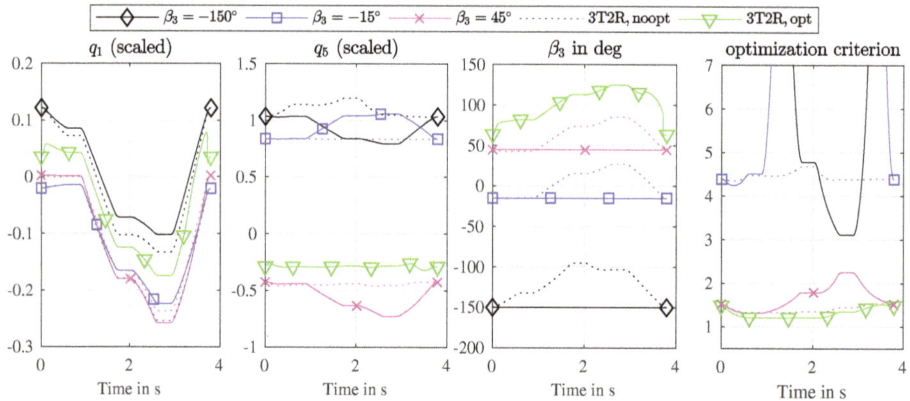

Figure 7. Results of the inverse kinematics with different settings for the trajectory of Figure 6.

6.2. Resolution of Functional Redundancy of a Parallel Robot in 3T2R Tasks

As elaborated in Section 4 and 5, the solution of the IKP for 3T2R and 3T3R tasks is necessary to solve the problem for parallel robots. Therefore, the trajectory evaluation for a 6UPS parallel robot in this section is preceded by the trajectory example for a serial link chain in the previous section. This robot belongs to the first class presented in Section 1.2, which is primarily addressed in this paper.

The robot has a Goughstructure [15] with symmetric alignment of the universal joint base couplings on a circle with radius $\|r_{0A_i}\| = 1\,\mathrm{m}$ and the spherical joint platform couplings on a circle with radius $\|r_{B_iE}\| = 0.4\,\mathrm{m}$. The initial pose was set to a center position $x_t^{\mathrm{T}} = [0,0,0.5\,\mathrm{m}]$ and the initial orientation x_r was set to zero, meaning an alignment of base and platform frame. The joint positions for each leg were defined to have the initial values $q_i^{\mathrm{T}} = [30°, -30°, 0.583\,\mathrm{m}, 0°, 30°, 60°]$ for the given initial platform pose to avoid switching $\pm\pi$ within the trajectory and to avoid gimballocksingularities. The joint limits were set around the resulting zero position to $\pm 0.5\,\mathrm{m}$ for the prismatic joint and $\pm 60°$ for all single revolute joints representing the universal and spherical joints. The values are higher than typical values for real robots to emphasize the effect of the nullspace movement in a bigger simulated workspace of the robot. The settings for the IK solver are similar to those in Section 6.1, since both cases regard solving the IKP for a trajectory.

The time evolution of platform pose and optimization criteria is depicted in Figure 8. The reference trajectory can be seen at the platform position in Figure 8a and the platform orientation expressed in XYZ Euler angles (β_1-β_3) relative to the base frame in Figure 8b. The IKP is solved using two different methods: Only solving the IKP for the legs separately, called "ser. IK" in Figure 8 and solving the IKP for all legs together, called "par. IK" in Figure 8. Both methods perform an optimization with only h_2 of (45), as justified in Section 6.1. The first approach only performs this optimization according to Section 3 for the first leg using the 3T2R method and then solves the IKP for all other legs with the 3T3R method. The second approach uses the optimization for all legs together according to the 3T2R method from Section 4. This results in improved values for the performance criteria depicted for h_1 in Figure 8c and for h_2 in a logarithmic scale in Figure 8d. Since the first approach does not regard the limits of the following legs, the optimization criterion gives high values indicating many joint limit violations. The second approach only shows peaks at $t = 1.5\,\mathrm{s}$ in Figure 8d that result from a joint position getting near to the limit, but not exceeding it. For the practical implementation, the computation time is only weakly influenced by the selection of the method, since calculating the (pseudo-)inverse for six 5×6 and 6×6 matrices or one 35×36 matrix does not present a challenge for current computing hardware. Therefore, the "par. IK"method should be preferred.

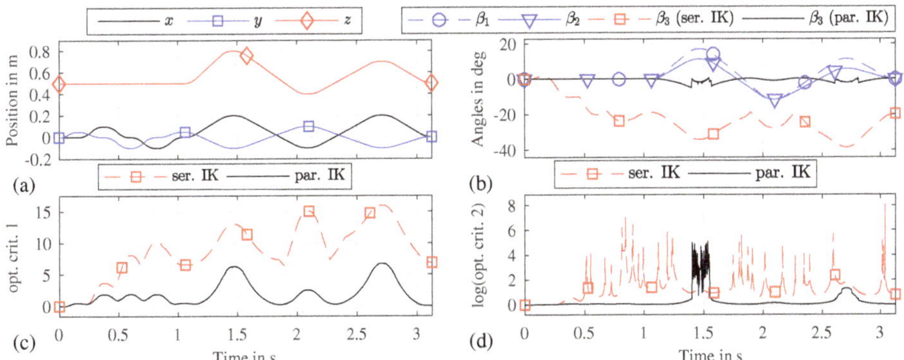

Figure 8. Results of the inverse kinematics of a 6UPS robot in a 3T2R task. (**a**) platform positions, (**b**) platform orientation in Euler angles, (**c**,**d**) optimization criteria $h_1(q)$ and $h_2(q)$.

6.3. Statistic Results for the Inverse Kinematics of Serial Link Chains

To emphasize the generality of the presented approach, the inverse kinematics is solved for a set of 309 serial kinematic chains with six joints. This set of six-DoF kinematics is generated by permutations of their Denavit–Hartenberg parameters and is reduced with the isomorphism detection of [34] to a minimal set, representing all possible six-DoF serial kinematics with full mobility. The approach is similar to the results of the evolutionary morphology of parallel robot leg chains of [11]. In contrast to the trajectory evaluations in the previous sections focusing on nullspace movement, the inverse kinematics is solved in this section for arbitrary reachable poses of the serial chain in its individual workspace. Therefore, different settings proved to be necessary, since for point-to-point movements, intermediate steps may be outside of the joint limits. In the trajectory case, the initial value for the IKP of the continuous trajectory is always very close to the desired pose of the next trajectory sample. Preventing the algorithm completely from leaving the allowed joint positions reduces the IK success rate. Therefore, the damping term for limit violation was not used in this evaluation, resulting in a constant $K_{\mathrm{Lim}}(q) = 1$ in (47). To reach again an allowed configuration when approaching the goal pose from intermediate steps with limit violations, the combined criterion $h_3(q)$ from (46) with $K_{h1} = 0.99$ and $K_{h2} = 0.01$ was used. Further empirically determined values for all different serial chains were the damping coefficients $K_T = 0.6$ and $K_N = 0.01$ in (47). Since these values provide good results for all serial chains with random geometric parameters and for random configurations, they can be regarded as a good choice generally.

To create a general evaluation case, the poses for testing the IK algorithm were generated by the forward kinematics of 50 different joint configurations of the chains uniformly distributed between the joint limits of $\pm\pi$ for revolute joints and $\pm 0.5\,\mathrm{m}$ for prismatic joints. Additionally, the Denavit–Hartenberg parameters were set to 50 different sets of uniformly distributed parameters between 0 and 1 meters or radians resulting in 2500 combinations for each of the 309 chains in total. The initial value q^0 for the solution of the IKP of (47) was set to random values from a uniform distribution within the joint limits. The inverse kinematics was calculated for the full pose with the 3T3R algorithm and only using the pointing direction together with the resolution of functional redundancy in the 3T2R algorithm. A maximum of 15 tries with random initial values was allowed to search for a solution of the IKP within the limits. After that, five more tries were allowed to find a solution violating the limits, but presenting a solution of the IKP to be able to distinguish the two cases, which allows further reasoning on the functionality and possible improvements. A success of the IK is defined as a solution within the joint limits.

The aggregated results are presented as histograms in Figure 9 for different settings of the algorithm. The histograms show that for the worst case in 3T2R (3T3R) tasks, the success rate is 87% (6%), marked by the position of the first bars in Figure 9a,c. These results can be vastly improved by setting the initial guess q^0 within 20% (w. r. t. the joint limit range) around the pose, from which the desired end-effector pose has been calculated. This improves the worst success rate of all kinematic chains to 98% for 3T2R (Figure 9b) and 95% for 3T3R tasks (Figure 9d).

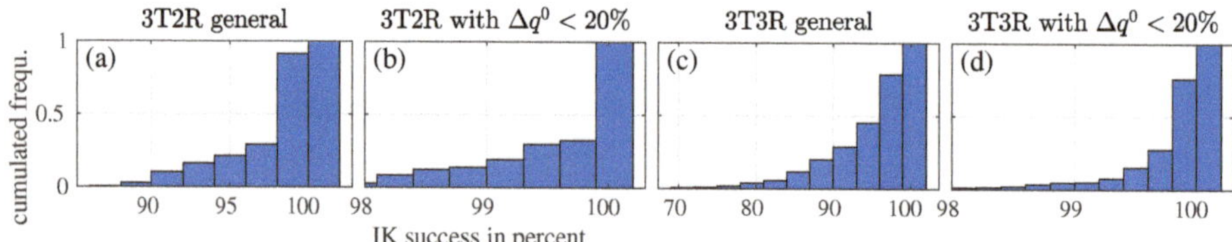

Figure 9. Histograms with cumulated frequency of the IK success for all kinematic chains with different settings: 3T2R tasks (**a,b**) vs. 3T3R tasks (**c,d**) and arbitrary initial value (**a,c**) vs. initial near goal pose.

A detailed investigation on the success rates of all possible serial chains is performed in Figure 10.

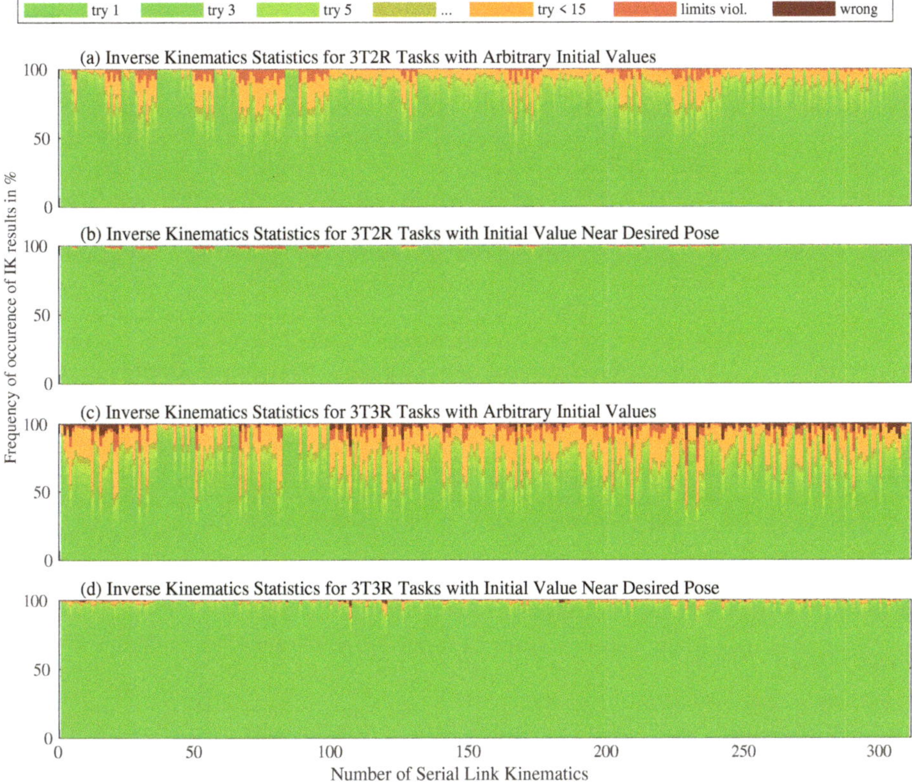

Figure 10. Detailed Statistics of the success of the inverse kinematics algorithm for 3T2R tasks (**a**,**b**) and 3T3R tasks (**c**,**d**). The Success of the IK solver is shown in different shades of green for increasing numbers of required tries. Different initial values q^0 are distinguished in (**a**,**c**) and (**b**,**d**).

The 309 serial kinematics are sorted according to their number of revolute joints and are listed on the horizontal axis of the figure: They contain three revolute joints up to no. 98, four Rjoints up to no. 240, and five Rjoints up to no. 301. The eight structures from 302 to 309 with six Rjoints differ in the parallelism of their joint axes. The first 240 kinematic chains with more than one prismatic joint can be seen as a rather academic example and are listed for the sake of completeness. The most prominent chains are the UPSchain from Section 6.2 at no. 266 and the six-DoF industrial robot from Section 6.1 at no. 309. Each bar represents the stacked relative frequency of the IK result state in percent for one kinematic chain. The result state is defined as the number of tries or the success. All bars add up to 100%, which corresponds to the 2500 configurations per chain. Beginning at the bottom, the number of tries necessary for the solution of the IKP is marked with colors from green to orange. Only cases with a violation of the limits (bright red) or wrong position (dark red) correspond to a failure of the algorithm, which has been addressed in the analysis of Figure 9, representing an aggregated form of Figure 10. The subfigures a–d of Figure 10 correspond to the ones in Figure 9. It can be observed that the quality of the results is clustered according to the kinematic groups. Structures with at most one Pjoint show a considerably better performance of the algorithm with a worst success rate of 97.16% for five Rjoints and 99.36% for six Rjoints for the 3T2R case (a), which can be seen at the very small red top parts of the bars in the corresponding range of the diagram. The worse performance of the 3T3R algorithm, mostly caused by limit violations, can be explained by joints changing their configuration, i.e., from "elbow up" to "elbow down", which causes limit violations but does not affect the 3T3R IK.

7. Discussion

A general kinematics model for parallel robots was introduced to solve the inverse kinematics problem for any kind of parallel robot in tasks with one redundant rotational DoF (3T2R). The prize of the generality of the approach is the increased size of the inverse-kinematics matrices, which is 10×11 for robots with a simple UPS structure instead of 6×6 for the non-redundant kinematics and grow up to 35×36 for general task redundant parallel robots with full mobility. This makes symbolic calculations of the kinematic matrices impossible, allowing only studies on mobility, singularities and other properties of the Jacobian matrix based on numeric calculations. The application of the proposed method can therefore be seen mainly in finding optimal trajectories for task redundant parallel robots in milling or drilling scenarios regarding stiffness, dexterity or joint limits. Due to the performance of the method demonstrated at exemplary cases, an online implementation is possible but has to be proved in future works to converge in real-time conditions for specific machines. The generality of the approach allows using it in a combined structural and dimensional synthesis sketched in Figure 2, extending the purely structural synthesis of parallel robot kinematics from [11,31] to a dimensional synthesis of all structures as shown in [34] for serial robots.

Author Contributions: conceptualization, M.S. and S.T.; methodology, M.S.; software, M.S.; validation, M.S.; formal analysis, M.S.; writing—original draft preparation, M.S.; writing—review and editing, M.S., S.T., T.O.; supervision, S.T. and T.O.; project administration, T.O.; funding acquisition, S.T. and T.O.

Funding: The financial support from the Deutsche Forschungsgemeinschaft (German Research Foundation, DFG) under grant number OR 196/33-1 is gracefully acknowledged.

Conflicts of Interest: The authors declare no conflict of interest.

Abbreviations

The following abbreviations are used in this manuscript:

PKM	parallel kinematic machine (parallel robot)
IKP	inverse kinematics problem
DoF	degrees of freedom
xTyR	x translational and y rotational degrees of freedom

Appendix A. Mathematical Symbols for Reciprocal Euler Angles in Inverse Kinematics

The following appendix contains additional detailed information about the kinematic constraint formulation of this paper. Appendix A.1 contains a mathematical proof for the properties of reciprocal Euler angles in inverse kinematics, which is only outlined in equ. 18 of [5]. The relations of the partial derivatives to the geometric and analytic Jacobian are derived in Appendixes A.2 and A.3. The matrix operations for the partial derivatives are replicated from [5] in Appendix A.4 and the contents of the single partial derivatives are given in Appendix A.5 to facilitate the understanding and implementation by the reader.

Appendix A.1. Proof for the Properties of Reciprocal Euler Angles

This section derives the effect of the reciprocity of Eulerangles at the example of the kinematics description of Section 3 and the frames of Figure 4b: An end-effector orientation $\beta = x_r$ gives the rotation matrix

$$^{D}\boldsymbol{R}_E(\beta, \boldsymbol{q}) = {}^{0}\boldsymbol{R}_D^{\mathrm{T}}(\beta)\,{}^{0}\boldsymbol{R}_E(\boldsymbol{q}), \tag{A1}$$

which rotates vectors from the actual end-effector frame $\mathscr{F}_E$ to the desired or platform end-effector frame $\mathscr{F}_D$. Using the XYZ Euler angles and exploiting the properties of SO(3) rotation matrices yields

$$^{0}\boldsymbol{R}_D^{\mathrm{T}}(\beta) = \boldsymbol{R}_z(-\beta_3)\boldsymbol{R}_y(-\beta_2)\boldsymbol{R}_x(-\beta_1), \tag{A2}$$

as introduced in (5). With an additional rotation $-\delta$ around the z-axis for the desired orientation, the resulting new Euler angles β' are

$$\beta_1' = \beta_1, \quad \beta_2' = \beta_2, \quad \beta_3' = \beta_3 - \delta. \tag{A3}$$

The additional rotation corresponds to the tool axis defined in Section 3 and leads to a new residual orientation error expressed as a rotation matrix

$$\begin{aligned}
{}^{D}R_E(\beta',q) &= {}^{0}R_D^{\mathrm{T}}(\beta')\,{}^{0}R_E(q) \\
&= \left({}^{0}R_D(\beta)R_z(-\delta)\right)^{\mathrm{T}}{}^{0}R_E(q) \\
&= R_z(\delta)\,{}^{0}R_D^{\mathrm{T}}(\beta)\,{}^{0}R_E(q) \\
&= R_z(\delta)\,{}^{D}R_E(\beta,q).
\end{aligned} \tag{A4}$$

The first residual orientation error from (A1) corresponding to β is defined as a rotation matrix

$$
{}^{D}R_E(\beta,q) = \begin{bmatrix} n_x & o_x & a_x \\ n_y & o_y & a_y \\ n_z & o_z & a_z \end{bmatrix} \tag{A5}
$$

and as a ZYX Euler angle representation

$$
\alpha = \begin{bmatrix} \alpha_1 \\ \alpha_2 \\ \alpha_3 \end{bmatrix} = \begin{bmatrix} \mathrm{arctan2}\left(n_y, n_x\right) \\ \mathrm{arctan2}\left(-n_z, \sqrt{a_z^2 + o_z^2}\right) \\ \mathrm{arctan2}\left(o_z, a_z\right) \end{bmatrix}. \tag{A6}
$$

Using the sign-aware operator $\mathrm{arctan2}(y,x)$ instead of $\arctan(y/x)$ allows angles to be in $(-\pi, +\pi]$, removes ambiguities and provides global differentiability. The second residual corresponding to β' only differs regarding the additional rotation δ. Combining (A4) and (A5) leads to

$$
{}^{D}R_E(\beta',q) = \begin{bmatrix} n_x' & o_x' & a_x' \\ n_y' & o_y' & a_y' \\ n_z' & o_z' & a_z' \end{bmatrix} = \begin{bmatrix} C_\delta\, n_x - S_\delta\, n_y & C_\delta\, o_x - S_\delta\, o_y & C_\delta\, a_x - S_\delta\, a_y \\ C_\delta\, n_y + S_\delta\, n_x & C_\delta\, o_y + S_\delta\, o_x & C_\delta\, a_y + S_\delta\, a_x \\ n_z & o_z & a_z \end{bmatrix},
$$

where $C_\delta = \cos(\delta)$ and $S_\delta = \sin(\delta)$. The ZYX Euler angles from this rotation matrix are

$$
\alpha' = \begin{bmatrix} \alpha_1' \\ \alpha_2' \\ \alpha_3' \end{bmatrix} = \begin{bmatrix} \mathrm{arctan2}\left(n_y', n_x'\right) \\ \mathrm{arctan2}\left(-n_z', \sqrt{a_z'^2 + o_z'^2}\right) \\ \mathrm{arctan2}\left(o_z', a_z'\right) \end{bmatrix} = \begin{bmatrix} \mathrm{arctan2}\left((C_\delta\, n_y + S_\delta\, n_x),(C_\delta\, n_x - S_\delta\, n_y)\right) \\ \mathrm{arctan2}\left(-n_z, \sqrt{a_z^2 + o_z^2}\right) \\ \mathrm{arctan2}\left(o_z, a_z\right) \end{bmatrix}, \tag{A7}
$$

where δ only influences the first component α_1'. This allows the conclusion that β_3 only influences α_1 and results in the dependencies

$$\alpha_1' = \alpha_1'(q, \beta_1, \beta_2, \beta_3) \tag{A8}$$
$$\alpha_2' = \alpha_2'(q, \beta_1, \beta_2) = \alpha_2 \tag{A9}$$
$$\alpha_3' = \alpha_3'(q, \beta_1, \beta_2) = \alpha_3 \tag{A10}$$

with the consequences for the kinematic modeling of robots in 3T2R tasks described in Section 3.

Appendix A.2. Relation of the Geometric Jacobian and the Partial Derivative of the Rotation Matrix

To be able to calculate the gradient matrices of Section 5, the very uncommon partial derivative of the end-effector rotation matrix ${}^{0}\!R_E$ with respect to the joint coordinates q_1 of the kinematic chain is needed as term "III" in (34) and (38). It can either be derived with computer algebra systems from the analytic expression of the rotation matrix, or by first devising the time derivative

$$
{}^{0}\dot{R}_E(q_1,\dot{q}_1) = \frac{\mathrm{d}}{\mathrm{d}t}{}^{0}\!R_E(q_1) = \frac{\partial}{\partial q_1}{}^{0}\!R_E(q_1)\dot{q}_1 \tag{A11}
$$

and then performing the derivative w.r.t $\dot{q}_1$. Using the operators of Appendix A.4 on the relation between rotation matrices and angular velocities leads to the formulation

$$
\frac{\partial}{\partial q_1}{}^{0}\!R_E(q_1) = \frac{\partial}{\partial \dot{q}_1}{}^{0}\dot{R}_E = \frac{\partial}{\partial \dot{q}_1}\overline{\overline{\prod}}\left(\overline{S}(\omega_E),{}^{0}\!R_E\right) \quad \text{with} \quad \omega_E = J_{\omega,1}(q_1)\dot{q}_1, \tag{A12}
$$

where the end-effector angular velocity ω_E can be expressed with the rotational part $J_{\omega,1}$ of the geometric Jacobian. It is used as the argument of the cross product matrix operator in stacked form

$$
\overline{S}(\omega) = \begin{pmatrix} 0 & \omega_z & -\omega_y & -\omega_z & 0 & \omega_x & \omega_y & -\omega_x & 0 \end{pmatrix}^{\mathrm{T}} \quad \text{with} \quad \omega = \begin{pmatrix} \omega_x & \omega_y & \omega_z \end{pmatrix}^{\mathrm{T}}. \tag{A13}
$$

Applying again the chain rule for differentiation of the function composition of the three functions $\overline{\overline{\prod}}$ (matrix product), S (cross product matrix) and ω_E (linear relation of velocities), as practiced in Section 5, yields

$$
\frac{\partial}{\partial q_1}{}^{0}\!R_E = \underbrace{\frac{\partial}{\partial \overline{S}}\overline{\overline{\prod}}\left(\overline{S},{}^{0}\!R_E\right)}_{\mathrm{I}\in\mathbb{R}^{9\times 9}} \underbrace{\frac{\partial}{\partial \omega}\overline{S}(\omega)}_{\mathrm{II}\in\mathbb{R}^{9\times 3}} \underbrace{J_{\omega,1}(q_1)}_{\mathrm{III}\in\mathbb{R}^{3\times\dim(q_1)}} \in \mathbb{R}^{9\times\dim(q_1)}. \tag{A14}
$$

The first factor is (A28), where the contents of ${}^{0}\!R_E$ are inserted. The second factor is a sign matrix which can be derived directly from (A13) and only contains $0/-1/1$. The third factor is the rotational part of the geometric Jacobian. This completes the calculation of the gradient matrix (34) for the Euler angles orientation residual, which was first introduced, but not pursued further in [7].

Appendix A.3. Relation of the Inverse- and Direct-Kinematics Matrices to the Analytic Jacobian

The properties of the gradient matrices from Section 5 can be illustrated further by comparing them to the analytic Jacobian of one kinematic chain, which gives the platform or end-effector velocity

$$
\dot{x} = J_1\dot{q}_1, \qquad \begin{bmatrix} \dot{x}_t \\ \dot{x}_r \end{bmatrix} = \begin{bmatrix} J_{t,1}(q_1) \\ J_{r,1}(q_1) \end{bmatrix}\dot{q}_1. \tag{A15}
$$

Also using the differential form (29) for the first kinematic leg chain of the parallel robot gives

$$
\frac{\mathrm{d}}{\mathrm{d}t}\Phi_1(q_1,x) = \Phi_{1,\partial q_1}(q_1,x)\dot{q}_1 + \Phi_{1,\partial x}(q_1,x)\dot{x} = 0 \tag{A16}
$$

and results reorganized to the form of (A15) with (32) and (35) written component-wise to

$$
\begin{bmatrix} \dot{x}_t \\ \dot{x}_r \end{bmatrix} = -\begin{bmatrix} -1 & 0 \\ 0 & \Phi_{r,1,\partial x_r}^{-1} \end{bmatrix}\begin{bmatrix} \Phi_{t,1,\partial q_1} \\ \Phi_{r,1,\partial q_1} \end{bmatrix}\dot{q}_1 = \begin{bmatrix} \Phi_{t,1,\partial q_1} \\ -\Phi_{r,1,\partial x_r}^{-1}\Phi_{r,1,\partial q_1} \end{bmatrix}\dot{q}_1. \tag{A17}
$$

By equating coefficients of (A15) and (A17) the relations

$$\Phi_{1,\partial q_1}(q_1) = -\Phi_{1,\partial x}(q_1, x)J_1(q_1), \tag{A18}$$

$$\Phi_{t,1,\partial q_1}(q_1) = J_{t,1}(q_1) \quad \text{and} \tag{A19}$$

$$\Phi_{r,1,\partial q_1}(q_1, x) = -\Phi_{r,1,\partial x_r}(q_1, x)J_{r,1}(q_1) \tag{A20}$$

can be obtained between the gradient matrices and the analytic Jacobian J_1 of the serial leg chain. The dependency on q_1 and x has been added to highlight the main requirement, namely the zero equality condition of (A16): (A18) and (A20) only hold if the orientation residual is zero. In the inverse kinematics procedure of Section 5.3 the residual at step k in (40) is unequal to zero, which means that the Equation (A20) does not hold in this case. The translational part is unaffected, as can be seen by the missing argument x in (A19).

Appendix A.4. Matrix Operations for Partial Derivatives

To simplify the calculations of the gradient matrices of the residuals in Section 5, operators for matrices are replaced by operators for vectors, to avoid differentiating matrices or w.r.t. matrices which would require multi-dimensional tensors. The column operator $\overline{R}$ for rotation matrices R to stack the coordinate systems unit vectors $n, o, a \in \mathbb{R}^3$ vertically instead of horizontally is defined as

$$\overline{R}(R) = \begin{bmatrix} n \\ o \\ a \end{bmatrix} \in \mathbb{R}^9 \quad \text{with} \quad R = \begin{bmatrix} n & o & a \end{bmatrix} = \begin{bmatrix} n_x & o_x & a_x \\ n_y & o_y & a_y \\ n_z & o_z & a_z \end{bmatrix} \in SO(3) \tag{A21}$$

to avoid differentiating matrices or w.r.t. matrices. The special properties of the $SO(3)$ group are not exploited and the operator can be used for $\mathbb{R}^{3\times3}$ as well. Matrix multiplication is expressed with the matrix product operator $\overline{\Pi}$ such that

$${}^1\overline{R}_3 = \overline{\Pi}\left({}^1\overline{R}_2, {}^2\overline{R}_3\right) = \overline{R}({}^1R_3) \quad \text{with} \quad {}^1R_3 = {}^1R_2\,{}^2R_3. \tag{A22}$$

The transposition operator P_T is a 9×9 permutation matrix such that

$${}^2\overline{R}_1 = P_T\,{}^1\overline{R}_2 = \overline{R}({}^1R_2^T) = {}^1\overline{R}_2^T \in \mathbb{R}^9 \quad \text{with} \quad {}^2R_1 = {}^1R_2^T \in SO(3) \quad \text{and} \quad {}^1\overline{R}_2 = \overline{R}({}^1R_2). \tag{A23}$$

Writing ${}^1\overline{R}_2^T$ instead of $P_T\,{}^1\overline{R}_2$ serves for the clarity of the expressions (34,36,38,39) and overloads the transposition operator for $\mathbb{R}^9$ noted with the bar.

Appendix A.5. Contents of the Partial Derivatives

The single expressions derived in Section 5 can be calculated with low computational effort from the definition of the XYZ and ZYX Euler angles from (5), (21) and (A6). With $\overline{R} = [n_x, n_y, n_z, o_x, o_y, o_z, a_x, a_y, a_z]^T$ the gradient "I" in (34,36,38,39) for ZYX angles becomes

$$\frac{\partial \alpha(\overline{R})}{\partial \overline{R}} = \begin{bmatrix} -\dfrac{n_y}{n_x^2+n_y^2} & \dfrac{n_x}{n_x^2+n_y^2} & 0 & 0 & 0 & 0 & 0 & 0 & 0 \\[2ex] 0 & 0 & -\sqrt{a_z^2+o_z^2} & 0 & 0 & \dfrac{n_z o_z}{\sqrt{a_z^2+o_z^2}} & 0 & 0 & \dfrac{n_z a_z}{\sqrt{a_z^2+o_z^2}} \\[2ex] 0 & 0 & 0 & 0 & 0 & \dfrac{a_z}{a_z^2+o_z^2} & 0 & 0 & -\dfrac{o_z}{a_z^2+o_z^2} \end{bmatrix} \tag{A24}$$

and the reciprocal gradient "IV" in (36) for XYZ angles yields

$$\frac{\partial \overline{R}(\beta)}{\partial \beta} = \begin{bmatrix} 0 & -S_2\,C_3 & -C_2\,S_3 \\ C_1\,S_2\,C_3 - S_1\,S_3 & S_1\,C_2\,C_3 & -S_1\,S_2\,S_3 + C_1\,C_3 \\ S_1\,S_2\,C_3 + C_1\,S_3 & -C_1\,C_2\,C_3 & C_1\,S_2\,S_3 + S_1\,C_3 \\ 0 & S_2\,S_3 & -C_2\,C_3 \\ -C_1\,S_2\,S_3 - S_1\,C_3 & -S_1\,C_2\,S_3 & -S_1\,S_2\,C_3 - C_1\,S_3 \\ -S_1\,S_2\,S_3 + C_1\,C_3 & C_1\,C_2\,S_3 & C_1\,S_2\,C_3 - S_1\,S_3 \\ 0 & C_2 & 0 \\ -C_1\,C_2 & S_1\,S_2 & 0 \\ -S_1\,C_2 & -C_1\,S_2 & 0 \end{bmatrix} \tag{A25}$$

with $C_i = \cos(\beta_i)$, $S_i = \sin(\beta_i)$. The property

$$\left(\frac{\partial \beta}{\partial \overline{R}}\right)\left(\frac{\partial \overline{R}(\beta)}{\partial \beta}\right) = \mathbf{1} \in \mathbb{R}^{3\times 3} \tag{A26}$$

can be used to test the implementation, if (A24, A25) are defined for the same Eulerangle convention.

The gradient of the matrix product (A22) w.r.t. the second factor used in (34)/II, (36)/III, (38)/II and (39)/III is

$$\frac{\partial}{\partial \overline{R}_2}\overline{\prod}\,(\overline{R}_1, \overline{R}_2) = \begin{bmatrix} R_1 & 0 & 0 \\ 0 & R_1 & 0 \\ 0 & 0 & R_1 \end{bmatrix} \tag{A27}$$

and to complete the enumeration the gradient w.r.t. the first factor used in (A14)/I is

$$\frac{\partial}{\partial \overline{R}_1}\overline{\prod}\,(\overline{R}_1, \overline{R}_2) = \begin{bmatrix} \mathrm{diag}(n_x) & \mathrm{diag}(o_x) & \mathrm{diag}(a_x) \\ \mathrm{diag}(n_y) & \mathrm{diag}(o_y) & \mathrm{diag}(a_y) \\ \mathrm{diag}(n_z) & \mathrm{diag}(o_z) & \mathrm{diag}(a_z) \end{bmatrix}, \tag{A28}$$

where $n_x, n_y, \ldots$ are the entries of R_2 and the diagmatrices are 3×3. By transposing the elements of the matrix product (A22), only the first form (A27) had to be used in Section 5.

References

1. Baron, L. A joint-limits avoidance strategy for arc-welding robots. Presented at the International Conference on Integrated Design and Manufacturing in Mechanical Engineering, Montreal, QC, Canada, 16–19 May 2000; note: the paper is not part of the conference proceedings published by Springer. Available online: https://www.researchgate.net/profile/Luc_Baron (accessed on 2 August 2019).
2. Huo, L.; Baron, L. Kinematic inversion of functionally-redundant serial manipulators: Application to arc-welding. *Trans. Can. Soc. Mech. Eng.* **2005**, *29*, 679–690. [CrossRef]
3. Žlajpah, L. On orientation control of functional redundant robots. In Proceedings of the IEEE International Conference on Robotics and Automation (ICRA), Singapore, 29 May–3 June 2017; pp. 2475–2482. [CrossRef]
4. Léger, J.; Angeles, J. Off-line programming of six-axis robots for optimum five-dimensional tasks. *Mech. Mach. Theory* **2016**, *100*, 155–169. [CrossRef]
5. Schappler, M.; Tappe, S.; Ortmaier, T. Resolution of Functional Redundancy for 3T2R Robot Tasks using Two Sets of Reciprocal Euler Angles. In *Advances in Mechanism and Machine Science, Proceedings of the 15th IFToMM World Congress on Mechanism and Machine Science, Kraków, Poland, 30 June–4 July 2019*; Springer: Berlin, Germany; 2019; Volume 73, pp. 1701–1710. [CrossRef]
6. Yoshikawa, T. Analysis and control of robot manipulators with redundancy. In *Robotics Research: The 1st International Symposium*; MIT Press: Cambridge, MA, USA, 1984; pp. 735–747; ISBN 0262022079.
7. Goldenberg, A.; Benhabib, B.; Fenton, R. A complete generalized solution to the inverse kinematics of robots. *IEEE J. Robot. Autom.* **1985**, *1*, 14–20. [CrossRef]

8.   Zhu, W.; Qu, W.; Cao, L.; Yang, D.; Ke, Y. An off-line programming system for robotic drilling in aerospace manufacturing. *Int. J. Adv. Manuf. Technol.* **2013**, *68*, 2535–2545. [CrossRef]

9.   Guo, Y.; Dong, H.; Ke, Y. Stiffness-oriented posture optimization in robotic machining applications. *Robot. Comput. Integr. Manuf.* **2015**, *27*, 367–376. [CrossRef]

10.   Tale-Masouleh, M.; Gosselin, C. Singularity analysis of 5-RPUR parallel mechanisms (3T2R). *Int. J. Adv. Manuf. Technol.* **2011**, *57*, 1107–1121. [CrossRef]

11.   Gogu, G. Solid Mechanics and Its Applications. In *Structural Synthesis of Parallel Robots, Part 1: Methodology*; Springer: Berlin, Germany; 2008; Volume 866.

12.   Huang, T.; Liu, H.; Chetwynd, D. Generalized Jacobian analysis of lower mobility manipulators. *Mech. Mach. Theory* **2011**, *46*, 831–844. [CrossRef]

13.   Merlet, J.P.; Perng, M.W.; Daney, D. Optimal trajectory planning of a 5-axis machine-tool based on a 6-axis parallel manipulator. In *Advances in Robot Kinematics*; Springer: Berlin, Germany, 2000; pp. 315–322.

14.   Hong, K.S.; Kim, J.G. Manipulability analysis of a parallel machine tool: application to optimal link length design. *J. Robot. Syst.* **2000**, *17*, 403–415. [CrossRef]

15.   Merlet, J.P. Solid mechanics and its applications. In *Parallel Robots*, 2nd ed.; Springer: Berlin, Germany, 2006; Volume 128.

16.   Zhang, D. *Parallel Robotic Machine Tools*; Springer: Berlin, Germany, 2009.

17.   Wang, J.; Gosselin, C.M. Kinematic analysis and singularity representation of spatial five-degree-of-freedom parallel mechanisms. *J. Robot. Syst.* **1997**, *14*, 851–869. [CrossRef]

18.   Zhang, D.; Gosselin, C.M. Kinetostatic modeling of N-DOF parallel mechanisms with a passive constraining leg and prismatic actuators. *J. Mech. Des.* **2001**, *123*, 375–381. [CrossRef]

19.   Zheng, K.J.; Gao, J.S.; Zhao, Y.S. Path control algorithms of a novel 5-DOF parallel machine tool. In Proceedins of the IEEE International Conference on Mechatronics and Automation, Niagara Falls, ON, Canada, 29 July–1 August 2005; Volume 3, pp. 1381–1385. [CrossRef]

20.   Gao, J.S.; Sun, H.; Zhao, Y.S. The primary calibration research of a measuring limb in 5-UPS/PRPU parallel machine tool. In Proceedings of the IEEE International Conference on Intelligent Mechatronics and Automation, Chengdu, China, 26–31 August 2004; pp. 304–308. [CrossRef]

21.   Liu, X.; Xu, Y.; Yao, J.; Xu, J.; Wen, S.; Zhao, Y. Control-faced dynamics with deformation compatibility for a 5-DOF active over-constrained spatial parallel manipulator 6PUS–UPU. *Mechatronics* **2015**, *30*, 107–115. [CrossRef]

22.   Wen, S.; Qin, G.; Zhang, B.; Lam, H.K.; Zhao, Y.; Wang, H. The study of model predictive control algorithm based on the force/position control scheme of the 5-DOF redundant actuation parallel robot. *Robot. Auton. Syst.* **2016**, *79*, 12–25. [CrossRef]

23.   Mbarek, T.; Nefzi, M.; Corves, B. Prototypische Entwicklung und Konstruktion eines neuartigen Parallelmanipulators mit dem Freiheitsgrad fünf. *VDI-Berichte Nr. 1892* **2005**. Available online: https://www.igmr.rwth-aachen.de/index.php/en/rob-en/rob-pentapod-en (accessed on 20 June 2019). (In German)

24.   Schreiber, H.; Gosselin, C. Analyse et conception d'un manipulateur parallèle spatial à cinq degrés de liberté. In French. *Mech. Mach. Theory* **2003**, *38*, 535–548. [CrossRef]

25.   Gao, F.; Peng, B.; Zhao, H.; Li, W. A novel 5-DOF fully parallel kinematic machine tool. *Int. J. Adv. Manuf. Technol.* **2006**, *31*, 201. [CrossRef]

26.   Bär, G.F.; Weiß, G. Kinematic Analysis of a Pentapod Robot. *J. Geometry Graph.* **2006**, *10*, 173–182.

27.   Lin, W.; Li, B.; Yang, X.; Zhang, D. Modelling and control of inverse dynamics for a 5-DOF parallel kinematic polishing machine. *Int. J. Adv. Robot. Syst.* **2013**, *10*, 314. [CrossRef]

28.   Alagheband, A.; Mahmoodi, M.; Mills, J.K.; Benhabib, B. Comparative analysis of a redundant pentapod parallel kinematic machine. *J. Mech. Robot.* **2015**, *7*, 034502. [CrossRef]

29.   Gogu, G. Fully-isotropic parallel manipulators with five degrees of freedom. In Proceedings of the 2006 IEEE International Conference on Robotics and Automation, Orlando, FL, USA, 15–19 May 2006; IEEE: Piscataway, NJ, USA, 2006; pp. 1141–1146. [CrossRef]

30.   Huang, Z.; Li, Q.C. General Methodology for Type Synthesis of Symmetrical Lower-Mobility Parallel Manipulators and Several Novel Manipulators. *Int. J. Robot. Res.* **2002**, *21*, 131–145. [CrossRef]

31.   Kong, X.; Gosselin, C.M. Type synthesis of 5-DOF parallel manipulators based on screw theory. *J. Robot. Syst.* **2005**, *22*, 535–547. [CrossRef]

32. Tale-Masouleh, M.; Saadatzi, M.H.; Gosselin, C.; Taghirad, H.D. A geometric constructive approach for the workspace analysis of symmetrical 5-PRUR parallel mechanisms (3T2R). In Proceedings of the ASME International Design Engineering Technical Conferences and Computers and Information in Engineering Conference, Montreal, QC, Canada, 15–18 August 2010; pp. 1335–1344. [CrossRef]
33. Cheng, L.; Wang, H.; Zhao, Y. Analysis and experimental investigation of parallel machine tool with redundant actuation. In *Intelligent Robotics and Applications, Proceedings of the International Conference on Intelligent Robotics and Applications, Wuhan, China, 15–17 October 2008*; Springer: Berin, Germany, 2008; pp. 179–188. [CrossRef]
34. Ramirez, D.; Kotlarski, J.; Ortmaier, T. Automatic generation of a minimal set of serial mechanisms for a combined structural—Geometrical synthesis. In Proceedings of the 14th IFToMM World Congress, Taipeh, Taiwan, 25–30 October 2015; [CrossRef]
35. Briot, S.; Khalil, W. *Dynamics of Parallel Robots*; Springer: Berlin, Germany, 2015.
36. Huo, L.; Baron, L. The joint-limits and singularity avoidance in robotic welding. *Ind. Robot.* **2008**, *35*, 456–464. [CrossRef]

Article

Efficient Closed-Form Task Space Manipulability for a 7-DOF Serial Robot

Gerold Huber * and Dirk Wollherr

Chair of Automatic Control Engineering (LSR), Department of Electrical and Computer Engineering, Technical University of Munich (TUM), 80333 Munich, Germany; dw@tum.de
* Correspondence: gerold.huber@tum.de

Received: 29 September 2019; Accepted: 23 November 2019; Published: 26 November 2019

Abstract: With the increasing demand for robots to react and adapt to unforeseen events, it is essential that a robot preserves agility at all times. While *manipulability* is a common measure to quantify agility at a given joint configuration, an efficient direct evaluation in task space is usually not possible with conventional methods, especially for redundant robots with an infinite number of Inverse Kinematic solutions. Yet, this is essential for global online optimization of a robot posture. In this work, we derive analytical expressions for a conventional 7-degrees of freedom (7-DOF) serial robot structure, which enable the direct evaluation of manipulability from a reduced task space parametrization. The resulting expressions allow array operation and thus achieve very high computational efficiency with vector-optimized programming languages. This direct and simultaneous calculation of the *task space manipulability* for large numbers of poses benefits many optimization problems in robotic applications. We show applications in global optimization of robot mounting poses, as well as redundancy resolution with global online optimization w.r.t. manipulability.

Keywords: manipulability; inverse kinematics function; kinematic optimization; redundant robot; 7-DOF; redundancy resolution

---

1. Introduction

It is a common requirement in robotic manipulation tasks to quantify the capabilities of a robot at a given pose. Having such a scalar measure allows comparison of different kinematic configurations in terms of the chosen metric, and can be considered at a path planning as well as at a control level. While these measures are usually defined in terms of a given joint configuration [1–5], the task of the robot is typically not given in this *joint space*. For a general robot the *task space* is usually defined in SE(3), i.e., the space of 3D poses consisting of translation and rotation. For many practical problems it is thus relevant to directly evaluate this measure w.r.t. a parametrization of SE(3) rather than the joints. This requires combining the evaluation of the Inverse Kinematic (IK) with the selected capability metric. But direct calculation of the IK is always robot-dependent and general analytic solutions are not possible. This is especially true for redundant robots that have more degrees of freedom (DOF) in joint space than in task space and thus admit an infinite number of IK solutions for a given end-effector pose. While analytic IK solutions are well known for conventional 6-DOF kinematics [6], for general robotic structures numeric IK solvers are applied. However, they require several iterations to find an approximated joint configuration for a given end-effector pose. This is sufficient for calculating single poses, but it is inefficient for optimization problem solvers that require evaluation of large numbers of poses. This especially prevents time-critical computation of global optima. Expressions that can be evaluated directly are thus superior for fast computation. While an analytical IK for a general robot structure does not exist, our work focuses on the most commonly used articulated 6- and 7-DOF robot serial kinematics. Yet, the 6 axis version can be viewed as a finite set of particular null space solutions of the 7-DOF.

1.1. Contribution

In this work, we develop a set of computationally efficient closed-form expressions to evaluate the *task space manipulability* of a 7-DOF serial robot structure, i.e., the mapping from a task space parametrization directly a manipulability measure. The main contributions of this work consist of:

1. a new parametrization of the state- and null space that results in concise IK expressions with symmetric structure in the individual components
2. analytical closed-form expressions from task space to manipulability measure w.r.t. joint limits, which allow *array operation* in vector-optimized programming languages. Note that array operation is also called *Vectorization* in e.g., *MATLAB*. It refers to the exploitation of Single Instruction Multiple Data (SIMD) instructions of modern Central Processing Unit (CPUs) and allows to operate on multiple data points simultaneously.
3. sensitivity analysis of manipulability in task space
4. real-time capable application for evaluating the task space manipulability of the entire null space, for globally optimal redundancy resolution w.r.t. manipulability of single poses and full trajectories on SE(3)

1.2. Related Work

For this concise review, we group previous work on the topic into the three areas: (1) performance measures in robotics, (2) direct methods for IK evaluation and algorithmic strategies on the velocity level, and (3) approaches for optimizing manipulability.

1.2.1. Performance Measures

Arguably the most common performance measure for robot structures is the *manipulability* measure defined by Yoshikawa [1]. It is proportional to the volume of an ellipsoid, spanned by directional capabilities of a kinematic structure to generate velocities in task space at a given joint configuration. It is purely kinematic and does not consider any dynamic components. Yoshikawa also proposed a dynamic manipulability ellipsoid [2] on the acceleration level, for cases where dynamic effects cannot be neglected. This formulation was improved by Chiacchio et al. [3] to correctly account for gravity. A new formulation of a dynamic manipulability ellipsoid that better depicts the real manipulator capabilities in terms of task space accelerations was proposed by Chiacchio [4].

Besides manipulability on the velocity and acceleration levels due to mere kinematic relations, it is essential for practical applications to also consider joint limits as constraints directly on the position level. Vahrenkamp et al. [5] extended Yoshkawa's basic manipulability, by directly integrating joint limit penalization into the definition of the kinematic velocity Jacobian. This is achieved via a joint limit potential function.

Bong-Huan Jun et al. [7] introduce a task-oriented manipulability measure. While Yoshikawa's original measure [1] denotes the manipulability of the whole manipulator system, [7] considers manipulability w.r.t. to sub-tasks that only affect parts of the task space, e.g., axis specific tasks. Karim Abdel-Malek and Wei Yu [8] proposed an alternative dexterity measure for robot placement that does not depend on explicit IK solutions. They analyze an augmented Jacobian matrix that does not only hold information about position and orientation, but also joint limits of the end-effector. It represents the reachable workspace with surface patches and is computationally very demanding.

Our work has the aim of developing closed-form solutions that allow efficient array operation. For this reason, the task space manipulability formulation developed in this work applies Yoshikawa's original measure from [1]. Because its definition uses a determinant to map the joints to a scalar metric, it thus allows expansion to a continuous polynomial expression for efficient evaluation.

1.2.2. Inverse Kinematics

The IK problem of serial robot structures can be solved very elegantly on the velocity level, due to the linear relation of joint and task space velocities. However, numeric integration of the resulting joint velocities to joint angles needs stabilization against numerical drift and thus results in an iterative scheme. Originally proposed by Wolovich and Elliot [9], this group of IK solvers is nowadays typically referred to as Closed-Loop Inverse Kinematic (CLIK) solvers. Colomé and Torras [10] give an overview of the most common CLIK solvers, with an additional experimental comparison in terms of convergence, numerical error, singularity handling, joint limit avoidance, and the capability of reaching secondary goals. Antonelli [11] conducted a stability analysis of priority-based kinematic CLIK algorithms for redundant kinematics. He provides sufficient conditions for the control gains. While different stabilization schemes for CLIK solvers are proposed, the choice of gain parameters used in the control structure is rarely addressed. In practice these parameters are often empirically tuned. Bjoerlykhaug [12] proposes the use of a genetic algorithm for optimizing the feedback gain used in CLIK solvers, in order to minimize iteration cycles and maximize accuracy. In an experimental evaluation, he achieved a 50% decrease in computation time through his feedback gain tuning. Reiter et al. [13] propose a strategy for finding higher-order time-optimal IK solutions for redundant robots. They lay out solutions for fourth-order time derivatives of joint trajectories, applying a multiple shooting optimization method. This higher-order continuous differentiability is especially important for application on elastic mechanisms.

Siciliano [14] gives a tutorial on early common online IK algorithms. He states the important features of a direct inverse kinematics function, i.e., repeatability, cyclicity, or cyclic behaviour, and online applicability. Shimizu et al. [15] outline an analytical IK computation for a 7-DOF serial robot. The approach directly parametrizes the end-effector pose with Cartesian coordinates for translation and a rotation matrix for orientation. However, the use of the 2-quadrant atan function, as opposed to the 4-quadrant atan2 function, results in two problems. For one, the entire task space is not covered, and two, it results in discontinuous joint functions w.r.t. the null space parameter and thus leads to discontinuous IK solutions and corresponding null space limitations. A similar strategy, but extended to the entire domain, is proposed by Faria et al. [16]. They propose a position-based IK solution for a 7-DOF serial manipulator with joint limit and singularity avoidance.

Besides approaches that use kinematic insight of a structure, several machine learning algorithms are also considered in the literature. A detailed review is beyond the scope of this work, but we want to give a concise overview of research activities. D'Souza et al. [17] apply a locally weighted projection regression to learn the IK of a 30-DOF humanoid robot. This maps the non-convex problem onto a locally convex problem that is suitable for direct learning. Tejomurtula and Kak [18], as well as Köker et al. [19], applied artificial neural networks for finding an IK mapping for 3-DOF robots and showed the feasibility of the problem using conventional error-backpropagation and Kohonen networks. Sariyildiz et al. [20] compare support vector regression and artificial neural networks for learning IK mappings of a 7-DOF serial robot. They find that support vector regression is less prone to local minima and requires very few training data. Genetic algorithms were already early applied by Parker et al. [21]. They pointed out low positioning accuracy, but emphasize its simplicity in application. Köker [22] proposes a hybrid approach combining Elman neural networks with genetic algorithms. He was able to significantly improve accuracy for IK solutions of a 6-DOF mechanism in comparison to pure neural networks. Very recently, Dereli et al. [23] proposed a strategy to apply quantum behaved particle swarm optimization for finding IK solutions of a 7-DOF serial robot.

The IK expressions developed in this work are similar to the analytical approaches in [15,16] in terms of parametrizing the null space as arm angle. However, the new task space parametrization that we introduce results in more concise and, more importantly, fully vectorizable expressions that allow efficient array operations. In contrast to existing approaches in the literature, this computational advantage makes our approach suitable for simultaneous evaluation of a large number of poses.

1.2.3. Optimizing Manipulability

In conventional industrial contexts, optimizing cycling time is always of interest. Several publications deal with this problem, e.g., Kamrani et al. [24] use the Response Surface Method [25] to optimize robot placement w.r.t. cycling time. Chan and Dubey [26], as well as Dariush et al. [27], use a projection method of the joint limit gradient potential function. This is used for local manipulability optimization on the velocity level. Dufour and Suleiman [28] present an approach of integrating the manipulability index into an optimization-based IK solver, by using linear approximations of the nonlinear manipulability measure with numeric gradient calculations at every time step. Jin et al. [29] mention the difficulty of real-time manipulability optimization that is related to a high computational burden since the manipulability is a non-convex function to the joint angles of a robotic arm. Due to the capability of *high-speed parallel distributed processing*, they propose an approach using dynamic neural networks in order to implement manipulability optimization in real-time. Conducting computer simulations, they show that the proposed method raises the manipulability by almost 40% on average compared to existing methods.

Besides local optimization of a given joint configuration, for many robotic tasks it is required to include manipulability as criteria for optimization of the whole trajectory. Lee [30] shows that a required motion can be approximated by a series of manipulability ellipsoids. Guilamo et al. [31] present an algorithm for trajectory generation that maximizes the volume of the manipulability ellipsoid. Yoshikawa [1] already observed that the optimal postures of various manipulators form the viewpoint of manipulability, and often show resemblance of those naturally taken by human arms. This motivates the idea of manipulability transfer using a *learning by demonstration* strategy that is introduced by Rozo et al. [32]. Their approach allows robots to learn and reproduce a continuous set of manipulability ellipsoids by an expert's demonstration. In order to encode and retrieve those ellipsoids, they apply Gaussian Mixture Models and Gaussian Mixture Regression. In Jaquier et al. [33] the same authors exploit tensor-based representation, to consider that manipulability ellipsoids lie on the manifold of symmetric positive definite matrices. Faroni et al. [34] present an approach that maximizes the average manipulability of the overall task. Their method is based on the optimization of a cost function that depends on various points along a predetermined path. In particular, if the task of the manipulator is known a priori, this approach provides global manipulability optimization.

An approach for directly quantifying manipulability of a redundant robot in task space is proposed by Zacharias et al. [35]. They introduce a capability map, to guide the decision on how to place a mobile robot relative to an object. It is a sampling-based approach, based on the manipulability index. While the approach reveals in which regions the robot is capable of grasping objects from different angles, the information of optimal approaching directions is lost.

The task space manipulability approach in this work enables for the first time global manipulability optimization with real-time capabilities, due to its efficient formulation.

1.3. Outline

The remainder of the paper is organized as follows. The problem of a closed-loop task space manipulation framework is outlined in Section 2. In Section 3, the derivation of all analytical mappings is explained. Evaluation and analysis of the resulting task space manipulability is discussed in Section 4 and applied in global optimization formulations in Section 5. We conclude the work and outline future directions of development in Section 6.

2. Problem Formulation

Given a n-DOF serial robot, its forward kinematics

$$\text{FK}: \quad \mathbb{R}^n \to \text{SE}(3) \times \mathbb{R}^{n-6}, \quad q \mapsto (z, \lambda) \tag{1}$$

maps the joints q onto the 3D end-effector pose z at a particular null space solution parametrized by λ. To quantify the capability of moving in the SE(3) task space at a given joint configuration q, a manipulability metric function

$$\text{M}: \quad \mathbb{R}^n \to \mathbb{R}^1, \quad q \mapsto \mu \tag{2}$$

is applied. A proper choice of parametrization for z and λ assures the existence of the inverse function

$$\text{IK}: \text{SE}(3) \times \mathbb{R}^{n-6} \to \mathbb{R}^n, \quad (z, \lambda) \mapsto \text{FK}^{-1}(z, \lambda) =: q. \tag{3}$$

We define the task space manipulability as the direct mapping

$$\text{M} \circ \text{IK}: \quad \text{SE}(3) \times \mathbb{R}^{n-6} \to \mathbb{R}^1, \quad (z, \lambda) \mapsto \mu \tag{4}$$

of a desired pose z in task space onto the manipulability measure μ, considering all null space solutions parametrized by λ. $(\text{M} \circ \text{IK})(z, \lambda)$ denotes the function composition $\text{M}(\text{IK}(z, \lambda))$. Figure 1 illustrates the task space manipulability for a certain end-effector pose z. Considering real-time critical online applications and feasibility of global optimization formulations, the development of the task space manipulability map can be broken down into three problems:

Problem 1: Find a parametrization of the task- and null space that exploits the kinematic structure for concise expressions.

Problem 2: Find closed-form expressions for all mappings from task space to manipulability that allow efficient array operation in vector-optimized programming languages.

Problem 3: Let $\mathcal{Q} \subset \mathbb{R}^7$ be the space of admissible joint configurations. Find an analytical expression of the range of the null space solutions $\Lambda(z) := \{\lambda \in \mathbb{R}^{n-6} \mid \text{IK}(z, \lambda) \in (Q)\}$, for which the inverse kinematics function $\text{IK}(\cdot, \lambda)$ results in an admissible joint configuration $q \in \mathcal{Q}$.

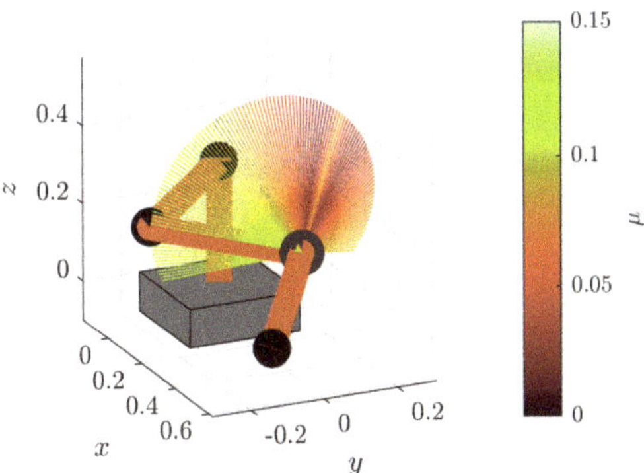

Figure 1. Illustration of the task space manipulability at a given end-effector pose. The null space of this 7-DOF S-R-S kinematics consists of the free elbow position (joint 4) along a circle. This position defines the direction of the forearm, i.e., the vector from the shoulder to the wrist. The colored fan shows all possible forearm poses with the corresponding manipulability color-coded from dark red (very bad) to light green (optimal). Colorless areas of the fan mark areas that violate joint constraints.

In this work, we investigate in detail the case of a 7-DOF serial robot kinematics in conventional Spherical-Revolute-Spherical (S-R-S) structure, such as the *KUKA LBR* series. In this context, S-R-S refers to a kinematic 7-DOF structure with alternating revolute joints, of which the rotation axes of the first and last 3 joints intersect. These two groups of intersecting axes behave kinematically like a spherical joint and are often referred to as shoulder and wrist. This type of kinematic structure leads to a 1-dimensional null space of solutions and thus $\lambda \in \mathbb{R}^1$.

3. Technical Approach

This section outlines the derivation of the closed-form task space manipulability for the considered special case of a 7-DOF serial robot kinematics. We first discuss the chosen manipulability mapping and possible reductions in joint space. Motivated by these reductions, we propose a task space projection onto a parameter space, which yields concise expressions for the IK. Figure 2 summarizes all developed mappings that are developed in this section. The section concludes with an analytic definition of the admissible null space at a given parameter end-effector pose.

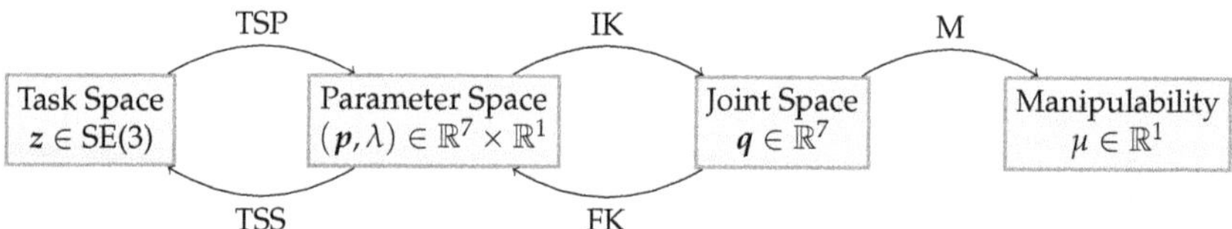

Figure 2. Relation of task space z, parameter space p, joint space q, and manipulability metric μ. The mappings are referred to as Task Space Projection (TSP) and Task Space Surjection (TSS), Forward Kinematic (FK) and Inverse Kinematic (IK), and Manipulability (M).

Notational Notes

Scalars are written in plain lower case, vectors in bold-face lower case. Matrices are bold-face upper case, while plain upper case symbols refer to coordinate frames, mathematical spaces, and sets.

For vector indices, we use the common anthropomorphic analogy of a human arm. We refer to the origin of the kinematic as base B, and to positions of joint 2, joint 4, and joint 6 as shoulder S, elbow E, and wrist W respectively. Body-fixed frames of the individual robot links are numbered 1 to 7 and relate to the bodies *after* the corresponding joints. The end-effector will be referred to as tool T.

Coordinate transformation matrices are written as A_{kj} with 2 indices and are read from right to left, e.g., A_{43} transforms the coordinate system from body-fixed frame of joint q_3 to joint q_4, whereas vector indices are read from left to right and their reference frame is written as left-hand side subscript. The notation $_B r_{SW}$ thus describes a vector r pointing from shoulder S to wrist W, expressed in base frame B. Cartesian base vectors of the coordinate systems are written as $\hat{x}$, $\hat{y}$, and $\hat{z}$. If a vector does not have a lower left index, it always refers to the base B.

3.1. Manipulability Measure

The central equation in robot kinematics is the linear forward velocity kinematic map

$$\dot{z}(q, \dot{q}) := J(q)\dot{q} \tag{5}$$

that relates general joint space velocities $\dot{q} \in \mathbb{R}^n$ to task space velocities $\dot{z} \in \mathbb{R}^6$, where the linear map $J(q) \in \mathbb{R}^{6 \times n}$ describes the velocity propagation from joint to task space at a given joint configuration $q \in \mathbb{R}^n$. It is defined by the kinematic chain and represents the derivative

$$J(q) = \frac{\partial \dot{z}(q, \dot{q})}{\partial \dot{q}}, \tag{6}$$

hence it is often referred to as *Robot Jacobian*.

Yoshikawa's manipulability measure [1], which we use in this work, is defined as

$$\mathrm{M}: \quad \mathbb{R}^n \to \mathbb{R}, \quad q \mapsto \sqrt{\det\left(J(q)J(q)^\top\right)} =: \mu \tag{7}$$

and is a measure, proportional to the volume of the velocity manipulability ellipsoid

$$\dot{q}^\top \dot{q} = 1 \tag{8a}$$
$$\dot{z}^\top (JJ^\top)^{-1}\dot{z} = 1. \tag{8b}$$

Note that (7) does not consider hardware-related joint limits. However, joint configurations that violate these constraints must not be considered.

Zlatanov et al. [36] explain that the forward velocity kinematic map (5) is not sufficient for exhaustive characterization of the singularities of a manipulator. Further, Staffetti et al. [37] show that many of these often-used manipulability indices are not invariant to change of reference frames, scale, or physical units. However, the big advantage of Yoshikawa's original manipulability metric is the fact that it can be expanded to a polynomial expression and thus qualifies for computationally efficient array operation. Further, derivatives can be calculated analytically. As outlined by Staffetti et al. [37], it is not a true metric for distance to a singularity but nonetheless serves as a relative comparison of manipulability qualities between joint configurations [38].

For a n-DOF serial robot kinematics, we refer to the $i = [1, n]$ absolute angular and translational velocities of the individual links, i.e., the velocity between the robot base B and the body-fixed frame of link i, as ω_{Bi} and v_{Bi}. Expressed w.r.t. the link frame i, the velocities of the kinematic chain are calculated with

$$_i\omega_{Bi} = A_{ip}\,_p\omega_{Bp} + _i\omega_{pi} \tag{9a}$$
$$_iv_{Bi} = A_{ip}\left(_pv_{Bp} + _p\omega_{Bp} \times _pr_{pi}\right), \tag{9b}$$

where $p = i - 1$ is the predecessor link of i and $(\times)$ denotes the cross product $\mathbb{R}^3 \times \mathbb{R}^3 \to \mathbb{R}^3$. In the following, *manipulability* refers to Yoshikawa's manipulability measure [1].

3.1.1. Reduction of First Joint

While Yoshikawa's manipulability measure is not invariant w.r.t. scale or physical units, it is in fact invariant to change of reference frames.

Proof. Given a vector of joint velocities $\dot{q}$ and task space velocities $\dot{z}$ w.r.t. to a reference frame A, the Jacobian matrix

$$_AJ(q) = \frac{\partial \,_A\dot{z}(q, \dot{q})}{\partial \dot{q}} \tag{10}$$

is used to define the manipulability index

$$_A\mu(q) = \sqrt{\det\left(_AJ(q)\,_AJ(q)^\top\right)}. \tag{11}$$

If this manipulability index is expressed in terms of a new reference frame B via the block transformation matrix

$$A_{BA}^{\mathrm{blk}}(q) = \begin{bmatrix} A_{BA}(q) & 0 \\ 0 & A_{BA}(q) \end{bmatrix}, \tag{12}$$

consisting of rotation matrices A_{BA}, the manipulability index reads

$$_B\mu(q) = \sqrt{\det\left(A_{BA}^{\mathrm{blk}}(q)\, _AJ(q)\left(A_{BA}^{\mathrm{blk}}(q)\, _AJ(q)\right)^\top \right)}. \tag{13}$$

Considering the fact that Euclidean transformation matrices have $\det(A) = 1$, we find

$$_B\mu(q) = \sqrt{\det\left(_AJ(q)\, _AJ(q)^\top \right)} = \,_A\mu(q) \tag{14}$$

i.e., the manipulability measure μ is invariant to change of reference frames. $\square$

If the reference frame is chosen to be fixed to any link after the first joint, it results in an expression for the manipulability measure that is independent of the first joint. This results from the fact that the first joint rotates the whole kinematic structure including the reference frame, but does not alter any geometric relations.

We consequently choose to formulate the Jacobian matrix w.r.t. to the end-effector frame, as this does not only lead to the independence of q_1, but also results in the most concise expression.

3.1.2. Reduction of Last Joint

For a special case of a 7-DOF serial kinematic, the parameter space of the manipulability can be further reduced. This special case consists of kinematic structures, whose origin of the end-effector frame lies on the rotation axis of the last joint q_n. The purely angular contribution of q_n does not alter the kinematic configuration but only rotates the reference frame and with it the manipulability ellipsoid. The shape of the ellipsoid is not affected and so q_n can also not influence the manipulability measure.

3.1.3. Closed-Form Expression

Exploiting these two reductions by formulating the $_TJ$ w.r.t. to the end-effector frame T and assuming the tool center point (TCP) along the last joint axis, it is possible to expand the entire determinant expression of the matrix $_TJ\,_TJ^\top \in \mathbb{R}^{6\times 6}$ from (7) to a symbolic polynomial expression using, e.g., *MATLAB Symbolic Math Toolbox™*. The advantage being that, unlike the original matrix expression, the polynomial form allows array operation in vector-optimized programming languages. This enables simultaneous evaluation of an entire set of joint configurations. The full manipulability function is listed in Appendix A.

3.2. Task Space Parametrization

The decision of choosing a parametrization for the SE(3) pose, as well as the 1D null space, is essential for the derivation of concise analytical formulations. We propose the following parameter requirements (PR) for a suitable parametrization in regard to the IK functions. The parameter set must

PR1: uniquely define the null space parameter for the entire space of SE(3).
PR2: result in a minimal number of parameters for the components of the IK vector map $p \mapsto q$.
PR3: allow direct application of the above-mentioned reductions.

Different approaches for null space parametrization were proposed in the literature. The redundancy is either directly parametrized by a redundant joint [39,40], or more commonly by a joint-independent arm angle [15,41]. Shimizu et al. [15] argued that joint-based parametrization is not suitable for the discussed 7-DOF S-R-S mechanism due to possible ambiguous results. Kreutz-Delgado et al. [41] define the arm angle as the angle between an arm and a reference plane. The arm plane is spanned by shoulder, elbow, and wrist locations. The reference plane is defined by a fixed vector and the vector from shoulder to wrist. Shimizu et al. [15] point out arithmetic singularities in the original definition whenever the two vectors are collinear. They enhance the robustness of the definition by defining the reference plane in terms of a particular solution $q_3 \overset{!}{=} 0$, which resembles the solution of conventional non-redundant 6-DOF mechanisms. While this definition is unique w.r.t. the conservative joint limits of their analyzed robot structure, it is ambiguous whenever the reduced non-redundant 6-DOF mechanism admits multiple configurations that result in the same end-effector pose.

In this work, we introduce a parametrization that fulfills all the above-discussed parameter requirements. Figure 3 illustrates the following discussion. Independent of a desired end-effector pose, positions of the base B and shoulder S are always stationary, where

$$_B r_{BS} := (l_B + l_1)\hat{z} \tag{15}$$

with link lengths of the base link l_B and the first link l_1. Additionally, defining a desired end-effector pose relative to the robot base in SE(3), consisting of $_B r_{BT}$ for translation and A_{TB} for orientation, determines not only the location of the tool-center-point T but also the wrist position

$$_B r_{BW} := {_B r_{BT}} - A_{B6}(l_6 + l_7 + l_T)\hat{z}, \tag{16}$$

with link lengths l_6 and l_7, and a potential tool length l_T. This wrist position is used for define the translational component of the end-effector pose z. The position $_B r_{SW}$ is parametrized by spherical coordinates $(r_{\text{ref}}, \gamma_{\text{ref}}, \beta_{\text{ref}})$ with coordinate plane $_B \hat{x}\hat{z}$, origin S and $_B \hat{z}$ as polar axis. The parameters are radius r_{ref}, longitudinal angle γ_{ref}, and azimuthal angle β_{ref}. Note that γ_{ref} and β_{ref} directly align with the rotation axis of q_1 and q_2. These two angles also define the reference frame R with

$$A_{RB}(\gamma_{\text{ref}}, \beta_{\text{ref}}) := A_y(\beta_{\text{ref}}) A_z(\gamma_{\text{ref}}). \tag{17}$$

The orientation is parametrized along a consecutive Euler angle sequence $Z \rightarrow Y' \rightarrow Z''$, which again corresponds to the sequence of the joint structure. However, instead of directly parametrizing A_{TB}, we parametrize the end-effector orientation with respect to the reference frame, i.e.,

$$A_{TR}(\gamma_{\text{EE}}, \beta_{\text{EE}}, \psi_{\text{EE}}) := A_z(\psi_{\text{EE}}) A_y(\beta_{\text{EE}}) A_z(\gamma_{\text{EE}}). \tag{18}$$

Regarding the stated parameter requirement PR2, this makes the IK functions of the wrist angles (q_5, q_6, q_7) as independent of the shoulder parameters $(r_{\text{ref}}, \gamma_{\text{ref}}, \beta_{\text{ref}})$ as possible, as will be seen in the IK Section 3.3.

The 1D null space is parametrized by the arm angle λ. In contrst to Shimizu et al. [15] we do not define the arm angle w.r.t. to the non-redundant solution $q_3 \overset{!}{=} 0$, but w.r.t. to the introduced reference frame R. Let λ be the arm angle, which defines a new frame L with

$$A_{LB}(\gamma_{\text{ref}}, \beta_{\text{ref}}, \lambda) := A_z(\lambda) A_{RB}(\gamma_{\text{ref}}, \beta_{\text{ref}}), \tag{19}$$

such that the negative frame base vector $(-_L \hat{x})$ points in direction of the elbow E. This uniquely defines the null space parameter as required in PR1. The full set of parameters is thus given with tuple $(p, \lambda) \in \mathbb{R}^6 \times \mathbb{R}$, consisting of the parameter vector

$$p := [r_{\text{ref}}, \gamma_{\text{ref}}, \beta_{\text{ref}}, \gamma_{\text{EE}}, \beta_{\text{EE}}, \psi_{\text{EE}}]^\top \tag{20}$$

and arm angle λ. The individual parameter range definitions are

$$
\begin{aligned}
r_{\text{ref}} &\in \left[\; r_{\text{ref}}^{\min}, \quad r_{\text{ref}}^{\max} \;\right] \\
\gamma_{\text{ref}} &\in \left[\; -\pi, \quad +\pi \;\right] \\
\beta_{\text{ref}} &\in \left[\; 0, \quad +\pi \;\right] \\
\gamma_{\text{EE}} &\in \left[\; -\pi, \quad +\pi \;\right] \\
\beta_{\text{EE}} &\in \left[\; 0, \quad +\pi \;\right] \\
\psi_{\text{EE}} &\in \left[\; -\pi, \quad +\pi \;\right] \\
\lambda &\in \left[\; -\pi, \quad +\pi \;\right]
\end{aligned}
\tag{21}
$$

and form the parameter space $\mathcal{P} \subset \mathbb{R}^7$. Note that the two parameters γ_{ref} and ψ_{EE} solely affect joints q_1 and q_7, which do not influence manipulability. The task space manipulability developed in this work can thus without loss of information be represented by the reduced parameter vector $p^{\text{red}} \in \mathcal{P}^{\text{red}} \subset \mathbb{R}^4$ consisting of

$$
p^{\text{red}} := [r_{\text{ref}}, \beta_{\text{ref}}, \gamma_{\text{EE}}, \beta_{\text{EE}}]^\top .
\tag{22}
$$

This complies with the stated requirement PR3. The presented parametrization is the fundamental core for the concise mappings developed in the remaining section.

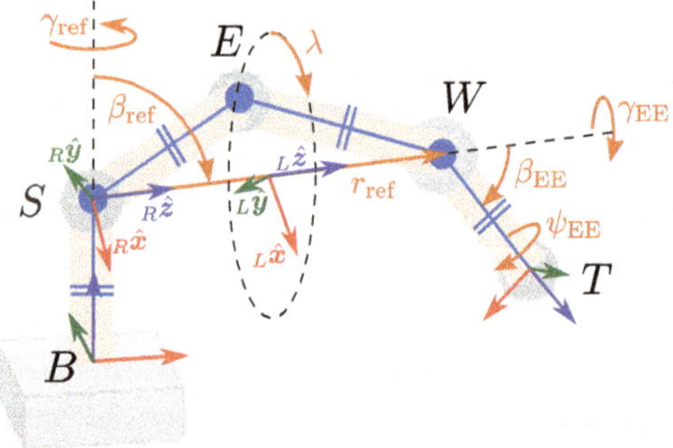

Figure 3. Parametrization of the Task Space. Positions of Base B and Shoulder S are fixed. Translation reference parameters $(r_{\text{ref}}, \gamma_{\text{ref}}, \beta_{\text{ref}})$ define the position of the Wrist W. The end-effector parameters $(\gamma_{\text{EE}}, \beta_{\text{EE}}, \psi_{\text{EE}})$ describe the rotation from reference frame R to tool frame T as consecutive $Z \to Y' \to Z''$ Euler angles. The null space is parametrized with λ. It defines the position of the elbow E via relative rotation between the elbow oriented frame L and frame R.

3.2.1. Task Space Projection

We refer to the extraction of the parameter vector $p = [r_{\text{ref}}, \gamma_{\text{ref}}, \beta_{\text{ref}}, \gamma_{\text{EE}}, \beta_{\text{EE}}, \psi_{\text{EE}}]^\top$ from a given end-effector pose $z \in \text{SE}(3)$ as *Task Space Projection*. Without loss of generality, we assume the pose $z \in \text{SE}(3)$ is described with Cartesian Coordinates (x, y, z) for translation ${}_B r_{BT}$ together with a Rotation matrix A_{TB} for orientation. As a reference matrix for extracting the parameter space angles $(\gamma_{\text{EE}}, \beta_{\text{EE}}, \psi_{\text{EE}})$ we state the rotation matrix for a general ZYZ Euler sequence

$$
A_{zyz}(\gamma, \beta, \psi) := A_z(\psi)\, A_y(\beta)\, A_z(\gamma) =
$$
$$
\begin{pmatrix}
c(\beta)\,c(\gamma)\,c(\psi) - s(\gamma)\,s(\psi) & c(\gamma)\,s(\psi) + c(\beta)\,c(\psi)\,s(\gamma) & -c(\psi)\,s(\beta) \\
-c(\psi)\,s(\gamma) - c(\beta)\,c(\gamma)\,s(\psi) & c(\gamma)\,c(\psi) - c(\beta)\,s(\gamma)\,s(\psi) & s(\beta)\,s(\psi) \\
c(\gamma)\,s(\beta) & s(\beta)\,s(\gamma) & c(\beta)
\end{pmatrix}
\tag{23}
$$

that shows that we can define a mapping $\mathrm{eul}_{ZYZ} : \mathrm{SE}(3) \to \mathbb{R}^3$ as

$$\mathrm{eul}_{ZYZ}: \quad \mathrm{SE}(3) \to \mathbb{R}^3, \quad A_{zyz} \mapsto \begin{bmatrix} \mathrm{atan2}\left(\left[A_{zyz}(z) \right]_{(3,2)} , \left[A_{zyz}(z) \right]_{(3,1)} \right) \\ \arccos\left(\left[A_{zyz}(z) \right]_{(3,3)} \right) \\ \mathrm{atan2}\left(\left[A_{zyz}(z) \right]_{(2,3)} , - \left[A_{zyz}(z) \right]_{(1,3)} \right) \end{bmatrix} =: \begin{bmatrix} \gamma \\ \beta \\ \psi \end{bmatrix} \tag{24}$$

that extracts the Euler angles from a rotation matrix in SE(3). The operator $[\,\cdot\,]_{(i,j)}$ returns the element at row i and column j of a matrix.

The Task Space Projection

$$\mathrm{TSP}: \quad \mathrm{SE}(3) \to \mathbb{R}^6, \quad z \mapsto p \tag{25a}$$

consists of the mappings

$$r_{\mathrm{ref}}(z) := \| {}_B r_{SW} \|_2 \tag{25b}$$

$$\beta_{\mathrm{ref}}(z) := \frac{\pi}{2} - \arctan \frac{\left[{}_B r_{SW} \right]_{(3)}}{\left[{}_B r_{SW} \right]_{(1)}} \tag{25c}$$

$$\gamma_{\mathrm{ref}}(z) := \mathrm{atan2}\left(\left[{}_B r_{SW} \right]_{(2)} , \left[{}_B r_{SW} \right]_{(1)} \right) \tag{25d}$$

$$\begin{bmatrix} \gamma_{\mathrm{EE}} \\ \beta_{\mathrm{EE}} \\ \psi_{\mathrm{EE}} \end{bmatrix}(z) := \mathrm{eul}_{ZYZ}\left(A_{7R}(z, \gamma_{\mathrm{ref}}, \beta_{\mathrm{ref}}) \right). \tag{25e}$$

With the shoulder-wrist vector

$$\begin{aligned} {}_B r_{SW} &:= {}_B r_{BR} - {}_B r_{BS} \\ &= {}_B r_{BT} - A_{B6}\, {}_6\hat{z}(l_6 + l_7 + l_T) - {}_B\hat{z}(l_B + l_1). \end{aligned} \tag{26}$$

and the rotation matrix

$$A_{7R}(z, \gamma_{\mathrm{ref}}, \beta_{\mathrm{ref}}) := A_{7T} A_{TB}(z) A_{BR}(\gamma_{\mathrm{ref}}, \beta_{\mathrm{ref}}), \tag{27}$$

derived from the desired task space pose. Rotation A_{7T} is the constant rotation matrix from body fixed frame of link 7 to the TCP frame.

3.2.2. Task Space Surjection

We refer to the inverse mapping, i.e., from the parameter vector p to the task space pose z, as Task Space Surjection (TSS)

$$\mathrm{TSS}: \quad \mathbb{R}^6 \to \mathrm{SE}(3), \quad p \mapsto z. \tag{28a}$$

The relations are given with

$${}_B r_{BT} := (l_B + l_1)\, {}_B\hat{z} + A_{BR}(\gamma_{\mathrm{ref}}, \beta_{\mathrm{ref}})\left({}_R\hat{z} r_{\mathrm{ref}} + A_{R7}(\gamma_{\mathrm{EE}}, \beta_{\mathrm{EE}}, \psi_{\mathrm{EE}})(l_6 + l_7 + A_{7T} l_T) \right) \tag{28b}$$

$$A_{TB} := A_{7T} A_{7R}(\gamma_{\mathrm{EE}}, \beta_{\mathrm{EE}}, \psi_{\mathrm{EE}}) A_{RB}(\gamma_{\mathrm{ref}}, \beta_{\mathrm{ref}}) \tag{28c}$$

using the established definitions in the previous sections.

3.3. Inverse Kinematics

In this section we derive closed-form expressions for the IK map. After discussing the choice of the default manipulator configuration, we derive the individual IK mappings of the robot joints. Corresponding to the S-R-S structure, we group the joints into shoulder angles $\{q_1, q_2, q_3\}$, the elbow angle $\{q_4\}$, and wrist angles $\{q_5, q_6, q_7\}$.

3.3.1. Manipulator Configuration

Due to the possible reconfiguration of the robot kinematics, i.e., whenever 3 revolute joint axes intersect in one point, with 2 being coaxial and the third being perpendicular to the links, there exists an alternative configuration

$$\text{FK}(\text{coaxial}_1, \text{perpendicular}, \text{coaxial}_2) = \text{FK}(\text{coaxial}_1 + \pi, -\text{perpendicular}_2, \text{coaxial}_2 + \pi) \tag{29}$$

that results in the same FK. In the 7-DOF S-R-S structure considered in this work, this is the case for the tuples (q_1, q_2, q_3), (q_3, q_4, q_5), and (q_5, q_6, q_7). Therefore, defining only the end-effector pose as well as the elbow position results in 8 possible configurations. Of course, it is important to derive an IK map that results in one specific configuration for the entire parameter space. The following derivation is designed to yield in a configuration as depicted in Figure 3 for the default case $q_1 = q_3 = q_5 = 0$. This is achieved by choosing the joint angle ranges

$$
\begin{aligned}
q_1 &\in \begin{bmatrix} -\pi, & +\pi \end{bmatrix} \\
q_2 &\in \begin{bmatrix} 0, & +\pi \end{bmatrix} \\
q_3 &\in \begin{bmatrix} -\pi, & +\pi \end{bmatrix} \\
q_4 &\in \begin{bmatrix} 0, & +\pi \end{bmatrix} \\
q_5 &\in \begin{bmatrix} -\pi, & +\pi \end{bmatrix} \\
q_6 &\in \begin{bmatrix} 0, & +\pi \end{bmatrix} \\
q_7 &\in \begin{bmatrix} -\pi, & +\pi \end{bmatrix}.
\end{aligned}
\tag{30}
$$

We refer to this definition as $\mathcal{Q}_{sc} \subset \mathbb{R}^7$, i.e., the space of joints in standard configuration.

3.3.2. Elbow Angles

The central geometric shape to express the arm portion of joints is the triangle $\overline{SEW}$ as depicted in Figure 3. It is fully defined by the parameter r_{ref}, as well as the robot related constant link lengths

$$r_{SE} := l_3 + l_4 \tag{31}$$

$$r_{EW} := l_5 + l_6. \tag{32}$$

The law of cosines in this triangle allows direct calculation of joint 4

$$r_{\text{ref}}^2 = r_{SE}^2 + r_{EW}^2 - r_{SE} r_{EW} \cos(\pi - q_4) \tag{33}$$

$$q_4(r_{\text{ref}}) := \pi - \arccos\left(\frac{r_{SE}^2 + r_{EW}^2 - r_{\text{ref}}^2}{2 r_{SE} r_{EW}}\right) \tag{34}$$

as well as the adjoint angles

$$r_{SE} = r_{\text{ref}}^2 + r_{EW}^2 - r_{\text{ref}} r_{EW} \cos(\theta_S) \tag{35}$$

$$\theta_S(r_{\text{ref}}) := \arccos\left(\frac{r_{\text{ref}}^2 + r_{EW}^2 - r_{SE}}{2 r_{\text{ref}} r_{EW}}\right) \tag{36}$$

and

$$r_{EW} = r_{\text{ref}}^2 + r_{SE}^2 - r_{\text{ref}}r_{SE}\cos\left(\theta_W\right) \tag{37}$$

$$\theta_W\left(r_{\text{ref}}\right) := \arccos\left(\frac{r_{\text{ref}}^2 + r_{SE}^2 - r_{EW}}{2r_{\text{ref}}r_{SE}}\right). \tag{38}$$

The latter are used to define alternative rotation frame compositions for the derivation of the remaining joints. See Figure 4 for an overview of the relations between all introduced coordinate frames.

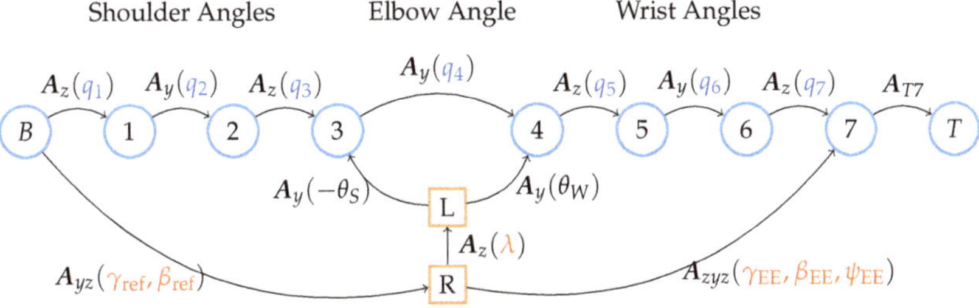

Figure 4. Reference frames and their relations. The blue frames B to T are fixed to the corresponding body-fixed coordinate systems of the robot links. Orange frames R and L are additional reference frames for the introduced parameter space. The arrows mark the rotations between the frames of reference.

3.3.3. Shoulder Angles

Reusing the ZYZ Euler sequence extraction function (24) makes it possible to directly define the IK function of the shoulder angles $\{q1, q2, q3\}$. The parameter-related frames R and L (cf. Figure 4) are used to compose the transformation matrix

$$A_{3B}(\boldsymbol{p}, \lambda) := A_y\left(-\theta_W\right) A_z\left(\lambda\right) A_{RB}\left(\beta_{\text{ref}}, \gamma_{\text{ref}}\right) \tag{39}$$

and extract

$$\begin{bmatrix} q_1 \\ q_2 \\ q_3 \end{bmatrix}(\boldsymbol{p}, \lambda) := \text{eul}_{ZYZ}\left(A_{3B}(\boldsymbol{p}, \lambda)\right). \tag{40}$$

3.3.4. Wrist Angles

Analogously to the shoulder angles, the wrist angles $\{q_5, q_6, q_7\}$ can be calculated by composing the transformation matrix

$$A_{74}(\boldsymbol{p}, \lambda) := A_{7R}\left(\gamma_{\text{EE}}, \beta_{\text{EE}}, \psi_{\text{EE}}\right) A_z(-\lambda) A_y\left(-\theta_S\right) \tag{41}$$

and extracting the wrist angles with

$$\begin{bmatrix} q_5 \\ q_6 \\ q_7 \end{bmatrix}(\boldsymbol{p}, \lambda) := \text{eul}_{ZYZ}\left(A_{74}(\boldsymbol{p})\right). \tag{42}$$

3.3.5. Overview

All closed-form expressions resulting from the IK mapping are fully listed in Appendix B. The parameter dependencies of the individual function components are

$$
\begin{aligned}
\text{IK}_1 &: \quad \mathbb{R}^4 \to \mathbb{R}^1, & (\theta_S(r_{\text{ref}}), \gamma_{\text{ref}}, \beta_{\text{ref}}, \lambda) &\mapsto q_1 & \text{(43a)} \\
\text{IK}_2 &: \quad \mathbb{R}^3 \to \mathbb{R}^1, & (\theta_S(r_{\text{ref}}), \beta_{\text{ref}}, \lambda) &\mapsto q_2 & \text{(43b)} \\
\text{IK}_3 &: \quad \mathbb{R}^3 \to \mathbb{R}^1, & (\theta_S(r_{\text{ref}}), \beta_{\text{ref}}, \lambda) &\mapsto q_3 & \text{(43c)} \\
\text{IK}_4 &: \quad \mathbb{R}^1 \to \mathbb{R}^1, & (r_{\text{ref}}) &\mapsto q_4 & \text{(43d)} \\
\text{IK}_5 &: \quad \mathbb{R}^3 \to \mathbb{R}^1, & (\theta_W(r_{\text{ref}}), \gamma_{\text{EE}}, \beta_{\text{EE}} - \lambda) &\mapsto q_5 & \text{(43e)} \\
\text{IK}_6 &: \quad \mathbb{R}^3 \to \mathbb{R}^1, & (\theta_W(r_{\text{ref}}), \gamma_{\text{EE}}, \beta_{\text{EE}} - \lambda) &\mapsto q_6 & \text{(43f)} \\
\text{IK}_7 &: \quad \mathbb{R}^4 \to \mathbb{R}^1, & (\theta_W(r_{\text{ref}}), \gamma_{\text{EE}}, \beta_{\text{EE}}, \psi_{\text{EE}} - \lambda) &\mapsto q_7 & \text{(43g)}
\end{aligned}
$$

and show the low dimensional dependency as required by PR2. Note that parameters γ_{ref} and ψ_{EE} do solely influence q_1 and q_7 resp., and thus do not influence manipulability. Further, in this formulation the shoulder and wrist joints result in equivalent mappings, with symmetrical assignments. Their relations are given as

$$
\begin{aligned}
\text{IK}_5 &= \text{IK}_3(\theta_W(r_{\text{ref}}), \beta_{\text{EE}}, \gamma_{\text{EE}} - \lambda) & \text{(44a)} \\
\text{IK}_6 &= \text{IK}_2(\theta_W(r_{\text{ref}}), \beta_{\text{EE}}, \gamma_{\text{EE}} - \lambda) & \text{(44b)} \\
\text{IK}_7 &= \text{IK}_1(\theta_W(r_{\text{ref}}), \psi_{\text{EE}}, \beta_{\text{EE}}, \gamma_{\text{EE}} - \lambda). & \text{(44c)}
\end{aligned}
$$

This is an interesting geometrical insight that results from the chosen parameter set.

3.4. Forward Kinematics

Although not used in the task space manipulability mapping, we state—for the sake of completeness—the forward mapping

$$
\text{FK}: \quad \mathbb{R}^7 \to \mathbb{R}^6 \times \mathbb{R}, \quad q \mapsto (p, \lambda) \tag{45}
$$

using the developed relations from the previous section on the IK problem. From the elbow angle q_4 and the relation (33), r_{ref} is mapped by

$$
r_{\text{ref}}(q_4) := \sqrt{r_{SE}^2 + r_{EW}^2 - r_{SE} r_{EW} \cos(\pi - q_4)}. \tag{46}
$$

The Euler angle extraction function (24) allows again a concise definition of the remaining mappings. The shoulder joints $\{q_1, q_2, q_3\}$ with the adjoint shoulder angle $\theta_S(r_{\text{ref}})$ from (35) parametrize

$$
\begin{bmatrix} \gamma_{\text{ref}} \\ \beta_{\text{ref}} \\ \lambda \end{bmatrix} := \text{eul}_{ZYZ}\left(A_{LB}(q, r_{\text{ref}})\right) \tag{47a}
$$

where

$$
A_{LB}(q, r_{\text{ref}}) := A_y(\theta_S(r_{\text{ref}})) A_z(q_3) A_y(q_2) A_z(q_1). \tag{47b}
$$

Analogously, the wrist joints $\{q_5, q_6, q_7\}$ and the adjoint wrist angle $\theta_W(r_{\text{ref}})$ from (37) define the end-effector parameters

$$
\begin{bmatrix} \lambda + \gamma_{\text{EE}} \\ \beta_{\text{EE}} \\ \psi_{\text{EE}} \end{bmatrix} := \text{eul}_{ZYZ}\left(A_{7L}(q, r_{\text{ref}})\right) \tag{48a}
$$

where

$$
A_{7L}(q, r_{\text{ref}}) := A_z(q_7) A_y(q_6) A_z(q_5) A_y(-\theta_W(r_{\text{ref}})). \tag{48b}
$$

The composition of rotations is in accordance with the structural relation depicted in Figure 4. This concludes the FK problem.

3.5. Admissible Parameter Space

The compact analytical expressions also allow solving analytically for an upper and lower bound of λ, given maximal joint angles $q_i^{\max}$. Let $\mathcal{Q} := \{q \mid q \in \mathcal{Q}_{\text{sc}}, |q_i| \leq q_i^{\max}\}$ be the space of admissible joint configurations. In this section, we determine the space of admissible parameters

$$
\mathcal{A} := \{(p, \lambda) \mid \text{IK}(p, \lambda) \in \mathcal{Q}\}. \tag{49}
$$

Recall the definition of the parameter vector $p := [r_{\text{ref}}, \gamma_{\text{ref}}, \beta_{\text{ref}}, \gamma_{\text{EE}}, \beta_{\text{EE}}, \psi_{\text{EE}}]^\top$ from (20). Only r_{ref} is of linear nature and thus has a limited range. The remaining parameters describe angles and hence need not be limited. While IK_4 directly relates joint limits of the elbow joint with the admissible range of r_{ref}, the null space parameter λ is related to all remaining joints. Each of which can potentially exclude partitions of the full range of λ. The set of admissible parameters $\mathcal{A}$ must consider all joint limits and results from the intersection

$$
\mathcal{A} = \bigcap_{i=1}^{n} \mathcal{A}_i, \tag{50}
$$

of the n individual joint-related portions.

3.5.1. Shoulder-Wrist Distance r_{ref}

Elbow joint 4 directly limits the parameter r_{ref}. Solving (43d) for r_{ref} gives

$$
r_{\text{ref}}(q_4) := \sqrt{r_{SE}^2 + r_{EW}^2 - 2r_{SE}r_{EW}\cos(\pi - q_4)} \tag{51}
$$

and defines the lower and upper bounds

$$
r_{\text{ref}}(q_4^{\max}) \leq r_{\text{ref}} \leq r_{\text{ref}}(0) \tag{52}
$$

with the upper boundary $r_{\text{ref}}(0)$ being the stretched out configuration of the robot. This defines

$$
\mathcal{A}_4 := \left\{ (p, \lambda) \in \mathcal{P} \;\middle|\; \sqrt{r_{SE}^2 + r_{EW}^2 - 2r_{SE}r_{EW}\cos(\pi - q_4^{\max})} \leq r_{\text{ref}} \leq r_{SE} + r_{EW} \right\} \tag{53}
$$

as the admissible parameter set w.r.t. joint 4.

3.5.2. Null Space Parameter λ

All remaining joints, i.e., shoulder joints $\{q_1, q_2, q_3\}$ and wrist joints $\{q_5, q_6, q_7\}$, limit parts of the null space parameter λ. The 4-quadrant $\text{atan2}(\cdot)$ functions from (43), however, are difficult to symbolically rewrite in terms of λ due to there piecewise definition. To circumvent this, we further introduce IK mappings that calculate the absolute joint angles. We define the extraction map of absolute values of the Euler sequence $|\operatorname{eul}_{ZYZ}| : \operatorname{SE}(3) \to \mathbb{R}^3_+$ as

$$|\operatorname{eul}_{ZYZ}| : \quad \operatorname{SE}(3) \to \mathbb{R}^3_+, \quad A_{zyz} \mapsto \begin{bmatrix} \arccos\left(\dfrac{[A_{zyz}(z)]_{(3,1)}}{\sin\left(\arccos\left([A_{zyz}(z)]_{(3,3)}\right)\right)}\right) \\ \arccos\left([A_{zyz}(z)]_{(3,3)}\right) \\ \arccos\left(\dfrac{-[A_{zyz}(z)]_{(1,3)}}{\sin\left(\arccos\left([A_{zyz}(z)]_{(3,3)}\right)\right)}\right) \end{bmatrix} =: \begin{bmatrix} |\gamma| \\ |\beta| \\ |\psi| \end{bmatrix} \tag{54}$$

which is used to find the absolute angles of the shoulder and wrist joints

$$\begin{bmatrix} |q_1 + \gamma_{\text{ref}}| \\ |q_2| \\ |q_3| \end{bmatrix} (p, \lambda) := |\operatorname{eul}_{ZYZ}|(A_{3B}(p, \lambda)) \tag{55a}$$

$$\begin{bmatrix} |q_5| \\ |q_6| \\ |q_7 + \psi_{\text{EE}}| \end{bmatrix} (p, \lambda) := |\operatorname{eul}_{ZYZ}|(A_{74}(p, \lambda)) \tag{55b}$$

analogously to the mapping eul_{ZYZ} from the previous Section 3.3. See Appendix C for the full definition of the absolute valued IK functions. Note that the mappings admit the same symmetrical assignments between the shoulder and wrist portion as the actual IK mapping discussed in Section 3.3.5.

Due to the concise formulations of the IK (55a), all functions can be solved for the null space parameter λ. By substituting the joint parameters with their respective limit, closed-form expressions are formed that deliver s_i candidates for lambda ranges

$$\lambda_i^{\lim} : \quad \mathbb{R}^7 \times \mathbb{R} \to \mathbb{C}^{s_i}, \quad (p, q_i^{\max}) \mapsto \lambda_i^{\lim}(p, q_i^{\max}) \quad \forall i \in [1,7] \setminus 4 \tag{56}$$

according to the $i = [1,7]$ joints. For q_2 and q_6, the respective middle joints of the shoulder and wrist angle tuples (q_1, q_2, q_3) and (q_5, q_6, q_7), we directly find $s_2 = s_6 = 2$ symmetric solutions for a positive and negative null space limit. However, solving the remaining mappings from IK (55a) for λ, results in more solution candidates. This results from the fact that, depending on the parameter configuration, these joints have the potential for cyclic behaviour for a linear increase in λ at a fixed pose (discussed in [15]). Joints q_3 and q_5 can thus reach up to $s_3 = s_5 = 4$ null space angles marking a joint limit. The first joint q_1 and last joint q_7 do also offer up to 4 critical values for λ, however, due to additional additive parameters γ_{ref} and ψ_{EE} resp., it is necessary to additionally consider solutions for $|-q_1 + \gamma_{\text{ref}}|$ and $|-q_7 + \gamma_{\text{ref}}|$. These solutions are evaluated with $\lambda_1^{\lim}(p|_{-\gamma_{\text{ref}}}, q_1^{\max})$ and $\lambda_7^{\lim}(p|_{-\psi_{\text{EE}}}, q_7^{\max})$. Consequently, $s_1 = s_7 = 8$ solution candidates for the first and last joint of the kinematic have to be considered.

Besides knowing the value of a critical limit, it is further essential for many applications to know if it expresses an upper or a lower limit. Similar to the approach in [16], the partial derivatives of the null space range mappings $\lambda_i^{\lim}$ w.r.t. the corresponding joint angle limit are used to characterize each limit candidate. For every $\ell \in \lambda_i^{\lim}$, the corresponding partial derivative is evaluated to decide

$$\ell \in \begin{cases} \text{is upper limit} & \text{if } \operatorname{sign}(\ell)\, \dfrac{\partial \lambda_i^{\lim}}{\partial q_i^{\max}} > 0 \\[2ex] \text{is lower limit} & \text{if } \operatorname{sign}(\ell)\, \dfrac{\partial \lambda_i^{\lim}}{\partial q_i^{\max}} < 0 \\[2ex] \text{is no limit} & \text{otherwise.} \end{cases} \tag{57}$$

In a second step, all solution candidates in $\lambda_i^{\lim}$ are tested for validity, to define the sets of actual upper and lower null space limit angles

$$\mathcal{L}_i^{\text{up}}(\boldsymbol{p}) := \left\{ \lambda \in \lambda_i^{\lim} \;\middle|\; \lambda \in \mathbb{R} \;\wedge\; |\text{IK}_i(\boldsymbol{p},\lambda)| = q_i^{\max} \;\wedge\; \text{sign}\,(\lambda)\,\frac{\partial \lambda}{\partial q_i^{\max}} > 0 \right\} \quad \forall i \in [1,7] \setminus 4 \quad (58a)$$

$$\mathcal{L}_i^{\text{low}}(\boldsymbol{p}) := \left\{ \lambda \in \lambda_i^{\lim} \;\middle|\; \lambda \in \mathbb{R} \;\wedge\; |\text{IK}_i(\boldsymbol{p},\lambda)| = q_i^{\max} \;\wedge\; \text{sign}\,(\lambda)\,\frac{\partial \lambda}{\partial q_i^{\max}} < 0 \right\} \quad \forall i \in [1,7] \setminus 4. \quad (58b)$$

These upper and lower limits form j pairwise ranges $\Lambda_{i,j}$ and define the remaining admissible parameter sets

$$\mathcal{A}_i := \left\{ (\boldsymbol{p},\lambda) \in \mathcal{P} \;\middle|\; \lambda \in \bigcup_j \Lambda_{i,j} \right\} \quad \forall i \in [1,7] \setminus 4 \,, \tag{59}$$

related to shoulder and wrist joints.

The full intersection set $\mathcal{A}$, as defined in (50), may consist of several separate regions. Directly evaluating all critical values of λ is especially interesting whenever planning a continuous path in task space. We apply the admissible parameter space in application Sections 5.1.3 and 5.2.2. All full function definitions of the limit candidates $\lambda_i^{\lim}$ are summarized in Appendix D.

4. Results

This section contains an evaluation of the task space manipulability framework developed in this work. We first give a run-time comparison to show the computational advantage of our closed-form expression in comparison to general numerical solutions. We show that uniform sampling in the new parameter space results in a superior probability distribution of the manipulability in comparison with direct sampling in joint space. Further, the sensitivity of the manipulability measure w.r.t. the parameters is analyzed.

4.1. Accuracy

Unlike numerical IK solvers that approximate the inverse mapping iteratively [42], or CLIK solvers [10–12] that converge to the solution from a control point of view, the analytical nature of our closed-form task space manipulability expression delivers exact results in a single iteration.

4.2. Run-Time Comparison

Complete evaluation of the closed-form IK and M mapping as single expressions allows automatic code generation of the symbolic expressions with e.g., the *MATLAB Coder*™ toolbox. These expressions allow array operations, or *vectorization* in *MATLAB*, such that a large number of solutions can be evaluated simultaneously. This leads to a significant computational boost, compared to algorithms that rely on matrix arithmetic and consequently have to sequentially evaluate multiple evaluations in programmatic loops. This property makes it further straightforward to calculate the task space manipulability of multiple samples on a powerful Graphics Processing Unit (GPU). The following run-time comparison was conducted in *MATLAB 2019a*, on a computer with Intel(R) Core(TM) i9-9900X CPU @ 3.50 GHz, 128 GB memory, and a *NVIDIA TITAN V* graphics card.

Besides different versions of our presented algorithm, we also tested the run-time of [15], representing typical analytical IK approaches in the literature, and the nonlinear optimization-based IK algorithm from the *Robotics System Toolbox*™ for *MATLAB*, representing iterative solver approaches. Figure 5 shows a run-time comparison of calculating the manipulability measures

$$\mu_n := (\text{M} \circ \text{IK})(\boldsymbol{p}_n), \quad \text{for } n = [1,N] \tag{60}$$

of N random samples $\boldsymbol{p}_n$.

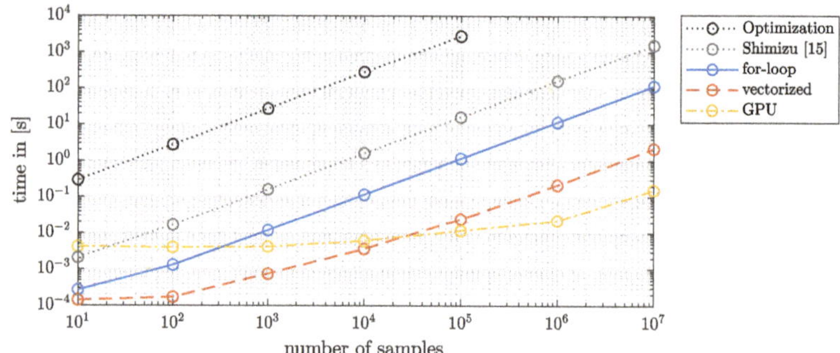

Figure 5. Run-time comparison of processing N poses w.r.t. their task space manipulability. Considered are the *MATLAB* robotics IK solver based on nonlinear optimization, the analytical IK solver by Shimizu et al. [15], and the presented approach in three versions: a conventional sequential loop structure, as well as vectorized evaluation on the central processing unit (CPU) and graphics processing unit (GPU).

As expected, the iterative optimization algorithm (applying BFGS Gradient Projection with solution tolerance 0.01) is the computationally most expensive solution method. It required an average of 37 iterations per pose and did not allow for direct selection of the arm angle. Two orders of magnitude faster and in addition producing exact inverse solutions are the analytical IK solvers found in the literature. They rely on matrix calculus and thus a for-loop structure for evaluation of multiple poses.

Our approach, which is entirely reduced to direct individual expressions, is over 10 times faster when implemented with the same conventional for-loop structure. Already for 200 evaluated samples, a simultaneous vectorized evaluation achieves another performance increase of factor 10. At the maximal evaluated amount of 10^7 samples, vectorization enables an even 50 times faster computation, compared to the implementation using for-loops. The advantage of calculating the task space manipulability on a GPU starts at an amount of 10^5 sample points. For a smaller number of samples, the overhead of initializing the data on the GPU does not pay off. Processing 10^7 samples, calculations on the GPU are 10 times faster then vectorized treatment on the CPU, and even 700 times faster than for conventional loop structures.Note that all time measurements include the generation of random samples on the CPU and GPU respectively.

Considering real-time application for a robot with a typical 1 kHz sampling rate, our approach allows evaluation of 1000 end-effector poses for their task space manipulability.

4.3. Sampling in Task Space

Not having to calculate the IK in an iterative fashion as done by CLIK solvers, evaluating manipulability directly in task space is computationally not much more expensive than directly calculating manipulability in joint space. However, choosing a different space for sampling random poses do have an influence on the probability distribution of resulting manipulability measures.

Before analyzing this difference, we first discuss the used sampling strategies. For a fair comparison, we cover the entire space without consideration of possible limits on the individual joints or parameters.

Let $u \in \mathbb{R}$ be a random number drawn from a uniform distribution in the range of $[0,1]$. Uniform sampling in joint space is straightforward with

$$q_i^{\text{uniform}} : \quad \mathbb{R} \to \mathbb{R}, \quad u \mapsto -\pi + 2\pi u \qquad \forall i \in [1,7] \tag{61}$$

due to the independence of its joints $q \in \mathbb{R}^7$.

For a random end-effector pose sample $(\boldsymbol{p}^{\mathrm{red}}, \lambda) = [r_{\mathrm{ref}}, \beta_{\mathrm{ref}}, \gamma_{\mathrm{EE}}, \beta_{\mathrm{EE}}, \lambda]^{\top}$ from the parameter space, one can choose the same strategy

$$p_1^{\mathrm{naive}} : \quad \mathbb{R} \to \mathbb{R}, \quad u \mapsto r_{\mathrm{ref}}^{\min} + (r_{\mathrm{ref}}^{\max} - r_{\mathrm{ref}}^{\min})u \tag{62a}$$

$$p_i^{\mathrm{naive}} : \quad \mathbb{R} \to \mathbb{R}, \quad u \mapsto -\pi + 2\pi u \quad \forall i \in [2,5] \tag{62b}$$

with respective scaling for the linear parameter r_{ref}. However, this *naive* form of sampling does not lead to a uniform distribution of samples in the task space SE(3), due to the interdependence of the coordinate components.

Recall that the first two parameters r_{ref} and β_{ref} describe translation in polar coordinates. Unlike in Cartesian coordinates, the base vectors are not constant. Consequently, direct uniform sampling of the radial coordinate r_{ref}, leads to sparser sampling further from the origin, due to the increasing circumference proportionally to r_{ref}. Proper uniform sampling of the translational part can be achieved by

$$p_1^{\mathrm{uniform}} : \quad \mathbb{R} \to \mathbb{R}, \quad u \mapsto \sqrt{\left(r_{\mathrm{ref}}^{\min}\right)^2 + \left(\left(r_{\mathrm{ref}}^{\max}\right)^2 - \left(r_{\mathrm{ref}}^{\min}\right)^2\right)u} \tag{63a}$$

$$p_2^{\mathrm{uniform}} : \quad \mathbb{R} \to \mathbb{R}, \quad u \mapsto -\pi + 2\pi u. \tag{63b}$$

An efficient method of uniform sampling on SO(3), i.e., 3D orientations, is proposed by Kuffner [43]. Uniform sampling of the individual angles of the Euler sequence results in a bias towards the polar regions of the unit sphere. He proposes to use an arctan function on the second angle to compensate for this bias. Uniform sampling of the end-effector orientation, parametrized by γ_{EE} and β_{EE}, is thus achieved with

$$p_3^{\mathrm{uniform}} : \quad \mathbb{R} \to \mathbb{R}, \quad u \mapsto -\pi + 2\pi u \tag{63c}$$

$$p_4^{\mathrm{uniform}} : \quad \mathbb{R} \to \mathbb{R}, \quad u \mapsto \arccos\left(1 - 2u\right). \tag{63d}$$

The last portion in our parameter tuple $(\boldsymbol{p}, \lambda)$ is the null space parameter λ that is independent and thus remains

$$p_5^{\mathrm{uniform}} : \quad \mathbb{R} \to \mathbb{R}, \quad u \mapsto -\pi + 2\pi u. \tag{63e}$$

Figure 6 illustrates the uniform sampling of the task space applying the uniform sampling strategy (63).

The above-discussed sampling strategies are now analyzed in conjunction with their respective mapping to the manipulability measure. Figure 7 shows the approximated cumulative distribution function (CDF) of manipulability resulting from 10^7 random samples. It shows that random sampling in joint space according to (61) is more likely to result in a joint configuration with poor manipulability of the robot. Uniform sampling in parameter space (63) produces much fewer joint configurations with poor manipulability, while at the same time more configurations with high manipulability. Naive sampling in parameter space (61) performs similarly good in the low manipulability section. However, it produces also fewer configurations with high manipulability. Considering a conventional 6-DOF robot, i.e., fixing the null space parameter λ to 0 or π, results in a slightly better probability density function (PDF) than for the discussed 7-DOF mechanism. This is a surprising result, as it is always argued that the redundancy improves manipulability. While it is true that the additional DOF has the potential to improve performance measures, poor exploitation might achieve the opposite.

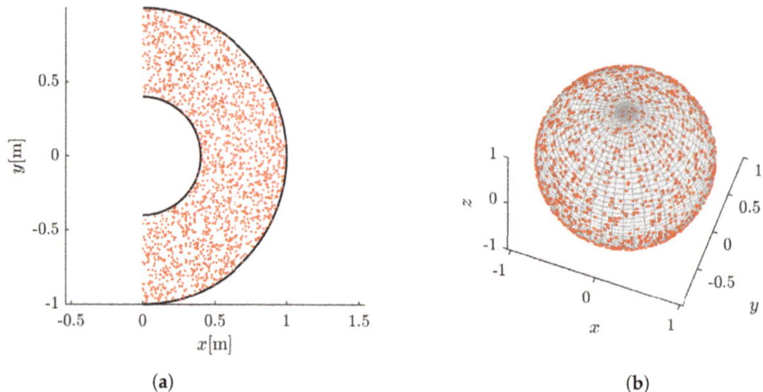

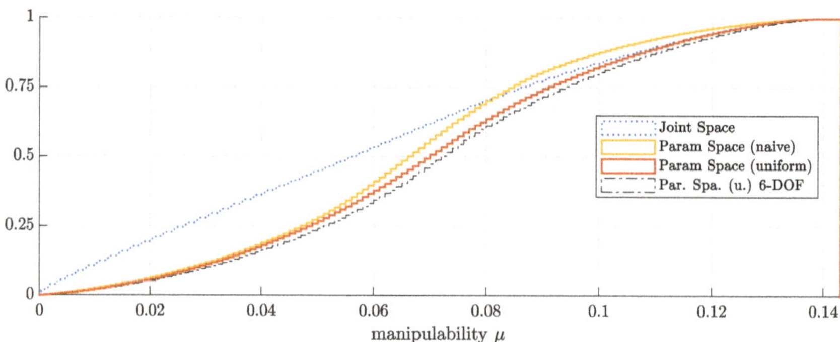

Figure 6. Uniform distributed sampling of the task space (2000 samples). (**a**) End-effector translation; (**b**) End-effector orientation.

Figure 7. Approximated cumulative distribution function (CDF) from a histogram of manipulability w.r.t. different sampling strategies (10^7 samples).

Kuhlemann et al. [44] showed in different use-cases that the seventh DOF of the *KUKA LBR iiwa* increased the average dexterity by 16% in comparison to a conventional 6 DOF *KUKA KR 10*. Both the shortcomings of the naive parameter sampling strategy and the apparent advantage of the 6-DOF mechanism are discussed in Section 4.4.4.

The average normalized manipulabilities achieved are 37% for uniform joint space sampling, 43% for naive parameter space sampling, and 50% for uniform sampling in parameter space. All numbers are w.r.t. the maximal encountered manipulability.

4.4. Parameter Sensitivity Analysis of Manipulability in Parameter Space

The sensitivity of the task space manipulability w.r.t. its parameters are analyzed by generating 10^7 random samples according to (63). These samples represent a uniform distribution of task space configurations. Figure 8 shows the bi-variate histograms of manipulability $\mu(\boldsymbol{p}^{\mathrm{red}}, \lambda)$ w.r.t. to the individual parameters.

Colors approximate the PDF of $\mu(r_{\mathrm{ref}}, \beta_{\mathrm{ref}}, \gamma_{\mathrm{EE}}, \beta_{\mathrm{EE}}, \lambda)$ at fixed values of the respective parameter. For all parameter values we find unimodal distributions, i.e., distributions with a single maxima.

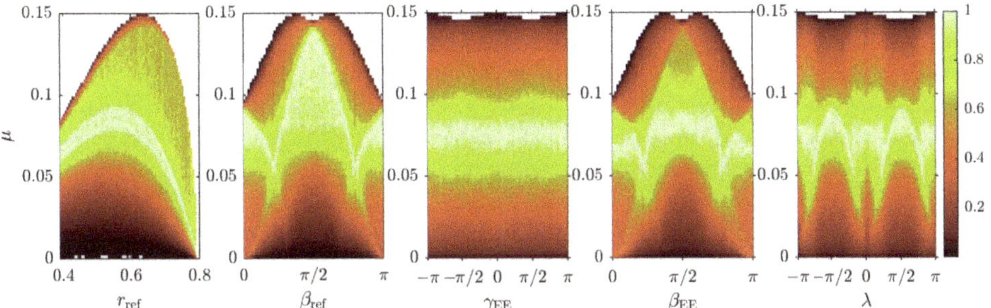

Figure 8. Bi-variate histograms of $\mu(r_{\text{ref}}, \beta_{\text{ref}}, \gamma_{\text{EE}}, \beta_{\text{EE}}, \lambda)$ w.r.t. to the individual parameters, based on 10^7 uniformly distributed parameter space samples. Colors are normalized along with the particular value of the parameter on the x-axis.

4.4.1. Translation Parameters r_{ref} and β_{ref}

The PDF of μ along the shoulder-wrist distance r_{ref} shows a preferred value of $0.57\,\text{m}$. Although a manipulability optimizing configuration cannot be found at this given value, the mode of the corresponding PDF, i.e., its local maxima, has the highest value of manipulability. Further, the probability of good manipulation is decreasing with r_{ref} towards the workspace singularity, i.e., a fully stretched arm of robot configuration.

The polar angle β_{ref} between the vertical and the shoulder-wrist reference vector has the highest manipulability mode at $\frac{\pi}{2}$ rad, although manipulability maximizing configurations are not found. For values approaching 0 and π rad, i.e., placing the wrist in line with the axis of base joint q_1, typically cause so-called *shoulder singularities* on conventional 6-DOF robots. While the 7-DOF kinematics do not necessarily result in a kinematic singularity, high manipulability is not possible either.

4.4.2. Orientation Parameters γ_{EE} and β_{EE}

The third parameter γ_{EE}, which describes a rotation around the shoulder-wrist vector, is the only one that seems to cause little variation in the manipulability PDF and does not allow a conclusion over a preferred configuration.

The consecutive rotation angle β_{EE} shows a similar influence as the reference angle β_{ref}. However, the mode of these PDFs is less prominent and tendentiously marks a lower manipulability.

4.4.3. Null Space Parameter λ

The null space parameter λ reveals that the highest manipulabilities can be found at $\lambda = \{0, \pm\pi\}$rad, i.e., the conventional upper and lower elbow configuration of 6-DOF kinematics. Although missing the absolute top manipulability poses, only small deviations of about ± 0.1 rad from these configurations result in a decrease of the manipulability mode of 25%, i.e., from 0.8 to 0.6. Better modes are found at $\lambda = \{\pm\frac{\pi}{2}\}$rad. Not only is their peak at a slightly higher manipulability of 0.85, but they are also less sensitive to a parameter change in λ. The latter is especially valuable for staying agile during unforeseen events.

4.4.4. Discussion of Manipulability in Different Sampling Strategies

The different sampling strategies discussed in Section 4.3 result in differences in the approximated CDFs, cf. Figure 7.

Naive vs. Uniformly Distributed Sampling

The difference between naive and uniform sampling solely affects parameters r_{ref} and β_{EE}. That is, the corresponding uniform sampling functions (63a) and (63d) correct the biases of the radial coordinate r_{ref} towards the origin, and the orientation towards the pole regions with azimuthal angle $\beta_{\mathrm{EE}} = \{0, \pi\}$rad, respectively. Consequently, these regions are sparser sampled in the uniformly distributed strategy. While this correction is negligible for the range of r_{ref} in this particular robot example, the improvement of the CDF towards better manipulability stems from a sparser sampling of the boundary regions of β_{EE}. Because exactly these boundaries lack high manipulability poses, as visible in the according bi-variate histogram in Figure 8.

6-DOF vs. 7-DOF Kinematics

According to Section 4.4.3, the apparent slight advantage of uniform distributed sampling of a conventional 6-DOF robot only holds for the over-all manipulability distribution illustrated in Figure 7. The parameter-specific histogram w.r.t. to the arm angle λ in Figure 8, on the other hand, reveals that the conventional 6-DOF configurations $\lambda = \{0, \pm\pi\}$rad do have a good manipulability distribution, but $\lambda = \{\pm\frac{\pi}{2}\}$rad configurations are preferable. A 7-DOF kinematics hence not only enables agile adaptation of the kinematic structure, but also contains arm angles that have a better PDFs of manipulability than its 6-DOF counterpart. At the same time, other arm angles show higher variability in the histogram and are more prone to decrease performance. An increase in manipulability by the additional DOF thus relies on a well-conceived utilization of such.

4.5. Number of Local Optima

While the analysis shown in the previous section gives insight in the probability distribution of the manipulability measure, it does not allow conclusions on how manipulability changes along the null space. Table 1 lists the number of local optima for a given end-effector pose. It shows that 80% of the robot poses do not have a unique manipulability maximizing null space solution, but up to 4 distinct optima.

Table 1. Distribution of local optima among 10^7 samples.

# optima	1	2	3	4
percentage	20%	41%	27%	12%

5. Applications

Two application directions that benefit from the closed-form expressions of the task space manipulability are outlined in this section. First, we demonstrate how global optimization problems can be formulated that profit from massive multi-start point pre-evaluation. Second, we propose a novel way of real-time redundancy resolution on the position level, which enables global manipulability optimization of single poses as well as for provided end-effector trajectories in SE(3).

5.1. Optimal Robot Placement

The analytic results from the previous Section 3 allow formulating interesting questions in terms of optimization problems. We consider the problem of optimal placement of the robot.

5.1.1. Best Overall Robot Configuration

The most basic optimization problem we considere is the question of finding the best overall robot configuration w.r.t. to manipulability. Mathematically, this problem can be stated as an unconstrained optimization problem

$$\underset{q}{\text{maximize}} \quad \mu(q) \tag{64}$$

directly finding the optimal joint configuration w.r.t. the manipulability measure. The global optimum is found with a multi-start strategy [45], where random samples are drawn from the admissible parameter space $\mathcal{P}$ and used as starting points for local optimizations. Figure 9, left side, shows the results of such a global optimization process with 1000 starting points. Note that the same problem can be formulated in *parameter space* and does yield the same result. All optimization iterations result in one of 8 equally good global optima, which can be reduced to 4 solutions due to symmetry of the shoulder joint. They further describe configurations in the pure xz-plane with $\lambda \in \{0, \pm 180\}°$. This is equivalent to the configurations achievable by a conventional 6-DOF robot.

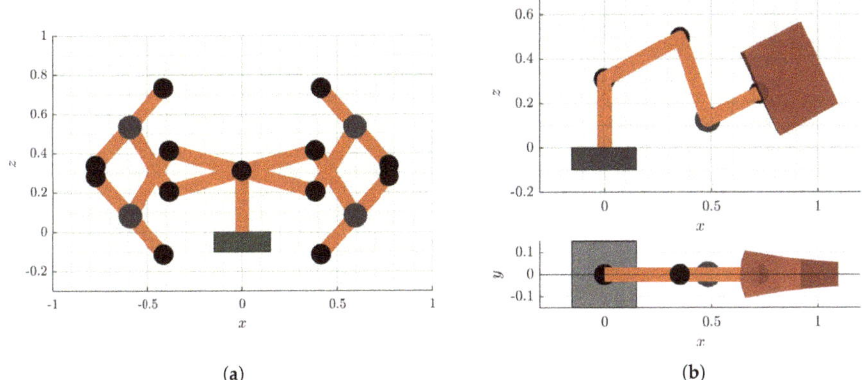

(a)   (b)

Figure 9. Results of the task space manipulability optimization of a robot mounting pose. (**a**): Overall best robot configuration. There are a total of 8 global optima with equal manipulability $\mu_{\max} = 0.143$. From 1000 random initial starting points, 83% of the optimization runs converged to one of the global optima. (**b**): Optimizing relative pose w.r.t. a workspace envelope of size $(\Delta x, \Delta y, \Delta z) = (0.4, 0.4, 0.3)$ m. Note that the cubic volume is projected onto the parameter space, hence the distortion in the illustration. The resulting configuration for pose z_0, again lies fully on the xz-plane. However, unlike the single best pose, only one single optimum is found.

5.1.2. Best Robot Configuration for Multiple Task Poses

In industrial settings, robots are often required to work at a certain number $i \in \mathbb{Z}^+$ of different task poses z_i. While the relative distances $\Delta z_i = z_i - z_1$ between this poses is defined, the optimal placement of the robot can be found by solving the optimization problem

$$\underset{z, \lambda}{\text{maximize}} \quad \sum_i (\mathrm{M} \circ \mathrm{IK} \circ \mathrm{TSP})(z + \Delta z_i, \lambda) \tag{65}$$

to find the relative pose z that maximizes the average manipulability of all i poses. Solving this problem directly, results in an infinite number of global poses. These solutions are rotationally symmetric around the base joint q_1 as well as the last joint q_7, as both these joints do not have an influence on the manipulability of the 7-DOF robot structure at consideration (discussed in Section 3.1).

The complexity of the optimization problem, as well as the number of global optima, can be drastically reduced by formulating the same problem in the lower dimensional parameter space

$$\underset{p, \lambda}{\text{maximize}} \quad \sum_i (\mathrm{M} \circ \mathrm{IK})(p + \Delta p_i, \lambda) \tag{66}$$

where $p_i = \mathrm{TSP}(z_i)$. The resulting optimal p can eventually be mapped to the corresponding task space parameter $z = \mathrm{TSS}(p)$. This result is useful for deciding on how to mount a robot relative to a given set of task poses z_i, or recalculating it online if task poses are time-variant and the robot structure is e.g., mounted on a mobile platform.

5.1.3. Optimizing Robot Mounting Positions Regarding a Workspace Envelope

In a modern scenario where robots are not only expected to repetitively execute the same tasks, a set of pre-defined task poses cannot always be formulated. But it is rather necessary for the robot to perform well in a defined workspace volume, e.g., given as a cubical volume $V = [-\Delta\frac{x}{2}, +\Delta\frac{x}{2}] \times [-\Delta\frac{y}{2}, +\Delta\frac{y}{2}] \times [0, +\Delta z]$. Due to all mappings involved in the task space manipulability being continuous, formulating a cost function for such a volume can be done using Fubini's theorem [46]. It allows calculation of the volume integral as triple integral. The objective for this optimization problem in task space reads

$$\underset{z_0, \lambda}{\text{maximize}} \iiint_V (\text{M} \circ \text{IK} \circ \text{TSP})(z_0 + z(x, y, z), \lambda)\, \mathrm{d}x\, \mathrm{d}y\, \mathrm{d}z \tag{67}$$

$$\text{subject to} \qquad \text{TSP}\,(z_0, +z(x, y, z), \lambda) \in \mathcal{A},$$

where the optimal task space volume origin z_0 needs to be found. This optimization can again be transformed to the lower dimensional parameter space

$$\underset{p_0, \lambda}{\text{maximize}} \iiint_V (\text{M} \circ \text{IK} \circ \text{TSP})(\text{TSS}(p_0) + z(x, y, z), \lambda)\, \mathrm{d}x\, \mathrm{d}y\, \mathrm{d}z \tag{68}$$

$$\text{subject to} \qquad \text{TSP}\,(\text{TSS}(p_0, \lambda) + z(x, y, z), \lambda) \in \mathcal{A}$$

with the condition that the whole Volume projected to parameter space must be within the set of admissible parameters. Figure 9, right side, shows the result of such a global optimization.

5.2. Redundancy Resolution

Solving for optimal robot poses online is essential for a robot to stay agile at all times. We demonstrate how the task space manipulability expressions developed in this work can be applied for real-time global manipulability optimization of single poses as well as full trajectories. The following run-time evaluations were conducted in *MATLAB 2019a*, on a computer with Intel(R) Core(TM) i3-7100 CPU @ 3.9 GHz and 32 GB memory.

5.2.1. Redundancy Resolution for Global Manipulability Optima

Approaches typically found in the literature focus on local optimization of manipulability based on local gradient information. Analysis of the number of existing local optima from Section 4.5, however, revealed that only 20% of end-effector poses have a unique global optimum. The computational advantage of our approach permits evaluating the manipulability of many poses simultaneously. Given a current robot pose z, our framework makes it possible to not only locally improve manipulability, but solve

$$\underset{\lambda}{\arg\max} \quad (\text{M} \circ \text{IK} \circ \text{TSP})(z, \lambda) \tag{69}$$

with a representative number of null space solution at a high resolution in real-time. Given the information of this *greedy optimization* strategy, the close-to-global optimum configuration can simply be picked. Solving for global optima in 0.25 ms at a resolution of 1° for λ enables application at typical robot sampling rates of 1 kHz.

Figure 10 shows manipulability of the full null space at a particular configuration.

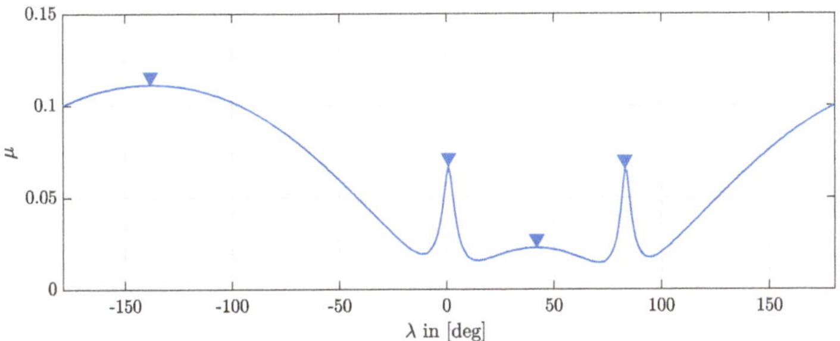

Figure 10. Multiple local optima of manipulability μ in the null space of $\boldsymbol{p}^{\text{red}} = [0.6, 0.7, 1.4, 0.7]^{\top}$.

This is an example of a pose with 4 local optima. If the current configuration of the robot is the solution for the given pose with the null space parameter $\lambda \in [0, 85]°$, a local optimization will only drive the redundancy resolution into a sub-optimal minima. In contrast, our approach allows finding the globally best configuration w.r.t. the admissible parameter space.

5.2.2. Optimizing Null Space Solution of Given End-Effector Trajectory

Several approaches can be found in the literature that maximize either the volume of a manipulability ellipsoid [31,47–49] or a predefined shape of the ellipsoid [33]. Yet all these approaches consider only local optimization.

Finding the best joint configuration for a given pose in task space simplifies to a 1D line search. However, given a full path in SE(3) it is also possible to find an optimal elbow trajectory that maximizes, e.g., the average manipulability while avoiding getting trapped in regions of poor manipulability. Note that a real manipulation task relies on a sophisticated path planner, capable of generating task-related paths that avoid obstacles while potentially fulfilling additional criteria. Knowledge about the task space manipulability, e.g., provided by our approach, may even be exploited by such a planner. This is, however, not the direct scope of this work. Instead, for a minimal working example, we use direct interpolation

$$\boldsymbol{p}(s) = s\,\boldsymbol{p}^{\text{start}} + (s - 1)\,\boldsymbol{p}^{\text{end}} \quad \text{with} \quad s = [0, 1] \tag{70}$$

between two poses as a simple path planner. Given are two random poses as depicted in Figure 11 to the left. On the right side of Figure 11, a contour plot of the manipulability of the full null space along the trajectory is shown. Red lines indicate not passable values in the null space due to joint limits, cf. Section 3.5. The blue line marks the trajectory that results from local optimization of manipulability. Note that at $s = 0.4$, the local optimization hits a joint limit of q_2. We stopped the line here, because it depends on a potential strategy for joint limit avoidance, which is not the scope of this work. A global optimization strategy that has predictive knowledge of the full null space development can exploit an initially sub-optimal path toward negative values of λ to circumvent the region of poor manipulability between $s = [0.6, 1]$. But this is usually not feasible in an online scenario with conventional global optimization strategies.

The computational advantage of our strategy, as seen in Figure 5, allows the computation of such a map with, e.g., a resolution of 100 steps in both parameters, s and λ, in under 5 ms. In combination with an online trajectory generator directly on SE(3), e.g., [50], this qualifies our task space manipulability approach to be used for predictive online manipulability optimization, e.g., with a receding horizon.

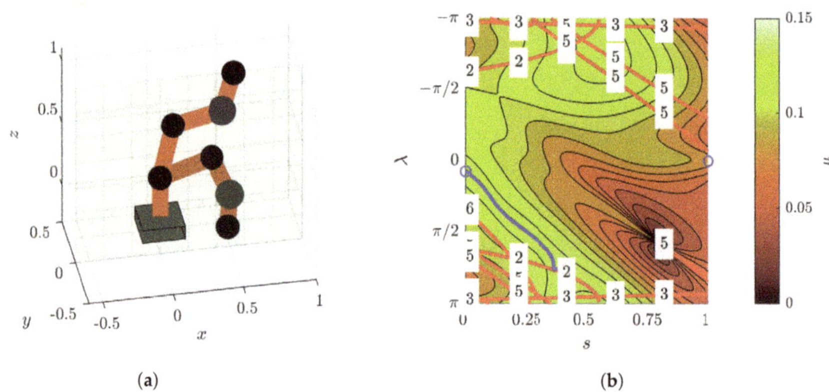

(a)   (b)

Figure 11. Null space manipulability over a parameter trajectory. The 3D plot (**a**) shows an exemplary start $p(s = 0)$ and end configuration $p(s = 1)$. The contour plot (**b**) shows the manipulability $\mu(p, \lambda)$ of the full null space. Red lines mark the limits $\lambda(p, q_{\max})$ of the admissible null space region. The numbers refer to the invoking joint. Blue circles mark desired $\lambda^d(s = 0)$ and $\lambda^d(s = 1)$, and the blue line marks a trajectory as it would be chosen by local optimization of $\lambda_{\max}(p)$.

6. Conclusions

Today's demand for adaptive and reactive robot behaviour requires sustaining the agility of a kinematic structure at all times. While manipulability is a common metric in robot research to quantify the capabilities of a robot at a given joint configuration, the robot task is directly defined in end-effector poses, which allows for multiple possible solutions. Unlike common metrics, which do not include the robot IK, a task space manipulability formulation is required to directly map an end-effector pose together with its null space solution onto the manipulability metric.

To achieve reactive robot behaviour, optimization of the null space at given poses must be performed online. In general, this requires efficient evaluation of a large number of configurations, especially in the case of redundant robots. In this work we developed a new closed-form approach for calculating manipulability directly from task space poses, for a redundant 7-DOF S-R-S serial robot kinematics. A novel parametrization of the task- and null space leads to concise IK, as well as admissible parameter mappings, which show symmetry in the structures of their individual expressions. Analysis of the resulting task space manipulability further revealed that the majority of end-effector poses do not have a unique, manipulability-maximizing null space solution. We thus argue that local optimization of the manipulability measure is not sufficient. A global optimization at high sampling frequencies, however, is not feasible with current approaches in the literature. The entire composition of the task space manipulability map proposed in this work allows for efficient array operations that can be exploited in vector-optimized programming languages, as well as GPU computing. Consequently, the simultaneous computation of a large number of poses in real-time is made possible. Our method, therefore, enables global online optimization of manipulability for single poses and even full SE(3) trajectories.

Future work will focus on further application development of our framework. Combining our task space manipulability approach with online planners opens an interesting field of predictive redundancy resolution for global manipulability optimization.

Author Contributions: conceptualization, G.H.; methodology, G.H.; software, G.H.; validation, G.H.; formal analysis, G.H.; investigation, G.H.; resources, D.W.; data curation, G.H.; writing–original draft preparation, G.H. and D.W.; writing–review and editing, G.H. and D.W.; visualization, G.H.; supervision, D.W.; project administration, D.W.; funding acquisition, D.W.,

Funding: The research leading to these results has received funding from the Horizon 2020 research and innovation programme under grant agreement №820742 of the project "HR-Recycler - Hybrid Human-Robot RECYcling plant for electriCal and eLEctRonic equipment".

Acknowledgments: The authors would like to thank Adam Horvath, Thomas Hartmann and Christian Ritter for discussions and initial implementations of the results.

Conflicts of Interest: The authors declare no conflict of interest. The funders had no role in the design of the study; in the collection, analyses, or interpretation of data; in the writing of the manuscript, or in the decision to publish the results.

Abbreviations

The following abbreviations are used in this manuscript:

CDF	cumulative distribution function
CLIK	Closed-Loop Inverse Kinematic
CPU	Central Processing Unit
DOF	degree of freedom
FK	Forward Kinematic
GPU	Graphics Processing Unit
IK	Inverse Kinematic
M	Manipulability
PDF	probability density function
PR	parameter requirements
SIMD	Single Instruction Multiple Data
S-R-S	Spherical-Revolute-Spherical
TCP	tool center point
TSP	Task Space Projection
TSS	Task Space Surjection

Appendix A. Manipulability

The full manipulability map M, discussed in Section 3.1, is given with $\mu(q) = \sqrt{\det\left({}_TJ\,{}_TJ^\top\right)}$. Note that the Jacobian matrices are formulated w.r.t. the tool frame T at the end-effector. The full symbolic expression for the determinant of the $\mathbb{R}^{6\times6}$ matrix results in the trigonometric polynomial

$$
\begin{aligned}
\mu(q)^2 := 2r_{SE}{}^2\,r_{EW}{}^2\left(\mathrm{c}\,(q_4)^2 - 1\right)\Bigg(& \\
+ r_{SE}{}^2\,\mathrm{c}\,(q_5)^2\,\mathrm{c}\,(q_6)^2\left(\mathrm{c}\,(q_2)^2 + \mathrm{c}\,(q_4)^2 - \mathrm{c}\,(q_2)^2\,\mathrm{c}\,(q_4)^2 - 1\right) & \\
+ r_{EW}{}^2\,\mathrm{c}\,(q_2)^2\,\mathrm{c}\,(q_3)^2\left(\mathrm{c}\,(q_4)^2 + \mathrm{c}\,(q_6)^2 - \mathrm{c}\,(q_4)^2\,\mathrm{c}\,(q_6)^2 - 1\right) & \\
+ \left(r_{SE}{}^2 + 2\,r_{SE}\,r_{EW}\,\mathrm{c}\,(q_4)\right)\left(\mathrm{c}\,(q_2)^2 + \mathrm{c}\,(q_6)^2 - \mathrm{c}\,(q_2)^2\,\mathrm{c}\,(q_6)^2 - 1\right) & \\
+ \left(r_{SE}{}^2\,\mathrm{s}\,(q_4)\,\mathrm{s}\,(q_6)\,\mathrm{c}\,(q_4)\,\mathrm{c}\,(q_5)\,\mathrm{c}\,(q_6) + r_{SE}\,r_{EW}\,\mathrm{s}\,(q_4)\,\mathrm{s}\,(q_6)\,\mathrm{c}\,(q_5)\,\mathrm{c}\,(q_6)\right)\left(1 - \mathrm{c}\,(q_2)^2\right) & \\
+ \Bigg(\left(r_{SE}\,r_{EW} + r_{EW}{}^2\,\mathrm{c}\,(q_4)\right)\mathrm{s}\,(q_2)\,\mathrm{s}\,(q_4)\,\mathrm{c}\,(q_2)\,\mathrm{c}\,(q_3) + r_{EW}{}^2\,\mathrm{c}\,(q_2)^2 - r_{EW}{}^2\Bigg)\left(1 - \mathrm{c}\,(q_6)^2\right) & \\
\Bigg). \quad (A1)
\end{aligned}
$$

Note that the manipulability measure μ does not depend on joints q_1 nor q_7. Further, the link lengths r_{BS} and r_{WT} do not affect manipulability.

Appendix B. Inverse Kinematic Functions from (43)

$$q_1(p,\lambda) := \gamma_{ref} + \text{atan2}\left(-s(\lambda)\,s(\theta_S),\, s(\beta_{ref})\,c(\theta_S) - c(\beta_{ref})\,c(\lambda)\,s(\theta_S)\right) \tag{A2}$$

$$q_2(p,\lambda) := \text{acos}\left(c(\beta_{ref})\,c(\theta_S) + c(\lambda)\,s(\beta_{ref})\,s(\theta_S)\right) \tag{A3}$$

$$q_3(p,\lambda) := \text{atan2}\left(s(\beta_{ref})\,s(\lambda),\, c(\lambda)\,s(\beta_{ref})\,c(\theta_S) - c(\beta_{ref})\,s(\theta_S)\right) \tag{A4}$$

$$q_4(p,\lambda) := \pi - \text{acos}\left(\frac{\frac{r_{EW}^2}{2} + \frac{r_{SE}^2}{2} - \frac{r_{ref}^2}{2}}{r_{EW}\,r_{SE}}\right) \tag{A5}$$

$$q_5(p,\lambda) := \text{atan2}\left(s(\gamma_{EE}-\lambda)\,s(\beta_{EE}),\, s(\beta_{EE})\,c(\theta_W)\,c(\gamma_{EE}-\lambda) - c(\beta_{EE})\,s(\theta_W)\right) \tag{A6}$$

$$q_6(p,\lambda) := \text{acos}\left(c(\beta_{EE})\,c(\theta_W) + s(\beta_{EE})\,s(\theta_W)\,c(\gamma_{EE}-\lambda)\right) \tag{A7}$$

$$q_7(p,\lambda) := \psi_{EE} + \text{atan2}\left(-s(\gamma_{EE}-\lambda)\,s(\theta_W),\, s(\beta_{EE})\,c(\theta_W) - c(\beta_{EE})\,s(\theta_W)\,c(\gamma_{EE}-\lambda)\right) \tag{A8}$$

Appendix C. Absolute Valued Inverse Kinematics Functions from (55a)

$$|q_1(p,\lambda)| := \gamma_{ref} + \text{acos}\left(\frac{s(\beta_{ref})\,c(\theta_S) - c(\beta_{ref})\,c(\lambda)\,s(\theta_S)}{\sqrt{1 - (c(\beta_{ref})\,c(\theta_S) + c(\lambda)\,s(\beta_{ref})\,s(\theta_S))^2}}\right) \tag{A9}$$

$$|q_2(p,\lambda)| := \text{acos}\left(c(\beta_{ref})\,c(\theta_S) + c(\lambda)\,s(\beta_{ref})\,s(\theta_S)\right) \tag{A10}$$

$$|q_3(p,\lambda)| := \pi - \text{acos}\left(\frac{c(\beta_{ref})\,s(\theta_S) - c(\lambda)\,s(\beta_{ref})\,c(\theta_S)}{\sqrt{1 - (c(\beta_{ref})\,c(\theta_S) + c(\lambda)\,s(\beta_{ref})\,s(\theta_S))^2}}\right) \tag{A11}$$

$$|q_4(p,\lambda)| := \pi - \text{acos}\left(\frac{\frac{r_{EW}^2}{2} + \frac{r_{SE}^2}{2} - \frac{r_{ref}^2}{2}}{r_{EW}\,r_{SE}}\right) \tag{A12}$$

$$|q_5(p,\lambda)| := \pi - \text{acos}\left(\frac{c(\beta_{EE})\,s(\theta_W) - s(\beta_{EE})\,c(\theta_W)\,c(\gamma_{EE}-\lambda)}{\sqrt{1 - (c(\beta_{EE})\,c(\theta_W) + s(\beta_{EE})\,s(\theta_W)\,c(\gamma_{EE}-\lambda))^2}}\right) \tag{A13}$$

$$|q_6(p,\lambda)| := \text{acos}\left(c(\beta_{EE})\,c(\theta_W) + s(\beta_{EE})\,s(\theta_W)\,c(\gamma_{EE}-\lambda)\right) \tag{A14}$$

$$|q_7(p,\lambda)| := \psi_{EE} + \text{acos}\left(\frac{s(\beta_{EE})\,c(\theta_W) - c(\beta_{EE})\,s(\theta_W)\,c(\gamma_{EE}-la)}{\sqrt{1 - (c(\beta_{EE})\,c(\theta_W) + s(\beta_{EE})\,s(\theta_W)\,c(\gamma_{EE}-la))^2}}\right) \tag{A15}$$

Appendix D. Admissible Null Space Parameter Functions from (56)

$$\lambda_1^{\lim}(\theta_S, \gamma_{\text{ref}}, \beta_{\text{ref}}, q^{\max}) :=$$

$$\left\{ \begin{array}{l} \pm\left(\pi - \text{acos}\left(\dfrac{\sqrt{s(\theta_S)^2 - s(\gamma_{\text{ref}} - q^{\max})^2\, s(\beta_{\text{ref}})^2} + c(\beta_{\text{ref}})\, s(\beta_{\text{ref}})\, c(\theta_S)\, s(\theta_S)\left(1 - c(\gamma_{\text{ref}} - q^{\max})^2\right)}{|s(\theta_S)|\,|c(\gamma_{\text{ref}} - q^{\max})|\, s(\theta_S)^2\left(s(\gamma_{\text{ref}} - q^{\max})^2\, s(\beta_{\text{ref}})^2 - 1\right)}\right)\right) \\[2em] \pm\left(\text{acos}\left(\dfrac{\sqrt{s(\theta_S)^2 - s(\gamma_{\text{ref}} - q^{\max})^2\, s(\beta_{\text{ref}})^2} + c(\beta_{\text{ref}})\, s(\beta_{\text{ref}})\, c(\theta_S)\, s(\theta_S)\left(c(\gamma_{\text{ref}} - q^{\max})^2 - 1\right)}{|s(\theta_S)|\,|c(\gamma_{\text{ref}} - q^{\max})|\, s(\theta_S)^2\left(s(\gamma_{\text{ref}} - q^{\max})^2\, s(\beta_{\text{ref}})^2 - 1\right)}\right)\right) \end{array}\right\} \tag{A16}$$

$$\lambda_2^{\lim}(\theta_S, \beta_{\text{ref}}, q^{\max}) := \left\{ \pm\,\text{acos}\left(\frac{c(q^{\max}) - c(\beta_{\text{ref}})\, c(\theta_S)}{s(\beta_{\text{ref}})\, s(\theta_S)}\right)\right\} \tag{A17}$$

$$\lambda_3^{\lim}(\theta_S, \beta_{\text{ref}}, q^{\max}) :=$$

$$\left\{ \begin{array}{l} \pm\left(\pi - \text{acos}\left(\dfrac{\sqrt{-c(\beta_{\text{ref}})^2 - c(q^{\max})^2\, c(\theta_S)^2 + c(q^{\max})^2 + c(\theta_S)^2} + c(\beta_{\text{ref}})\, c(\theta_S)\, s(\theta_S)\left(1 - c(q^{\max})^2\right)}{|c(q^{\max})|\, s(\beta_{\text{ref}})\left(s(q^{\max})^2\, s(\theta_S)^2 - 1\right)}\right)\right) \\[2em] \pm\left(\text{acos}\left(\dfrac{\sqrt{-c(\beta_{\text{ref}})^2 - c(q^{\max})^2\, c(\theta_S)^2 + c(q^{\max})^2 + c(\theta_S)^2} + c(\beta_{\text{ref}})\, c(\theta_S)\, s(\theta_S)\left(c(q^{\max})^2 - 1\right)}{|c(q^{\max})|\, s(\beta_{\text{ref}})\left(s(q^{\max})^2\, s(\theta_S)^2 - 1\right)}\right)\right) \end{array}\right\} \tag{A18}$$

$$\lambda_5^{\lim}(\theta_W, \beta_{\text{EE}}, q^{\max}) := \gamma_{\text{EE}} - \lambda_3^{\lim}(\theta_W, \beta_{\text{EE}}, q^{\max}) \tag{A19}$$

$$\lambda_6^{\lim}(\theta_W, \beta_{\text{EE}}, q^{\max}) := \gamma_{\text{EE}} - \lambda_2^{\lim}(\theta_W, \beta_{\text{EE}}, q^{\max}) \tag{A20}$$

$$\lambda_7^{\lim}(\theta_W, \gamma_{\text{EE}}, \beta_{\text{EE}}, q^{\max}) := \gamma_{\text{EE}} - \lambda_1^{\lim}(\theta_W, \gamma_{\text{EE}}, \beta_{\text{EE}}, q^{\max}). \tag{A21}$$

References

1. Yoshikawa, T. Manipulability of robotic mechanisms. *Int. J. Rob. Res.* **1985**, *4*, 3–9. [CrossRef]
2. Yoshikawa, T. Dynamic manipulability of robot manipulators. In Proceedings of the 1985 IEEE International Conference on Robotics and Automation, St. Louis, MO, USA, 25–28 March 1985; pp. 1033–1038.
3. Chiacchio, P.; Chiaverini, S.; Sciavicco, L.; Siciliano, B. Influence of gravity on the manipulability ellipsoid for robot arms. *J. Dyn. Syst. Meas. Contr.* **1992**, *114*, 723–727. [CrossRef]
4. Chiacchio, P. A new dynamic manipulability ellipsoid for redundant manipulators. *Robotica* **2000**, *18*, 381–387. [CrossRef]
5. Vahrenkamp, N.; Asfour, T.; Metta, G.; Sandini, G.; Dillmann, R. Manipulability analysis. In Proceedings of the 12th IEEE-RAS International Conference on Humanoid Robots (Humanoids 2012), Osaka, Japan, 29 November–1 December 2012; pp. 568–573.
6. Waldron, K.J.; Schmiedeler, J. Kinematics. In *Springer Handbook of Robotics*, 2nd ed.; Siciliano, B., Khatib, O., Eds.; Springer: Cham, Switzerland, 2016; pp. 11–36.
7. Jun, B.H.; Lee, P.M.; Kim, S. Manipulability analysis of underwater robotic arms on ROV and application to task-oriented joint configuration. *J. Mech. Sci. Technol.* **2008**, *22*, 887–894. [CrossRef]
8. Abdel-Malek, K.; Yu, W.; Yang, J. Placement of Robot Manipulators to Maximize Dexterity. *Int. J. Rob. Autom.* **2004**, *19*, 6–14. [CrossRef]
9. Wolovich, W.A.; Elliott, H. A computational technique for inverse kinematics. In Proceedings of the 23rd IEEE Conference on Decision and Control, Las Vegas, NV, USA, 12–14 December 1984; pp. 1359–1363.
10. Colomé, A.; Torras, C. Closed-loop inverse kinematics for redundant robots: Comparative assessment and two enhancements. *IEEE/ASME Trans. Mechatron.* **2014**, *20*, 944–955. [CrossRef]
11. Antonelli, G. Stability analysis for prioritized closed-loop inverse kinematic algorithms for redundant robotic systems. *IEEE Trans. Rob.* **2009**, *25*, 985–994. [CrossRef]
12. Bjoerlykhaug, E. A Closed Loop Inverse Kinematics Solver Intended for Offline Calculation Optimized with GA. *Robotics* **2018**, *7*, 7. [CrossRef]

13. Reiter, A.; Müller, A.; Gattringer, H. On Higher Order Inverse Kinematics Methods in Time-Optimal Trajectory Planning for Kinematically Redundant Manipulators. *IEEE Trans. Ind. Inf.* **2018**, *14*, 1681–1690. [CrossRef]

14. Siciliano, B. Kinematic Control of Redundant Robot Manipulators: A Tutorial. *J. Intell. Rob. Syst.* **1990**, *3*, 201–212. [CrossRef]

15. Shimizu, M.; Kakuya, H.; Yoon, W.K.; Kitagaki, K.; Kosuge, K. Analytical inverse kinematic computation for 7-DOF redundant manipulators with joint limits and its application to redundancy resolution. *IEEE Trans. Rob.* **2008**, *24*, 1131–1142. [CrossRef]

16. Faria, C.; Ferreira, F.; Erlhagen, W.; Monteiro, S.; Bicho, E. Position-based kinematics for 7-DoF serial manipulators with global configuration control, joint limit and singularity avoidance. *Mech. Mach. Theory* **2018**, *121*, 317–334. [CrossRef]

17. D'Souza, A.; Vijayakumar, S.; Schaal, S. Learning inverse kinematics. In Proceedings of the 2001 IEEE/RSJ International Conference on Intelligent Robots and Systems (IROS), Maui, HI, USA, 29 October–3 November 2001; Volume 1; pp. 298–303.

18. Tejomurtula, S.; Kak, S. Inverse kinematics in robotics using neural networks. *Inf. Sci.* **1999**, *116*, 147–164. [CrossRef]

19. Köker, R.; Öz, C.; Çakar, T.; Ekiz, H. A study of neural network based inverse kinematics solution for a three-joint robot. *Rob. Autom. Syst.* **2004**, *49*, 227–234. [CrossRef]

20. Sariyildiz, E.; Ucak, K.; Oke, G.; Temeltas, H.; Ohnishi, K. Support Vector Regression based inverse kinematic modeling for a 7-DOF redundant robot arm. In Proceedings of the 2012 International Symposium on Innovations in Intelligent Systems and Applications, Trabzon, Turkey, 2–4 July 2012; pp. 1–5.

21. Parker, J.K.; Khoogar, A.R.; Goldberg, D.E. Inverse kinematics of redundant robots using genetic algorithms. In Proceedings of the 1989 International Conference on Robotics and Automation (ICRA), Scottsdale, AZ, USA, 14–19 May 1989; pp. 271–276.

22. KöKer, R. A genetic algorithm approach to a neural-network-based inverse kinematics solution of robotic manipulators based on error minimization. *Inf. Sci.* **2013**, *222*, 528–543. [CrossRef]

23. Dereli, S.; Köker, R. A meta-heuristic proposal for inverse kinematics solution of 7-DOF serial robotic manipulator: quantum behaved particle swarm algorithm. *Artif. Intell. Rev.* **2019**, 1–16. [CrossRef]

24. Kamrani, B.; Berbyuk, V.; Wäppling, D.; Feng, X.; Andersson, H. Optimal usage of robot manipulators. In *Robot Manipulators*; IntechOpen: London, UK, 2010.

25. Myers, R.H.; Montgomery, D.C.; Anderson-Cook, C.M. *Response Surface Methodology: Process and Product Optimization Using Designed Experiments*, 4th ed.; John Wiley & Sons: Hoboken, NJ, USA, 2016.

26. Chan, T.F.; Dubey, R.V. A weighted least-norm solution based scheme for avoiding joint limits for redundant joint manipulators. *Trans. Rob. Autom.* **1995**, *11*, 286–292. [CrossRef]

27. Dariush, B.; Hammam, G.B.; Orin, D. Constrained resolved acceleration control for humanoids. In Proceedings of the 2010 IEEE/RSJ International Conference on Intelligent Robots and Systems (IROS), Taiwan, 18–22 October 2010; pp. 710–717.

28. Dufour, K.; Suleiman, W. On integrating manipulability index into inverse kinematics solver. In Proceedings of the 2017 IEEE/RSJ International Conference on Intelligent Robots and Systems (IROS), Vancouver, BC, Canada, 24–28 September 2017; pp. 6967–6972.

29. Jin, L.; Li, S.; La, H.M.; Luo, X. Manipulability optimization of redundant manipulators using dynamic neural networks. *IEEE Trans. Ind. Electron.* **2017**, *64*, 4710–4720. [CrossRef]

30. Lee, S. Dual redundant arm configuration optimization with task-oriented dual arm manipulability. *Trans. Rob. Autom. 1* **1989**, *5*, 78–97. [CrossRef]

31. Guilamo, L.; Kuffner, J.; Nishiwaki, K.; Kagami, S. Manipulability optimization for trajectory generation. In Proceedings of the IEEE International Conference on Robotics and Automation (ICRA), Orlando, FL, USA, 15–19 May 2006; pp. 2017–2022.

32. Rozo, L.; Jaquier, N.; Calinon, S.; Caldwell, D.G. Learning manipulability ellipsoids for task compatibility in robot manipulation. In Proceedings of the 2017 IEEE/RSJ International Conference on Intelligent Robots and Systems (IROS), Vancouver, BC, Canada, 24–28 September 2017; pp. 3183–3189.

33. Jaquier, N.; Rozo, L.D.; Caldwell, D.G.; Calinon, S. Geometry-aware Tracking of Manipulability Ellipsoids. In Proceedings of the Proceedings Robotics: Science and Systems, Pittsburgh, PA, USA, 26–30 June 2018.

34. Faroni, M.; Beschi, M.; Visioli, A.; Tosatti, L.M. A global approach to manipulability optimisation for a dual-arm manipulator. In Proceedings of the 2016 IEEE 21st International Conference on Emerging Technologies and Factory Automation (ETFA), Berlin, Germany, 6–9 September 2016; pp. 1–6.
35. Zacharias, F.; Borst, C.; Wolf, S.; Hirzinger, G. The capability map: A tool to analyze robot arm workspaces. *Int. J. Humanoid Rob.* **2013**, *10*, 1350031. [CrossRef]
36. Zlatanov, D.; Fenton, R.G.; Benhabib, B. Singularity analysis of mechanisms and manipulators via a velocity-equation model of the instantaneous kinematics. In Proceedings of the International Conference on Robotics and Automation (ICRA), San Diego, CA, USA, 8–13 May 1994; Volume 2, pp. 980–985
37. Staffetti, E.; Bruyninckx, H.; De Schutter, J. On the invariance of manipulability indices. In *Advances in Robot Kinematics*; Lenarčič, J., Thomas, F., Eds.; Springer: Dordrecht, The Netherlands, 2002; pp. 57–66.
38. Gotlih, K.; Troch, I. Base invariance of the manipulability index. *Robotica* **2004**, *22*, 455–462. [CrossRef]
39. Lee, S.; Bejczy, A.K. Redundant arm kinematic control based on parameterization. In Proceedings of the IEEE International Conference on Robotics and Automation (ICRA), Sacramento, CA, USA, 9–11 April 1991; pp. 458–465.
40. Tondu, B. A closed-form inverse kinematic modelling of a 7R anthropomorphic upper limb based on a joint parametrization. In Proceedings of the 2006 6th IEEE-RAS International Conference on Humanoid Robots, Genova, Italy, 4–6 December 2006; pp. 390–397.
41. Kreutz-Delgado, K.; Long, M.; Seraji, H. Kinematic analysis of 7-DOF manipulatorsiter. *Int. J. Rob. Res.* **1992**, *11*, 469–481. [CrossRef]
42. Goldenberg, A.; Benhabib, B.; Fenton, R. A complete generalized solution to the inverse kinematics of robots. *J. Rob. Autom.* **1985**, *1*, 14–20. [CrossRef]
43. Kuffner, J.J. Effective sampling and distance metrics for 3D rigid body path planning. In Proceedings of the IEEE International Conference on Robotics and Automation (ICRA), New Orleans, LA, USA, 26 April–1 May 2004; pp. 3993–3998.
44. Kuhlemann, I.; Jauer, P.; Ernst, F.; Schweikard, A. Robots with seven degrees of freedom: Is the additional DoF worth it? In Proceedings of the 2016 2nd International Conference on Control, Automation and Robotics (ICCAR), Hong Kong, China, 28–30 April 2016; pp. 80–84.
45. Dixon, L.C.W. The Global Optimization Problem. An Introduction. *Toward Global Optim.* **1978**, *2*, 1–15.
46. Fubini, G. Sugli integrali multipli. *Rend. Acc. Naz. Lincei* **1907**, *16*, 608–614.
47. Chiacchio, P. Exploiting redundancy in minimum-time path following robot control. In Proceedings of the 1990 American Control Conference, San Diego, CA, USA, 23–25 May 1990; pp. 2313–2318.
48. Chiu, S. Control of redundant manipulators for task compatibility. In Proceedings of the 1987 IEEE International Conference on Robotics and Automation (ICRA). Raleigh, NC, USA, 31 March–3 April 1987; Volume 4; pp. 1718–1724.
49. Somani, N.; Rickert, M.; Gaschler, A.; Cai, C.; Perzylo, A.; Knoll, A. Task level robot programming using prioritized non-linear inequality constraints. In Proceedings of the 2016 IEEE/RSJ International Conference on Intelligent Robots and Systems (IROS), Daejeon, Korea, 9–14 October 2016; pp. 430–437.
50. Huber, G.; Wollherr, D. An Online Trajectory Generator on SE(3) for Human Robot Collaboration. *Robotica* **2019**, accepted, doi:10.1017/S0263574719001619. [CrossRef]

 *robotics*

Article

Kineto-Elasto-Static Design of Underactuated Chopstick-Type Gripper Mechanism for Meal-Assistance Robot

Tomohiro Oka [1,*], Jorge Solis [2], Ann-Louise Lindborg [3,4], Daisuke Matsuura [1], Yusuke Sugahara [1] and Yukio Takeda [1]

[1] Department of Mechanical Engineering, Tokyo Institute of Technology, Tokyo 152-8552, Japan; matsuura.d.aa@m.titech.ac.jp (D.M.); sugahara.y.ab@m.titech.ac.jp (Y.S.); takeda.y.aa@m.titech.ac.jp (Y.T.)

[2] Department of Engineering and Physics, Karlstad University, 651 88 Karlstad, Sweden; jorge.solis@kau.se

[3] Camanio Care AB, 131 30 Nacka, Sweden; annlouise.lindborg@camanio.com or ann-louise.lindborg@mdh.se

[4] School of Innovation, Design and Engineering, Mälardalen University, 721 23 Västerås, Sweden

* Correspondence: tomohiro.oka.822@gmail.com; Tel.: +81-3-5734-2177

Received: 28 May 2020; Accepted: 28 June 2020; Published: 30 June 2020

Abstract: Our research aims at developing a meal-assistance robot with vision system and multi-gripper that enables frail elderly to live more independently. This paper presents a development of a chopstick-type gripper for a meal-assistance robot, which is capable of adapting its shape and contact force with the target food according to the size and the stiffness. By solely using position control of the driving motor, the above feature is enabled without relying on force sensors. The gripper was designed based on the concept of planar 2-DOF under-actuated mechanism composed of a pair of four-bar chains having a torsion spring at one of the passive joints. To clarify the gripping motion and relationship among the contact force, food's size and stiffness, and gripping position, kineto-elasto-static analysis of the mechanism was carried out. It was found from the result of the analysis that the mechanism was able to change its gripping force according to the contact position with the target object, and this mechanical characteristic was utilized in its grasp planning in which the position for the gripping the object was determined to realize a simple control system, and sensitivity of the contact force due to the error of the stiffness value was revealed. Using a three-dimensional (3D) printed prototype, an experiment to measure the gripping force by changing the contact position was conducted to validate the mechanism feature that can change its gripping force according to the size and the stiffness and the contact force from the analysis results. Finally, the gripper prototype was implemented to a 6-DOF robotic arm and an experiment to grasp real food was carried out to demonstrate the feasibility of the proposed grasp planning.

Keywords: mechanism design; meal-assistance robot; chopstick-type gripper; under-actuated mechanism; kineto-elasto-static analysis; grasp planning

---

1. Introduction

WHO (World Health Organization), or rather the WHOQOL (World Health Organization Quality of Life) group, has developed a test instrument to evaluate the quality of life, as a complement to a standard health assessment. The instrument had questions within the following domains: physical, psychological, level of independence, social relationships, environment, and spirituality/religion/personal beliefs. Within the domain "level of independence" WHO has the following facets: mobility, activities of daily living, dependence on medicinal substances and medical aids, dependence on non-medical

Robotics 2020, 9, 50; doi:10.3390/robotics9030050

substances (alcohol, tobacco, drugs), communication capacity, and work capability [1]. The independency and possibility to manage activities of daily living are taken into account as one of the important abilities.

Undernutrition is a serious problem amongst the elderly. For example, elderly people in Scandinavia have a high risk of malnutrition and eating difficulty is a risk factor for it according to Nyberg et al. [2]. Japan and Sweden share the same problem with the demographical development, and both countries are robot and technical-friendly. This gives a common goal to manage to take care of the elderly.

Up to now, several robotic systems for eating have been introduced. Handy 1 is an assisting robotic system, which is composed of a robotic arm that works with different trays, depending on the task. For eating, it has a tray with different compartments, where different kinds of food can be placed and the user chooses what compartment to take from. It was also designed to assist other activities than eating, such as drinking, and make-up application, which requires different kinds of detachable slide-on tray sections and end effectors [3]. Developments of robotic assistive eating devices with chopstick-type gripper have been done in Japan [4]. The mechanism in [4] is based on an industrial robot-like arm. The meal assistance robot "My Spoon" is composed of a manipulator arm with 5-DOFs and an end-effector that is controlled by a joystick. As the end effector has a spoon and a fork that together work as a pincer, it picks up food between the spoon and fork and lifts it up to the mouth and, when the user touches the spoon, the fork folds back to release the bite. The food is put into a box with four compartments [5,6]. The "Neater Eater" started as a 2-DOF arm robot with a spoon as the end-effector, which is moved by the user with a damping mechanism that absorbs tremor [7]. Now that the Neater Eater robotic V6 has been developed [8]. It takes the food by turning the plate and scrape towards the brim and can be controlled by a touch screen. The "Meal Buddy" has a 3-DOF robotic arm and a spoon as the end-effector, and it has three bowls for the food that are mounted on a board using magnets. It is possible to choose different ways of control devices [9]. "Obi" is one of the newest eating devices on the market. Obi's base shape has similarities to a drop and the arm is situated on the right side of the plate that has four compartments/bowls to put the food in. It is white and has 6-DOF. It is controlled by the user with two buttons, one for choosing which one of the four compartments with food to take from, and the other button for starting the grasping of food and lifting to the mouth. The LARM [10] clutched arm is driven by a single actuator from which the motion is transmitted to its joints with the help of gears and electromagnetic clutches. The arm has a parallelogram-based mechanism for the limb part, which drives the upper arm and forearm from the shoulder. Even though the kinematic design has been considerably simplified, the system can be only used for eating and the end-effector cannot adapt to the stiffness of the grasped food. An assistive robot for self-feeding that is capable of handling Korean food, including sticky rice, has been introduced in [11]. It is composed of a dual-arm manipulator with a total of 6-DOF (without the gripper). The first robotic arm (spoon-arm) uses a spoon to transfer the food from a container on a table to the user's mouth. The second robotic arm picks food up from a container and then puts it on the spoon of a spoon-arm. The level of cognitive load to the user increases and two different tools are required while eating due to the use of the dual-arm manipulation.

In most cases mentioned above, meal assistant robots/systems are basically composed of a robot arm to move the gripping tool of foods, such as spoon and chopsticks. From the point of view of real use of meal assistant robot for the promotion of independent life of elderly, it is very important to consider safety, how it looks, sounds, and so on, as well as the complexity of the mechanical design, the total cost of the system and power consumption.

Based on the background and discussions mentioned above, we have undertaken development/research in order to promote an independent life of elderly people by a multi grip tool to facilitate eating. To raise the standard of living of the elderly from the viewpoint of eating habits, Japan–Swedish industry-academia collaboration program began in 2017 [12]. In this project, we used the eating aid Bestic. The initiative to Bestic came from Sten Hemmingsson, who had the need for the product himself and who wanted to be able to continue eating independently. The product development is described in the report "Bestic An eating-aid for persons with little or no ability to move their arms" [13]. The product has continued developing with the influence of users [14]. Our research team aimed to expand its function by attaching a gripper instead of a spoon to Bestic developed at Camanio Care AB in Sweden.

We aimed to enhance the functionalities of Bestic in order to enable frail elderly to have an independent meal experience and reduce caregiver's burden. For this purpose, in this research, we aimed to further develop the current commercial version of Bestic for targeting current Bestics' users as well as frail and dependent elderly. Therefore, we focused on developing different kinds de-attachable multigrip tools that can passively adapt to the stiffness of the grasped object as well as integrating the vision system for enhancing the usability of the proposed system (e.g., vision-based control). In particular, in this research, we aimed to adapt the eating device Bestic to Japanese eating customs, and we developed a chopstick-type gripper that allows users to properly eat Japanese food as well as to perform other daily life activities.

Regarding multigrip tools, chopsticks are widely used because of its usefulness in picking up many kinds of foods and its very simple composition. Although using chopsticks looks like needing dexterous operation by human's fingers, since it can be used to do various manipulation of foods, such as picking up, cutting, sticking, and so on; it can actually be used easily once the purpose of use has been adequately narrowed. In this paper, we rather focus on the picking up operation of chopsticks and propose a reasonable design of a gripper mechanism that employs an under actuation principle to make the gripper adaptable against various size and stiffness of foods by solely using position control and without introducing force sensors. "Solely use" means that the gripper only has "open" or "close" states, and a control system only gives maximum or minimum positions to the actuator corresponding to each state. In order to calculate the configuration of the mechanism against the given property of the target food and the actuator input, and to figure out a suitable design of the gripper and grasp planning which derives the motor input angle and the contact position with the target object, in order to realize a simple control system, Kineto-elasto-static analysis was performed in this paper. Also, an effect of the stiffness error on the gripping force is investigated. An experiment using the prototype to measure the contact force was conducted to validate the modeling of the mechanism and the feature of the mechanism obtained by the kineto-elasto-static analysis. Finally, the gripper prototype was implemented in a 6-DOF robotic arm and the grasp planning experiment was carried out to demonstrate the feasibility of the proposed grasp planning.

2. Meal-Assistance Robot and Multi-Gripper

2.1. Composition of the Meal-Assistance Robot with Vision System and Multi-Gripper

The proposed system is composed of Bestic arm, a mini PC (NUC6i5SYK), a camera (Intel Real Sense SR300), an articulated arm with camera attachment (with passive joints), and different kinds of de-attachable multigrip tools, different kinds of de-attachable multigrip tools (e.g., spoon, chopstick-type, etc.), as shown in Figure 1. A vision-based feature extraction algorithm for estimating the location of the food on the plate has been developed by extracting the color feature [15]. After detecting the location of the object to be grasped, by means of a proposed distributed vision-based control system, the coordinates (x and y) are then transmitted to the control system of Bestic [15].

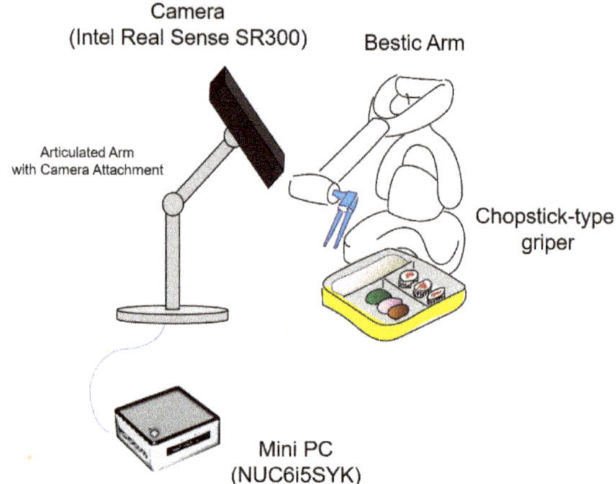

Figure 1. Proposed robot system.

2.2. Requirement to the Multi-Gripper

In our research, we consider the following major constraints to the multigrip tools in order to adapt the new functionalities to the present Bestic:

1. It shall only be powered by the motor that normally drives the spoon.
2. It shall be possible to grasp Japanese foods that are hard to manipulate with a spoon.
3. It shall be easy to clean.
4. It shall both be possible to pick up food on a plate and then release it in the mouth of a user.
5. It shall be possible to be taken off for replacement to different tools like a toothbrush for example.

In the following part of the paper, we will focus on the design and grasp planning of the chopstick-type gripper while taking into account these constraints explained above.

2.3. Design Requirements to the Chopstick-Type Gripper

The gripper for the meal-assistance robot is required to adjust its gripping force according to foods. Each food has its size, mass, and stiffness, which is a problem for the gripper to adjust the gripping force. Using a force sensor for controlling the gripping force leads to high cost and complexity of the system, which makes the robot unaffordable for users. In addition, it seems hard to prepare a precise database of physical property of foods for the control of the gripping force, because the number of food types is enormous, even if only Japanese food is considered. Thus, a chopstick-type gripper that enables adjusting the gripping force according to food's size and stiffness without any force sensors is required. In this paper, a mechanism that is based on the concept of under-actuation is proposed and analyzed to achieve passively adjust its gripping force while using an elastic element, and grasp planning is investigated.

3. Chopstick-Type Gripper Mechanism

Figure 2 shows the proposed mechanism. It is a planar 2-DOF mechanism composed of two four-bar closed loops, ABCD (Loop 1) and BEFC (Loop 2). The degree of freedom of this mechanism is calculated while using Gruebler's equation described as Equation (1).

$$F = 3(N - J - 1) + \sum_{i=1}^{J} f_i$$

$$= 3(7 - 8 - 1) + 8 = 2$$

(1)

where, N is the number of links, and J is the number of joints, and $f_i(i = 1 \sim J)$ is the degree of the freedom of the joint. In this case, all of joints are revolute pairs, thus $f_i = 1(i = 1 \sim J)$.

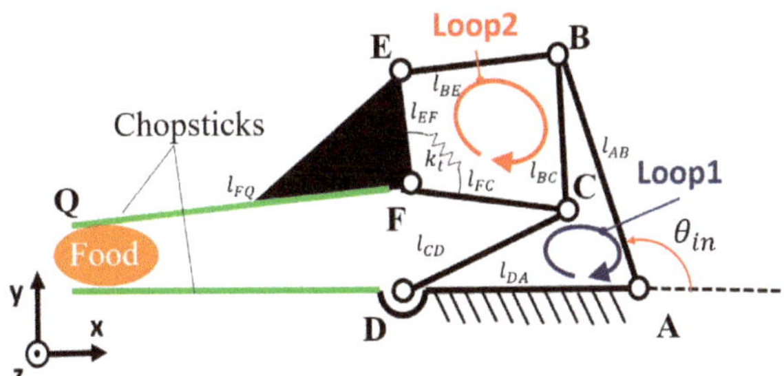

Figure 2. Proposed mechanism of the gripper.

A pair of chopsticks is comprised of the base link AD with an extension and the coupler link FQ. All joints A to F are revolute-type, and only one of them, A, is actuated. Thus, te entire mechanism is under-actuated [16–19], but, since a torsion spring is attached at joint F, connecting the links FC and FE, its configuration is statically determined once the actuator's input angle and external force on the end-effector (gripping force between chopsticks) are specified. When the chopsticks are not in contact with the food and no external force is applied, joint F is considered to remain at a neutral angle determined by the torsion spring, and the mechanism's configuration is thus only determined by the input angle of the actuator. In contrast, when the chopsticks contact with the food, the shape of Loop 2 changes according to the deformation of the torsion spring due to a deterministic motion of Loop 1. Thus, Loop 1 generates the path of the tip of the chopsticks part, and Loop 2 regulates the gripping force with an object to be grasped. Therefore, the gripping force is adjustable by the position control of the driving motor, and each of the loops can be separately designed. Additionally, the proposed mechanism has a feature to grasp an object at the tip part and adjust its force with deformation of the torsion spring, while the other underactuated mechanisms, such as [16–19], were designed to wrap around an object to grasp it.

4. Kineto-Elasto-Static Analysis of the Mechanism

This section addresses the kineto-elasto-static analysis of the proposed gripper mechanism. For the analysis, the model includes the contact point with the object to be grasped, and the object is modeled as a simple compression spring. The solution of the analysis is obtained from the condition that satisfies both the geometrical and mechanical relationship of the system. A simple computational scheme is needed in order to reduce the calculation time. In our previous paper [20], we presented an analysis scheme where L and ϕ are estimated from an assumed w and the assumed w is iteratively adjusted until the contact force is converged. However, it suffered a high computational cost. In the present paper, a new analysis scheme that needs less calculation cost is introduced, and the calculation performance is compared with our previous analysis scheme. Using the analysis, the relationship among the contact force of the gripper, the size and the stiffness of food, and the contact position is obtained to propose a grasp planning scheme.

4.1. Modeling of the Mechanism Including the Contact point

Under the assumption that food to be picked up is an elastic body, a kineto-elasto-static model, including the chopstick mechanism and food, is set as shown in Figure 3. In order to take the relative motion between the food and the chopsticks into account, a prismatic pair and a revolute pair are set at the contact point between the food and the end-effector of the mechanism on the link FQ, and the joint is set as P. A distance between joint F and joint P, and an angle of FP against the horizontal (x) axis are denoted as L and ϕ, respectively, and these parameters are set as unknown. f stands for the contact force. Its direction is considered to be normal to FQ, and the magnitude is denoted by a scalar f [N]. A target food is modeled as a vertical spring of which the length is set as w and the spring constant is set as K [N/m], and with its one end point fixed on the base link AD at $(x_P, 0)$. Subsequently, the position of P is set as (x_P, w), using h, which is set as the distance from the tip and the P in x-axis direction. The initial size and deformation of the food are denoted as w_o [mm] and Δw [mm], respectively. The angle θ_B between links BE and BC and the virtual external torque M_{EX}, which is added for the sake of convenience and its value should be zero, are introduced. In this paper, assumptions are made that the mechanism moves quasi-statically in the horizontal plane where the inertial and gravitational forces and viscosity of the food can be neglected. Additionally, as for the balance of forces between the contact force and the reaction force from the food, only the y-axis component is considered, and the x-axis component is ignored, since the one end of the spring is considered being fixed on the base link AD.

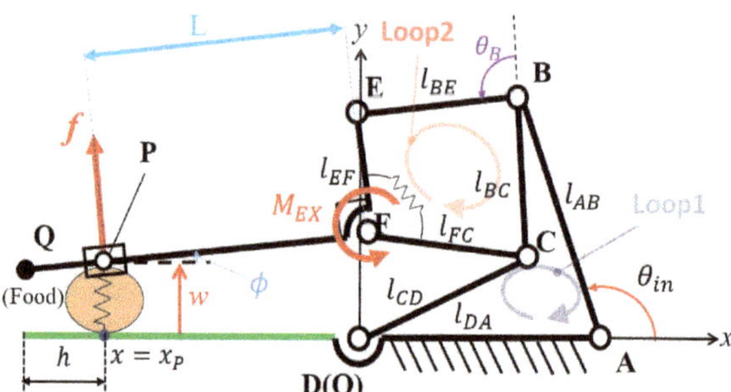

Figure 3. Modeling of the mechanism including the contact point.

4.2. Analysis Scheme Introducing a Virtual Torque

The analysis scheme is referenced to [21,22] in that an additional parameter of force is set to satisfy the mechanical condition. When the input motor angle θ_{in} is given, using one parameter, all of the position of the joints can be determined, since the whole system has 2-DOF. In the analysis scheme, the angle θ_B is set as the convergence parameter for determining the joint positions of the system.

Once the x-position (x_P), initial size w_o, and the stiffness K of the target food are given, the configuration of the mechanism and contact force for a given input angle θ_{in} can be obtained according to the procedure described below and are shown in Figure 4.

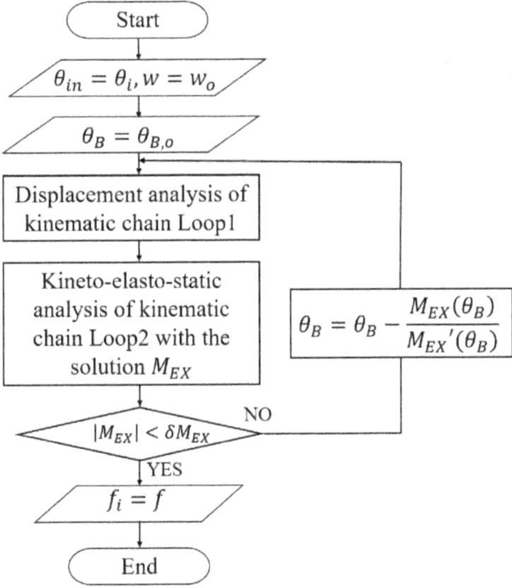

Figure 4. Flow chart of the analysis scheme.

1. With the input angle θ_{in}, the displacement of the kinematic Loop 1 is determined. Note that θ_i is the input angle of which value is between the given initial position to the given final position.
2. The x-position of the target x_P is given, same as the x-position of point P.
3. The statistic force analysis of the kinematic Loop 2 is carried out with joint positions described with the parameter θ_B. When considering the equilibrium of force and moment on each link, simultaneous equations that include the virtual external torque M_{EX} on joint F and the x-component and y-component forces at the joints B, C, E, and F as unknown are formulated. Subsequently, the value of M_{EX} can be calculated.
4. When the absolute value of M_{EX} is smaller than the threshold δM_{EX}, the calculation is terminated, and the value f can be described with the parameter θ_B at that time. When its value exceeds δM_{EX}, adjustment of the value of θ_B is done and the calculation is repeated back to step 3 until the process converges.

Let us think of Loop 2, which is a four-bar mechanism with 1-DOF. When the parameter θ_B is given, positions of joints E and F can be determined as functions of θ_B, $E = E(\theta_B), F = F(\theta_B)$. By using them, position of joint P can be described as Equation (2) under consideration of the geometric relations of Loop 2,

$$P = F(\theta_B) + \frac{L}{l_{EF}} R(\angle EFQ) \cdot (E(\theta_B) - F(\theta_B)) = \begin{bmatrix} P_x \\ P_y \end{bmatrix} = \begin{bmatrix} x_P \\ w \end{bmatrix} \tag{2}$$

where, $R(\angle EFQ)$ represents a rotational matrix in respect to $\angle EFQ$.

Because P_x is the same as the given x_P, L can be obtained from Equation (2) when considering the x components. L can therefore be considered to have another function with respect to θ_B, $L(\theta_B)$. Accordingly, the size of food w and position of P can also be obtained as the function of θ_B, $w = w(\theta_B)$ and $P = P(\theta_B)$, considering the y components of Equation (2). Additionally, the angle θ_{EFC} between links EF and FC and the angle ϕ are expressed as $\theta_{EFC} = \theta_{EFC}(\theta_B)$, $\phi = \phi(\theta_B)$. In this analysis scheme, it is set that the contact force from the gripper and the reaction force from the food are balanced in the y-direction, then their relationship is described as;

$$f \cos \phi = K(w_o - w).$$

(3)

From these geometrical and mechanical relations, the magnitude f and the torque τ_k of the torsion spring are obtained as;

$$f(\theta_B) = \frac{K(w_o - w(\theta_B))}{\cos \phi(\theta_B)},$$

$$\tau_k(\theta_B) = k_t(\theta_{EFC,initial} - \theta_{EFC}(\theta_B)).$$

(4)

From the free body diagram of Loop 2 that is shown in Figure 5, the equilibrium of the force and moment on each link is formulated, as follows.

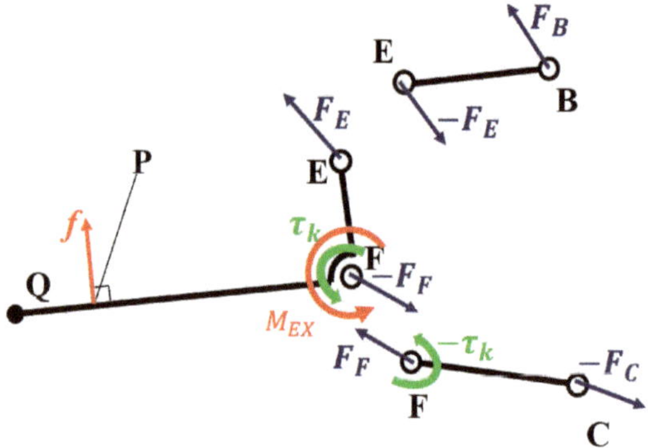

Figure 5. Free body diagram included virtual torque M_{EX}.

Link BE:

$$\left. \begin{array}{r} F_B - F_E = 0 \\ (E(\theta_B) - B(\theta_B)) \times (-F_E) = 0 \end{array} \right\}$$

(5)

Link EFQ:

$$\left. \begin{array}{r} F_E - F_F + f(\theta_B)e(\theta_B) = 0 \\ \tau_k(\theta_B) + (E(\theta_B) - F(\theta_B)) \times F_E - f(\theta_B)L(\theta_B) + M_{EX} = 0 \end{array} \right\}$$

(6)

Link FC:

$$\left. \begin{array}{r} F_F - F_C = 0 \\ -\tau_k + (C(\theta_B) - F(\theta_B)) \times (-F_C) = 0 \end{array} \right\}$$

(7)

From Equations (4) to (7), simultaneous equations are obtained as the following Equation (8).

$$A(\theta_B) \begin{bmatrix} F_B \\ F_E \\ F_F \\ F_C \\ M_{EX} \end{bmatrix} = b(\theta_B) \tag{8}$$

where,

$$A(\theta_B) = \begin{bmatrix} 1 & 0 & -1 & 0 & 0 & 0 & 0 & 0 & 0 \\ 0 & 1 & 0 & -1 & 0 & 0 & 0 & 0 & 0 \\ 0 & 0 & y_E(\theta_B) - y_B & -(x_E(\theta_B) - x_B(\theta_B)) & 0 & 0 & 0 & 0 & 0 \\ 0 & 0 & 1 & 0 & -1 & 0 & 0 & 0 & 0 \\ 0 & 0 & 0 & 1 & 0 & -1 & 0 & 0 & 0 \\ 0 & 0 & -(w(\theta_B) - y_F(\theta_B)) & x_P(\theta_B) - x_F(\theta_B) & 0 & 0 & 0 & 0 & 1 \\ 0 & 0 & 0 & 0 & 1 & 0 & -1 & 0 & 0 \\ 0 & 0 & 0 & 0 & 0 & 1 & 0 & -1 & 0 \\ 0 & 0 & 0 & 0 & 0 & 0 & y_C - y_F(\theta_B) & -(x_C - x_F(\theta_B)) & 0 \end{bmatrix}$$

$$b(\theta_B) = \begin{bmatrix} 0 & 0 & 0 & 0 & 0 & -\tau_k(\theta_B) + f(\theta_B)L(\theta_B) & 0 & 0 & \tau_k(\theta_B) \end{bmatrix}^T$$

By solving Equation (8), $M_{EX} = M_{EX}(\theta_B)$ is obtained. When the reaction force from the target food and torque of the torsional spring are balanced, value of M_{EX} should be zero. A numerical computation based on Newton-Raphson method [21] can solve this equation regarding $M_{EX}(\theta_B) = 0$.

4.3. Comparison of the Computational Efficiency

In Figure 6, the computational efficiency of the analysis scheme is compared with the scheme of [20]. Design parameter values, such as link lengths, are given in Table 1. By using the simulation based on the kinematic analysis, the initial position of the motor as 109 deg and the final position as 117.5 deg were decided. As the motor angle is 117.5 deg, the gripper is closed and the tip of the chopstick part (point Q) reaches the other tip of the chopstick part of link AD, when there is no target object between the chopstick parts. These parameters are obtained through the process that is explained in the following section.

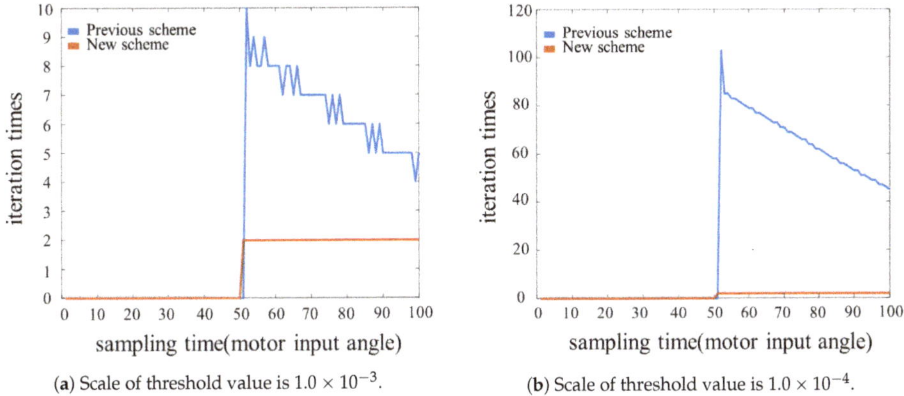

(a) Scale of threshold value is 1.0×10^{-3}.   (b) Scale of threshold value is 1.0×10^{-4}.

Figure 6. Comparison of iteration times among the analysis schemes.

Table 1. Parameters of the mechanism.

Parameter	Value	Parameter	Value
l_{AB}	31 mm	l_{BE}	24 mm
l_{BC}	24 mm	l_{EF}	12 mm
l_{CD}	24 mm	l_{FC}	22 mm
l_{DA}	34.5 mm	l_{FQ}	115.5 mm
$k_t = 3.52 \times 10^{-2}$ Nm/rad		$\angle$EFC $= \pi/2$ rad	

Figure 6 shows the iteration times against the sampling time (which corresponds to the input motor angle). The sampling time is set as 0 when the motor input angle is initial position (109 deg) and as 100 when the motor input angle reaches the final position (117.5 deg), and intermediate values are linearly ramped. The parameters used in the analysis methods are set as $w_o = 15$ mm, $K = 1.0 \times 10^3$ N/m, $h = 5$ mm. Note that δf is the threshold vale for the convergence judgement used in [20]. In order to match the scales of threshold value used in the two analysis schemes, in Figure 6a $\delta f = 1.0 \times 10^{-3}$N, $\delta M_{EX} = 1.0 \times 10^{-3}$ Nm are used, and in Figure 6b $\delta f = 1.0 \times 10^{-4}$ N, $\delta M_{EX} = 1.0 \times 10^{-4}$ Nm are used for the computations. Figure 6a shows that the iteration times in the new scheme are constant as 2, while the iteration times in the previous scheme are decreasing from 10 as the sampling time increases, and the new scheme's computational times are less than the previous scheme's ones. Figure 6b shows that the iteration times in the new scheme are constant as 2, likewise Figure 6a, while the iteration times in the previous scheme are decreasing from 100 as the sampling time increases, and the new scheme's computational times are less than the previous scheme's ones. These results are caused by that in the previous scheme the parameter w is changed by the fixed value, while in the new scheme the parameter θ_B is changed by Newton-Raphson method. Also, in the previous scheme, it is needed to solve the complex non-linear simultaneous equation, while, in new scheme, it is needed to solve the simple equation as Equation (8), and then it can be said that the cost of calculation of the new scheme is less than the one of the previous scheme. Therefore, the new analysis scheme is better than the previous scheme with respect to the computational efficiency. The new analysis scheme is used in the following part of this paper.

4.4. Numerical Example

Numerical examples are shown, in which the analysis scheme that is described in the previous section is applied. The results are obtained by a numerical software MATLAB. Design parameter values, such as link lengths, are given in Table 1, and δM_{EX} and $\delta \theta_B$ are set as $\delta M_{EX} = 1.0 \times 10^{-10}$ Nm, $\delta \theta_B = 1.0 \times 10^{-10}$ rad, respectively.

Figure 7 shows a result of the magnitude of contact force f against the initial food size w_o and the food stiffness K when the input angle reaches the preliminarily determined target value ($\theta_{in} = 117.5$ deg). In this analysis, ranges of w_o and K are set as 1 mm to 20 mm and 0.01×10^3 N/m to 1.0×10^3 N/m, respectively. The numerical calculation was carried out with 100 divisions on each parameter. From the result, it is figured out that the bigger initial size and stiffness result in the larger magnitude of the contact force. The contact force seems to be constant for a wide range of stiffness among the range of w_o between 2 mm and 4 mm. In the range of w_o between 4 mm and 11 mm, the contact force increases according to the increments of the stiffness. Among the range of w_o between 11 mm and 20 mm, while the contact force increases sharply within the range of K up to around 0.3×10^3 N/m, it increases smoothly in the other range of K. Additionally, the magnitude of force f increases monotonically as the value of the initial food size w_o increases.

Figure 8 shows the relation between contact force f [N], food's stiffness K [N/m] and the contact point h [mm]. At this time, the input angle of the driving motor θ_{in} is set as 117.5 deg, and the initial size of food w_o is varied as 5 mm, 10 mm, 15 mm, 20 mm. From these results, it is figured out that the magnitude of force f increases monotonically as the value of the distance from the tip h [mm] increases. In other words, it can be said that the larger distance between the contact point and the tip of the chopstick part, the larger force the mechanism outputs. Additionally, it can be said that the gripper mechanism is able to change its gripping force actively by changing the position of the contact point, and this mechanism's feature can be utilized for a grasping planning in the following section.

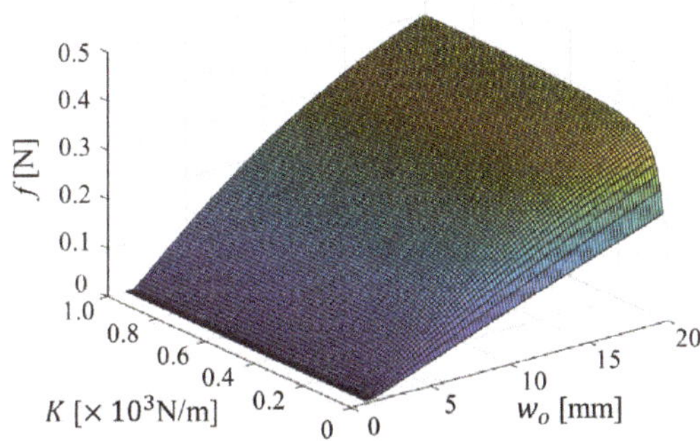

Figure 7. Relation between contact force f and food's size w_o, food's stiffness K as $\theta_{in} = 117.5$ deg.

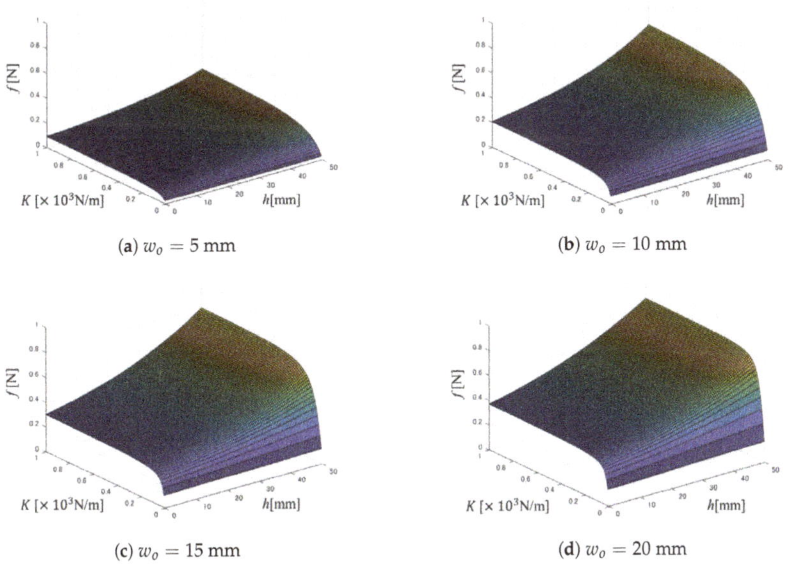

(**a**) $w_o = 5$ mm

(**b**) $w_o = 10$ mm

(**c**) $w_o = 15$ mm

(**d**) $w_o = 20$ mm

Figure 8. Contact force related to the stiffness and the contact position as $\theta_{in} = 117.5$ deg.

5. Grasp Planning Considering the Characteristic of the Mechanism

This section addresses grasp planning utilizing the feature of the mechanism, that is the mechanism can change the contact force with changing the contact position. The proposed grasp planning derives the input motor angle and the contact position to grasp up food in order to realize a simple control system.

5.1. Grasp Planning Algorithm

The grasping of the chopstick-type gripper is modeled in order to know how much force is needed to pick up a food. The proposed mechanism is a planar mechanism, and the gripping of the target food is established by the equilibrium of forces between the gravitational force of the food and the friction forces from the gripper, as shown in Figure 9. Subsequently, the equilibrium of the forces is described as Equations (9) and (10) under the assumption that the two friction forces $F_{friction}$ are equal. In this case, m, g, μ are set as the mass of the food, the gravitational acceleration, and the coefficient of static friction, respectively.

$$2F_{friction} = mg \tag{9}$$

$$F_{friction} \leq \mu f_{gripping} \tag{10}$$

Thus, to pick up the food, the gripping force is described as Equation (11) with the safety factor $s(> 1)$ for the maximum friction force.

$$f_{gripping} = s\frac{mg}{2\mu} \tag{11}$$

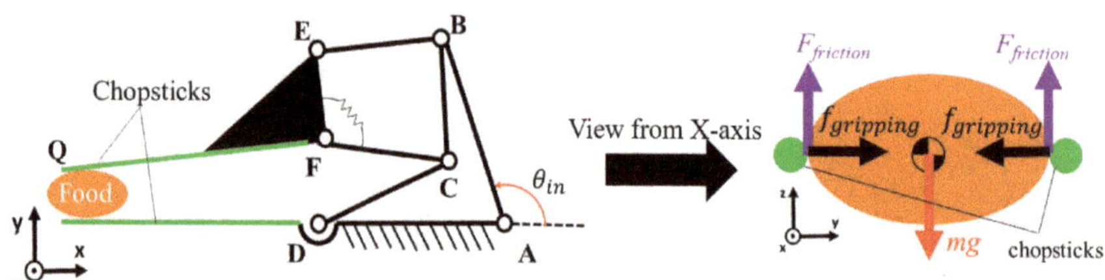

Figure 9. Simple modeling of the grasping.

In the following, the gripping force to pick up food will be obtained by the position control of the driving motor of the gripper mechanism and the end-effector of the manipulator in order to implement a simple control system, and the grasp planning is proposed based on it. In other words, the grasp planning aims to obtain the required force in order to pick up the food without feedback control.

The grasp planning algorithm is proposed from the results obtained from the analysis. In the paper, the grasp planning is regarded as the planning to decide the contact position and the input motor angle for the gripper to grasp the food. Figure 10 shows the algorithm of the proposed grasp planning. For the grasp planning, it is considered that the magnitude of the contact force $f(h) = f\cos\phi$ monotonically increases with respect to the contact position h when the stiffness of the food is known, as Figure 8 shows. Additionally, the range of contact point is set as $0 \leq h \leq h_{limit}$. The procedure of grasp planning is summarized, as follows.

1. For the given w_o, K, m and μ, the gripping force to pick up food, $f_{gripping}$, is calculated by Equation (11).
2. The motor input angle θ_{in} is given, and the function $f(h)$ is derived using the analysis.
3. It is determined if the solution h exists, which satisfies $f(h) = f_{gripping}$ in the assumed contact range ($0 \leq h \leq h_{limit}$). When the solution h exists in the range, the h is obtained by solving the equation $f(h) = f_{gripping}$. When the solution h does not exist, the motor input angle is updated and the calculation goes back to step.2.

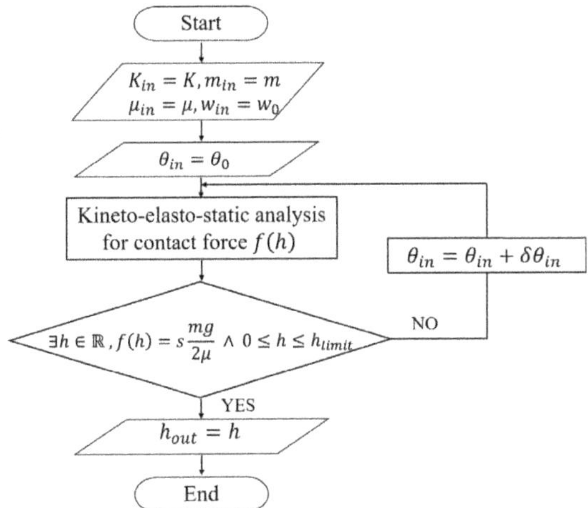

Figure 10. Flowchart of the grasp planning based on the analysis.

5.2. Case Study of the Grasp Planning

Case studies to decide the input motor angle and the contact position by the proposed grasp planning are described. For the examples, the safety factor is set as $s = 2.0$, the update value of the input motor angle is set as $\delta\theta_{in} = 1$ deg, and the range of the contact is set as $0 \leq h \leq 50$ mm.

(case.1) As $K = 0.5 \times 10^3$ N/m, $w_o = 20$ mm, $m = 0.025$ kg, $\mu = 0.40$

From Equation (11), the required force for grasping is $f_{gripping} = 0.613$ N. From Figure 11a, when the input motor angle is $\theta_{in} = 117.5$ deg, h which satisfies $f(h) = f_{gripping}$ exists, and the solution is $h = 33.43$ mm. Figure 11b shows that the calculation is converged.

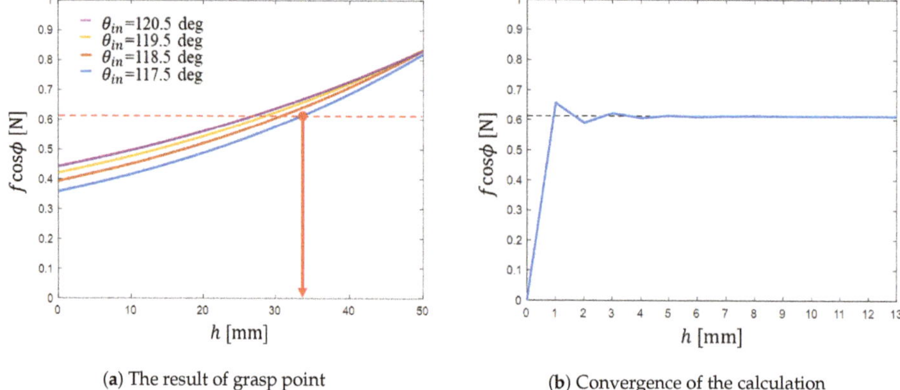

(**a**) The result of grasp point

(**b**) Convergence of the calculation

Figure 11. The result of grasp planning of case.1.

(**case.2**) As $K = 0.1 \times 10^3$ N/m, $w_o = 10$ mm, $m = 0.020$ kg, $\mu = 0.35$

From Equation (11), the required force for grasping is $f_{gripping} = 0.420$ N. From Figure 12a, when the input motor angle is $\theta_{in} = 117.5$ deg, h, which satisfies $f(h) = f_{gripping}$ does not exist, then the input motor angle is updated. From Figure 12a, when the input motor angle is $\theta_{in} = 118.5$ deg, the h which satisfies $f(h) = f_{gripping}$ exists, and the solution is $h = 40.89$ mm. Figure 12b shows that the calculation is converged.

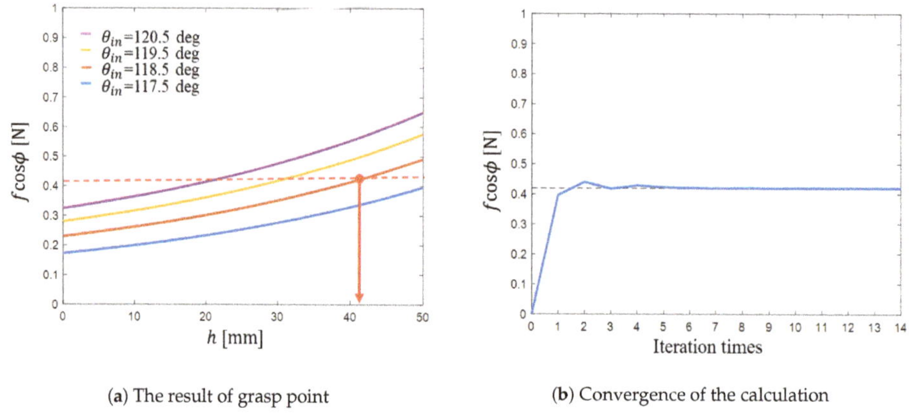

(**a**) The result of grasp point

(**b**) Convergence of the calculation

Figure 12. The result of grasp planning of case.2.

In this paper, it is assumed that the parameters of the food such as dimensions, mass and stiffness are obtained from the database based on the measurement by the vision system. Then, the error of these parameter values is inevitable. Here, let us investigate the effect of the error of stiffness value, which seems the most difficult to get precise value among the parameters, on the gripping performance through examples. The error of the stiffness K is considered under an assumption that its value includes maximum error by 20 % of its nominal value. Two cases of nominal stiffness values: $K = 0.5 \times 10^3$ N/m and $K = 0.1 \times 10^3$ N/m for $\theta_{in} = 117.5$ deg, and $h = 5$ mm are considered. Taking into consideration the maximum 20 % error in stiffness, we obtained that the contact forces vary between 0.314 N (-0.94%) and 0.319 N ($+0.63$ %) (nominal value: 0.317 N) for $K = 0.5 \times 10^3$ N/m, and between 0.267 N (-3.96 %) and 0.285 N ($+2.52$ %) (nominal value: 0.278 N) for $K = 0.1 \times 10^3$ N/m, respectively. From these results, it is known that the sensitivity of the stiffness error on the contact force error is low while the stiffness for soft food is more sensitive to the contact force than for hard food. Therefore, it can be said that the proposed mechanism can achieve a stable gripping under the existence of the estimated stiffness value error. In the case where a quite sensitive force control is required to handle a very delicate food, in order to avoid hurting the food, a feedback control system, such as an impedance control system, may be applied. Even in such a case, a low-cost control system may be constructed by adding an angular displacement sensor at joint F to measure the spring force, which is based on the advantage of the proposed mechanism.

6. Design, Prototyping and Experiment

6.1. Design of the Prototype

The design process for determining the parameter values as shown in such as Table 1 is described. As for the design of the mechanism, the property of the food is set as Table 2. First of all, the whole size of the mechanism is determined. The length of chopstick is set on 100 mm considering on the length of real chopsticks, and the mechanism should be small enough to set as the end-effector of the meal assistance robot. In this case, the total size of two loops was set to be about 40 mm × 40 mm. Subsequently, the length of each link of Loop 1($l_{AB}, l_{BC}, l_{CD}, l_{DA}$) was determined so that the path of the chopstick part is close to the actual movement of chopsticks. Next, the length of each link of Loop 2 (l_{BE}, l_{EF}, l_{FC}) is determined, so that the larger the size of the food is, the lager the contact force outputs. At this time, the parameters were set, so that the contact force monotonously increases according to the increase of the motor input angle θ_{in} when the stiffness of food K is constant and the size is changed. In addition, the range of the motor input angle θ_{in} is set so that the tip of the chopstick part matches when the maximum θ_{in} is given in the initial loop with the grasp planning algorithm in Figure 10 when there is no target object. Finally, the spring constant of the torsion spring installed k_t was determined. The used torsion spring is a linear spring, and the order of the contact force is determined by the value of the spring constant. From Equation (11) and Table 2, the order of the required gripping force was determined and, with the safety factor $s = 2.0$, the maximum force was determined as $f_{gripping} = 0.98$ N, then the spring constant was determined. The prototype was fabricated using 3D printer as shown in Figure 13. The prototype has a double supported structure, and the actuator used for the prototype was VS-12M servo (Vigor Precision Ltd.).

Table 2. Parameter of target food.

Mass m [g]	$0 < m \leq 30$
Initial width w_o [mm]	$5 \leq w_o \leq 20$
Friction coefficient μ [-]	$0.3 \leq \mu$
Stiffness $K [\times 10^3$ N/m]	$0.01 \leq K \leq 1.0$

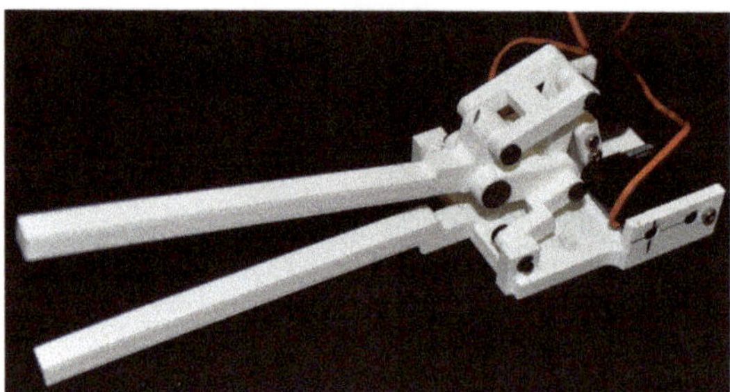

Figure 13. Gripper prototype fabricated using 3D printer (bird's eye view).

6.2. Experiment Using the Gripper Prototype with Changing the Contact Point

An experiment to measure the contact force with changing the contact position and the stiffness of the contacted object was carried out in order to validate the modeling of the mechanism including the contact point and the results of the kineto-elasto-static analysis that the gripper can change its contact force according to the size and the stiffness of food, and the contact position. In the experiment, the reaction force from the food ($K\Delta w$ [N]) was measured through a force gauge with an attachment including a compression spring, and the measured values were compared with the theoretical result of the analysis ($f \cos \phi$ [N]) based on Equation (3). Figure 14 shows the experimental setup with the gripper prototype to measure the contact force from link EFQ. The experimental setup was composed of the gripper prototype, a linear guide with a measure, and a force gauge with an attachment having a compression spring. The input angle of the motor was controlled by PWM control using a microcomputer Arduino Mega. The resolution of the motor angle was 0.1 deg. The linear guide enabled the gripper to change the contact point. The displacement was measured by the scale alongside the linear guide. The attachment of the force gauge is shown in Figure 15, and it was fabricated with a three-dimensional (3D) printer. This attachment reproduced the characteristic of the food having an stiffness and the modeling of the mechanism, as shown in Figure 3. The force gauge was DS2-20N (IMADA), and the resolution was 0.01 N. The compression spring in the attachment was replaced to change the spring constant, and the linear bushes inside the attachment made the contact point move smoothly. The experiment was carried out with changing the spring constant, using five kinds of spring, of which spring constant: 0.05×10^3 N/m, 0.1×10^3 N/m, 0.3×10^3 N/m, 0.5×10^3 N/m, and 1.0×10^3 N/m. Additionally, the contact point was changed in the area as $0 \text{ mm} \leq h \leq 50 \text{ mm}$, and the contact force was measured at every 10 mm of h. As for the input angle of the driving motor of the gripper, the input was set as $\theta_{in} = 117.5, 118.5, 119.5,$ and 120.5 deg. In the experiment, the measured values were set as the average of the five times measurement values.

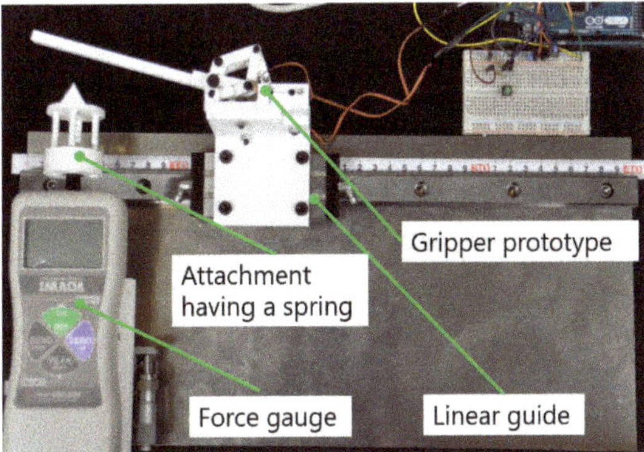

Figure 14. Experimental setup to measure the contact force with changing the contact position.

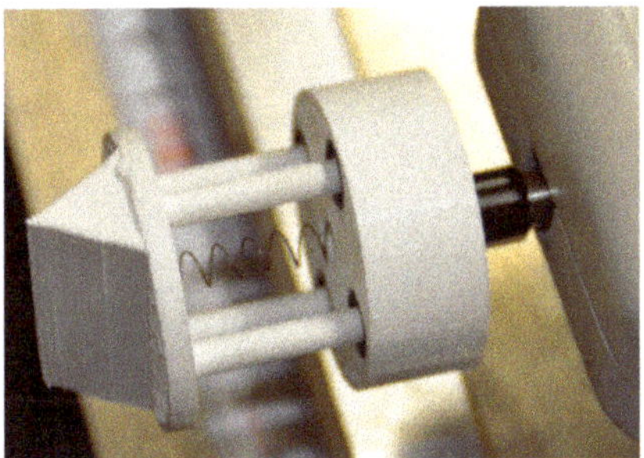

Figure 15. Attachment of the force gauge.

Figure 16 shows examples of the experimental results to measure the contact force with comparing the results of the theoretical analysis. In the figures, the theoretical value was calculated by $f \cos \phi$ [N], and the measured value was obtained by $K\Delta w$ [N] in Equation (3).The maximum difference between the measured value and the theoretical value was 0.06 N when $w_o = 15$ mm, $K = 1.0 \times 10^3$ N/m, $\theta_{in} = 120, h = 40$ mm. The main reason of these differences is considered to come from the x-component of the contact force $f \sin \phi$. In the analysis, the x-component of the contact force $f \sin \phi$ was neglected. The value is small when the value ϕ is small, so the influence seems small. However, when the ϕ becomes big, the influence of x-component of the contact force $f \sin \phi$ cannot be neglected, and the force seems to affect the movement of the attachment of the force gauge. When considering that the resolution of the force gauge was 0.01 N, the deviations between the theoretical values and measured values were small when the spring constant was set as $K = 0.05 \times 10^3$ N/m, $K = 0.1 \times 10^3$ N/m and $K = 0.5 \times 10^3$ N/m. Thus, it can be said that the analysis is validated through the experiment.

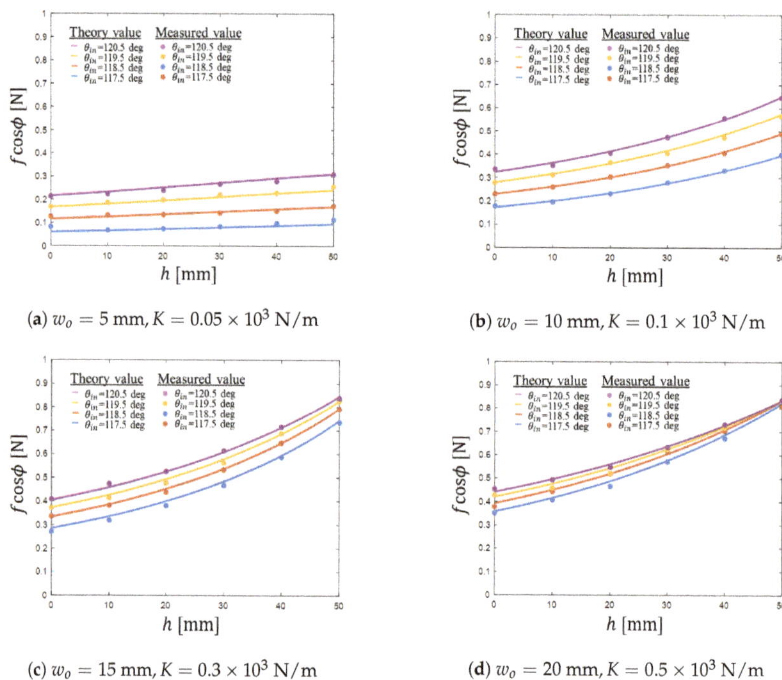

(**a**) $w_o = 5$ mm, $K = 0.05 \times 10^3$ N/m

(**b**) $w_o = 10$ mm, $K = 0.1 \times 10^3$ N/m

(**c**) $w_o = 15$ mm, $K = 0.3 \times 10^3$ N/m

(**d**) $w_o = 20$ mm, $K = 0.5 \times 10^3$ N/m

Figure 16. Comparisons between the experimental value and the theory value (Example).

6.3. Experiment of the Grasp Planning with 6-DOF Robot Arm

An experiment using 6-DOF robot arm was carried out in order to demonstrate the feasibility of the grasp planning proposed in the previous section. Figure 17 shows the experimental set up, which is composed of 6-DOF robot arm (LR Mate 200iD/4S, FANUC) and the gripper prototype, which is the same as the one used for the experiment in the previous section. The gripper prototype was implemented to the robot arm with the connected part which was fabricated by a 3D printer. In this experiment, the displacement of the gripper was controlled by the manual controller of the robot arm, and the orientation of the gripper was kept constant (the value of the orientation of the end-effector was set as $roll = 33, pitch = -67, yaw = -176$). The motor angle of the gripper was controlled through Arduino Mega, which the same one used in the previous experiment.

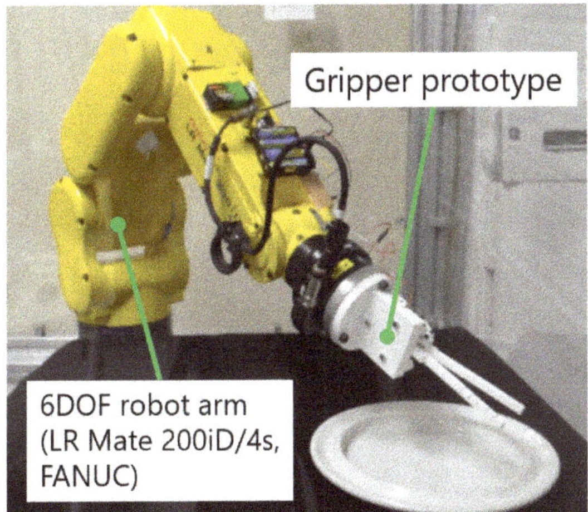

Figure 17. Experimental setup with 6-DOF robot arm

In the experiment, the target food was sushi roll, as shown in Figure 18, and the width to be grasped was measured as 16.6 mm, and the mass was measured as $m = 0.0131$ kg. From [23], the stiffness of sushi roll was obtained as $K = 0.31 \times 10^3$ N/m, and the viscosity of the food was neglected in the experiment. Additionally, the coefficient of friction was roughly set as $\mu = 0.4$ referencing [24]. From Equation (11), the required force for grasping was $f_{gripping} = 0.320$ N. From Figure 19a, when the input motor angle is $\theta_{in} = 117.5$ deg, h, which satisfies $f(h) = f_{gripping}$ exists, and the solution was obtained as $h = 2.10$ mm. Figure 19b shows that the calculation is converged.

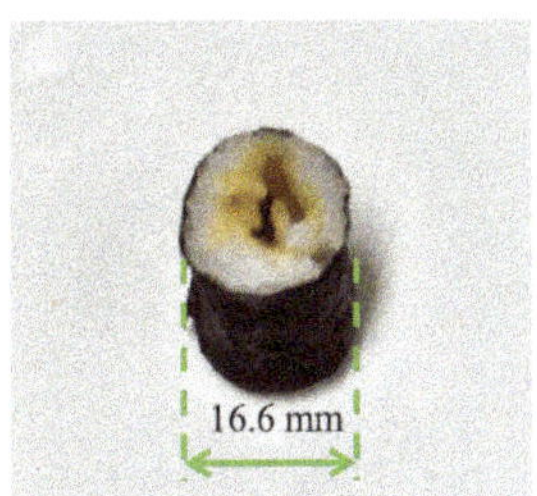

Figure 18. Target food.

Figure 20 shows the grasping experiment using the 6-DOF robot arm. From the grasp planning, the input parameters were obtained as $\theta_{in} = 117.5$ deg and $h = 2.1$ mm, and these parameters were used in the experiment. Note that the actual contact point was not point contact, and the grasping was carried out at the area where $0 \leq h \leq 10$ mm. From the figure, it was observed that the gripper successfully picked up the target food. From the result, the proposed grasp planning was confirmed to be effective under the experimental condition.

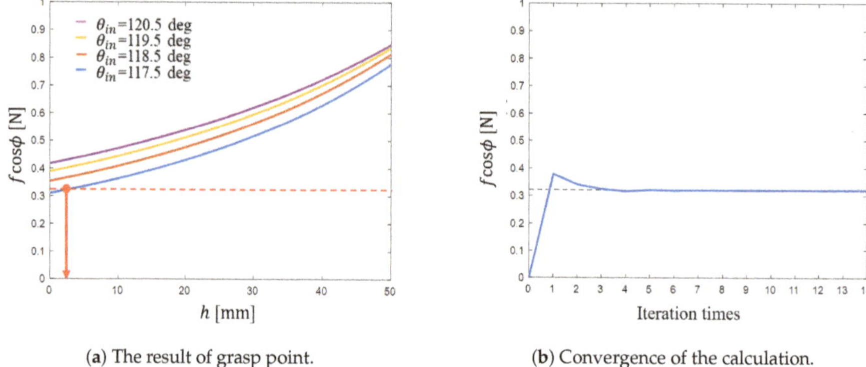

(**a**) The result of grasp point.

(**b**) Convergence of the calculation.

Figure 19. The result of grasp planning.

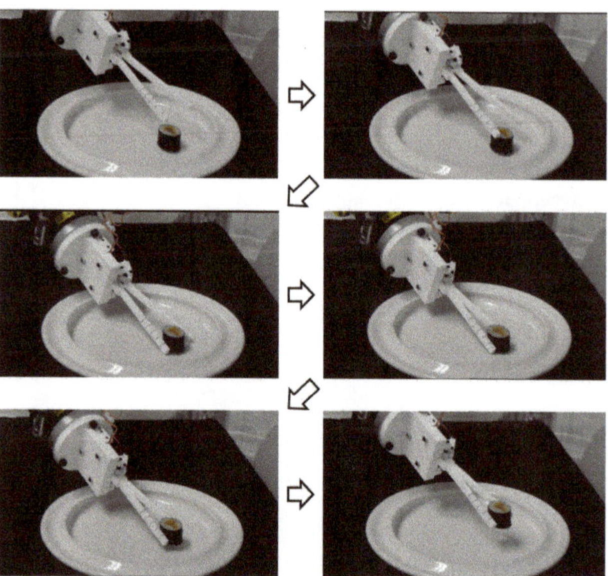

Figure 20. Grasping test with 6-DOF robot arm.

7. Conclusions

In this paper, a chopstick-type gripper mechanism based on the concept of under-actuation and solely use of position control, which is capable of adapting its shape and contact force according to size and stiffness of target foods, was proposed. Modeling of the mechanism, including the contact with target food having stiffness, has been done in order to design a practical gripper. Based on this model, an analysis scheme based on iterative calculations of kineto-elasto-static analysis has been formulated and shown to be improved through comparison with the other analysis scheme of our previous work from the point of view of the computational efficiency. Based on the result of analysis, it is revealed that the proposed mechanism is able to adjust its contact force according to the size and stiffness of target foods, and that the gripper mechanism is able to change its gripping force according to the contact position with the target object, which is utilized for the grasp planning that determines the contact position suitable for its grasping. Through examples, it has been revealed that the contact force of the proposed mechanism is less sensitive against the stiffness error. While using the gripper prototype, the contact force was measured by a force gauge with an attachment having a spring with changing the contact point with a linear guide. From the result of the experiment, the modeling of mechanism in the kineto-elasto-static analysis and the mechanism's feature that the mechanism can change its contact force according to the size and stiffness and the contact position with the object was validated. Additionally, using the 6-DOF robot arm, the grasping test of a real food was conducted utilizing the proposed grasp planning. The gripper was able to lift up the food using the input parameters obtained by the grasp planning, and the feasibility of the grasp planning was confirmed. For future works, experiments to grasp other kinds of food will be carried out in order to decide the scope of the application of the gripper. Additionally, a control system utilizing some feedback signal without implementing complex and expensive instruments based on the impedance control will be introduced to achieve more stable and appropriate gripping a wide variety of foods. Furthermore, in addition to the gripper mechanism and its control system, future work includes the design of the gripper with different shape of the end-effector for more practical development.

Author Contributions: Conceptualization, T.O.; Investigation, T.O.; Methodology, T.O.; Project administration, J.S., A.-L.L. and Y.T. Resources, T.O., Supervision, D.M. and Y.S.; Validation, Y.T.; Writing—original draft, T.O.; Writing-Review & Editing, J.S., A.-L.L., D.M., Y.S. and Y.T. All authors have read and agreed to the published version of the manuscript.

Funding: This work was carried out as a part of the SICORP under the responsibility of the Japan Science and Technology Agency (JST) and was supported in part by JSPS KAKENHI JP17H03162.

Conflicts of Interest: The authors declare no conflict of interest.

References

1. The WHOQOL Group. The World Health Organization quality of life assessment (WHOQOL): Development and general psychometric properties. *Soc. Sci. Med.* **1998**, *46*, 1569–1585. [CrossRef]
2. Nyberg, M.; Olsson, V.; Pajalic, Z.; Örtman, G.; Andersson, H.S.; Blücher, A.; Wendin, K.; Westergren, A. Eating difficulties, nutrition, meal preferences and experiences among elderly: A literature overview from a Scandinavian context. *J. Food Res.* **2015**, *4*, 22–37. [CrossRef]
3. Topping, M.; Smith, J. The development of Handy 1, a rehabilitation robotic system to assist the severely disabled. *Ind. Rob.* **1998**, *25*, 316–320 [CrossRef]
4. Yamasaki, A.; Fukushima, M.; Masuda, R. Trial manufacture of the meal assistance robot using chopsticks and control of grasp force for foods. *Rob. Soc. Jpn.* **2012**, *30*, 917–923. [CrossRef]
5. Soyama, R. The development of meal-assistance robot 'My Spoon'. In Proceedings of the 8th International Conference on Rehabilitation Robotics, Taejon, Korea, 23–25 April 2003; pp. 88–91.
6. Secom. My spoon. Available online: https://www.secom.co.jp/english/myspoon/ (accessed on 25 January 2020).
7. Michaelis, J. Introducing the neater eater. *Action Res.* **1988**, *6*, 2–3.
8. NeaterSolutions. Neater Eater. Available online: https://neater.co.uk/neater-eater/ (accessed on 25 January 2020).
9. Products Incorporated 2003, R. Meal buddy. Available online: http://www.richardsonproducts.com/mealbuddy.html (accessed on 25 January 2020).
10. Copilusi, C.; Ceccarelli, M. An application of LARM clutched arm for assisting disabled people. *Int. J. Mech. Control* **2015**, *16*, 57–66.
11. Song, W.K.; Kim, J. Novel assistive robot for self-feeding. *Rob. Syst. Appl. Control Program.* **2012**, *1*, 43–60.
12. EU-Japan Centre for Industrial Cooperation. Japan-Sweden Academia-Industry International Collaboration. Available online: https://www.eu-japan.eu/news/bilateral-call-japan-sweden-academia-industry-international-collaboration (accessed on 25 January 2020).

13. Norén A-L. Bestic: An Eating-Aid for Persons with Little or No Ability to Move Their Arms. Master's Thesis, Chalmers University of Technology, Gothenburg, Sweden, 2005.

14. Lindborg, A.L.; Lindén, M. Development of an eating aid - from the user needs to a product. *Stud. Health Technol. Inf.* **2015**, *211*, 191–198.

15. Solis, J.; Karlsson, C.; Ogenvall, M.; Lindborg, A.L.; Takeda, Y.; Zhang, C. Development of a vision-based feature extraction for food intake estimation for a robotic assistive eating device. In Proceedings of the 2018 IEEE 14th International Conference on Automation Science and Engineering (CASE). Munich, Germany, 20–24 August 2018; pp. 1105–1109.

16. Birglen, L.; Gosselin, C.M. Kinetostatic analysis of underactuated fingers. *IEEE Trans. Rob. Autom.* **2004**, *20*, 211–221. [CrossRef]

17. Stavenuiter, R.A.; Birglen, L.; Herder, J.L. A planar underactuated grasper with adjustable compliance. *Mech. Mach. Theory* **2017**, *112*, 295–306. [CrossRef]

18. Fukaya, N.; Asfour, T.; Dillmann, R.; Toyama, S. Development of a five-finger dexterous hand without feedback control: The TUAT/Karlsruhe humanoid hand. In Proceedings of the 2013 IEEE/RSJ International Conference on Intelligent Robots and Systems, Tokyo, Japan, 3–7 November 2013; pp. 4533–4540.

19. Ueno, T.; Oda, M. Development of an index finger for the dexterous hand for space. *J. Rob. Soc. Jpn.* **2010**, *28*, 349–359. [CrossRef]

20. Oka, T.; Matsuura, D.; Sugahara, Y.; Solis, J.; Lindborg, A.; Takeda, Y. Chopstick-type Gripper Mechanism for Meal-Assistance Robot Capable of Adapting to Size and Elasticity of Foods. In *Mechanism Design for Robotics*; Springer: Cham,Switzerland, 2018; pp. 284–292.

21. Suwa, T.; Iwatsuki, N.; Ikeda, I. Kinematic Analysis and synthesiS of Flexible Mechanism Composed of Underactuated Mechanism Constrained with Elastic Element. In Proceedings of the Mechanical Engineering Congress, Hokkaido, Japan, 14 September 2015; S1170203.

22. Iwatsuki, N.; Kotte, T.; Morikawa, K. Simultaneous control of the motion and stiffness of redundant closed-loop link mechanisms with elastic elements. *J. Mech. Sci. Technol.* **2010**, *24*, 285–288. [CrossRef]

23. Sakamoto, N.; Yuya, M.; Higashimori, M.; Kaneko, M. An Optimum Design for Handling a Visco-elastic Object Based on Maxwell Model. *J. Rob. Soc. Jpn.* **2007**, *25*, 166–172. [CrossRef]

24. Editorical Commitee of "Shokuhin-kagaku-binran". *Shokuhin-Kagaku-Binran*; Kyoritsu Shuppan Co, Ltd.: Tokyo, Japan, 1978; p. 265. (In Japanese)

Article

A Planar Parallel Device for Neurorehabilitation

Jawad Yamine [1], Alessio Prini [2], Matteo Lavit Nicora [2], Tito Dinon [2], Hermes Giberti [3] and Matteo Malosio [2,*

[1]   Politecnico di Milano, Via la Masa 1, 20156 Milano, Italy; jawad.yamine@mail.polimi.it
[2]   Istituto di Sistemi e Tecnologie Industriali Intelligenti per il Manifatturiero Avanzato,
     Consiglio Nazionale delle Ricerche, via G. Previati 1/E, 23900 Lecco, Italy; alessio.prini@stiima.cnr.it (A.P.);
     matteo.lavit@stiima.cnr.it (M.L.N.); tito.dinon@stiima.cnr.it (T.D.)
[3]   Dipartimento di Ingegneria Industriale e dell'Informazione, Università di Pavia, Via ferrata 5,
     27100 Pavia, Italy; hermes.giberti@unipv.it
*   Correspondence: matteo.malosio@cnr.it; Tel.: +39-0341-235-0204

Received: 15 October 2020; Accepted: 27 November 2020; Published: 3 December 2020

Abstract: The patient population needing physical rehabilitation in the upper extremity is constantly increasing. Robotic devices have the potential to address this problem, however most of the rehabilitation robots are technically advanced and mainly designed for clinical use. This paper presents the development of an affordable device for upper-limb neurorehabilitation designed for home use. The device is based on a 2-DOF five-bar parallel kinematic mechanism. The prototype has been designed so that it can be bound on one side of a table with a clamp. A kinematic optimization was performed on the length of the links of the manipulator in order to provide the optimum kinematic behaviour within the desired workspace. The mechanical structure was developed, and a 3D-printed prototype was assembled. The prototype embeds two single-point load cells to measure the force exchanged with the patient. Rehabilitation-specific control algorithms are described and tested. Finally, an experimental procedure is performed in order to validate the accuracy of the position measurements. The assessment confirms an acceptable level of performance with respect to the requirements of the application under analysis.

Keywords: parallel kinematic architecture; kinematic optimization; rehabilitation robotics; assist-as-needed control algorithms

---

1. Introduction

Stroke is one of the main causes of long-term disability worldwide and the most common in Western countries [1]. The number of patients having difficulties in performing daily-living activities due to physical disabilities is constantly increasing, making the availability of therapists and caregivers more and more inadequate and, therefore, creating an unmet market need.

Robotic devices for neurorehabilitation have been widely investigated, developed and introduced in the market to offer a valid alternative to conventional therapy and fill the constantly growing gap between supply and demand [2,3]. Since the invention of the MIT-Manus [4], robot-assistance, force-feedback and force-based control are sought after features of neurorehabilitation devices [5], enabling them to sense the patient's interaction with the robot, react accordingly and adapt the level of physical assistance provided. Most of the proposed robots are technically advanced, but are relatively expensive and designed for clinical settings, which makes it hard for patients to afford such treatment. There are also examples of commercial general-purpose industrial manipulators, properly equipped with force-based control algorithms, exploited in rehabilitation scenarios [6–8]. They can be very flexible and useful for testing purposes but, on the other side, they are inherently relatively expensive with respect to rehabilitation budget requirements.

Focusing on upper-limb rehabilitation, a number of low-cost rehabilitation devices are currently available, typically passive or passively gravity-balanced [9]. Nevertheless, the lack of actuation, of an assist-as-needed support and of haptic capabilities, preclude them to be effectively used by patients with low/medium motion capabilities. The development of low-cost rehabilitation devices also meets the need of low-income countries where the healthcare system is lacking and the medical personnel is insufficient. In these countries, where even hospitals cannot afford expensive mechanical devices, the challenge is to conceive and develop low-cost and easily-replicable systems for rehabilitation, as far as possible.

Some tabletop actuated devices have been specifically developed with the aim of satisfying economic and installation requirements in out-of-clinic environments. These solutions often rely on reduced complexity and optimized costs by limiting the number of degrees of freedom with respect to complex rehabilitation devices, such as exoskeletons [10–12], in order to partially meet the affordability requirements. However, strictly reducing the number of degrees of freedom of exoskeletons can sometimes lead to drawbacks. The authors of [13] developed an interesting elbow rehabilitation device; but, since the architecture is not supported or constrained to a fixed structure, the device weighs on the shoulder of the patient with a consequent lack of rehabilitation for that specific body part.

The large majority of tabletop devices are constituted by rigid links and joints. Nevertheless, it is worth to mention the existence of alternative solutions. CUBE is a tabletop cable-driven device enabling 3D-movements of the upper limb [14]. Despite its peculiar and interesting kinematic architecture, it does not provide a steady support for the hand in spatial movements since its end-effector is constrained only by two groups of three wires. MOTORE is an interesting mobile robot for upper-limb rehabilitation, but the need of resting completely the forearm on the device can constrain the upper arm excessively and lead to a high elevation angle of the elbow [15].

Alternative solutions can mobilize the upper arm for specific movements, but do not allow a wide movement of the upper limb, both in terms of shoulder and elbow. For instance, Nam et al. developed a portable device, capable of mobilizing the pronosupination of the forearm, unusual capability for tabletop devices [16]. However, its kinematic structure does not allow the rehabilitation of the upper limb in an extensive range of motion, since it does not provide any mobilization of the shoulder.

Moreover, planar movements are largely used for upper-limb neurorehabilitation and they represent the basis for interesting works such as the one proposed by Zadravec et al., in which a solution to model the planar movement trajectory formation [17] is suggested. By referring to articulated kinematic structures, it is possible to highlight a characteristic shared among different devices. The human body is inherently symmetric with respect to the sagittal plane. Nevertheless, several devices are characterized by a non-symmetric structure that could cause kinetostatic performance and manipulability ellipses to also be asymmetric with respect the sagittal plane. Asymmetrical kinematic structures produce asymmetrical shapes of the manipulability ellipsis, leading to an asymmetric kinetostatic behaviour for right-handed and left-handed patients. This is true for the kinematic structure of several rehabilitation devices, such as MIT-Manus [4], Braccio di Ferro [18] and NURSE [19].

Focusing on symmetrical kinetostatic behaviour with respect to the sagittal plane, some paradigmatic devices can be found. Some of them exploit Cartesian kinematic architectures, both serial and parallel. Wu et al. developed an admittance-controlled Cartesian serial kinematic architecture [20], while Zollo et al. proposed a planar orthogonal parallel rehabilitation device [21]. Both these devices are characterized by an inherent isotropic kinetostatic behaviour. However, in the opinion of the authors, such architectures are relatively cumbersome and complex and would not allow an effective commercial exploitation, especially in low-budget rehabilitation scenarios. An additional solution is provided by the H-MAN [22], a differential-based isotropic planar device for upper-limb rehabilitation. Although the authors consider its design outstanding, the goal of this work was to develop a device able to exploit extensively the range of motion of the upper limb, without leading to a relatively bulky structure. In these terms, the notable architecture of H-MAN would have resulted in a big and not straightforwardly portable device if properly scaled to allow large movements of the upper limb, mainly because of its Cartesian structure.

The aim of the present work was to present a rehabilitation device, namely PLANarm2, developed specifically to achieve an acceptable compromise in terms of (1) workspace symmetry with respect to the sagittal plane, (2) relatively large workspace, (3) portability and (4) affordability (Figure 1). The well-known 5R planar kinematic chain was considered a promising solution [23]. It is a matter of fact that this architecture has already been adopted to realize the haptic device developed by Klein et al. [24]. Starting from the parametric model of the 5R kinematics, link lengths of PLANarm2 have been optimized to have good kinematic performances in the large majority of its workspace, properly dimensioned to overlap the range of motion of the upper limb. Its symmetric kinematic structure is inherently characterized by a symmetrically distributed kinetostatic behaviour with respect to the sagittal plane. Moreover, in order to reduce the total cost of the device, it has been designed to be clamped quite easily on a standard table and to facilitate both portability and fast installation inside already furnished environments. As opposed to the device described in [24], which is characterized by a self-supported manipulandum, the PLANarm2 manipulandum slides on a table or an a desk, whose surface supports the gravitational load. The links of the parallel structure only transmit horizontal forces, limiting bending loads. This allowed the device to be realized by additive manufacturing techniques with plastic material, in line with the affordability requirement.

The paper is organized as follows: the kinematic architecture is presented in Section 2; the mechanical design and its optimization are described in Section 3; the main components of the prototype and a brief cost analysis are reported in Section 4; the control framework is presented in Section 5; results of an experimental assessment are outlined in Section 6; conclusions are drawn in Section 7.

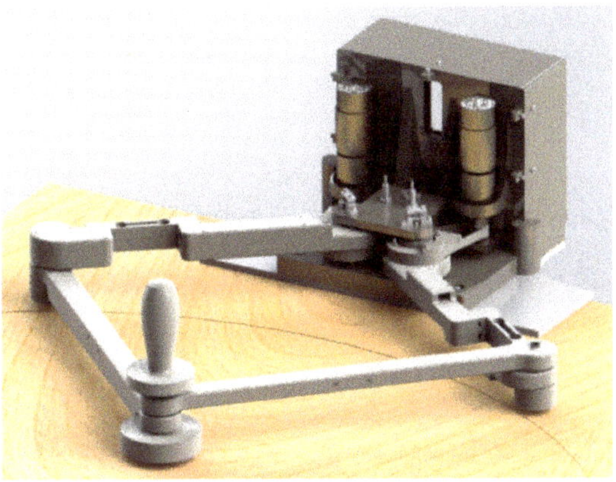

Figure 1. 3D model of the prototype.

2. Kinematics

The forward and inverse kinematics presented in this section were developed to provide a less general but more efficient formulation than the one in [25]. The model proposed in the mentioned work provides the solution to the inverse kinematics problem for each of the four a possible configurations depicted in Figure 2. In addition, the forward kinematic problem leads to two solutions, one for the up-configuration and one for the down-configuration. In order to reduce the computational burden, the model presented in the following pages has been developed specifically for the configuration of interest, which is configuration (a) in the up-configuration.

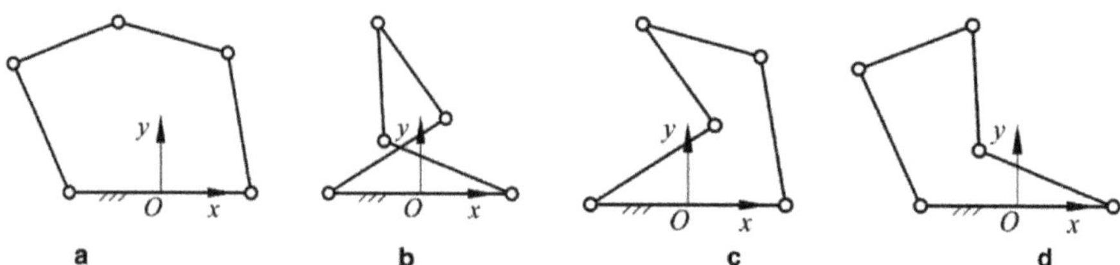

Figure 2. Four configurations of the planar 5R parallel architecture. (**a**): "+ −", (**b**): "− −", (**c**): "− +", (**d**): "+ −", by denoting the convex (+) or the concave (−) configuration of the left and the right elbow joints, respectively.

2.1. Architecture

The device described in this paper is a 2-DOF parallel kinematic manipulator. It is characterized by a structure made up of four links and a fixed frame connected by five revolute joints. The main reason for the choice of this kind of closed-loop architecture was the possibility of placing both motors on a fixed base. Thanks to this solution, the robot is characterized by a relatively high stiffness and lower moving masses if compared to serial manipulators, therefore providing higher dynamic performances, a lighter structure and, potentially, better positioning accuracy.

With reference to the generic planar parallel five-bar mechanism depicted in Figure 3, the end-effector $P(x, y)$ is connected to the base by two legs, each of which consists of three revolute joints and two links. Joints A_1 and A_2 are connected to the base where they are actuated. The joints at the other end of each actuated link are denoted as B_1 and B_2. A fixed global reference system $O - xy$ is located in the midpoint of the segment $\overline{A_1 A_2}$ with the y axis normal to $\overline{A_1 A_2}$ and the x axis directed along $\overline{A_1 A_2}$. The mechanism is characterized by a symmetric structure where $\overline{OA_1} = \overline{OA_2} = R_3(r_3)$, $\overline{A_1 B_1} = \overline{A_2 B_2} = R_1(r_1)$ and $\overline{B_1 P} = \overline{B_2 P} = R_2(r_2)$. The notation $R_i(i = 1, 2, 3)$ represents the link lengths with dimensions while $r_i(i = 1, 2, 3)$ represents dimensionless lengths of the links. Given:

$$D = \frac{R_1 + R_2 + R_3}{3} \tag{1}$$

One can obtain the three non-dimensional parameters:

$$r_1 = R_1/D, \quad r_2 = R_2/D, \quad r_3 = R_3/D \tag{2}$$

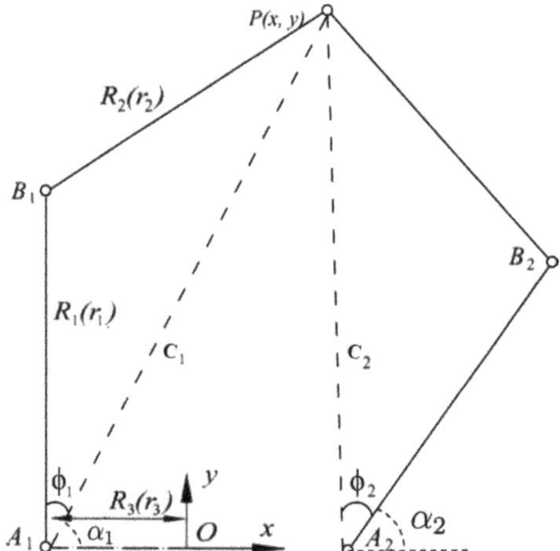

Figure 3. The planar 5R parallel mechanism.

It is important to stress the fact that such an architecture is characterized by four possible configurations "*a*, *b*, *c* and *d*", as shown in Figure 2. However, only configuration "*a*" will be considered in the scope of this paper. Moreover, on the basis of the singularity analysis done in [25], the following constraints must be applied:

1. $r_2 > r_1 + r_3$ in order to avoid the uncertainty singularity where $B_1 P B_1$ is extended.
2. $r_1 > r_3$ and $r_2 > r_3$ in order to have a manipulator with a surrounded workspace

2.2. Inverse Kinematics

The joint variables $\theta = [\theta_1, \theta_2]^T$ are expressed as a function of the end-effector position $P = [x, y]^T$ using the following inverse kinematic equations:

$$\begin{bmatrix} \theta_1 \\ \theta_2 \end{bmatrix} = \begin{bmatrix} \alpha_1 + \arccos\left(\frac{r_1^2 - r_2^2 + (\sqrt{(x-r_3)^2 + y^2})^2}{2r_1\sqrt{(x-r_3)^2 + y^2}} \right) \\ \alpha_2 - \arccos\left(\frac{r_1^2 - r_2^2 + (\sqrt{(x+r_3)^2 + y^2})^2}{2r_1\sqrt{(x+r_3)^2 + y^2}} \right) \end{bmatrix} \tag{3}$$

where

$$\alpha_1 = \begin{cases} \arctan\left(\frac{y}{x-r_3} \right) & \text{if } \arctan\left(\frac{y}{x-r_3} \right) \geq 0 \\ \arctan\left(\frac{y}{x-r_3} \right) + \pi, & \text{otherwise} \end{cases} \tag{4}$$

$$\alpha_2 = \begin{cases} \arctan\left(\frac{y}{x+r_3} \right) & \text{if } \arctan\left(\frac{y}{x+r_3} \right) \geq 0 \\ \arctan\left(\frac{y}{x+r_3} \right) + \pi, & \text{otherwise} \end{cases} \tag{5}$$

2.3. Forward Kinematics

The forward kinematic relations are derived using the variables described in Figure 4. Regarding the notation, m is the midpoint of segment $\overline{B_1 B_2}$, β is the angle between segment $\overline{B_1 B_2}$ and the x-axis, d is the distance between the end-effector P and segment $\overline{B_1 B_2}$, a represents the distance from m to B_2 and, finally, γ is the angle between segment $\overline{B_1 B_2}$ and $\overline{B_2 P}$.

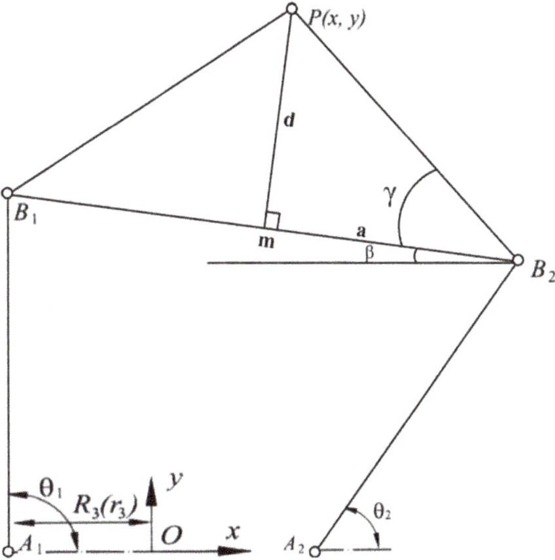

Figure 4. Forward kinematics scheme for up-configuration.

The end-effector position can be calculated as:

$$\mathbf{P} = \begin{bmatrix} m_x + d\cos(\frac{\pi}{2} + \beta) \\ m_y + d\sin(\frac{\pi}{2} + \beta) \end{bmatrix} \tag{6}$$

where:

$$m_x = \frac{r_1(\cos(\theta_1) + \cos(\theta_2))}{2} \tag{7}$$

$$m_y = \frac{r_1(\sin(\theta_1) + \sin(\theta_2))}{2} \tag{8}$$

$$d = r_2\sin(\cos^{-1}(\frac{\sqrt{(r_1(\cos(\theta_1) - \cos(\theta_2)) - 2r_3)^2 + (r_1(\sin(\theta_1) - \sin(\theta_2)))^2}}{2r_2})) \tag{9}$$

$$\beta = \arctan\left(\frac{r_1(\sin(\theta_1) - \sin(\theta_2))}{(r_1(\cos(\theta_1) - \cos(\theta_2)) - 2r_3}\right) \tag{10}$$

2.4. Jacobian

Differentiating Equation (6) with respect to time and rearranging the terms one can obtain:

$$\mathbf{J} = \begin{bmatrix} J_{11} & J_{12} \\ J_{21} & J_{22} \end{bmatrix} \tag{11}$$

where:

$$J_{11} = -\frac{r_1s_1}{2} - 2\cos(C)\frac{Ar_1c_1 - Br_1s_1}{D} - \frac{E}{B^2}(Br_1c_1 + Ar_1s_1) \tag{12}$$

$$J_{12} = -\frac{r_1s_2}{2} + 2\cos(C)\frac{Ar_1c_2 + Br_1s_2}{D} - \frac{E}{B^2}(-Br_1c_2 + Ar_1s_2) \tag{13}$$

$$J_{21} = -\frac{r_1c_1}{2} - 2\sin(C)\frac{Ar_1c_1 - Br_1s_1}{D} + \frac{E'}{B^2}(Br_1c_1 + Ar_1s_1) \tag{14}$$

$$J_{22} = -\frac{r_1c_2}{2} + 2\sin(C)\frac{Ar_1c_2 + Br_1s_2}{D} + \frac{E'}{B^2}(-Br_1c_2 + Ar_1s_2) \tag{15}$$

in which:

$$s_i = \sin(\theta_i), \quad i = 1, 2 \tag{16}$$

$$c_i = \cos(\theta_i), \quad i = 1, 2 \tag{17}$$

$$A = r_1s_1 - r_1s_2 \tag{18}$$

$$B = 2r_3 + r_1c_1 + r_1c_2 \tag{19}$$

$$C = \frac{\pi}{2} + \tan^{-1}(\frac{A}{B}) \tag{20}$$

$$D = 8r_2\sqrt{1 - \frac{B^2 + A^2}{4r_2^2}} \tag{21}$$

$$E = \frac{r_2\sin(C)}{1 + \frac{A^2}{B^2}}\sqrt{1 - \frac{A^2 + B^2}{4r_2^2}} \tag{22}$$

$$E' = \frac{r_2\cos(C)}{1 + \frac{A^2}{B^2}}\sqrt{1 - \frac{A^2 + B^2}{4r_2^2}} \tag{23}$$

3. Mechanical Design

3.1. Workspace

The theoretical reachable workspace for upper arm neurorehabilitation in Cartesian coordinates was defined in [26] through a transformation from articular to Cartesian space, performed using the direct kinematics of the human arm. The inclusive theoretical platform was defined as the union between the workspace defined for minimum limb lengths and the workspace defined for the maximum limb lengths. The resulting workspace is identified by an ellipse with centre c = [0, 513.5] mm, minor axis = 222 mm and major axis = 502.75 mm.

Since the population under study in [26] was right-handed, the authors of that research centred the reachable workspace at x = 55.75 mm. Consequently, the y-axis of PLANarm2 has been translated in order to have it aligned with the centre of the reachable workspace, as shown in Figure 5a. Since the manipulator is designed to be home based, it will be installed on a regular home table or desk. An average sized table is assumed to have a length of, at least, 1500 mm and a width of about 800 mm. Furthermore, the patient must be located at a distance of 200 mm away from the table, as shown in Figure 5b.

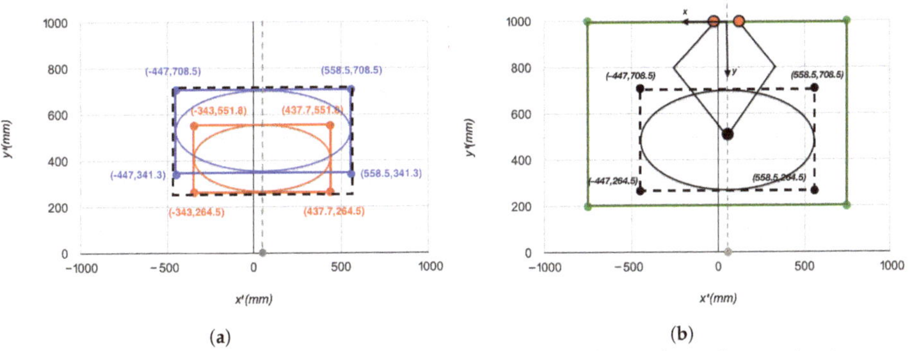

(a)   (b)

Figure 5. (a) Sketch showing the total reachable workspace by combining the workspace of patients with minimum limb lengths and those with maximum limb lengths. (b) Sketch showing the placement of the rehabilitation device in accordance with the reachable workspace.

3.2. Kinematic Optimization

The link lengths have been optimized in order to provide the best kinetostatic performance. The lower r_3 is, the larger the theoretical workspace is [25]. The maximum workspace is obtained when the joints connected to the ground are coaxial ($r_3 = 0$). However, due to mechanical constraints, the lowest possible value of R_3 was chosen to be equal to 45 mm. Based on the reachable workspace, it was sufficient to choose $R_1 + R_2 = 800$ mm. Finally, in order to determine the values of R_1 and R_2, the minimum stiffness and isotropy were optimized over the radial direction. Both indexes are radially symmetric [23] and therefore they are plotted against the y-direction in Figures 6 and 7. The interval of interest is y = [300 mm, 700 mm], which includes the reachable workspace.

1. I_2: The minimum singular value (minimum stiffness)

     The I_2 index corresponding to the minimum singular value is defined as:

$$I_2 = \sigma_{min} = \sqrt{\lambda_{min}} \tag{24}$$

The greater the minimum singular value, the greater the minimum rigidity of the machine. The minimum singular value was plotted against the radial direction of the manipulator for different ratios of r_1/r_2 and the results are shown in Figure 6.

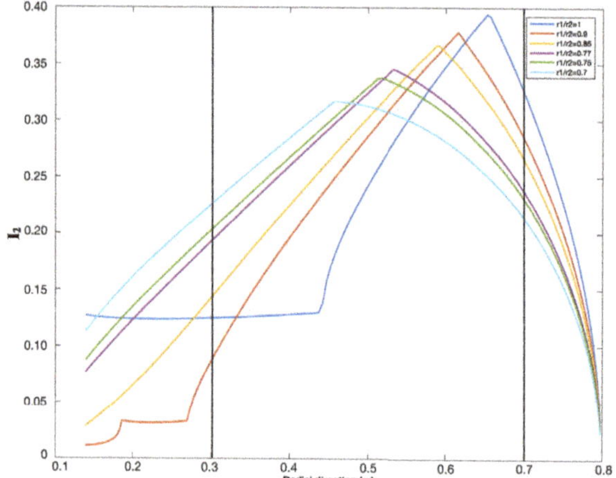

Figure 6. Plot of the minimum singular value over the radial direction for different values of R_1/R_2.

2. I_1: Conditioning number (Isotropy)

    The conditioning number is a measure of the isotropy of the manipulator from the rigidity point of view. It is defined as the ratio between the maximum and minimum singular values of the Jacobian:

$$I_1 = cond(J) = \sqrt{\frac{\lambda_{max}}{\lambda_{min}}} = \frac{\sigma_{max}}{\sigma_{min}} \tag{25}$$

The closer this ratio is to 1, the more consistent the stiffness of the machine will be along the main directions. The conditioning over the workspace is radially symmetric; therefore, in order to understand the behaviour of the conditioning index when changing the length of the link, the conditioning index of the manipulator was plotted against the radial direction (y-direction) for different values of R_1/R_2. The results are shown in Figure 7.

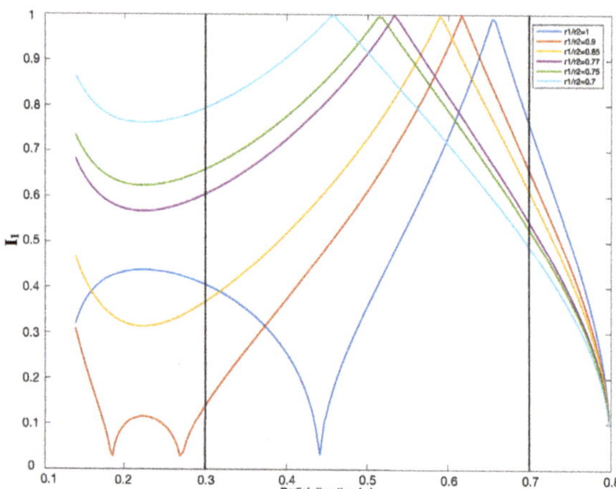

Figure 7. Conditioning index over the radial direction for different values of R_1/R_2.

With reference to Figures 6 and 7, both performance indexes are optimal in the reachable workspace, which lies between 300 mm and 700 mm (as defined in Section 3.1) when $R_1/R_2 = 0.77$. Finally, the obtained link lengths are as follows:

$$R_1/R_2 = 0.77, \; R_1 + R_2 = 800\,\text{mm} => R_1 = 348\,\text{mm}, \; R_2 = 452\,\text{mm}, \; R_3 = 45\,\text{mm}$$

After applying the optimized link lengths, the conditioning index and the manipulability force ellipses were plotted on the workspace and the results are shown in Figure 8.

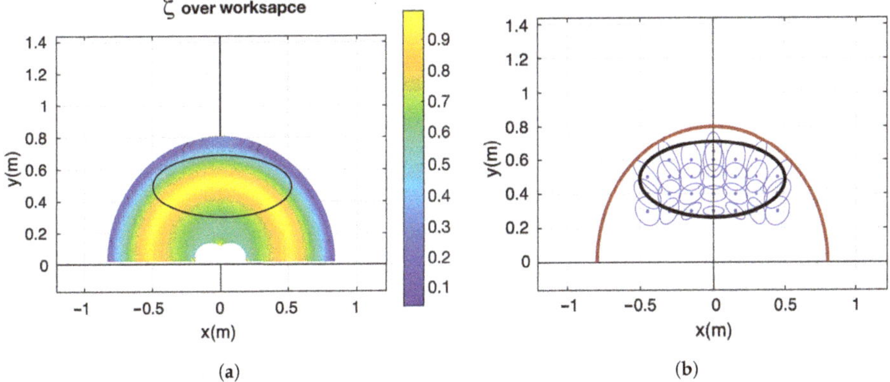

Figure 8. (**a**) Plot of conditioning number over the workspace. (**b**) Plot of the manipulability force ellipses over the workspace.

The conditioning index is larger than 0.6 on the whole workspace and higher than 0.75 on 90% of the workspace. Better manipulability is therefore achieved if compared to the 3DOM architecture [24], which provides a conditioning index larger than 0.2 on the whole workspace, higher than 0.33 on 97% of the workspace and higher than 0.5 on 84% of the workspace. The manipulability force ellipses also reflect an acceptable manipulability index due to their not extremely elongated shape. Moreover, the ellipses plots highlight the symmetric distribution of manipulability, demonstrating an identical kinetostatic performance for right-handed and left-handed users.

3.3. Kinetostatics

The selection of the desired specifications for the actuators was based on the maximum torque and maximum velocity required on the actuated joints.

1. Maximum torque:
   The robot target is 28 N, as the one of the MIT-MANUS [27], taken as a reference value for its considerable clinical exploitation. This force is translated to joints A_1 and A_2 on the basis of Equations (26)–(28).

$$\mathbf{T} = \mathbf{J}^T \cdot \mathbf{F} \tag{26}$$

$$\mathbf{T} = \begin{bmatrix} t_1 \\ t_2 \end{bmatrix} \tag{27}$$

$$\mathbf{F} = \begin{bmatrix} f_x \\ f_y \end{bmatrix} \tag{28}$$

where T is the torque matrix, t_1 and t_2 are the torques transferred to θ_1 and θ_2, respectively, J is the Jacobian, F is the force matrix and f_x and f_y are the forces exerted by the patient in the x-direction and y-direction, respectively. Based on Equations (26)–(28), the maximum torque translated to the actuated joints is calculated to be equal to 11.2 Nm.

2. Maximum velocity:

Considering common neurorehabilitation exercises, the maximum velocity required at the end-effector is assumed to be lower than 0.5 m/s in the Cartesian space. In fact, Krebs et al. states, with experiments, that the tangential velocities for circular movements performed by stroke patients is below 0.5 m/s [28]. They also present linear velocities for point-to-point movements lower than 0.25 m/s. The corresponding angular velocity on the joints depends on the configuration of the manipulator and it is maximal when the minimum singular value of the Jacobian is minimal. Accordingly, the maximum angular velocity needed on the actuated joints was calculated to be equal to 4 rad/s or 38 RPM.

4. Prototype

4.1. Description

On the basis of the considerations reported in the previous sections, the PLANarm2 prototype was developed and assembled. The mechanical assembly of the manipulator is composed of five subsystems: base, motors, transmission, links and end-effector. A proper mechatronic design has to consider that the choice of all the components must be carried out keeping in mind the expected behaviour of the final controlled device.

Impedance and its dual admittance control are today part of the state-of-the-art in physical Human–Robot Interaction (pHRI) and essential control strategies for rehabilitation devices. Impedance control requires a direct force/torque control [29] and backdrivable motors are therefore preferred. High torque and low velocity needed for this application, as reported in Section 3.3, clash with the characteristics of electrical motors that in general express high velocity and low torque. High torque and low velocity electrical motors (i.e., torque motors) are available on the market, but they are generally expensive, and not suitable for the low-cost device described. PLANarm2 is moved by two 24 V motors (EMG49 model from Robot-electronics) equipped with a non-backdrivable 49:1 gearbox (resulting in a no-load speed of 143 rpm and a stall torque of 19.6 Nm) further reduced by a 3:1 pulley belt transmission connected to the corresponding link.

The links have been designed to embed a Cantilever Beam load cell (Model 830, Richmond Industries Ltd., Reading, UK) measuring the torque transmitted through the links actuated by motors. This solution allows to evaluate straightforwardly the torque by measuring the shear force at the cantilever sensor and multiplying it by the length of the link. This motor choice reflects in the impossibility to use an impedance control algorithm in favour of an admittance strategy, which requires force sensing and good position/velocity control. Each actuator is equipped with an incremental encoder sensor with a final resolution on the link rotation of 0.00213 rad. The low-level controller, in line with the affordability nature of the device, is represented by an Arduino DUE board (Atmel SAM3X8E based on a ARM Cortex-M3) that, in conjunction with two VNH5019 motor drivers, provides full control of the robot's movements. The Arduino board, the motor drivers and other electronic elements required to operate, are installed on a PCB and mounted on the device. Closed control loops are processed directly by the Arduino board and not by additional commercial motor drivers. This choice has been made in order to exploit the low cost and versatility of a general purpose microcontroller unit (MCU).The MCU controller communicates with a PC through a serial RS-232. PC will be responsible for the high-level control logic and GUI for robot control.

Proximity sensors are used to detect the end stroke of each arm, as a reference for the incremental encoders. The large majority of the components have been 3D-printed. The complete mechanical structure is shown in Figure 9.

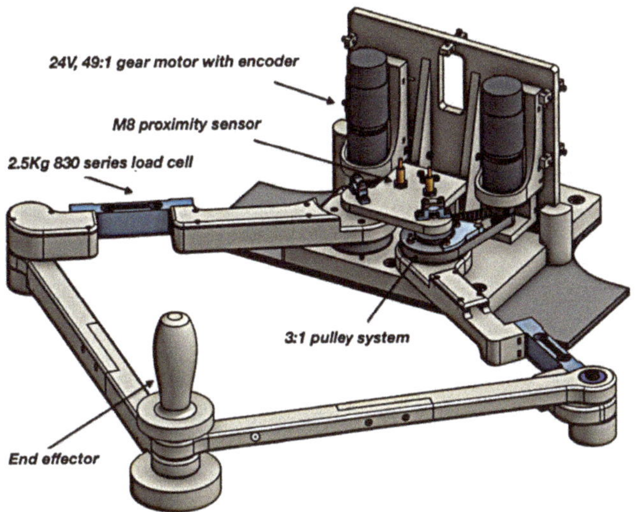

Figure 9. 3D model of the prototype.

4.2. Cost Estimation

The device is designed specifically for being an affordable rehabilitative device with a low-volume production. Given the complexity of some parts of the device and its intrinsic prototype nature, the most flexible and suitable technology for this production volume is Additive Manufacturing (AM) thanks to the ability to run a production without the long term investment in specific tooling and the flexibility on implementing any layout adjustment, upgrade or customization.

The device has been optimized to be produced out of polymeric materials on a Fused Deposition Modelling (FDM) AM machine. The specific machine used to print the device is a Stratasys F370 printer and the used materials are Stratasys ABS and Stratasys QSR soluble support material both in the 1.75 mm filament diameter. The print project consists of 4 different print trays for a total of 97 printing hours, 2292 cm^3 of building material and 446 cm^3 of support material, with additional 25 h of washing time (most of them performed while the machine was printing other subsequent trays). The magnitude of the building cost could be roughly estimated with the cost of building material (Stratasys ABS cost: 0.18 EUR/cm^3 (in 2020)) added to the cost of support material (Stratasys QSR cost: 0.19 EUR/cm^3 (in 2020)) used along the fabrication and is approx € 501. The concept of affordability has been employed for the selection of essential components like electric motors and drivers, load cells and electronics components too, bringing the cost of bought material to a rough total of € 720. For the assembly of the structure, one single operator was able to perform the whole operation during a single working day time with no specific tools and with a few other components like standard metric screws, nuts, ball bearings and pulleys, for an additional rough cost of € 100. Two additional days were required to assemble the electrical and electronic assembly and wiring, for a rough cost of € 200. Regarding the aforementioned observations, with a total estimated cost of € 1521, PLANarm2 could be considered an affordable device for a limited production run.

5. Control

When designing a rehabilitation device, mechatronic and control aspects are equally important. Following [30], it is possible to divide the existing control strategies for neurorehabilitation devices into three main branches: assistive, intended to help patients perform certain movements; corrective, intended to help patients improve their movement accuracy; and resistive, intended to further challenge the patient's capabilities. The authors decided to make available all these control strategies for the user of the PLANarm2 device. In particular, the assistive mode is realized both by a passive *trajectory_controller* and by an active *admittance_controller*, the corrective mode is introduced through a so-called *tunnel_controller* and the resistive mode is implemented as a particular case of the admittance control. In order to develop an effective and modular control architecture, the authors decided to leverage the functionalities of the *ros_control* package [31], available within the Robotic Operative System (ROS) framework [32]. With reference to Figure 10, the control structure is made up of four main components: a controller manager, the set of available controllers, a hardware interface and the real controlled robot. The *controller manager* is responsible for handling the controllers implemented in the system; it activates, deactivates and switches them depending on the user's command. Once a specific *controller* is activated, it has access to the current state of the robot and, depending on its internal algorithm, it can use that information to compute the next command to be sent to the robot. This back and forth data transmission is made possible by the *hardware interface* component of the control architecture, in charge of interfacing the software portion of the system with the hardware one through the *read()* and *write()* methods. For this specific application, the *hardware_interface* has also been equipped with a dummy transmission performing the transformation between Cartesian space and actuator space so that commands can be computed more intuitively referring to the Cartesian reference frame.

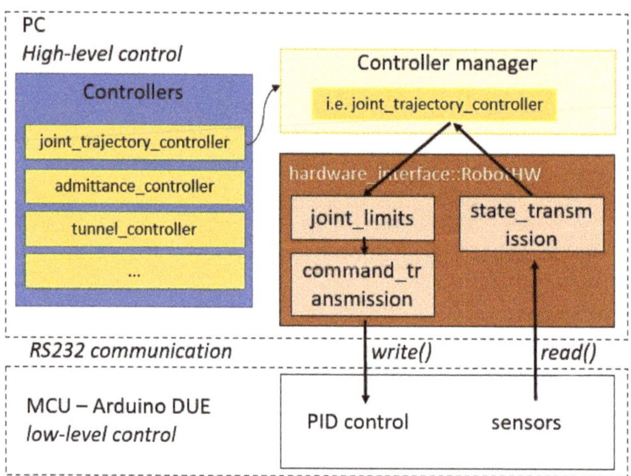

Figure 10. Control architecture for PLANarm2.

As from Figure 10, the control structure could be split in low-level control and high-level control. The low-level portion of the control architecture is represented by the PID loops for position and velocity control running on the Arduino DUE board. These capabilities are often built-in for commercial robotic devices but, in this case, given the use of a general purpose Arduino DUE for cost-effectiveness and flexibility reasons, they must be redesigned from scratch. The high-level portion is implemented on a PC to exploit the *ros_control* capabilities.

5.1. Low-Level Control

A fundamental requirement for the implementation of the PID loops is to guarantee a fixed control time step. This has been achieved by equipping the micro-controller with ChibiOS [33], an efficient open-source Real Time Operative System (RTOS) specifically designed for embedded applications. Using ChibiOS, it is possible to guarantee a time step of 1 ms for measure and control, while ensuring a 200 Hz communication with PC via serial interface.

As highlighted in Section 4, the chosen actuators are sold with an embedded incremental encoder for precise position measurement. However, no velocity sensor was installed on the motors and therefore the speed value had to be estimated using the PID loop depicted in Figure 11.

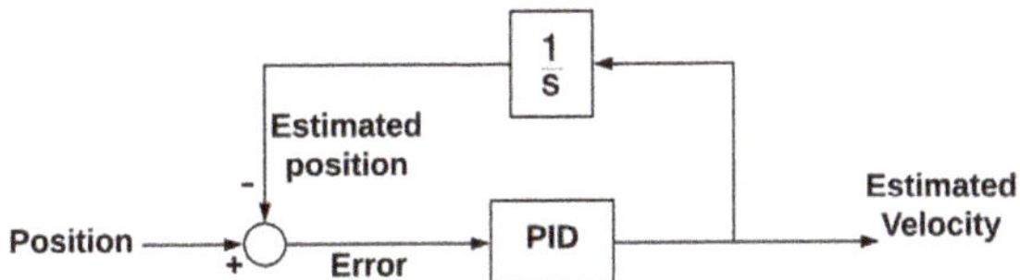

Figure 11. Feedback loop of the velocity estimator.

The performance of the velocity estimator was then analysed in terms of frequency response and the corresponding Bode diagram is reported in Figure 12.

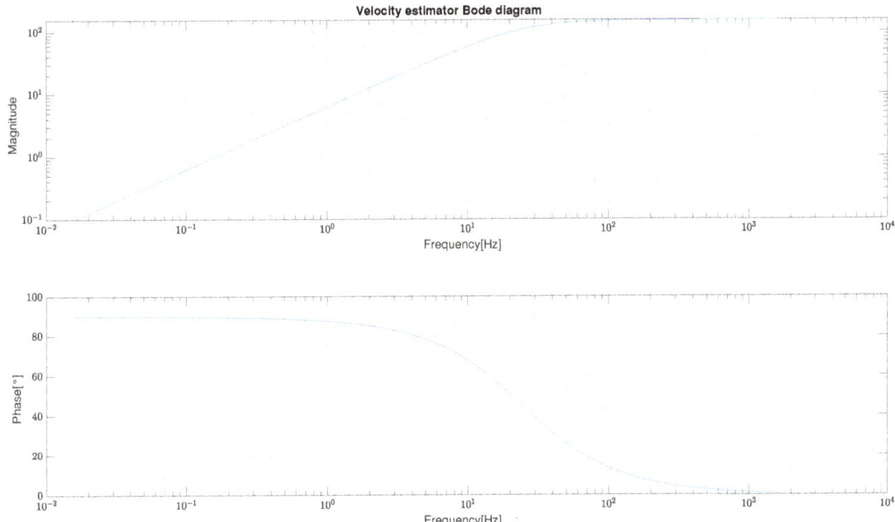

Figure 12. Bode diagram of the velocity estimator.

The *velocity_controller* was then realized using a common PID loop and the speed estimator just introduced. Similarly, the *position_controller* was implemented by encapsulating the *velocity_controller* within an additional PID loop, as schematized in Figure 13.

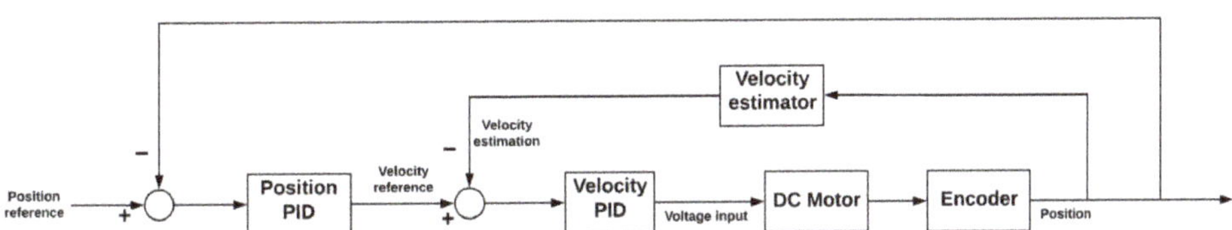

Figure 13. Feedback loop of position controller.

With reference to Figure 14, the response of the *position_controller* to a step input is characterized by a 5% overshoot, a settling time of 0.21 s and a steady state error lower than 0.5%, which is considered acceptable for the application of interest.

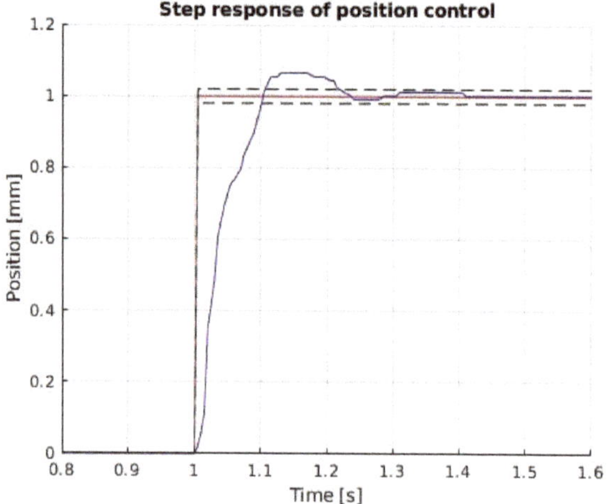

Figure 14. Step response of position controller.

5.2. High-Level Control

Now, on top of the basic robot functionalities just presented, it is possible to implement the rehabilitation-specific controllers introduced at the beginning of this section.

5.2.1. Trajectory Controller

The *trajectory_controller* can be used to perform passive rehabilitation exercises. Since it is a common tool, the authors decided to exploit the so-called *joint_trajectory_controller* [34], available as part of the *ros_control* package. This controller takes as input trajectories specified as a set of waypoints to be reached at specific time instants and attempts to execute them as well as the mechanism allows. The interpolation between waypoints can be performed using linear, cubic or quintic 1D splines, depending on the level of continuity that has to be guaranteed. For this specific project, the authors chose to specify a desired position, velocity and acceleration for each waypoint and then used quintic splines interpolation to ensure continuity at the acceleration level. Thanks to the trajectory controller, PLANarm2 is capable of following any path that lays within the workspace of the robot with the performances achieved by the position controller, described in Section 5.1.

5.2.2. Admittance Controller

Starting from Hogan's work [29], indirect force control strategies such as impedance and its dual admittance control can be considered the most proper and efficient way to control a robot interacting with its environment. As highlighted in [30], impedance and admittance control are also the simplest and probably most used way to carry out an assistance-as-needed control in robotic neurorehabilitation. The possibility to change on-line their parameters, and therefore the robot's behaviour, also allows to sophisticate the algorithm in several ways. As reported in Section 4, in order to guarantee the device's simplicity and affordability, it is not possible to realize a direct effort control. This seems to clash with the need to realize a haptic device and, for this reason, the authors choose a strategy similar to the one described in [35]. Given a reference force $F_r(t)$, coming from the digital environment connected to the device, it is possible to control the motors with a velocity reference (v_r) obtained through a PI control loop over the force error F_e, where $F_e(t) = F_r(t) - F_m(t)$ with F_m being the measured force. For the sake of simplicity, Equation (29) has been written only for one of the controlled joints:

$$v_r = \frac{1}{D_e} \cdot F_e(t) + K_i \cdot \int_0^t F_e(t')dt' \tag{29}$$

The proportional parameter in Equation (29) is called $\frac{1}{D_e}$ to highlight that the transparency felt by the user will increase while D_e, that can be associated to a virtual damping, decreases. A proper choice of these parameters must also take into account the disturbance rejection.

5.2.3. Tunnel Controller

Corrective rehabilitation is proven effective when aiming to improve motion coordination. To provide this functionality, the authors decided to develop a so-called *tunnel_controller*, similar to what is presented in [36]. The controller takes as input a predefined trajectory and builds a virtual tunnel of user-defined width around it. The patient is allowed to move freely along the path and, whenever the tunnel's boundaries are exceeded, a restoring force is produced in order to correct the undesired movement. A schematic representation of this concept is reported in Figure 15.

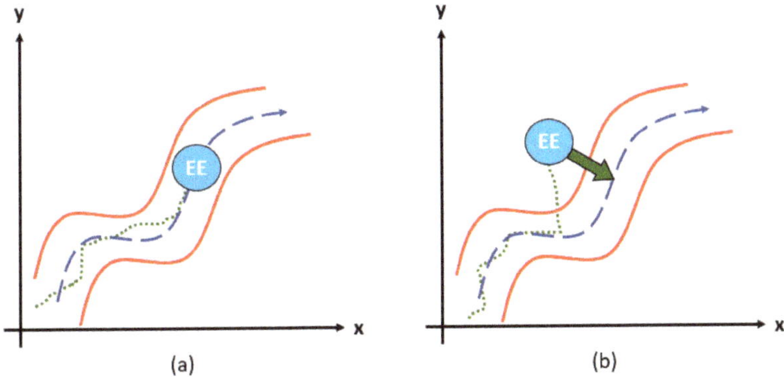

Figure 15. Schematic tunnel representation. On the left (**a**), the end-effector moves freely within the tunnel. On the right (**b**), a restoring force brings the end-effector back within the tunnel.

Differently from the *trajectory_controller*, for which input trajectories are time-parametrized, the *tunnel_controller* requires paths expressed in terms of curvilinear abscissa s. In order to guarantee coherence with the other controllers, a method that automatically transforms a time-parametrized trajectory into its corresponding s-version has been implemented so that the same computed trajectory can be applied to all the available controllers. In addition, a new coordinate system $(\vec{t},\vec{n})$ has been defined on the trajectory $f(s)$ at any instant, denoting by $\vec{t}$ and $\vec{n}$ the tangential and the normal vectors, respectively, where $\vec{n} \times \vec{t} = \vec{x} \times \vec{y}$, as shown in Figure 16.

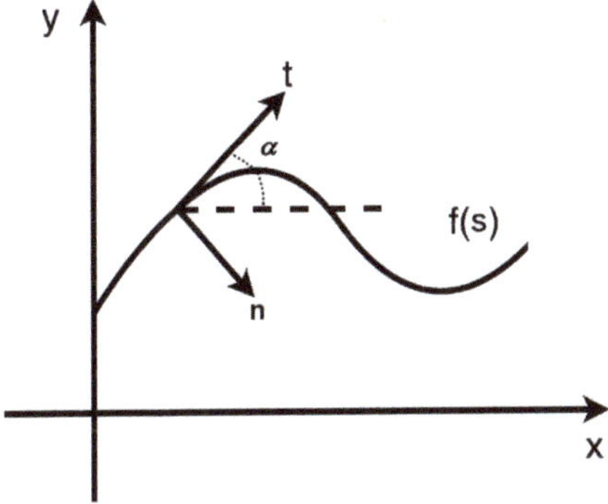

Figure 16. Reference frame for tunnel control.

The patient's force on the end-effector is projected from the Cartesian reference frame to the new reference frame according to the instantaneous slope α of the requested trajectory. Then, the controller's basic working principle is similar to the one of the *admittance_controller*. For every control cycle, the normal distance n_{ee} between desired and actual position of the end-effector, with respect to the given trajectory, is calculated. If that distance is smaller than the user-defined tunnel half-width W, tangential and normal measured forces are given as input to a high-level PI loop set with a reference of 0N. On the contrary, if the end-effector is detected outside said tunnel, the force F_{refN} used as reference for the PI loop related to the normal direction is computed as in Equation (30), where K_v represents the stiffness of the virtual spring responsible for the generation of the corrective force.

$$F_{refN} = n_{ee} \cdot K_v \tag{30}$$

The effect of this approach is that the patient is allowed to move freely inside the virtual tunnel but, whenever the boundaries are exceeded, a virtual spring generates a corrective force that compensates the error and guides the end-effector back inside the tunnel. On top of this, an acceleration limit has been implemented within the controller's logic for safety reasons: if any spasm or sudden movement of the patient occurs, it can be absorbed.

6. Experimental Assessment

This section presents the results of the experimental assessments performed on the developed prototype. The objective of these experimental tasks is to carry out functional tests able to confirm the goodness of the mechatronic project and control structure chosen. Improvements on the algorithms presented in this work are already under consideration by the authors. Notice that the performance of the *trajectory_controller* is directly connected to the results obtained for the low-level *position_controller* reported in Section 5.1 and therefore not reported here for brevity. However, data collected for the *admittance_controller* and the *tunnel_controller* together with an analysis on the accuracy of the position measurements are discussed in detail hereafter.

6.1. Admittance Controller Validation

The admittance controller was tested on the PLANarm2 prototype. The algorithm has been implemented starting from Equation (29). The final implemented algorithm is slightly different from the ideal case since, for instance, the noise affecting the measured force must be considered. For this reason, the signal coming from the sensors is processed with a simple exponential filter. This filter can be expressed by the formula:

$$F_m(n) = F_m(n-1) \cdot (1 - \alpha) + F(n) \cdot \alpha \tag{31}$$

In this case, α is taken as $\alpha = 1 - e^{-dt \cdot 2\pi f_{cutoff}}$ with f_{cutoff} the ideal cut off frequency. This filter was chosen because of its simplicity and functionality. Its frequency response is represented in Figure 17, showing a magnitude >70% before the cutoff frequency.

Different kinds of filters (n-order filters) are under consideration of the authors in order to improve the performances in terms of admittance readiness. Starting from the filtered force measures, the PI control loop (following Equation (29)) is implemented on the high-level control hardware, running at 200 Hz. The *admittance_controller*'s performance is presented in terms of frequency response function between measured force (F_m) and measured velocity on a single axis (y axis considering the reference presented in Figure 3) in a particular configuration ($x = 0$, $y = 0.5$). Similar results could be obtained for the perpendicular axis.

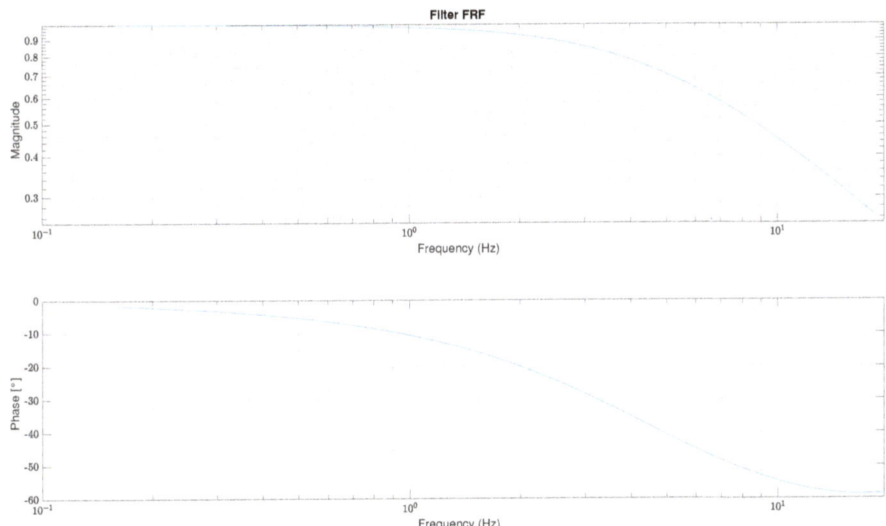

Figure 17. Exponential filter response function with $f_{cutoff} = 5\,\text{Hz}$.

In Figure 18, the frequency response function (FRF) for the admittance/force-tracking control is depicted. The graph shows a coherence >80% until 5 Hz of frequency, meaning that the output measured can be considered related to the input. Phase is quite constant until 7 Hz and shows a small delay for the frequency range between 0 and 7 Hz. The magnitude trend begins with a fall due to the pole in the origin and the stabilizes around 7 Hz.

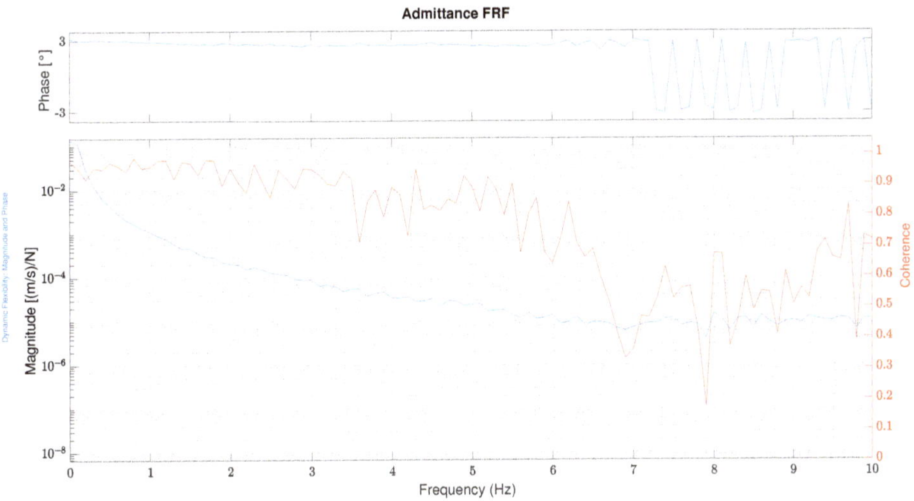

Figure 18. Admittance frequency response function (FRF) with $D_e = 20$ and $k_i = 0.01$.

6.2. Tunnel Controller Validation

As explained in Section 5.2, as long as the end-effector remains within the tunnel's boundaries, the *tunnel_controller* is based on the same working principle of the *admittance_controller*. For this reason, the assessment of the behaviour of PLANarm2 in those conditions is redundant and is not discussed here. On the other hand, it is interesting to see what happens whenever the end-effector is guided against the mentioned boundaries. The controller was set up with the following parameters: $K_p = 0.05$ and $K_i = 0.01$ for the force tracking loop, $K_v = 500$ N/m and $W = 0.01$ m for the virtual stiffness and the half-width of the tunnel, respectively. Figure 19 reports the data collected while moving the end-effector along a certain predefined trajectory. The top plot represents the normal distance from the given trajectory against time together with an indication of the tunnel's boundaries (black dashed line). It can be easily noticed that during the experimental run the end-effector was driven outside of those boundaries a few times. The plot in the middle depicts the trend of the force exerted on the end-effector in the direction normal to the desired trajectory, while the bottom plot reports the trend of the corrective force produced by the virtual spring. As shown, the end-effector is free to move within the tunnel boundaries with the same performances highlighted for the *admittance_controller*. However, as soon as the end-effector is driven against the tunnel boundaries, the force required to further increase the normal distance from the trajectory rises due to the corrective force generated by the virtual spring. It is worth mentioning that, as can be seen in Figure 19, a certain degree of discontinuity in the force produced by the virtual environment has been maintained. This choice is justified by the fact that thanks to the discontinuity itself, the patient can intuitively "feel" the contact with the tunnel's boundaries and try to autonomously correct its motion.

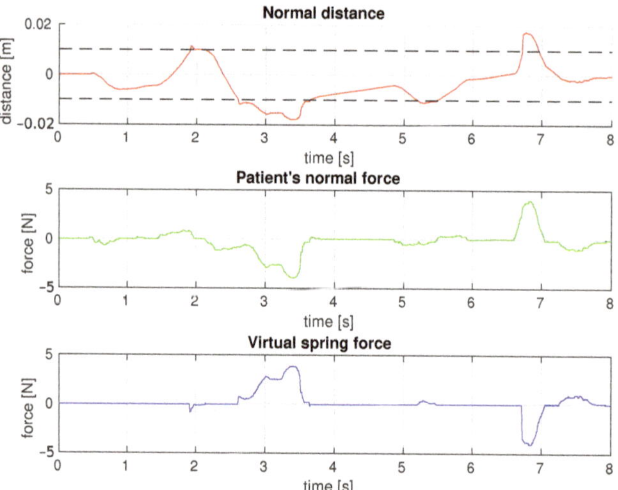

Figure 19. The top plot shows the normal distance from the desired trajectory against time, the middle plot shows the normal force applied on the end-effector against time and the bottom plot shows the force produced by the virtual spring.

6.3. Position Measurement Accuracy

Each actuated joint is driven, through a proper transmission system, by a motor with an embedded incremental encoder. The measured position is then transformed to the joint space by multiplying by the gear ratio of the transmission system (3:1). Finally, the position in the joint space reference is converted to the Cartesian space reference. This procedure of obtaining the position in the Cartesian reference, along with the inaccuracies generated in the embedded encoder, contribute to the generation of a measurement error. In order to quantify this measurement error, a test to measure the accuracy of the device was performed.

6.3.1. Test Bench

In order to measure the actual position of the end-effector, a Vicon marker-based motion capture system was used. The system was setup with 10 cameras tracking the motion of reflective trackers installed on the device. A table was placed within the area under the scope of the cameras and the PLANarm2 device was installed on it. Then, three markers were installed on the planar manipulator as shown in Figure 20: two were placed on the base to act as reference frames, and one was installed on the end-effector to track its position.

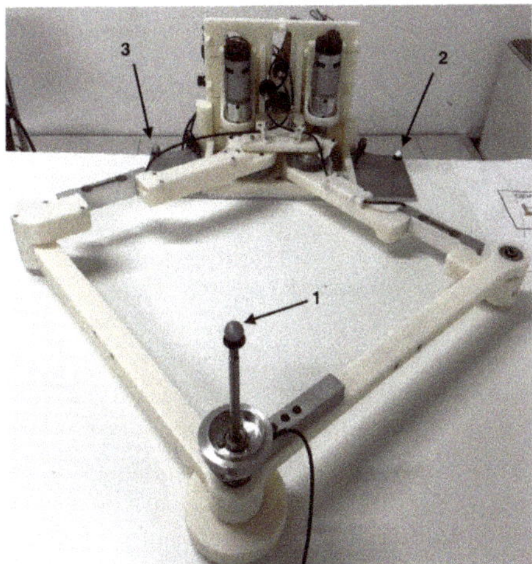

Figure 20. Placement of the markers on the prototype for the experimental procedure.

6.3.2. Data Analysis

After recording the position of the end-effector using the Vicon Nexus software, the results were plotted against the measurements taken by the encoder, along a generic, irregularly-shaped trajectory.

As it can be seen from Figure 21, there exists a small error when comparing the position taken from the motion capture cameras and the position recorded from encoder. This error arises from the combination of different factors, including encoder uncertainties as well as mechanical measurements inaccuracies related to the lengths of the links. Moreover, mechanical backlash is another source of error, as can be noticed in the inversion of the motion. However, the maximum error recorded when comparing the two results was 0.02 m and the average was 0.009 m, considered acceptable for the final application.

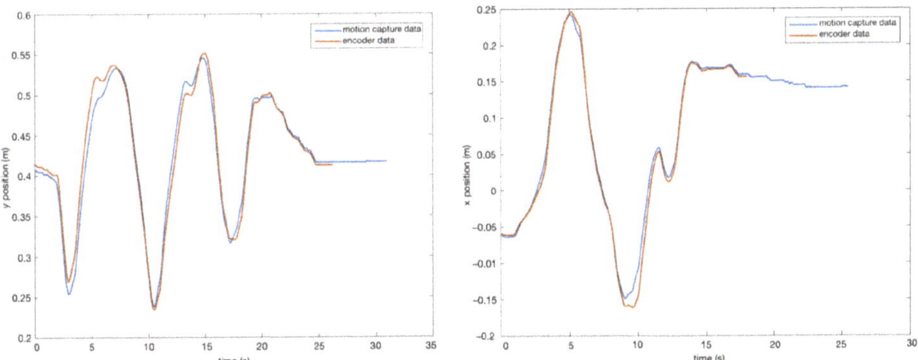

Figure 21. Position of motion capture and encoder vs. time: x-position (**right**) and y-position (**left**).

7. Conclusions

As a result of the increasing number of patients suffering disabilities due to stroke, many research groups have proposed devices aimed at facilitating the rehabilitation process. However, most of these devices are technically advanced and designed for clinical use. This paper presents the prototype of an affordable device for upper-limb neurorehabilitation based on a planar five-bar parallel kinematic mechanism.

The optimal link lengths were obtained by optimizing the conditioning index and the minimum singular value of the Jacobian over the workspace. Components were chosen starting from kinematic and dynamic evaluations as well as on the desired performances. A 3D-printed prototype was presented and the main components and characteristic were analysed. Different kinds of controllers were implemented in order to verify the effectiveness of the prototype and the goodness of the design. Both active and passive controllers were tested and the measured performances showed a good dynamic behaviour. In order to validate the measurements of the end-effector position, a test procedure was followed. The position of the end-effector was recorded using motion capture cameras and compared to the measurements obtained from the encoders. It was shown that the measurements taken by the encoders are accurate enough for the target application.

Next steps will include more refined admittance and assist-as-needed control algorithms, starting from the obtained results and considering new improvements, in order to assist the patients in performing the required tasks according to their capabilities. Finally, a graphic user interface is being implemented in order provide a visual feedback to the patient while performing rehabilitation tasks.

Author Contributions: Conceptualization, J.Y. and M.M.; data curation, J.Y., A.P. and M.L.N.; formal analysis, J.Y.; funding acquisition, M.M.; investigation, J.Y., A.P. and M.L.N.; methodology, J.Y.; resources, T.D. and M.M.; software, J.Y., A.P. and M.L.N.; supervision, H.G. and M.M.; validation, J.Y., A.P. and M.L.N.; visualization, J.Y., A.P. and M.L.N.; writing—original draft, J.Y.; writing—review and editing, J.Y., A.P., M.L.N and M.M. All authors have read and agreed to the published version of the manuscript.

Funding: This work was partially funded by Fondazione Cariplo within the EMPATIA@Lecco project, Rif. 2016-1428 Decreto Regione Lombardia 6363 del 30/05/2017.

Acknowledgments: The authors would like to thank: João Carlos Dalberto and Roberto Bozzi for supporting the mechanical and electrical realization of the prototype; Alessandra Tafaro and Roberta Pozzi for the administrative support of the project.

Conflicts of Interest: The authors declare no conflict of interest. The funders had no role in the design of the study; in the collection, analyses, or interpretation of data; in the writing of the manuscript, or in the decision to publish the results.

References

1. Millán, M.; Dávalos, A. The Need for New Therapies for Acute Ischaemic Stroke. *Cerebrovasc. Dis.* **2006**, *22* (Suppl. S1), 3–9. [CrossRef]
2. Maciejasz, P.; Eschweiler, J.; Gerlach-Hahn, K.; Jansen-Troy, A.; Leonhardt, S. A survey on robotic devices for upper limb rehabilitation. *J. Neuroeng. Rehabil.* **2014**, *11*, 3. [CrossRef] [PubMed]
3. Zhang, X.; Yue, Z.; Wang, J. Robotics in Lower-Limb Rehabilitation after Stroke. *Behav. Neurol.* **2017**, *2017*, 1–13, [CrossRef] [PubMed]
4. Krebs, H.; Hogan, N.; Volpe, B.; Aisen, M.; Edelstein, L.; Diels, C. Overview of clinical trials with MIT-MANUS: A robot-aided neuro-rehabilitation facility. *Technol. Health Care* **1999**, *7*, 419–423. [CrossRef] [PubMed]
5. Marchal-Crespo, L.; Reinkensmeyer, D.J. Review of control strategies for robotic movement training after neurologic injury. *J. Neuroeng. Rehabil.* **2009**, *6*, 20, [CrossRef] [PubMed]
6. Toth, A.; Fazekas, G.; Arz, G.; Jurak, M.; Horvath, M. Passive robotic movement therapy of the spastic hemiparetic arm with REHAROB: report of the first clinical test and the follow-up system improvement. In Proceedings of the 9th International Conference on Rehabilitation Robotics (ICORR), Chicago, IL, USA, 28 June–1 July 2005; pp. 127–130. [CrossRef]

7. Malosio, M.; Pedrocchi, N.; Tosatti, L.M. Robot-assisted upper-limb rehabilitation platform. In Proceedings of the 2010 5th ACM/IEEE International Conference on Human-Robot Interaction (HRI), Osaka, Japan, 2–5 March 2010; pp. 115–116. [CrossRef]

8. Li, G.; Cai, S.; Xie, L. Cooperative Control of a Dual-arm Rehabilitation Robot for Upper Limb Physiotherapy and Training. In Proceedings of the 2019 IEEE/ASME International Conference on Advanced Intelligent Mechatronics (AIM), Hong Kong, China, 8–12 July 2019; pp. 802–807. [CrossRef]

9. Prochazka, A. Passive Devices for Upper Limb Training. In *Neurorehabilitation Technology*; Springer: London, UK, 2012; pp. 159–171.

10. Rehmat, N.; Zuo, J.; Meng, W.; Liu, Q.; Xie, S.Q.; Liang, H. Upper limb rehabilitation using robotic exoskeleton systems: A systematic review. *Int. J. Intell. Robot. Appl.* **2018**, *2*, 283–295, [CrossRef]

11. Qassim, H.M.; Wan Hasan, W.Z. A Review on Upper Limb Rehabilitation Robots. *Appl. Sci.* **2020**, *10*, 6976, [CrossRef]

12. Gull, M.A.; Bai, S.; Bak, T. A Review on Design of Upper Limb Exoskeletons. *Robotics* **2020**, *9*, 16, [CrossRef]

13. Zuccon, G.; Bottin, M.; Ceccarelli, M.; Rosati, G. Design and Performance of an Elbow Assisting Mechanism. *Machines* **2020**, *8*, 68, [CrossRef]

14. Cafolla, D.; Russo, M.; Carbone, G. CUBE, a Cable-driven Device for Limb Rehabilitation. *J. Bionic Eng.* **2019**, *16*, 492–502, [CrossRef]

15. Avizzano, C.A.; Satler, M.; Cappiello, G.; Scoglio, A.; Ruffaldi, E.; Bergamasco, M. MOTORE: A mobile haptic interface for neuro-rehabilitation. In Proceedings of the 2011 RO-MAN, Atlanta, GA, USA, 31 July–3 August 2011; pp. 383–388. [CrossRef]

16. Nam, H.S.; Hong, N.; Cho, M.; Lee, C.; Seo, H.G.; Kim, S. Vision-Assisted Interactive Human-in-the-Loop Distal Upper Limb Rehabilitation Robot and its Clinical Usability Test. *Appl. Sci.* **2019**, *9*, 3106, [CrossRef]

17. Zadravec, M.; Matjačić, Z. Planar arm movement trajectory formation: An optimization based simulation study. *Biocybern. Biomed. Eng.* **2013**, *33*, 106–117, [CrossRef]

18. Casadio, M.; Sanguineti, V.; Morasso, P.G.; Arrichiello, V. Braccio di Ferro: A new haptic workstation for neuromotor rehabilitation. *Technol. Health Care* **2006**, *14*, 123–142. [CrossRef] [PubMed]

19. Chaparro-Rico, B.D.M.; Cafolla, D.; Ceccarelli, M.; Castillo-Castaneda, E. NURSE-2 DoF Device for Arm Motion Guidance: Kinematic, Dynamic, and FEM Analysis. *Appl. Sci.* **2020**, *10*, 2139. [CrossRef]

20. Wu, Q.; Chen, B.; Wu, H. Adaptive Admittance Control of an Upper Extremity Rehabilitation Robot With Neural-Network-Based Disturbance Observer. *IEEE Access* **2019**, *7*, 123807–123819. [CrossRef]

21. Zollo, L.; Accoto, D.; Torchiani, F.; Formica, D.; Guglielmelli, E. Design of a planar robotic machine for neuro-rehabilitation. In Proceedings of the IEEE International Conference on Robotics and Automation, ICRA 2008, Pasadena, CA, USA, 19–23 May 2008; pp. 2031–2036. [CrossRef]

22. Campolo, D.; Tommasino, P.; Gamage, K.; Klein, J.; Hughes, C.M.; Masia, L. H-Man: A planar, H-shape cabled differential robotic manipulandum for experiments on human motor control. *J. Neurosci. Methods* **2014**, *235*, 285–297. [CrossRef]

23. Giberti, H.; Cinquemani, S.; Ambrosetti, S. 5R 2dof parallel kinematic manipulator—A multidisciplinary test case in mechatronics. *Mechatronics* **2013**, *23*, 949–959. [CrossRef]

24. Klein, J.; Roach, N.; Burdet, E. 3DOM: A 3 Degree of Freedom Manipulandum to Investigate Redundant Motor Control. *IEEE Trans. Haptics* **2014**, *7*, 229–239. [CrossRef]

25. Liu, X.J.; Wang, J.; Pritschow, G. Kinematics, singularity and workspace of planar 5R symmetrical parallel mechanisms. *Mech. Mach. Theory* **2006**, *41*, 145–169. [CrossRef]

26. Corona-Acosta, I.P.; Castillo-Castaneda, E. Dimensional Synthesis of a Planar Parallel Manipulator Applied to Upper Limb Rehabilitation. In *Multibody Mechatronic Systems*; Ceccarelli, M., Hernández Martinez, E.E., Eds.; Springer International Publishing: Cham, Switzerlands, 2015; pp. 443–452.

27. Krebs, H.I.; Ferraro, M.; Buerger, S.P.; Newbery, M.J.; Makiyama, A.; Sandmann, M.; Lynch, D.; Volpe, B.T.; Hogan, N. Rehabilitation robotics: Pilot trial of a spatial extension for MIT-Manus. *J. Neuroeng. Rehabil.* **2004**, *1*, 5. [CrossRef]

28. Krebs, H.I.; Aisen, M.L.; Volpe, B.T.; Hogan, N. Quantization of continuous arm movements in humans with brain injury. *Proc. Natl. Acad. Sci. USA* **1999**, *96*, 4645–4649. [CrossRef] [PubMed]

29. Hogan, N. Impedance Control: An Approach to Manipulation: Part I—Theory. *J. Dyn. Syst. Meas. Control* **1985**, *107*, 1–7. [CrossRef]

30. Proietti, T.; Crocher, V.; Roby-Brami, A.; Jarrassé, N. Upper-Limb Robotic Exoskeletons for Neurorehabilitation: A Review on Control Strategies. *IEEE Rev. Biomed. Eng.* **2016**, *9*, 4–14. [CrossRef] [PubMed]

31. Chitta, S.; Marder-Eppstein, E.; Meeussen, W.; Pradeep, V.; Tsouroukdissian, A.; Bohren, J.; Coleman, D.; Magyar, B.; Raiola, G.; Lüdtke, M.; et al. Ros_control: A generic and simple control framework for ROS. *J. Open Source Softw.* **2017**, *2*, 456. [CrossRef]

32. Quigley, M. ROS: An open-source Robot Operating System. In Proceedings of the ICRA 2009, Kobe, Japan, 12–17 May 2009.

33. ChibiOS Free Embedded RTOS. Available online: http://www.chibios.org/ (accessed on 18 November 2020).

34. Joint Trajectory Controller. Available online: http://wiki.ros.org/joint_trajectory_controller (accessed on 18 November 2020).

35. Seraji, H. Adaptive admittance control: An approach to explicit force control in compliant motion. In Proceedings of the 1994 IEEE International Conference on Robotics and Automation, San Diego, CA, USA, 8–13 May 1994; Volume 4, pp. 2705–2712. [CrossRef]

36. Ding, B.; Ai, Q.; Liu, Q.; Meng, W. Path Control of a Rehabilitation Robot Using Virtual Tunnel and Adaptive Impedance Controller. In Proceedings of the 2014 Seventh International Symposium on Computational Intelligence and Design, Hangzhou, China, 13–14 December 2014; Volume 1, pp. 158–161. [CrossRef]

Publisher's Note: MDPI stays neutral with regard to jurisdictional claims in published maps and institutional affiliations.

Article

Kinematic Optimization for the Design of a Collaborative Robot End-Effector for Tele-Echography [†]

Alessandro Filippeschi [1,2,*], Pietro Griffa [3] and Carlo Alberto Avizzano [1,2]

1 TeCIP Institute, Scuola Superiore Sant'Anna, Via Moruzzi 1, 56124 Pisa, Italy; c.avizzano@santannapisa.it
2 Department of Excellence in Robotics and AI, Scuola Superiore Sant'Anna, 56127 Pisa, Italy
3 ETH Zurich, Rämistrasse 101, 8092 Zürich, Switzerland; griffap@student.ethz.ch
* Correspondence: a.filippeschi@santannapisa.it; Tel.: +39-0508-82307
† This paper is an extended version of our paper published in Griffa, P.; Filippeschi, A.; Avizzano, C.A. Kinematic Optimization for the Design of a UR5 Robot End-Effector for Cardiac Tele-Ultrasonography. In Proceedings of The 3rd International Conference of IFToMM Italy, held online, 9–11 September 2020.

Abstract: Tele-examination based on robotic technologies is a promising solution to solve the current worsening shortage of physicians. Echocardiography is among the examinations that would benefit more from robotic solutions. However, most of the state-of-the-art solutions are based on the development of specific robotic arms, instead of exploiting COTS (commercial-off-the-shelf) arms to reduce costs and make such systems affordable. In this paper, we address this problem by studying the design of an end-effector for tele-echography to be mounted on two popular and low-cost collaborative robots, i.e., the Universal Robot UR5, and the Franka Emika Panda. In the case of the UR5 robot, we investigate the possibility of adding a seventh rotational degree of freedom. The design is obtained by kinematic optimization, in which a manipulability measure is an objective function. The optimization domain includes the position of the patient with regards to the robot base and the pose of the end-effector frame. Constraints include the full coverage of the examination area, the possibility to orient the probe correctly, have the base of the robot far enough from the patient's head, and a suitable distance from singularities. The results show that adding a degree of freedom improves manipulability by 65% and that adding a custom-designed actuated joint is better than adopting a native seven-degrees-freedom robot.

Keywords: design synthesis; kinematic optimization; telemedicine; human robot interaction

Citation: Filippeschi, A.; Griffa, P.; Avizzano, C.A. Kinematic Optimization for the Design of a Collaborative Robot End-Effector for Tele-Echography. *Robotics* **2021**, *10*, 8. https://doi.org/10.3390/robotics10010008

Received: 12 November 2020
Accepted: 26 December 2020
Published: 1 January 2021

Publisher's Note: MDPI stays neutral with regard to jurisdictional claims in published maps and institutional affiliations.

1. Introduction

The aging of the population makes the need for medical examinations increase every year. The available specialists are insufficient to meet this need and this shortage will worsen in the forthcoming years. Tele-medicine is a viable solution to cope with this trend and to serve areas far from hospitals.

Tele-medicine services available or under development in many of the WHO (World Health Organization) countries are typically focused on sharing examination results among specialists such as in the case of tele-radiology, tele-pathology, tele-dermatology, and tele-psychiatry [1]. However, advances in robotic and computer graphics technologies fostered the development of robotic telemedicine systems. Examples include endoscopy [2], ultrasonography [3,4] and palpation [5–7]. Among these examinations, ultrasonography (USG) is one of the most important to make a decision on a patient's need to be directed to a specialist.

In the early robotic systems designed for telemedicine (e.g., [3,8]), robotic arms were designed on purpose to place the USG probe on the patient's body. In recent years, the availability of affordable robotic arms, e.g., from Universal Robots (Energivej 25 DK-5260, Odense, Denmark) and Franka Emika (Infanteriestraße 19, 80797, Munich, Germany), has enabled the possibility to drastically reduce the costs of such systems. The most

155

used commercial robotic arm for tele-USG is the UR5 from Universal Robots, which has 6 DoFs (degrees of freedom). However, to the authors' knowledge, none of the proposed systems have investigated the possibility to add a degree of freedom to ease the remote manipulation of the probe.

In this paper, we study the advantages of adding such a DoF either to an end-effector to be mounted on the UR5 robot or by considering a COTS 7-DoFs arm such as the Panda by Franka Emika. We define a force manipulability metric based on the USG task and, based on this metric, we optimize the design of the end-effector in three cases: first, the end-effector is mounted on the UR5 robot and it has no DoFs with regard to the robotic arm's tip; second, one DoF is add by a rotational joint whose axis is perpendicular to the probe axis; and, third, the end-effector is mounted on the Panda robot with no additional DoFs. This work follows a preliminary study presented in [9], in which the general Yoshikawa manipulability index was adopted and the Franka Emika Panda robot was not considered. Concerning this work, we adopt an optimization metric more specific to the task, obtaining significantly different results.

The paper firstly introduces USG and current approaches to tele-USG. In Section 3, after a definition of the requirements, a model of the patient is defined and the robotic arms along with the probe are presented. This is followed by a discussion of the performance metrics that could be adopted in the design of the end-effector. The same section reports the target metric, the constraints, and the formulation of the optimization problem. Section 4 reports the details of the implementation of the problem. Section 5 reports the results of the study and their discussion. Section 6 concludes the paper.

2. Background

2.1. Ultrasonography

Ultrasonography is a technique in which sound waves are sent towards the human body, whose tissues reflect them and whose echoes are used to make a picture, called sonogram. In ultrasonography, the sonographer holds the probe of the USG machine and places it on the patient's body. This probe emits the ultrasound waves and records the reflected waves. The resulting signal is sent to a machine which reconstructs the sonogram.

The successful reconstruction of the sonogram depends on the correct positioning of the probe. The sonographer must find the correct window where the ultrasound beam is most effective to reconstruct the target and the correct orientation(s) to have the meaningful images to formulate a diagnosis. Ultrasonography is applied to several medical examinations, including those targeting the heart (echocardiography), on which this paper focuses. In echocardiography (ECG), the sonographer places the probe on the user's chest at five anatomical locations to obtain the five standard windows, i.e., suprasternal, left parasternal, right parasternal, apical, and subcostal [10] (see Figure 1).

During the examination, the sonographer places the probe on one of these locations and applies a wrench that makes the probe pin about the sought contact point until obtaining the desired acoustic window. Then, the sonographer records one or more ultrasound images and moves the probe to the next location.

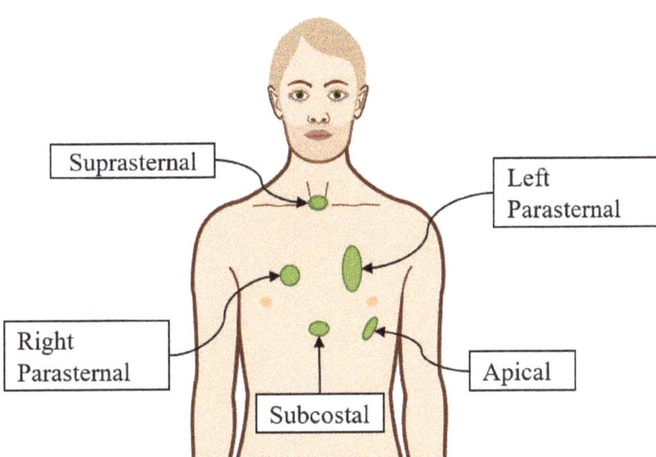

Figure 1. Five locations to obtain the five windows of echocardiography.

2.2. Tele-USG Systems

Several systems for Tele-USG have been developed in the last two decades. Some of them allow the specialist to fully control the pose of the probe at the patient site, others feature a mechanical rig that places the probe correctly on the patient torso, whereas the specialist remotely controls the orientation of the probe. A review on the Tele-USG systems was recently presented by Adams et al. [11]. In the following, we recall the main projects that have been developed throughout the years, with a final focus on those that use a robotic arm at the patient site to enable full control of the pose of the probe. An example of tele-USG system which features force feedback is described in [12,13]. The system includes a 6 DoFs robot composed of an orientable pantograph and an end-effector that allows for 3D positioning of the probe and a reasonable decoupling of translational and rotational DoFs. More recently, a complete tele-USG system was developed within the European project OTELO [14]. The system includes a 6 DoFs robot at the patient's site and a 6 DoFs haptic interface at the expert site. The robot at the patient site is custom-designed and has PPRRRP kinematics, with the three R joints in a wrist configuration. In this line, Arbeille et al. [3] developed a tele-USG system which works over a satellite link to make available echography examination for astronauts (TERESA project [15]). In their system, the patient and the expert sites are linked by a videoconference system. At the patient's site, a non-specialist operator places the robot (ESTELE) on the patient. This robot, purposely designed for USG, is composed of a rigid structure that is placed on the patient and that hosts an RRR spherical wrist manipulator that orients the probe on the patient's body. The following ARTIS project (European Space Agency contract number ESA No. 21210/07/NL/HE) further developed this concept, resulting in a simpler robot (the kinematic optimization is reported in [16]) at the patient site, which still needs to be held by an assistant. In recent years, the MELODY system [17] stemmed from the TERESA and ARTIS projects and has been commercialized by AdEchoTech. In this latter system, the robot at the patient site is held by a passive manipulator that balances the device's weight and helps the assistant to place the robot on the patient's torso. A similar approach has been adopted for the FASTele system [18], which aims at a prompt intervention during an emergency. In addition, in this case, the robot at the patient site allows for partial control of the probe. In particular, the robot has PRRP kinematics where the joint axis is parallel to the patient's longitudinal axis, the two R joints determine the probe's main axis orientation, and the latter P joint is directed along the probe's axis and features two springs to keep the contact with the patient's body. Differently from the OTELO system, this robot allows only for small displacements of the probe, whereas the gross motion from one anatomical location to the other is carried out by a non-specialist, under video surveillance of the expert.

In the TER tele-USG system [19], a different slave robot at the patient site is proposed. This robot decouples the gross positioning of the probe from its pose fine refinement on the patient's body. This robot has a kinematic structure composed of two parts: The first is parallel, with a ring that has a planar motion and that is actuated by either four McKibben artificial muscles or four DC motors trough four belts in an antagonistic configuration. The second is serial and composed of a wrist and a prismatic joint for fine-tuning of the probe position along its axis.

In the ReMeDi tele-examination system [8], a 6 DoFs serial arm is mounted on a mobile base to allow the robot to achieve a correct position around the patient and to allow the sonogrpaher to place the probe correctly from remote using a multimodal diagnostician user interface [4].

This latter system, as well as the many recent tele-USG systems, adopts a robotic arm with serial kinematics at the patient's site. Moreover, most of these recent systems use commercial robotic arms to improve the feasibility and to make both the development of such systems and the final product more affordable. For example, the system proposed in [20] uses a Viper s650 arm, whereas the system developed by Mathur et al. uses a KUKA LWR arm [21]. The most used commercial arm is thus far the Universal Robot UR5 [22–25]. The main reasons are the compliance with the ISO 10218-1:2006, which makes it usable as a collaborative robot, and its affordability. At the same time, its kinematic structure makes its workspace cover almost all the patient's torso. The Panda arm from Franka Emika has similar advantages. It has 7 DoFs, but it has a smaller workspace and lower joint torques limits with regard to the UR5 arm. Tele-USG setups the uses the Panda arm include the works presented by Sandoval et al. [26] and Kaminski et al. [27]. In the former, the probe is attached to the robot's flange using a compliant prismatic joint with variable stiffness for the patient's safety, whereas, in the latter, specifically aimed at Thyroid USG, the probe is rigidly attached to the end effector of the arm.

Thanks to the aforementioned advantages, most of these systems use a rigid end effector, which is attached to the last link of the arm, to hold the probe. None of these systems investigated the possibility to add one actuated DoF to the end-effector to optimize the dexterity of the robot at the patient's site in all the positions required for echocardiography. In particular, we study the effect of adding one DoF to the end-effector of the UR5 robot. Moreover, we compare this solution to the adoption of the Panda arm, which has natively 7 DoFs, is cost-effective, and brings advantages similar to the UR5 robot.

3. End-Effector Design Optimization

3.1. Requirements

In the USG examination task, the robot at the patient site must reach any target point on the torso and guarantee to move without collisions between any two anatomical locations. Therefore, the workspace of the robot has to include the torso. Moreover, the robot has to allow the sonographer to change the probe's orientation while exerting the contact wrench. According to experimental measurements acquired within the ReMeDi project, the main component of the contact force is along the probe long axis and equals 12 ± 3 N, whereas the remaining two components are approximately equal to 5 ± 1 N. The torque component along the probe long axis is approximately 0.02 Nm, whereas the remaining two components are equal to 0.5 ± 0.3 Nm. In addition to the capability to maintain the contact wrench, it is important to guarantee that the robot's configurations are far from singularities when the probe moves from one location to another over the patient. For the safety and the acceptability of the system, it is indeed not desirable that the manipulator makes weird and fast movement when leaving one contact point to reach the next. Finally, we treat the problem as quasi-static, focusing only on the contact phases. In fact, the time spent in any anatomical location is much higher than the time needed to move from one location to another. Therefore, it is not worth maximizing the speed of the end-effector between two locations.

Model of the Patient and Target of the ECG

The definition of a patient's model is a difficult task for two main reasons: first, patients can have very differently shaped and sized bodies. Second, patients can lie on the back or on the side. In this paper, we make some simplifications to make the problem tractable. First, we adopt an average-sized chest, and we approximate its shape to an elliptic cylinder aligned to the human longitudinal axis, whose semi-axis lengths are 26 and 15 cm, respectively. Second, we assume that the patient lies on their back, which is largely the most common case in ECG. Third, we assume the heart as a particle target H placed inside the chest. Fourth, we define five target points $E_i, i = 1, \ldots 5$ on the chest to represent the probe positions where the standard windows are sought (see Section 2.1). Finally, we sample the chest using 16 points $P_i, i = 1, \ldots, 16$ equally distributed in the longitudinal and lateral directions in the coronal plane, and we extend the samples set by adding points E_i to P_i, thus having a total of 21 points. Figure 2 shows the model of the chest along with the heart, the E_i targets, and the sample points P_i along with the reference frame Σ_s used to define the pose of the patient. The origin of Σ_s is S, which is at the base of the throat, the y_S axis is directed towards the head in the longitudinal direction, whereas the z_S axis points upwards. Points E_i and P_i are defined in Σ_S.

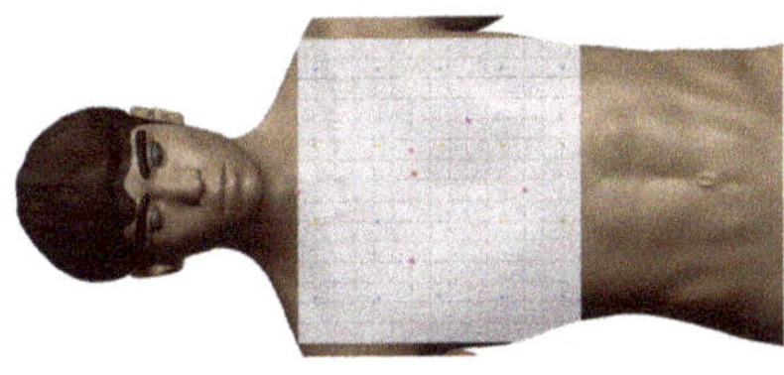

(a) Patient and model of her/his chest

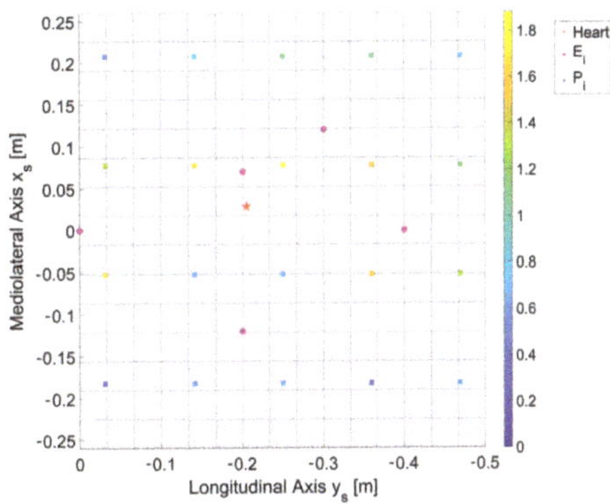

(b) Coronal view of the chest model along with heart, target points E_i and sample points P_i, whose color represent the respective λ_i according to the colorbar

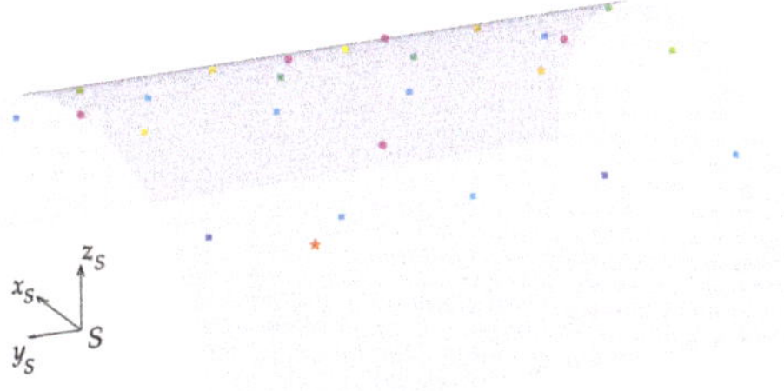

(c) 3D view of the model of the chest along with Σ_S frame.

Figure 2. The mathematical model of the patient. The heart is represented as a red asterisk, whereas target points E_i are plotted in orange and the sample points P_i in green.

Based on the aforementioned definitions, the target of the ECG is defined as placing the USG probe on the patient in every point P_i while pointing the axis of the probe towards the heart H. This task sets five of the six DoFs of the probe. In fact, the rotation of the probe about its axis remains free. However, to use the arm in a teleoperation setting, the doctor needs to have full control of the pose of the probe. Therefore, in Section 3.6, the orientation of the lateral axis of the probe will also be set to replicate commonly used orientations of the probe.

3.2. UR5 and Panda Kinematics

This section introduces the kinematic models of the three adopted arms. The Universal Robots UR5 is a 6 DoFs manipulator which includes revolute joints only. Its kinematics is similar to an anthropomorphic arm, with the noticeable difference that the last three R joints are not arranged in a spherical wrist fashion, so that all six joints contribute to both the translational and rotational motion of the end-effector [28]. A description of the kinematics of the robot is reported in Figure 3a, whereas Table 1 reports the Denavit–Hartenberg parameterization of the robot. In Figure 3a, Σ_6 and Σ_7 are the frames attached to the end-effector without and with additional joint, respectively.

The Franka Emika Panda is a 7 DoFs manipulator composed of eight links connected by revolute joints. Its kinematics includes a full shoulder, an elbow, and a wrist, thus being similar to a human arm. As for the UR5, the wrist is not spherical, hence all joints contribute to both the position and orientation of the end-effector. The robot is shown in Figure 3, whereas Table 1 reports the Denavit–Hartenberg parameters as reported by the manufacturer. For the Panda robot, the same notation of the UR5 with additional DoF holds.

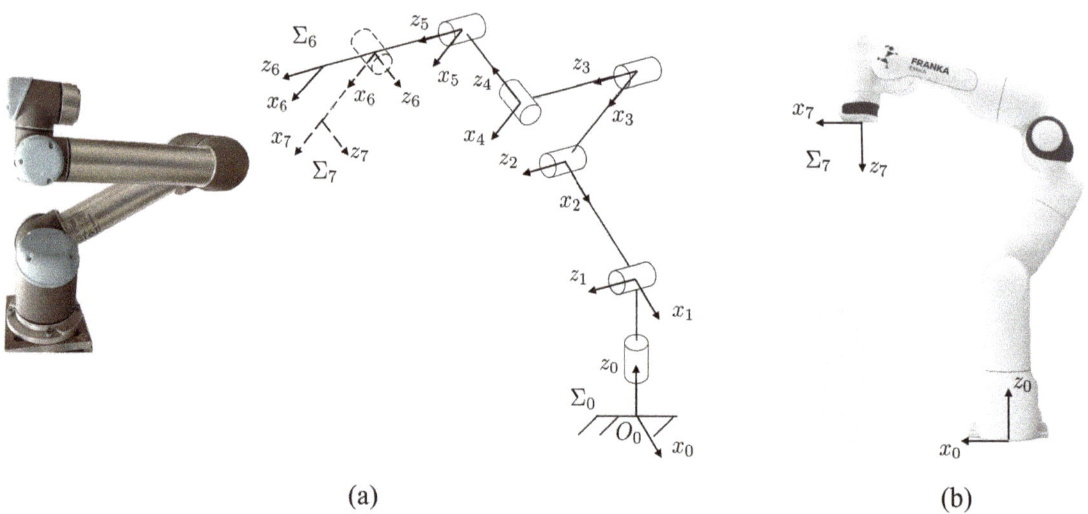

Figure 3. (**a**) The Universal Robot UR5 robot and the adopted kinematic model without and with additional joint. (**b**) The Panda robot from Franka Emika along with the first and the last frames, Σ_0 and Σ_7, respectively.

$\mathbf{q} = [q_1, \dots q_n]^T$ and $\dot{\mathbf{q}} = [\dot{q}_1, \dots \dot{q}_n]^T$ are the joint angles vector and the vector of their time derivatives, respectively. The geometric Jacobians J of the three manipulators satisfies:

$$\begin{bmatrix} \mathbf{v}_C \\ \omega_p \end{bmatrix} = J\dot{\mathbf{q}} \tag{1}$$

where $\mathbf{v}_C$ is the velocity of the end-effector, i.e., the center of the surface of the probe that goes in contact with the patient's skin; ω is the angular velocity of the end-effector; $n = 6$ for the UR5 robot; and $n = 7$ for the remaining two arms.

Table 1. Denavit–Hartenberg parameters for the definition of the kinematics of the two robots. The seventh frame of the UR5 with additional joint is drawn as separate from the sixth for clarity of representation, even though their origins are superimposed. * UR5 without additional joint; ** UR5 with an additional joint; d_6 is a design parameter which is part of the optimization.

		UR5		
link	a_i	α_i	d_i	θ_i
1	0	$\pi/2$	0.0895	θ_1
2	−0.4250	0	0	θ_2
3	−0.3922	0	0	θ_3
4	0	$\pi/2$	0.1091	θ_4
5	0	$-\pi/2$	0.0946	θ_5
6 *	0	0	0.0823	θ_6
6 **	0	$-\pi/2$	d_6	θ_6
7	0	0	0	θ_7

		Panda		
link	a_i	α_i	d_i	θ_i
1	0	0	0.333	θ_1
2	0	$-\pi/2$	0	θ_2
3	0	$\pi/2$	0.316	θ_3
4	0.0825	$\pi/2$	0	θ_4
5	−0.0825	$-\pi/2$	0.384	θ_5
6	0	$\pi/2$	0	θ_6
7	0.088	$\pi/2$	0.107	θ_7

3.3. End-Effector Kinematics

Echocardiography probes (see Figure 4) have a longitudinal axis z_p in their prominent direction and a second perpendicular axis x_p that defines the ultrasound plane π_p (we do not explicitly consider the case of 3D ECG, for which a similar treatment could be proposed). ECG probes are typically symmetric with regard to π_p and to a plane perpendicular to π_p which passes through z_p. The image obtained by ultrasound is a circular sector Γ which lies in π_p and is symmetric with regard to z_p. In the recent ECG probes, the x_p axis can be rotated via software. Therefore, it is defined in the software reference configuration, in which the rotation of the ultrasound plane with regard to z_p is 0. This typically means that x_p lies in one of the two symmetry planes of the probe. Additionally, we define two points , C and G on the z_p axis (see Figure 4). The former is the centroid of the contact surface of the probe with the patient. The latter is the point at which the probe is rigidly attached to the end-effector.

The centroid C is assumed to lie on the z_p axis. Given the symmetry planes of the probe and that the image Γ is symmetric with regard to z_p, it is reasonable to assume that the sonographer operates to have $C \in z_p$ in order to have the best image. Moreover, from the requirements, we have forces in the order of 5N in the x_p and y_p directions, whereas the torque along z_p is nearly 0. Therefore, C cannot lie far from z_p. We also assume that $\left\| \overrightarrow{GC} \right\| = 4$ cm, which can be easily obtained by design for the most common ECG probes. The point G lies on z_p by the design of the end-effector. A different choice would produce unnecessary torques due to $\mathbf{w}$, which the attachment to the-effector would have to balance. This design constraint does not reduce the generality of the approach. In fact, point G can be arbitrarily placed in the frame of the end-effector, and we define $\mathbf{p}_G = [x_G y_G z_G]^T$ the position of G in the end-effector frame Σ_n.

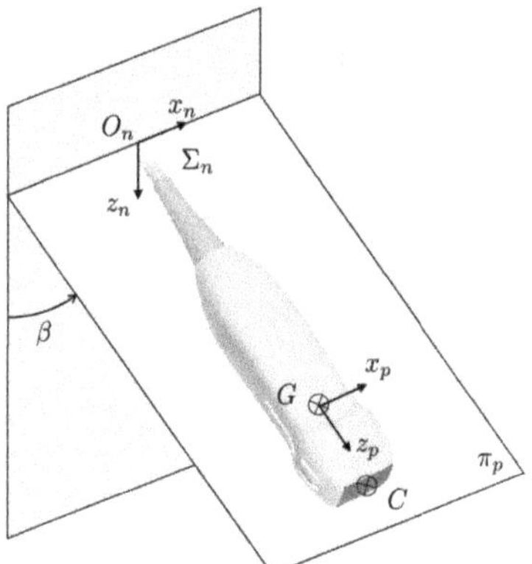

Figure 4. Echocardiographic probe along with the definition of its kinematic variables.

Whereas 7-DoFs manipulators offer a kinematic redundancy that can be exploited to optimize the orientation of the probe with joints in the proximity of the patient, the UR5 does not have this possibility. Therefore, in absence of a seventh DoF, we decided to add a further design variable β, which is the fixed angle by which frame Σ_6 is rotated about x_6 to obtain the final orientation of the probe (see Figure 4). The pose of the probe with regard to Σ_n is then obtained by means of the homogeneous transformation matrix

$$T_p = \begin{bmatrix} 1 & 0 & 0 & x_G \\ 0 & \cos\beta & -\sin\beta & y_G \\ 0 & \sin\beta & \cos\beta & z_G \\ 0 & 0 & 0 & 1 \end{bmatrix} \tag{2}$$

where β is a design variable for the UR5 without additional DoF, whereas it is used in the other two cases to properly orient z_p, i.e., $\beta = \pi/2$ for the UR5 with additional DoF, and $\beta = 0$ for the Panda robot.

3.4. Optimization Variables

The design of the end-effector includes four main variables: the position of G with regard to the terminal link of the robot, and the inclination of the plane π_p with regard to the terminal link x_n axis, i.e., the fixed angle β for the UR5 without additional joint.

In addition to these variables, we include the position of the patient with respect to the robot base in the optimization problem. To do that, we consider that the patient is steady with respect to the global frame Σ_0 attached to the base of the robot (see Figure 3) and that the pose of its reference frame Σ_s with regard to Σ_0 is defined by

$$T_s = \begin{bmatrix} \mathbf{I} & \mathbf{p}_s \\ \mathbf{0} & 1 \end{bmatrix} \tag{3}$$

where $\mathbf{p}_s = [x_s y_s z_s]^T$. Therefore, when there is no additive DoF in the UR5 robot, the seven optimization variables are contained in the array $\mathbf{x}_A = [x_s y_s z_s x_G y_G z_G \beta]$. In the other two cases, the optimization variables included in the array $\mathbf{x}_B$ are six: $\mathbf{x}_B = [x_s y_s z_s x_G y_G z_G]$. To simplify the notation, when possible, we refer to $\mathbf{x}$ to mean the optimization array.

3.5. Objective Function

The experimental results reported in Section 3.1 set a target for the tele-USG task that will be exploited to define the objective function of the design optimization. The main objective of the design is overall good manipulability of each arm, especially in the areas that surround the target points E_i.

In the literature, several manipulability measures have been proposed to compare different manipulators and to optimize their design [29,30]. In particular, when the capability of exerting wrenches is concerned, Patel et al. [29] moved from the classical Yoshikawa's analysis based on the calculation of the force ellipsoids to define task-specific indices. These indices take into account the wrench $\mathbf{w}$ that the arm has to produce at the end-effector.

In the simplest case in which no specific cost hierarchy is to be applied to the components of the manipulator torque vector τ and all the components of the wrench $\mathbf{w}$ are equally important, the Yoshikawa manipulability index provides a well-recognized measure of local manipulability, which can be easily extended by integration to a global index of manipulability. However, in the case in which the joint torque limits are not the same for all joints, and when there are preferable directions in the wrench space for the wrench exertion, a better choice to evaluate the manipulability was initially proposed by Bicchi et al. [31] and elaborated for the force analysis in [32]. In this latter work, the authors proposed the following metric:

$$\eta = \frac{\mathbf{w}^T W_u \mathbf{w}}{\tau^T W_\tau \tau} \tag{4}$$

where W_u and W_τ are positive definite and allow for applying weights to the different components of $\mathbf{w}$ and τ.

Other approaches to the optimization of manipulability are based on the wrench requirements and are task-specific indexes. These include the task-dependent performance index [33], in which the author proposed to minimize the weighted sum of the normalized differences between the transmission ratios desired for the task, i.e., the task ellipsoid semi-axis lengths, and the actual transmission ratios, i.e., the force ellipsoid semi-axis lengths. Though computationally simple, the application of the method to this problem requires careful tuning of the weights to account for the different units of the wrench components and to include the different cost of the manipulator actuators in terms of torque. Therefore, Equation (4) is selected as the metric.

To set the optimization problem as a minimization problem, the metric

$$\mu = \frac{1}{\eta} = \frac{\tau^T W_\tau \tau}{\mathbf{w}^T W_u \mathbf{w}} \tag{5}$$

is adopted and elaborated for the purpose of this study. The matrix W_t is introduced to account for the different capabilities of the joint actuators. It is a diagonal matrix whose elements are the inverse of the maximum joint torques provided by the manufacturer: $W_t = \mathrm{diag}(1/t_{m_1}, \ldots, 1/t_{m_n})$, $\quad i = 1 \ldots n$. In the case of the additional joint of the UR5, $t_{m_n} = t_{m_{n-1}}$. The matrix W_u is diagonal as well, and it accounts for the task defined in Section 3.1: $W_u = \mathrm{diag}(w_1, \ldots, w_6)$, where $w_1 = 12, w_2 = 5, w_3 = 5, w_4 = 0.5, w_5 = 0.5, w_6 = 0.02$ are obtained from the requirements reported in Section 3.1. This choice allows us to take in consideration the physical units both in the joint and the task space.

From the statics of the manipulator, we have

$$\tau = J^T \mathbf{w}, \tag{6}$$

Since the singularity configurations of the manipulators are avoided in the optimization process, J is always full rank, with $\dim \mathcal{R}(J) \geq 6$, where $\mathcal{R}(J)$ is the range of J. This means that all wrenches can be obtained using active torques and that no wrench applied to the end-effector can be balanced by a null vector in the torque space. By putting Equation (6) into Equation (5), we obtain

$$\mu = \frac{\mathbf{w}^T J W_t J^T \mathbf{w}}{\mathbf{w}^T W_u \mathbf{w}} \tag{7}$$

which is a generalized Rayleigh quotient that we want to minimize. Since both W_t and W_u are diagonal, the minimization of μ is straightforward and reduces to

$$\tilde{\mu} = \arg\min_{\sigma} W_u^{-1} J W_t J^T \mathbf{w} = \sigma \mathbf{w} \tag{8}$$

which only requires the computation of the eigenvalues of the matrix $W_u^{-1} J W_t J^T$. The local index $\tilde{\mu}_i$ is evaluated at each patient target point P_i to compute a global manipulability index

$$\mathcal{M} = \sum_{i=1}^{21} \lambda_i \tilde{\mu}_i \tag{9}$$

The mentioned

$$d_i = \arg\min_{d_j} d_j = \left\| \overrightarrow{P_i E_j} \right\| \, j = 1 \ldots 5, \tag{10}$$

$\tilde{E}_i$, and the corresponding target point, λ_i, are calculated as follows

$$\lambda_i = v_i \left(1 - e^{\frac{-1}{20 d_i}} \right) \quad \begin{cases} v_i = 1.5 & \text{for } \tilde{E}_i = E_5 \\ v_i = 3 & \text{otherwise} \end{cases} \tag{11}$$

that implies $\lambda_i \in [0.3, 3]$.

3.6. Constraints

The first constraint of the problem is the accomplishment of the task while keeping the manipulator far from singularities. In the definition of the task, the rotation of the probe around z_p is not left as an optimization variable, unnecessarily increasing the optimization domain. Instead, five $\tilde{x}_i \, i = 1 \ldots 5$ orientations for the probe lateral axis x_p are defined for each of the E_i points so that the orientation of the plane π_p is nearly correct when in the reference configuration. For each point P_i, the orientation $\tilde{x}$ of the nearest target point E is selected. For each point P_i, an inverse kinematics problem is then solved, i.e., the vector $\mathbf{q}_i$, which defines the configuration of the manipulator is calculated by imposing that

$$C = P_i + \mathbf{p}_S$$
$$z_p \parallel \overrightarrow{P_i H} \tag{12}$$
$$x_p \parallel \tilde{x}_i$$

For the solution of the inverse kinematics, six joint configurations spanning the joint space within the joint limits are defined as starting points. These starting points are used to solve the inverse kinematics numerically by means of the weighted damped least-squares method described in [34]. For each P_i, three outcomes are possible: First, no solution is found. In this case, the optimization array $\mathbf{x}$ is penalized by setting $\tilde{\mu}_i = 10^8$, which makes it not competitive against feasible solutions. Second, one solution $\mathbf{q}_i$ is found. In this case, the condition number of the manipulator $\kappa(\mathbf{q}_i)$ is computed and compared against the acceptability threshold $\tilde{\kappa}$, which is set by approaching each manipulator to a singularity configuration. If $\kappa(\mathbf{q}_i) \geq \tilde{\kappa}$, the array $\mathbf{x}$ is penalized by setting $\mu_i = 10^8$. Finally, if more solutions are found, they are filtered by the condition number criterion. Then, they are ranked according to $\tilde{\mu}_i$ and the best one is selected to contribute to $\mathcal{M}$. This procedure does not guarantee that the best solution (in terms of $\tilde{\mu}_i$) of the inverse kinematics is found, but the proposed method was tested in several configurations to check that at least the known multiple solutions (e.g., elbow-up vs. elbow-down) of the inverse kinematics were obtained in the target points for some positions of the patient with regard to the base. A thorough exploration requires, at least for the 7 DoFs manipulators, at each optimization step, the solution of a nested optimization problem, which would make the time to find a solution to increase exponentially, without guaranteeing to find an optimal solution.

The second set of constraints (Equation (13)) is imposed for feasibility, safety, and acceptability reasons: the base cannot be placed far from the patient, because part of the patient chest would be outside of the manipulator workspace. Moreover, the base cannot be too close to the patient's head or between her/his legs.

$$\mathbf{p}_{sm} \leq \mathbf{p}_s \leq \mathbf{p}_{sM} \tag{13}$$

The third constraint set 14 is added to limit the size of the end-effector, to avoid it being hard to control, possibly making the joint torques insufficient to obtain $\mathbf{w}$, expensive, and cumbersome to realize, as it would be bigger and bulkier due to the higher stress.

$$\mathbf{p}_{Gm} \leq \mathbf{p}_G \leq \mathbf{p}_{GM} \tag{14}$$

A fourth constraint limits the probe orientation. This is a limit for the angle with which the probe is mounted, in the case of a rigid end-effector, and a limit on θ_7 in the other case.

$$\begin{cases} \beta_m \leq \alpha \leq \beta_M & 6\,\text{DoFs} \\ \theta_{7m} \leq \theta_7 \leq \theta_{7M} & 7\,\text{DoFs} \end{cases} \tag{15}$$

Finally, it is necessary to guarantee that the robot avoids any possible collision in its motion around the patient's body. We did not include this constraint in the optimization problem, thus speeding its solution up, but we verified ex-post (see Section 4) that, for some trajectories around the patient, there were no collisions.

4. Implementation

The whole optimization was implemented in Matlab R2019A except for the collision avoidance that was modeled in Gazebo. Peter Corke's Robotics Toolbox [35] was used for the definition of the manipulator, the computation of the Jacobians J and the solution of the inverse kinematics using the "ikine" method of the SeialLink class. The robots were created starting from the existent models of the UR5 and Panda arms as new instances of the SerialLink class. The end-effector was then implemented as a tool of the SerialLink object.

The optimization problem was implemented using the Matlab Optimization Toolbox and the MultiStart algorithm, which allows a thorough exploration of the optimization domain and the exploitation of parallel computing. The MultiStart algorithm samples uniformly the optimization domain within the bounds and runs local solvers to find local minima. After all start points are evaluated, the algorithm compares the local solvers' solution to return a "global" minimum. In our implementation, after a pilot trial to evaluate the order of magnitude of $\mathcal{M}$, we set the "FunctionTolerance" parameter, which is used to compare the minima of $\mathcal{M}$, to 10^{-7}, and the "XTolerance" parameter, which is the minimum distance between two $\mathbf{x}$ points to be considered as separated, to 10^{-4}. The selected solver is "fmincon", which is a gradient-based method that exploits an interior point algorithm suitably designed to account for both equality and inequality constraints. The torque limits of the UR5 arm are $t_{m1} = t_{m2} = t_{m3} = 150$ Nm and $t_{m4} = t_{m5} = t_{m6} = 28$ Nm. The torque limits of the Panda arm are $t_{m1} = t_{m2} = t_{m3} = t_{m4} = 87$ Nm and $t_{m5} = t_{m6} = t_{m7} = 12$ Nm.

Constraints defined in Equations (13)–(15) were set as bounds of the optimization array components. In particular, regarding Equation (13), we set $\mathbf{p}_{sm} = [0.35, 0.10, -0.40]$ m. The first component ensures that the base of the robot is outside the patient. The second component guarantees that the base is at least 3 cm from the neck along the patient's longitudinal axis. The latter represents the vertical displacement of the patient with regard to the base. It was set after pilot trials that showed increasing difficulties for the arm to reach all the target points when moving the patient too much below the robot. We set $\mathbf{p}_{sM} = [0.90, 0.60, 0.10]$ m. These maximum values were set after pilot trials, as the arm encountered increasing difficulties to reach all target points. Regarding the position of probe with regard to the arm flange, we limited the range in the plane of the arm terminal flange to ± 0.15 m in both directions. Larger displacements would lead to a cumbersome end-effector, with a relevant effect on the magnitude of the wrench at the end-effector that the arm can balance. The limits in the direction perpendicular to the flange were set to 0.05 and 0.30 m. The former gives sufficient room for the design of the end effector (0 means that the probe penetrates into the arm's flange), while the latter value was set after pilot trials. Finally, we set $\theta_{7m} = \beta_m = -3\pi/4$ and $\theta_{7M} = \beta_M = 3\pi/4$. These values allow the probe to point towards the arm's terminal flange, and they were retained sufficient to explore the reasonable configurations. The constraints defined in Equation (12) were embedded in the computation of the objective function (see Algorithm 1).

The optimal array $\tilde{\mathbf{x}}$ minimizes $\mathcal{M}$ while satisfying the constraints. Algorithm 1 synthesizes the whole optimization procedure to find $\tilde{\mathbf{x}}$.

Algorithm 1: Optimization algorithm

Data: sample points P_i, weights λ_i
Result: optimal
create the manipulator MN;
set W_u, W_t, $\mathcal{M} = 0$;
compute P_i, λ_i;
compute inverse kinematics starting points $\mathbf{q}_0$;
set bounds in MultiStart according to Equations (13)–(15);
while *optimization stop criteria in MultiStart are not met* **do**
    set $\mathcal{M} = 0$;
    get x from MultiStart;
    if *MN == UR5* **then**
        update MN tool: MN.tool(x_G, y_G, z_G, β);
    else if *MN == UR5 + 1 DoF* **then**
        update d_6 in Denavit Hartenberg table: $d_6 = z_G$;
        update MN tool: MN.tool(x_G, y_G);
    else if *MN == Panda* **then**
        update MN tool: MN.tool(x_G, y_G, z_G);
    end
    compute $P_i + \mathbf{p}_S$;
    for $i \leftarrow 1$ **to** 21 **do**
        compute target pose T_i according to constraints (12);
        for $j \leftarrow 1$ **to** 6 **do**
            $\mathbf{q}_{i,j} = $ MN.ikine$(T_i, \mathbf{q}_{0_j})$;
            set $\tilde{\mu}_{i,j} = 10^8$;
            if *exist* $\mathbf{q}_{i,j}$ **then**
                for $k \leftarrow 1$ *to number inverse kinematics solutions* **do**
                    $J_{i,j} = $ MN.jacobe$(\mathbf{q}_{i,j,k})$ (compute the Jacobian);
                    compute $\kappa_{i,j,k}$;
                    if $\kappa_{i,j,k} < \tilde{\kappa}$ **then**
                        compute $\sigma_{i,j,k,l} = \text{eig } W_u^{-1} J_{i,j} W_t J_{i,j}^T$;
                        $\tilde{\mu}_{i,j,k} = \min_l \sigma_{i,j,k,l}$;
                  end
                end
                $\tilde{\mu}_{i,j} = \min_l \sigma_{i,j,l}$;
            end
        end
        $\tilde{\mu}_i = \min_j \tilde{\mu}_{i,j}$;
        $\mathcal{M} = \mathcal{M} + \tilde{\mu}_i$;
    end
    return $\mathcal{M}$ to MultiStart;
end

Finally, to verify the eventual occurrence of collisions when moving around the patient's chest, and possibly avoid them, the manipulators comprehensive of all their elements (robotic arm, the box representing the end-effector, and body of the patient) were modeled in ROS-Gazebo. Simulations were run imposing the trajectories of interest among points P_i and the collision among the elements were identified.

5. Results

The results of the optimization for the three manipulators are reported in Table 2. In addition to $\mathcal{M}$, this table reports the ratio

$$\epsilon = \max \tilde{\mu}_i / \min \tilde{\mu}_i \tag{16}$$

that allows an evaluation of the uniformity of the arm's behavior across the patient's torso.

Feasible solutions have been found for each manipulator. None of the components of the x array is on the boundaries set by Equations (13)–(15), making us confident that the choice of the boundaries has not excluded optimal solutions nearby the boundaries.

The obtained optimal vectors reported in Table 2 were used to calculate the objective metric $\tilde{\mu}$ at the target points, thus enabling a detailed evaluation of the behavior of the arms. To have a richer representation of the performance of the arm, 48 target points equally spaced on the patient's torso were used. Figure 5 shows the objective metric μ evaluated at these points. With regard to the arms' performance, we report in Table 2 the ratio between the maximum and minimum value of μ calculated in these target points.

The simulations run in Gazebo showed that it is possible to reach every P_i without colliding with the patient or having self collisions.

Table 2. Optimization results for the three manipulators. UR5+1 stands for Ur5 with additional joint.

Robot	x_S [m]	y_S [m]	z_S [m]	x_G [m]	y_G [m]	z_G [m]	β [deg]	$\mathcal{M}$ [10^{-4}]	ϵ
UR5	0.55	−0.23	0.11	0.017	−0.019	0.161	−65.9	5.56	8.54
UR5 + 1	0.73	−0.33	0.20	0.037	0.005	0.123	-	1.97	3.23
Panda	0.39	−0.17	0.31	0.023	0.071	0.146	-	5.28	14.1

Discussion

The results of the optimization, reported in Table 2, show that adding a DoF to the UR5 robot improves the performance index significantly (65%). The overall value of $\mathcal{M}$ and the smaller value of ϵ support this statement.

The same benefits are not present when using the Panda arm. This is likely due to the smaller size of the arm and the limited range of motion of the joints. These limited the feasible optimization domain, especially regarding $\mathbf{p}_s$, thus limiting the possibility to achieve the performance level of the UR5 with one additional DoF. This statement is also consistent with the large value of ϵ (see Table 2 and Figure 5c). Figure 5c shows that $\tilde{\mu}$ is relatively high along a line directed as x_s is passing through the arm base and around this line in the vicinity of the base (except for a second line parallel and next to this one), whereas $\tilde{\mu}$ increases far from the arm base. A possible explanation, which is also supported by the relatively high value of $\sqrt{x_G^2 + y_G^2}$, is that the optimization process was forced to keep the base close to the patient to guarantee that all targets could be reached. This makes the arm assume very compact configurations in the proximity of the base, with detrimental effects on the manipulability even in the presence of kinematic redundancy.

Differently from the Panda, the UR5 has a large workspace, which covers the patients' torso with a good margin. However, it is not redundant. The large size and a larger optimization domain (seven variables instead of six) made ϵ much smaller than for the Panda. Figure 5a shows that, apart for one target point, $\tilde{\mu}$ is pretty uniform. If that point is removed, the ratio drops to 4.82, which is much closer to the case of UR5 with an additional joint than to the case of Panda arm. Curiously, the overall performance $\mathcal{M}$ is similar to the Panda arm.

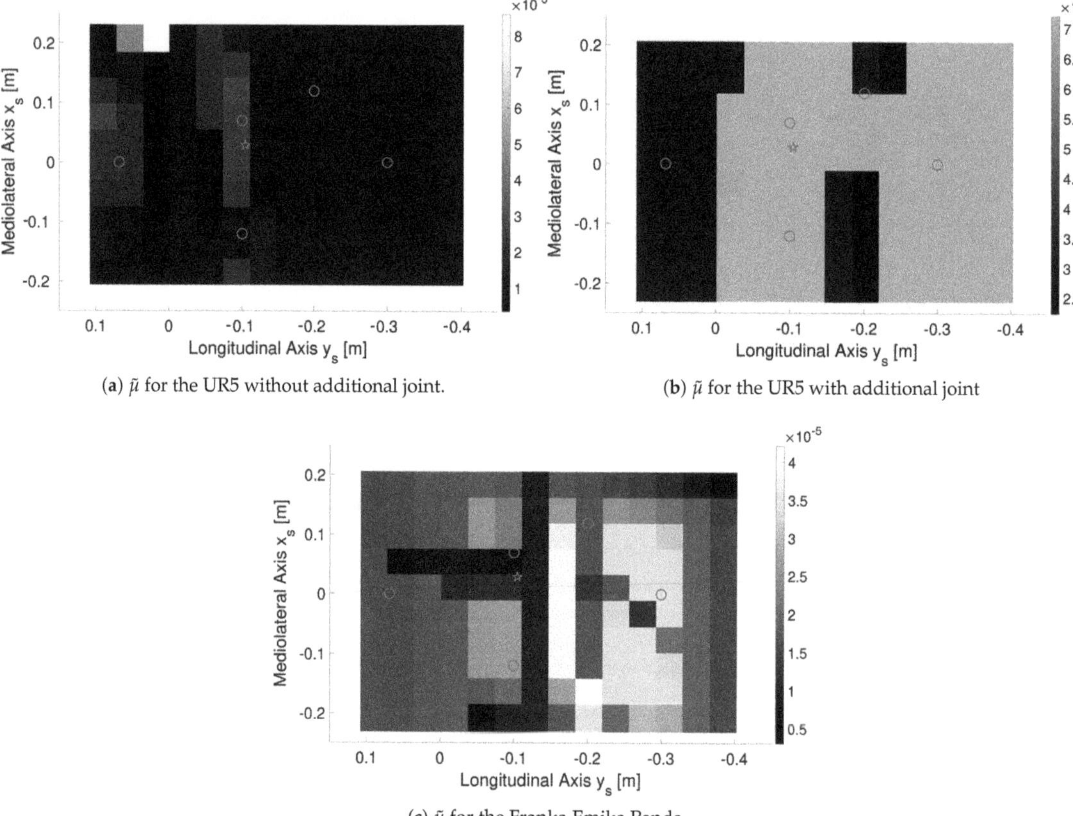

(a) $\bar{\mu}$ for the UR5 without additional joint.

(b) $\bar{\mu}$ for the UR5 with additional joint

(c) $\bar{\mu}$ for the Franka Emika Panda

Figure 5. The target metric $\bar{\mu}$ calculated at the target points for the three arms. The color represents μ as indicated in the color bars besides the plots. The color scale is different for the three arms. Circles represent points E_i, whereas the star represents the heart.

The UR5 with additional joint has a very uniform behavior with a noticeably better overall performance. The combined effects of redundancy and large workspace made the optimization process free of the limitation encountered in the case of the Panda. In the comparison against the UR5 without additional joint, we note that, in the optimization, the angle β and the variable θ_7 play a similar role from the point of view of the kinematics. However, β is fixed at each optimization step, whereas θ_7 can vary at each target point. Apart from the reported advantage for the UR5 in the proposed optimization, we highlight that the optimality of β and therefore of the performance of the arm are limited to these optimization settings. When moving to the real case, it is likely that having a seventh DoF instead of an optimal β will increase the performance difference between the two arms.

The decision to introduce an additional, custom degree of freedom improves the efficiency in the accomplishment of the task, but it also introduces several complications that need to be addressed appropriately. First, this joint will have to be actuated, and this will require the design of an actuated joint. It will have to be housed in such a way that the motion is easily transmissible, but does not affect the mechanical performance of the end-effector and the proper execution of the examination. It also becomes necessary to design a special fixing to mount the probe and to leave a sufficient room to allow the correct movement of the probe. Moreover, the addition of an actuated joint complicates the path towards a successive certification of the device. Despite these complications in the design of the end-effector, in our opinion, the workspace limitations of the Panda arm make it worthwhile to consider adding a seventh DoF to the UR5 robot..

Finally, we make two comments on this study. First, the study is conceived primarily for teleoperation of the robot at the patient site, but the obtained results can be equally applied to a robot at the patient site that carries out USG examination with some degrees of autonomy. Second, the discussion follows uniquely the kinematic optimization, without taking into account practical issues related to the teleoperated control of the arm. These aspects will have to be taken into account in the successive step of designing the whole teleoperated robotic system.

6. Conclusions

In this paper, we present the optimization of the design of an end-effector for tele-echocardiography, while comparing a non-redundant solution against two redundant ones. We show that the introduced redundancy allows for a noticeable improvement of manipulability, and that a custom design of an actuated joint to be added to an existing 6 DoFs manipulator is better than adopting a native 7 DoFs Panda arm.

Author Contributions: Conceptualization, A.F. and C.A.A.; methodology, A.F. and C.A.A.; software, A.F. and P.G.; validation, A.F. and P.G.; formal analysis, A.F. and P.G.; writing—original draft preparation, A.F. and P.G.; writing—review and editing, C.A.A. and A.F.; supervision, A.F. and C.A.A.; project administration, C.A.A. and A.F.; and funding acquisition, C.A.A. All authors have read and agreed to the published version of the manuscript.

Funding: This research received no external funding.

Institutional Review Board Statement: Not applicable.

Informed Consent Statement: Not applicable.

Data Availability Statement: Data available in a publicly accessible repository that does not issue DOIs. This data can be found here: https://www.dropbox.com/s/vl3ch1ndu7bwet7/codeUSGkinopt.rar?dl=0.

Conflicts of Interest: The authors declare no conflict of interest.

References

1. World Health Organization. *Telemedicine: Opportunities and Developments in Member States: Report on the Second Global Survey on eHealth*; World Health Organization: Geneva, Switzerland, 2010.
2. Valdastri, P.; Simi, M.; Webster, R.J., III. Advanced technologies for gastrointestinal scopy. *Annu. Rev. Biomed. Eng.* **2012**, *14*, 397–429. [CrossRef] [PubMed]
3. Arbeille, P.; Capri, A.; Ayoub, J.; Kieffer, V.; Georgescu, M.; Poisson, G. Use of a robotic arm to perform remote abdominal telesonography. *Am. J. Roentgenol.* **2007**, *188*, W317–W322. [CrossRef] [PubMed]
4. Filippeschi, A.; Brizzi, F.; Ruffaldi, E.; Jacinto Villegas, J.M.; Landolfi, L.; Avizzano, C.A. Evaluation of diagnostician user interface aspects in a virtual reality-based tele-ultrasonography simulation. *Adv. Robot.* **2019**, *33*, 840–852. [CrossRef]
5. Endo, T.; Kawasaki, H.; Mouri, T.; Doi, Y.; Yoshida, T.; Ishigure, Y.; Koketsu, K. Five-fingered haptic interface robot: HIRO III. In Proceedings of the World Haptics 2009—Third Joint EuroHaptics Conference and Symposium on Haptic Interfaces for Virtual Environment and Teleoperator Systems, Salt Lake City, UT, USA, 18–20 March 2009; pp. 458–463.
6. Coles, T.R.; John, N.W.; Gould, D.; Caldwell, D.G. Integrating haptics with augmented reality in a femoral palpation and needle insertion training simulation. *IEEE Trans. Haptics* **2011**, *4*, 199–209. [CrossRef] [PubMed]
7. Filippeschi, A.; Brizzi, F.; Ruffaldi, E.; Jacinto, J.M.; Avizzano, C.A. Encountered-type haptic interface for virtual interaction with real objects based on implicit surface haptic rendering for remote palpation. In Proceedings of the 2015 IEEE/RSJ International Conference on Intelligent Robots and Systems (IROS), Hamburg, Germany, 28 September–2 October 2015; pp. 5904–5909.
8. Arent, K.; Cholewinski, M.; Domski, W.; Drwiega, M.; Jakubiak, J.; Janiak, M.; Szczesniak-Stanczyk, D. Selected Topics in Design and Application of a Robot for Remote Medical Examination with Use of Ultrasonography and Auscultation from the Perspective of the ReMeDi Project. *J. Autom. Mob. Robot. Intell. Syst.* **2017**, *11*, 82–94.
9. Griffa, P.; Filippeschi, A.; Avizzano, C.A. Kinematic Optimization for the Design of a UR5 Robot End-Effector for Cardiac Tele-Ultrasonography. In *The International Conference of IFToMM ITALY*; Springer: Berlin/Heidelberg, Germany, 2020; pp. 423–430.

10. Mitchell, C.; Rahko, P.S.; Blauwet, L.A.; Canaday, B.; Finstuen, J.A.; Foster, M.C.; Horton, K.; Ogunyankin, K.O.; Palma, R.A.; Velazquez, E.J. Guidelines for performing a comprehensive transthoracic echocardiographic examination in adults: recommendations from the American Society of Echocardiography. *J. Am. Soc. Echocardiogr.* **2019**, *32*, 1–64. [CrossRef] [PubMed]

11. Adams, S.J.; Burbridge, B.; Obaid, H.; Stoneham, G.; Babyn, P.; Mendez, I. Telerobotic Sonography for Remote Diagnostic Imaging: Narrative Review of Current Developments and Clinical Applications. *J. Ultrasound Med.* **2020**. [CrossRef] [PubMed]

12. Salcudean, S.; Bell, G.; Bachmann, S.; Zhu, W.H.; Abolmaesumi, P.; Lawrence, P.D. Robot-assisted diagnostic ultrasound–design and feasibility experiments. In *International Conference on Medical Image Computing and Computer-Assisted Intervention*; Springer: Berlin/Heidelberg, Germany, 1999; pp. 1062–1071.

13. Abolmaesumi, P.; Salcudean, S.E.; Zhu, W.H.; Sirouspour, M.R.; DiMaio, S.P. Image-guided control of a robot for medical ultrasound. *IEEE Trans. Robot. Autom.* **2002**, *18*, 11–23. [CrossRef]

14. Delgorge, C.; Courreges, F.; Bassit, L.A.; Novales, C.; Rosenberger, C.; Smith-Guerin, N.; Bru, C.; Gilabert, R.; Vannoni, M.; Poisson, G.; et al. A tele-operated mobile ultrasound scanner using a light-weight robot. *IEEE Trans. Inf. Technol. Biomed.* **2005**, *9*, 50–58. [CrossRef] [PubMed]

15. Vieyres, P.; Poisson, G.; Courrèges, F.; Mérigeaux, O.; Arbeille, P. The TERESA project: From space research to ground tele-echography. *Ind. Robot. Int. J.* **2003**, *30*, 77–82. [CrossRef] [PubMed]

16. Nouaille, L.; Smith-Guérin, N.; Poisson, G.; Arbeille, P. Optimization of a 4 dof tele-echography robot. In Proceedings of the 2010 IEEE/RSJ International Conference on Intelligent Robots and Systems, Taipei, Taiwan, 18–22 October 2010; pp. 3501–3506.

17. Avgousti, S.; Panayides, A.S.; Christoforou, E.G.; Argyrou, A.; Jossif, A.; Masouras, P.; Novales, C.; Vieyres, P. Medical telerobotics and the remote ultrasonography paradigm over 4g wireless networks. In Proceedings of the 2018 IEEE 20th International Conference on e-Health Networking, Applications and Services (Healthcom), Ostrava, Czech Republic, 17–20 September 2018; pp. 1–6.

18. Ito, K.; Sugano, S.; Iwata, H. Portable and attachable tele-echography robot system: FASTele. In Proceedings of the 2010 Annual International Conference of the IEEE Engineering in Medicine and Biology, Buenos Aires, Argentina, 31 August–4 September 2010; pp. 487–490.

19. Vilchis, A.; Troccaz, J.; Cinquin, P.; Masuda, K.; Pellissier, F. A new robot architecture for tele-echography. *IEEE Trans. Robot. Autom.* **2003**, *19*, 922–926. [CrossRef]

20. Chatelain, P.; Krupa, A.; Navab, N. Optimization of ultrasound image quality via visual servoing. In Proceedings of the 2015 IEEE International Conference on Robotics and Automation (ICRA), Seattle, WA, USA, 26–30 May 2015; pp. 5997–6002.

21. Mathur, B.; Topiwala, A.; Schaffer, S.; Kam, M.; Saeidi, H.; Fleiter, T.; Krieger, A. A semi-autonomous robotic system for remote trauma assessment. In Proceedings of the 2019 IEEE 19th International Conference on Bioinformatics and Bioengineering (BIBE), Athens, Greece, 28–30 October 2019; pp. 649–656.

22. Mathiassen, K.; Fjellin, J.E.; Glette, K.; Hol, P.K.; Elle, O.J. An ultrasound robotic system using the commercial robot UR5. *Front. Robot. AI* **2016**, *3*, 1. [CrossRef]

23. Fang, T.Y.; Zhang, H.K.; Finocchi, R.; Taylor, R.H.; Boctor, E.M. Force-assisted ultrasound imaging system through dual force sensing and admittance robot control. *Int. J. Comput. Assist. Radiol. Surg.* **2017**, *12*, 983–991. [CrossRef] [PubMed]

24. Finocchi, R.; Aalamifar, F.; Fang, T.Y.; Taylor, R.H.; Boctor, E.M. Co-robotic ultrasound imaging: A cooperative force control approach. In Proceedings of the Medical Imaging 2017: Image-Guided Procedures, Robotic Interventions, and Modeling, Orlando, FL, USA, 11–16 February 2017.

25. Geng, C.; Xie, Q.; Chen, L.; Li, A.; Qin, B. Study and Analysis of a Remote Robot-assisted Ultrasound Imaging System. In Proceedings of the 2020 IEEE 4th Information Technology, Networking, Electronic and Automation Control Conference (ITNEC), Chongqing, China, 12–14 June 2020; Volume 1, pp. 389–393.

26. Sandoval, J.; Laribi, M.A.; Zeghloul, S.; Arsicault, M.; Guilhem, J.M. Cobot with Prismatic Compliant Joint Intended for Doppler Sonography. *Robotics* **2020**, *9*, 14. [CrossRef]

27. Kaminski, J.T.; Rafatzand, K.; Zhang, H.K. Feasibility of robot-assisted ultrasound imaging with force feedback for assessment of thyroid diseases. In Proceedings of the Medical Imaging 2020: Image-Guided Procedures, Robotic Interventions, and Modeling, Houston, TX, USA, 15–20 February 2020.

28. Kebria, P.M.; Al-Wais, S.; Abdi, H.; Nahavandi, S. Kinematic and dynamic modelling of UR5 manipulator. In Proceedings of the 2016 IEEE International Conference on Systems, Man, and Cybernetics (SMC), Budapest, Hungary, 9–12 October 2016; pp. 004229–004234.

29. Patel, S.; Sobh, T. Manipulator performance measures-a comprehensive literature survey. *J. Intell. Robot. Syst.* **2015**, *77*, 547–570. [CrossRef]

30. Kelaiaia, R.; Company, O.; Zaatri, A. Multiobjective optimization of a linear Delta parallel robot. *Mech. Mach. Theory* **2012**, *50*, 159–178. [CrossRef]

31. Bicchi, A.; Melchiorri, C.; Balluchi, D. On the mobility and manipulability of general multiple limb robots. *IEEE Trans. Robot. Autom.* **1995**, *11*, 215–228. [CrossRef]

32. Bicchi, A.; Prattichizzo, D. Manipulability of cooperating robots with passive joints. In Proceedings of the 1998 IEEE International Conference on Robotics and Automation (Cat. No. 98CH36146), Leuven, Belgium, 20–20 May 1998; Volume 2, pp. 1038–1044.

33. Chiu, S.L. Kinematic characterization of manipulators: An approach to defining optimality. In Proceedings of the 1988 IEEE International Conference on Robotics and Automation, Leuven, Belgium, 20–20 May 1988; pp. 828–833.

34. Nakamura, Y.; Hanafusa, H. Inverse Kinematic Solutions With Singularity Robustness for Robot Manipulator Control. *J. Dyn. Syst. Meas. Control.* **1986**, *108*, 163–171. [CrossRef]
35. Corke, P. *Robotics, Vision and Control: Fundamental Algorithms in MATLAB® Second, Completely Revised*; Springer: Berlin/Heidelberg, Germany, 2017; Volume 118.

Article

Development of an Automatic Robotic Procedure for Machining of Skull Prosthesis

Kevin Castelli, Marco Carnevale and Hermes Giberti *

Dipartimento di Ingegneria Industriale e dell'Informazione, Università degli Studi di Pavia, Via A. Ferrata 5, 27100 Pavia, Italy; kevin.castelli@unipv.it (K.C.); marco.carnevale@unipv.it (M.C.)
* Correspondence: hermes.giberti@unipv.it; Tel.: +39-0382-9852-55

Received: 26 October 2020; Accepted: 11 December 2020; Published: 14 December 2020

Abstract: The project presented in this paper develops within the field of automation in the medical-surgical sector. It aims at automating the process for the realization of prosthetic devices for the skull in cranioplasty, following a craniotomy intervention for brain tumor removal. The paper puts emphasis on the possibility to create the prosthetic device in run-time during the surgery, in order to ease the work that surgeons have to do during the operation. Generally, a skull prosthesis is realized before the day of the intervention, based on the plan of the medical operation, on the results of computed tomography, and through image processing software. However, after the surgery is performed, a non-negligible geometrical uncertainty can be found between the part of the skull actually removed and the cut planned during the preliminary analysis, so that the realized prosthesis (or even the skull, at worse) may need to be retouched. This paper demonstrates the possibility to introduce a fully automated process in a hospital environment, to manufacture in runtime the prosthetic operculum, relying on the actual geometry of the incision of the skull detected during the intervention. By processing a 3D scan of the skull after the craniectomy, a digital model of the prosthesis can be created and then used as an input to generate the code to be run by a robotic system in charge of the workpiece machining. Focusing on this second step, i.e., the manufacturing process, the work describes the way the dimensions of the raw material block are automatically selected, and the way robot trajectories for milling operation are automatically generated. Experimental validation demonstrates the possibility to complete the prosthesis within the surgery time, thus increasing the accuracy of the produced prosthesis and consequently reducing the time needed to complete the operation.

Keywords: automation in surgery; robotic machinery; G-code generation

---

1. Introduction

Automation is showing a relevant development in medicine, particularly in the field of surgical robotics and the surgical sector in general. In many cases, laboratories have been completely automated to satisfy requirements from blood sample to elaboration of clinical reports. Robotic systems can precisely execute actions being directly commanded by a doctor or through local systems, whereas, to authors' knowledge, remotely controlled systems or robots capable of working in a complete unmanned way are not yet available. In the context of Industry 4.0 [1], the development of automation in the medical field is also changing the properties of medical devices exploited into operating rooms and in daily assistance of patients. A revolution is happening, involving both the production of prosthesis and their management, which leads to a consistent saving of time and money. To this regard, additive manufacturing is one of the many applications which are catching on. 3D printing allows to create custom made prosthesis [2] or even organs [3] based on computer-aided design (CAD) systems.

Some of the main technologies adopted for the realization of prosthetic devices are three-dimensional printing method (3DP), stereolithography, fused deposition modelling (FDM), and selective laser sintering (SLS) [4]. As a further step, in the last two decades the possibility to customize the geometry of the prosthesis for each patient has been consolidated in many fields, cranioplasty [5,6] and maxillo–facial surgery [7–9], for instance. Patient-specific implants reduce the time of surgery compared to the case in which the shaping is carried out during the surgery, and they are likely to reduce the revision rate after surgery [10]. A test case is reported in [11], in which 40 operations for mandibular reconstruction were compared. Twenty out of forty were carried out with the conventional technology (i.e., strips adapted in the operating theatre), the remaining through digital surgery, where the implant was prepared in advance based on computed tomography (CT) data. Results demonstrates that 3D printing enabled a saving of 33 min in the reconstruction, two hours in the operation, and the hospital stay was reduced of over three days. The overall saving for each operation was estimated to be 3450 euros.

Another kind of approach for the manufacturing of prosthesis is based on computer numerical control (CNC) of metals [12,13]. In [13] the authors investigate the possibility to manufacture a skull prosthesis through computer numerical control (CNC), based on a digital model of the skull reconstructed through computed tomography data. The starting raw material adopted is a titanium plate. As a matter of fact, manufacturing through machining reduces the production time compared to additive manufacturing. Researches have also been carried out in the field of CNC machinery of femoral head prosthesis, to increase the accuracy of the machined geometry and to shorten production times. In this latter case, stainless steel is exploited as the raw material [14].

In the field of craniectomy, the surgeon commonly plans the feasible cutting lines before performing the operation, based on the analysis of CT results or other data (e.g., magnetic resonance). The image from the CT is then converted to a three-dimensional model through dedicated software, very specific and expensive, so that the entire procedure is generally outsourced. The needed prosthesis can then be designed based on the 3D model, to generate a preoperatively customized implant that properly fits the patient's skull [15]. The computer-aided design (CAD) model of the prosthesis is then transformed into a suitable file format for rapid prototyping machines (e.g., standard tessellation language (STL)), so that two solid prototypes of both the prosthesis and the skull can be generated. This second step allows the surgeon to verify, before the day of the intervention, if the designed prosthesis properly fit the skull, and to carry out any necessary correction before the final prosthesis is manufactured. Finally, the last step consists in the realization of the actual prosthesis that will be implanted in the patient cranium. In the standard procedure described above, the surgery is performed after the prosthetic device has been created, so that any difference between the planned cutting line and the actual part of the skull removed would result in an unfit or maladjusted prosthesis [15]. This implies the surgeon is required to perform milling operations on the operculum, extending the time required in the operating room, and increasing the risk of mistakes or of achieving imprecise results. In the worst case, even the cut in the skull must be retouched to make the prosthesis fit and to complete the operation.

To overcome these issues, this paper proposes and develops a renewed procedure to realize the skull prosthesis in real-time during the surgery, through an automatic process based on machining of a plastic material. The possibility of machining the prosthesis would allow to reduce the production time, making the latter compatible with the duration of the surgery, and to extend the range of workable materials, compared to additive manufacturing. The possibility of machining a radio–transparent material would, indeed, be a very relevant outcome, since it would facilitate, as an example, post-operation radiological exams and radiotherapy without the need to remove the prosthesis from the cranium. In [16] the authors compare the clinical performances of custom-made prefabricated polymethyl methacrylate (PMMA) prosthesis with the performance of prosthesis molded with the same material during the surgery (i.e., intra-operatively), pointing out that the higher operating time required for intra-operative molding led to higher blood loss and infection rate. In this regard, the procedure presented in our work, which enables the automatic robot programming and machining of a plastic prosthesis, is aimed at reducing the fabrication time of an intra-operatively fabricated prosthesis, accomplishing to the need of a short fabrication time.

The paper is meant to represent a further step towards the possibility of carrying out the machining procedure during the surgery through a dedicated robotic system, which could be adopted in a clinical environment with more flexibility (of layout and of applicability) compared to a standard milling machine.

The proposed process is engineered to minimize the need of intervention of medical personnel, not necessarily qualified or trained to use a robotic system and relative software. To these aims, a new perspective is presented: immediately after the incision, the hole left by the craniotomy is scanned by a member of the medical team, following the on-video instructions. Acquired data are then matched with the topography of the patient skull, previously taken before the surgery. A digital model of the prosthesis is then generated and converted in a suitable digital format (e.g., STL), to be adopted for the automatic generation of the code for a robotic manufacturing system (i.e., trajectory planning). The paper deals with the automatic procedure for the machinery of the prosthetic device, relying on the digital model of the prosthesis as a starting point, whereas the image processing software generating the digital model has been engineered by a partner company and is not part of the work here described.

The preliminary phase of the established algorithm consists in the automatic selection of the dimension of the raw material block and in the definition of the optimal orientation of the workpiece inside the raw material volume. After this phase, the software can output on the screen in the surgery room instructions on the dimensions of the block of raw material to be selected from the ones available in stock, and how to properly install it on the machine. The second step of the automatic procedure consists in the trajectory planning, defining the trajectories to be followed by the milling tool to carry out the machinery.

The described procedure would allow to complete the prosthesis within the time required by surgeon to complete the operation, increasing the accuracy of the produced prosthesis and consequently reducing the surgery time.

The paper is organized as follows: in Section 2 the main aspects of craniotomy surgery are recalled; standard procedures and new perspective for the manufacturing process are compared and commented. Section 3 illustrates the way the STL file model of the prosthesis is processed to automatically identify the dimensions of the raw material block which minimize the waste material, and Section 4 describes the algorithm to generate the machining procedure and the trajectories for a robotic manufacturing system. Finally, the developed software is validated in Section 5, in which a five degree-of-freedom (DOF) milling machine is exploited for the machinery of a plastic prosthesis, based on the code automatically generated through the proposed algorithm. Finally, conclusions are drawn is Section 6. The quality of the produced specimens and the compatibility of machinery time with surgery duration demonstrate the possibility to introduce a fully automated process in a hospital environment.

2. Standard and New Perspectives for Craniotomy and Cranioplasty

This section analyses the steps related to the bone excision (i.e., craniotomy) and implant placement (i.e., cranioplasty) which are performed every time a surgeon must access the brain to remove tumors. The proposal and the perspective of the new procedure is then outlined.

In the standard procedure the only craniotomy usually takes 4–6 h. Before the surgery, the surgeon analyses the case to be examined and defines the way to design the incision, to carry out the surgery in the best possible way, both aesthetically and functionally. The digital procedure to realize the prosthesis model is generally carried out by external companies, enabling the production of a prototype which allows the surgeon to verify, before the day of the intervention, if the designed prosthesis properly fits the skull. Eventually, the final prosthesis to be implanted is manufactured to be ready for the day of the surgery.

During the intervention, the surgeon draws the contour line on the shaved scalp where the cut must be done, using a plastic operculum realized as a reference shape. It is important that the margins on the skull and those of the final prosthesis are perfectly fitting, which is facilitated by the fact that the plastic operculum has smaller dimensions compared to the final prosthesis. The skull is then cut following the trace, but after the incision the dimensions and inclinations of the extremities of the skull cut and of the final prosthesis might slightly differ. The prosthesis might need to be reshaped accordingly to fit the actual hole, requiring further finishing operations. In the worst case even the patient's skull must be readjusted. All these activities require high precision, and sterile rooms, including the environment, tools, and materials.

In the novel procedure proposed in this paper, the prosthesis manufacturing process is engineered to be carried out in real-time during the surgery, minimizing the interaction with medical personnel. It is conceived to be an automated process, receiving as input the digital model of the part to be manufactured (e.g., STL format) and directly outputting the final prosthesis to be implanted, just requiring simple post manufacturing activities.

The steps of the proposed procedure are as follows:

- The hole left by the craniotomy is scanned and data are matched with the topography of patient skull identified before the surgery. A digital model of the prosthesis is created and saved in STL file format (as already mentioned, this process has been engineered by a partner company specialized in the field, and it is not discussed in the present paper).
- The prosthesis 3D model, still oriented at this stage with the inclination resulting from the coordinate system of the skull scan (see Figure 1a), must be re-oriented to properly fit the workpiece raw block to be machined. The latter is automatically selected among the number of standards available in stock at the hospital, based on the dimension of the prosthesis to be realized and after an iterative procedure in which the prosthesis is reoriented to minimize the quantity of scraps.
- The path and trajectories to be followed by the tool center point (TCP) of the mill need to be computed and generated, with the constraint that inner and outer surfaces (defined by the R_{int} and R_{ext} radii of Figure 1) are manufactured one side at once (see Section 4). The task of limiting robot movements inside a specific workspace might also be needed [17].
- The code instructions to command a robotic system with the designed trajectory must be generated and uploaded on a at least four DOF machine (plus one more DOF for tool rotation) (Section 5).
- Once the prosthesis has been finished, it could be removed, sterilized, and directly used by the surgeon. At most, burs should be removed with tools already available in the surgery room.

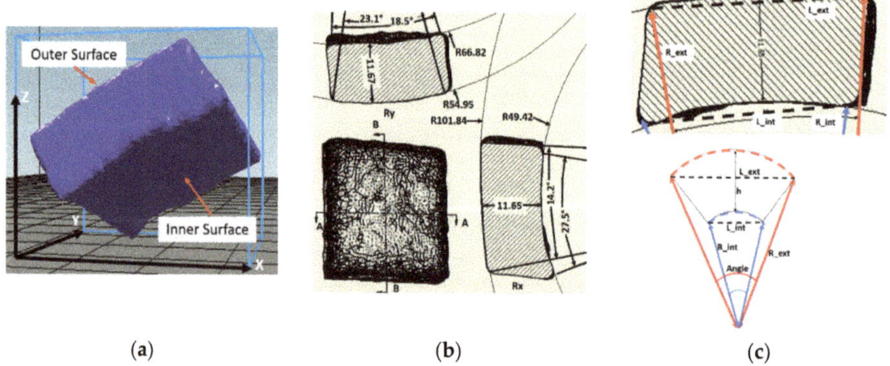

(a)   (b)   (c)

Figure 1. Geometrical parameters for the description of the operculum geometry. (**a**) Digital model (**b**) Main sections. (**c**) Considered dimensions.

Figure 1a reports an example of the digital model of a prosthesis and Figure 1b the corresponding draft on the XY plane, also representing two sections (AA and BB) in the two main directions.

The dimensions of each prosthesis are classified, as defined in Figure 1c, by considering the radii (external R_ext and internal R_int), the lengths (external L_ext and internal L_int) and thicknesses along the X and Y direction, for both the external and internal surfaces.

As an example, the geometrical results of 25 opercula (deriving from five skulls), representative of the overall population of prosthesis according to the medical experts involved in the activity, are summarized as statistical results in Table 1.

Table 1. Dimensional analysis of a set of 25 prostheses.

	Mean	**Max**	**Mode**	**std**
Rx_ext	82.4	196	50	38.78
Rx_int	85.08	500	47	93.532
Ry_ext	78.8	214	50	45.038
Ry_int	100	2848	43	554.9
hx	9.8	12	9	1.5
hy	10	12	12	1.5
	mean	**max**	**mode**	**min**
Lx_ext	29.68	45	30	19
Lx_int	30.76	47	31	20
Ly_ext	27.6	38	25	19
Ly_int	28.12	50	32	17

3. Automatic Selection of the Raw Material Block

The three main criteria for the selection of the starting block for the manufacturing of the workpiece are as follows:

- Processing waste shall be as minimal as possible.
- The starting block must be such as to allow the prosthesis manufacturing in a time shorter than surgery duration. In order to contain the manufacturing time, the inner and outer surface of the prosthesis (see Figure 1a for nomenclature) are machined one side at a time;
- A region for gripping must be considered in the lower-center area of the raw block, in addition to the material volume needed for the workpiece.

The digital model of the prosthesis to be manufactured shall undergo few preliminary operations before being exploited for the automatic selection of the starting raw block. After the model file is read, non-manifold tests are carried out at first (e.g., aimed at identifying self-intersecting geometries or open surfaces), followed by mesh reconstruction when needed. Since the non-manifold check is likely to be time consuming due to the size of the mesh, redefined points are removed.

The algorithm exploited for the identification of the best starting block dimensions is first in charge of defining the orientation of the prosthesis digital model within this raw block, to better fit its dimensions and to minimize the volume to be machined, thus reducing material waste. Figure 1a shows an example of the orientation the original model is likely to have at the beginning of this reorientation procedure.

The process starts with the definition of the reference system, whose origin is placed in the center of the rectangle circumscribing the model projection in the XY plane. Figure 2 represents the way the algorithm works: The prosthesis volume is firstly projected on the XY plane (projection marked with white shape, Figure 2a), and a plane bounding box is then evaluated, defining the minimum dimensions of the block required in this XY plane (i.e., black box, Figure 2a). The difference between the area of the prosthesis projection and the area of the bounding box is then evaluated through the application of the Gauss formula [18], quantifying in such a way the amount of material to be machined in this plane and, thus, the wasted material.

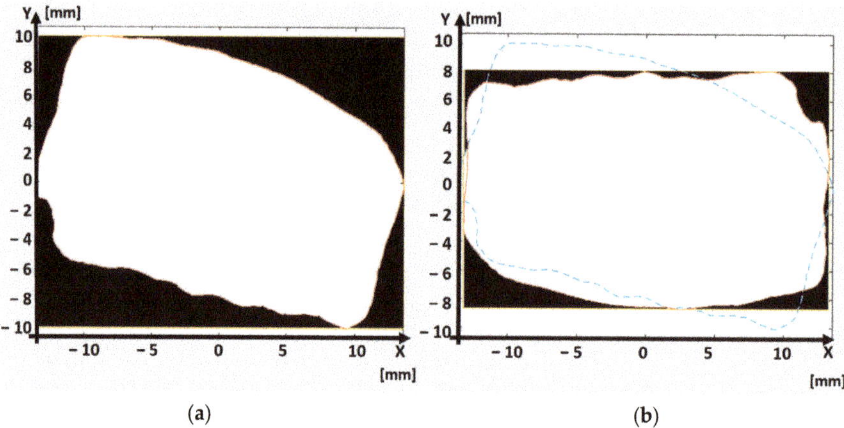

Figure 2. Orientation of the workpiece on a single plane through minimization of residual area between the workpiece projection and the bounding box. (**a**) Initial orientation. (**b**) Final orientation.

An iterative cycle allows then minimizing the difference between the prosthesis projection and the bounding box area: the 2D contour is iteratively turned with predefined steps of angular increase, around one axis at once (i.e., axis Z when considering projection on XY plane). For each step, a new bounding box is evaluated, and the difference between the prosthesis projection area and the bounding box area is updated. Once the rotation range selected in the software is completely investigated (e.g., 180 degrees), the Z-axis angular position is established. During the minimization, a constraint is set to keep the longest side of the bounding box oriented along the x-axis direction and the shortest side along the y-axis direction of the reference frame, which will be exploited in the procedure for the generation of the tool center point (TCP) trajectories as described in the following Section 4.

Figure 2b shows an example of the oriented workpiece at the end of this minimization procedure. The dashed blue line represents the original orientation as in Figure 2a, whereas the white contour represents the final orientation of the workpiece. The dimensions of the bounding box in this final configuration are smaller than those of Figure 2a, meaning that a smaller raw block can be adopted. Moreover, the area included between the bounding box and the prosthesis projection (black area in the figure), representing the amount of material to be machined, is much smaller in Figure 2b than Figure 2a.

After having oriented the workpiece in the XY plane, the procedure for the minimization of the residual area is repeated first in the XZ and finally in the YZ planes. The prosthesis 3D geometry is projected into one plane at once, and the areas of the projected geometry and of the corresponding bounding box are evaluated through Gauss formula.

Table 2 reports, as an example, the values of residual areas obtained for 6 sample of prosthesis, out the 25 analyzed. For each of the reported samples (i.e., 1a, 1b, 2a, 2b, 5a, 5b) the algorithm execution is carried out on XY, XZ, and YZ planes in sequence. After operating on each plane, the residual area achieved in the previous planes are slightly changed, so that the overall procedure is repeated with further runs until no significant variations are encountered with respect to the previous one.

Table 3 reports, for the same cases as in Table 2, the machinery time which would be required to realize the prosthesis in the original placement of the 3D digital model (i.e., the one oriented as the skull geometry), after the first run of the algorithm for orientation, and so on. It can be observed how the orientating procedure significantly reduces the machinery time right after the first run. Non-negligible differences can also be observed between run 1 and the end of run 3.

Table 2. Example of residual areas at the end of the orientation procedure.

Sample	Projection on Plane	Area after 1st Run (mm^2)	Area after 2nd Run (mm^2)	Area after 3rd Run (mm^2)
1a	XY	49.4843	40.4213	40.4213
	XZ	49.4086	49.2327	49.2327
	YZ	29.5412	26.1143	26.1143
1b	XY	30.2398	30.2398	30.2398
	XZ	64.6581	64.6581	64.6581
	YZ	41.8535	41.8535	41.8535
2a	XY	74.3664	69.4032	62.4382
	XZ	118.9427	81.0414	81.0965
	YZ	30.1888	26.9894	27.1190
2b	XY	392.1400	255.4181	252.2044
	XZ	487.2517	365.7807	365.2796
	YZ	171.7894	131.7848	133.7236
5a	XY	164.1555	164.1555	164.1555
	XZ	160.6684	160.6684	160.6684
	YZ	198.8549	198.8549	198.8549
5b	XY	204.3723	121.0910	111.9669
	XZ	397.2964	385.9294	385.6223
	YZ	114.1317	89.7570	79.7622

Table 3. Machinery time (MT) required to produce the prosthesis in the original orientation and after the orientating procedure.

Sample	MT Original Orientation (min)	MT after 1st Run (min)	MT after 2nd Run (min)	MT after 3rd Run (min)
1a	78	44	41	41
1b	72	37	37	37
2a	94	53	50	48
2b	149	108	100	99
5a	189	90	90	90
5b	171	98	80	79

Figure 3 represents the final orientation in the XY, XZ, and YZ plane, respectively, for the cases reported in Tables 2 and 3.

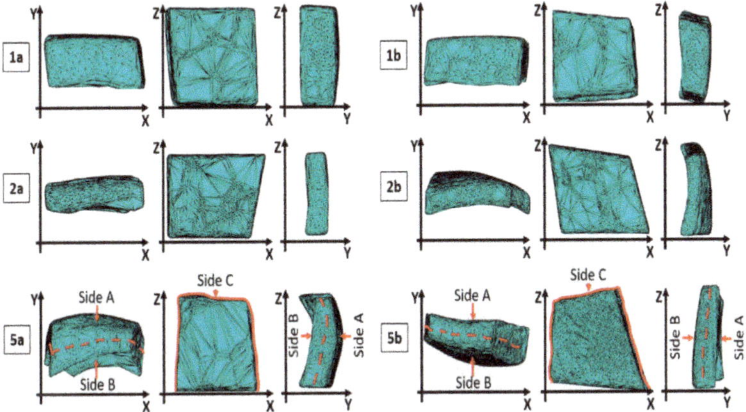

Figure 3. Examples of final orientation of the prosthetic opercula.

At the end of the execution of the algorithm, the digital model of the prosthesis is suitably oriented within a rectangular parallelepiped, whose dimensions give indications for the selection of the starting block of raw material. Before giving the ultimate indication and to allow for the block selection, the dimensions of the raw material block are increased by adding a gap in the z direction, in the x direction, and for the entire thickness of the raw block along y-direction. These dimensions are configurable parameters in the procedure (e.g., z_gap_block and x_gap_block), selectable on the basis of the dimensions of the gripping device and on the cutting tool exploited in the robotic cell. This additional material is added to create a margin at the top and bottom of the workpiece during the contouring phase, needed to grip the workpiece and to prevent the milling cutter from overrunning the prosthesis surface.

After the final dimensions of the starting raw block have been identified, the developed software can output on the screen in the operating room the numeric code of the raw block to be select from the ones available in stock (selected from a database) and give indications on how to properly install it on the machine.

4. Software for Generating the Tool Centre Point (TCP) Trajectories

This section describes the way TCP trajectories and robot commands are automatically generated starting from the oriented digital model obtained through the algorithm described in the previous section. Compared to previous research works dealing with the issue of automated computer-aided process planning (ACAPP) [19], the present application does not require the automatic selection of a specific cutting tool [20], which can be preliminary identified, once and for all, during the setup phase of the cutting parameters (see Section 5) and based on the material adopted. Moreover, the absence of specific features like pockets or holes in the geometry of the prosthesis (see Figure 1a) allows the development of a trajectory planning algorithm as simple as possible, compared to other state-of-art works [21]. The algorithm shall not indeed be designed to work with general sculptured surfaces [20], but it must be a special-purpose code specifically designed to work with a dedicated geometry. This enables the possibility to obtain a very stable process, avoiding in any case any manned procedure for checks, which could not be acceptable in the considered application.

4.1. Steps of the Machinery Procedure

The generation of the tool path starts by cutting the oriented 3D model with planes, parallel to the XY plane (see Figure 4) and suitably spaced. The spacing of the layers can be set according to the depth of each desired tool pass (h_layer). The slicing returns a contour of the workpiece per each layer, identified by finding the intersection between the slicing layer and the prosthesis model. An example of the obtained contours is represented in Figure 4. The adopted algorithm is a self-developed algorithm for conventional slicing (i.e., plane-triangle intersection combined with loop closure), already exploited for other manufacturing applications [22].

After the contours are generated, the algorithm identifies three separate portions of the prosthesis, so that the process of material removal can be divided in three phases: the first one is related to the machining of the prosthesis outer side (referenced as side A, see Figure 3); the second one is related to the machining of the inner side (referenced as side B, see Figure 3); the third one, referenced as side C (see Figure 3), is related to the remaining contour, as described in the following. The procedure for the identification of these three parts is based on the vectors normal to each segment constituting the contours of Figure 4, evaluated through the cross product between the unitary vector defining the orientation of each segment (following the contour in a clockwise manner) and the unit vector normal to each slicing plane. The angular position associated to each normal vector can then be exploited to identify the quadrant in which each vector lays. The procedure is described in Figure 5: if the angle defining the normal orientation is included in the first or second quadrants, the corresponding segment is associated to side A. If the angle is included in the third or fourth quadrants, the corresponding segment is associated to side B.

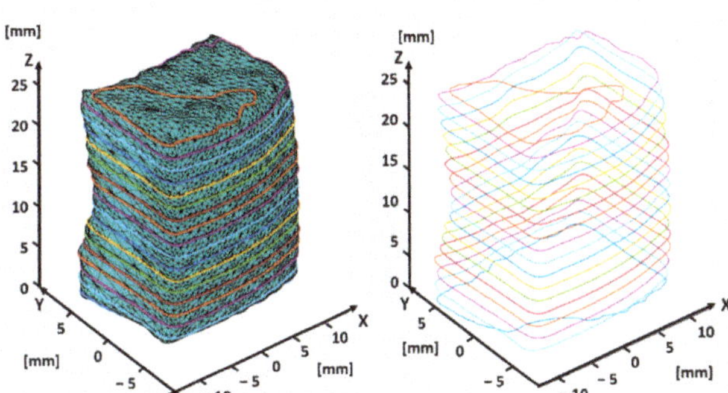

Figure 4. Closed profiles obtained through the slicing procedure of the oriented prosthesis.

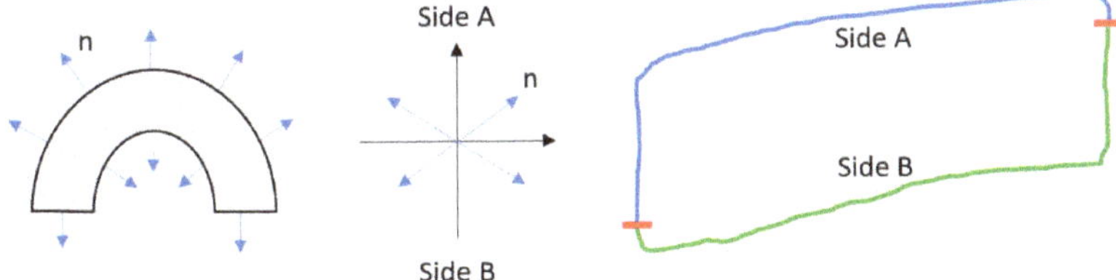

Figure 5. Identification of side A and side B of the workpiece through normal vectors analysis.

After identifying sides A and B, the volume of material to be removed can be recognized for each of the two sides, as reported in Figure 6.

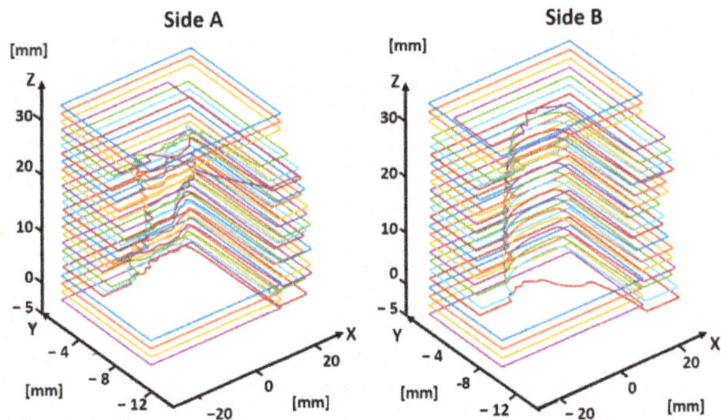

Figure 6. Volume of material to be removed corresponding to sides A and B.

The process for material removal is conceived to be carried out with the processing of side A first, and of side B afterwards. When starting the machining, the tool is in the home position, with the workpiece platform oriented to show the side A to the milling cutter. The TCP is positioned in the center of the region where the block will be chiseled, coincident with the origin of the local reference frame. After finishing side A, the tool exits the stock, allowing the platform to be rotated and side B to be machined.

Figure 7 reports the orientation of the tool with respect to the workpiece reference frame (X_{SW}, Y_{SW}, Z_{SW}).

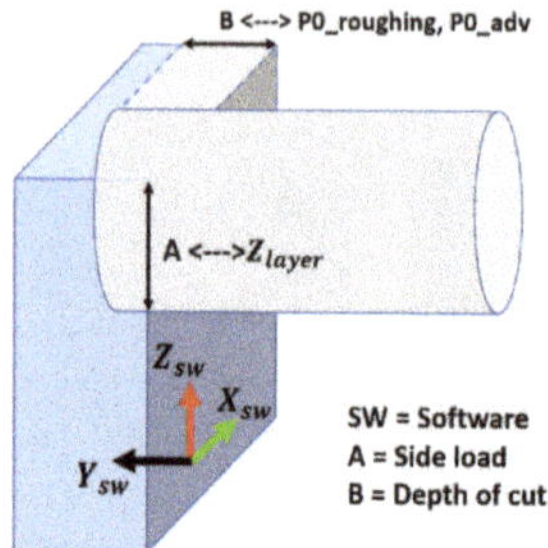

Figure 7. Orientation of the tool with respect to the workpiece reference frame (i.e., X_{SW}, Y_{SW}, Z_{SW}).

To guarantee that the workpiece is kept steady during the entire machinery, a narrow strip of material is maintained around the entire prosthesis during the machinery of sides A and B, so that it remains connected to the stock. To obtain this strip, the milling tool has to remove the material along the y-axis up until a certain depth (see Figure 8A,B). A representation of the "holding strip" is reported in the scheme of Figure 8C. It is referred to as side C. Its geometry is computed automatically in accordance with the geometry of the prosthesis, to make sure that there will be no chance of an unsteady placement of the part or even premature separation of the workpiece from the raw block, which would corrupt the quality of the final part. This machining approach increases the reliability of the overall process.

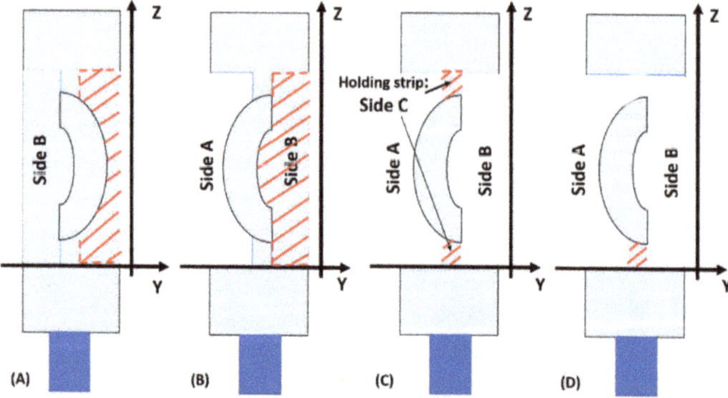

Figure 8. Schematic view of the machining steps. (**A**) Machinery of Side A. (**B**) Machinery of Side B. (**C**) Holding strips left after the machinery of sides A and B (i.e., Side C). (**D**) Separation of the prosthesis from the stock.

The third and last phase of the machinery procedure then consists in the removal of this anchoring strip (Figure 8D), to completely separate the prosthesis from the stock. The width of the holding strip is set automatically by the software so that it can be machined in one single pass, following the tool path obtained by projecting the prosthesis geometry in the XZ plane, as described in the Section 4.2. In order to guarantee the highest possible rigidity of the to-be-machined part along the toolpath, a residual strip of material is left at its bottom at the end of this job, so that the prosthesis is not completely detached from the block through the robotic milling procedure. A subsequent job will be in charge of separating the prosthesis and bring it to the sterilization chamber.

4.2. Generation of Toolpath Trajectories

The generation of the toolpath trajectories [23] for sides A and B is carried out by intersecting the volume of material to be removed, identified in Figure 6, with planes parallel to the XZ reference plane. The distance between subsequent planes depends on the type of pass to be generated, which can either be a roughing or smoother pass. In all the zones where the cutting plane parallel to XZ does not intersect the operculum, a roughing pass can be performed, allowing a great time reduction. On the contrary, when the plane intersects the geometry of the prosthesis, smoother passes are implemented.

The intersection of the mentioned planes with the prosthesis geometry is detected by evaluating the distances between each plane and the points belonging to the prosthesis contours, previously identified and reported in Figure 4: if the oriented distances between the plane under analysis and two subsequent points show the same sign, both the points lay on the same side with respect to the plane. On the other hand, if the distances show opposite sign, an intersection between the segment and the plane is identified.

The designed toolpath starts from a point generated at a certain distance (tool clearence) in the mill direction of advancement (i.e., −Ysw direction in Figure 7) from the first plane cutting the workpiece. The algorithm function then looks for the closest point, which can either lay on the same plane (i.e., same pass) or on the following plane along the axial direction (i.e., next pass), generating in such a way the entire tool path. Figure 9 reports an example of the final trajectories. When dealing with the points belonging to the prosthesis contour, the algorithm shall avoid milling part of the operculum [24]: it is able to discern the sequence of points found on the same plane, automatically retracting the cutter and jumping the prosthesis zone, to re-enter the workpiece where needed. The exemplary toolpath trajectories highlight the first and second roughing wipe, and the thin wipes adopted when the cutting planes intersect the prosthesis geometry.

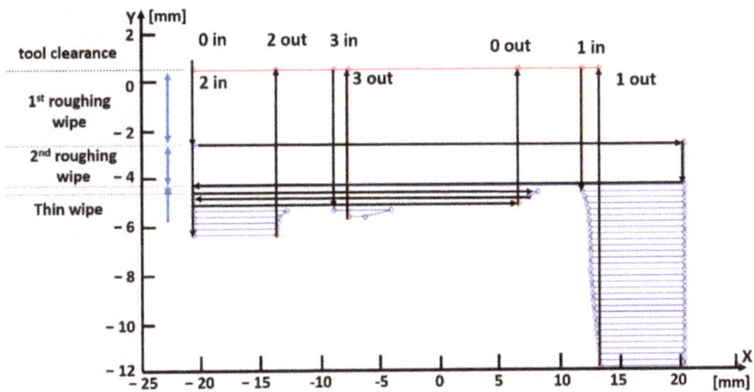

Figure 9. Example of toolpath trajectories generated.

Figure 10 shows the trajectory result for a section parallel to XY plane, highlighting the machining of sides A and B. At the end of machining of sides A and B a last trajectory is defined by linking all the points of the prosthesis contour, to define a last finishing pass to reduce the quantity of residual burr on the prosthesis surface.

The tool trajectories for the machining of side C are finally identified by projecting the 3D model of the holding strip onto the XZ plane and reconstructing the profile of the contour by using alpha shape boundary detection. An example of this projection is reported in Figure 11: the path to be followed by the TCP is constructed using the projected contour (i.e., the orange continuous line in Figure 11), its normal unitary vectors and the bit radius. The final toolpath is represented with a dashed line in the figure (labelled as tool path).

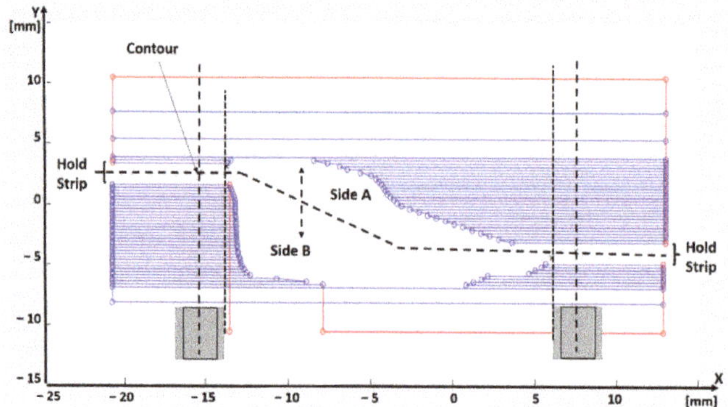

Figure 10. Example of tool path trajectories for machining side A and B.

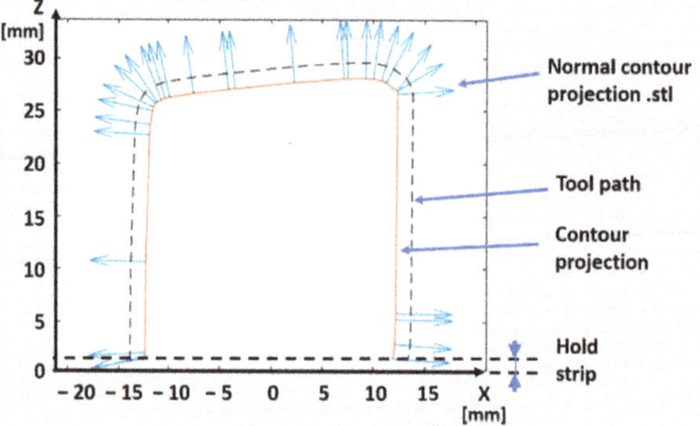

Figure 11. Projection on the XZ plane for the definition of the side C tool path.

4.3. Generation of the Machine Code

A further section of the developed code automatically generates the code instruction needed to command a robotic system with the designed TCP trajectory (e.g., a G-code when the robotic system adopted is able to read it). It is composed of three phases, listed as follows:

- Referring the trajectories to the machine reference system.
- Check for required corrections, to make sure the points are still compliant with the machine limits and check for machining time. As for the latter, the software checks the expected machining time and compare it the maximum allowed, depending on the surgery time. The estimation of the machining time is done by summing the duration of each linear interpolation obtained by connecting two consecutive points of the final toolpath, assuming a constant velocity. This approach is chosen for its computational simplicity: the contribution of the acceleration times has been discarded under the assumption that the motion planner is governed by a look ahead algorithm that tends to guarantee the commanded speed throughout the toolpath (hence improving the quality of the subtractive process), making the time contribution related to acceleration phases marginal.
- Writing the robot code: A code function autonomously translates the arrays of points defining the computed toolpath into machine instructions. This routine takes as input the points of side A, side B and side C obtained from the above described procedure and returns the machine code file. As for example, in the case of G-code, this file starts with information related to the setup of the machine and to the local reference frames that the machine will have to use for side A. The trajectories are then automatically printed in the text as G-code functional blocks: for instance, movements that engage the mill with the stock are defined as *"G1 X<Px> Y<Py> Z<Pz> F<Feed1>"*

(where *Px*, *Py* and *Pz* are the coordinates to be reached with a linear interpolation, and *Feed1* is the target TCP velocity defined in the next section). The repositioning movements of the mill use *G0* instead of *G1*. This operation is executed also for the side B, after having run a rotation of 180° of platform holding the workpiece. Lastly, the contour is appended to the file. It then ends with *go-to-home* position and with *turn-the machine off*.

5. Experimental Tests

In order to validate the trajectory planning software, an available CNC machine has been used as if it were a robot, feeding the controller with a G-code automatically output by the developed algorithm, starting from the digital model of a reference prosthesis having dimensions Rx_ext = 252 mm, Rx_int = 238 mm, Ry_ext = 255 mm, Ry_int = 244 mm. For the tests described in this section, a five-axis CNC milling machine was available (Pocket NC V2, Belgrade, MT, USA, Resolution: 6.10 microns (XYZ), 0.01° (A and B); Speed: 1524 m/min (XYZ), 40°/s; Spindle speed: 2–10 kRPM). Experimental tests have been carried out to check the software behavior and to set the most suitable parameters and hardware for the machining the selected material (the latter is not mentioned here, since the entire process is under the patenting process by an industrial company partner of this work).

5.1. Setting of Working Parameters

The process parameters are listed in Figure 12. The top of Figure 12a represents a side view of the block under machinery, whereas the bottom of Figure 12b the top view. None of the parameters indicated in the figure, exploited by the software to generate the toolpath, must be initialized by the operator (i.e., the surgeon), who should be exempted as much as possible from technical evaluations about the machinery process. For this reason, the experimental tests described in this paragraph are aimed at identifying the best parameters for the selected material, to be pre-set in the final software actually used in the hospital to operate the machinery.

- Miller diameter: d_m [mm]
- Cut movement speed: Feed1 [mm/min]
- Free movement speed: Feed0 [mm/min]
- Initial layer height: h_layer0
- Layer height: h_layer

- Safe distance tool-workpiece: tool_clearance
- Dimension of the strip per layer: p0_adv
- Roughing pass: p_roughing
- Thin pass: p_adv

- External margin along X: X_gap_block
- External margin along Z: Z_gap_block
- Starting block thickness X: X_block
- Starting block thickness Y: Y_block

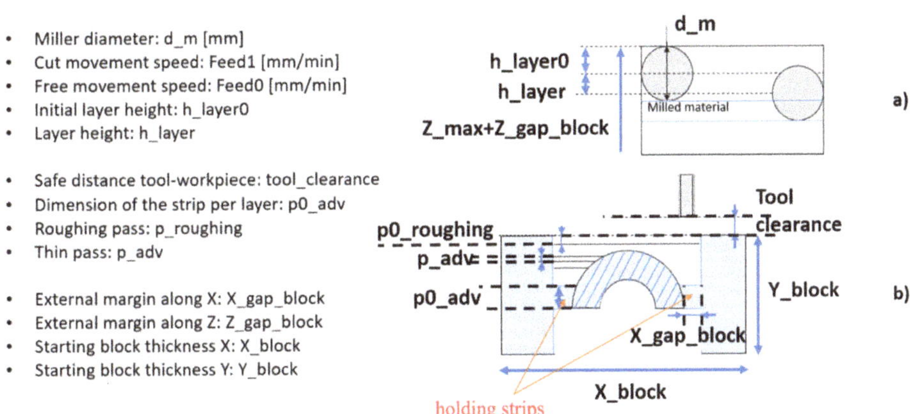

Figure 12. Process parameters. (**a**) Side view. (**b**) Top view.

In the experimental campaign carried out, the process parameters are grouped into two categories, the first one enclosing those that are kept constant during the tests, the second one including parameters that can be tuned to improve the machining operation.

The parameters kept constant during the experiments are the mill diameter (*d_m*, see Figure 12a), the first side load (*h_layer0*) always equal to half of the mill diameter, the gap left between the mill and the raw block to assure a safe and fast repositioning of the TCP with respect to the workpiece surface (*Tool_clearance*), and the speed related to this movement (named *Feed0*, since it is related to G0 blocks in the G-code). Additionally, the width of holding strip that constrains the part to the stock until the final contouring is carried out (*p0_adv*) and the depth of roughing passes (*p_roughing*), also related to the mill diameter. Furthermore, the parameters (*X_gap_block* and *Z_gap_block*) mentioned in Section 3, used to define of the overall dimensions of the raw block, are not changed during the experiments.

On the other hand, the parameters directly related to the chip removal have been varied during the testing phase, having a direct influence on the quality of the final implant. They are the spindle speed (*S*), the side load (*h_layer*, see Figure 12a), the depth of cut (*p_adv*, see Figure 12b) and the speed of the TCP (named *Feed1*, since it is related to G1 blocks). The optimal set for these parameters depends on the material to be machined and on the characteristics of tool used.

The prosthetic plates are machined from a polymeric material (not mentioned here in detail, since the entire process is under patenting process by an industrial company partner of this work). As indicated by the mill producer, two types of setting can be used as a starting point in the fine-tuning process based on the plastic hardness, indicated as hard and soft plastic. They are indicated in Table 4.

Table 4. Parameters recommended by the milling producer for two type for soft and hard plastic.

	Speed (RPM)	Feed per Tooth [mm]	Side Load (h_layer)	Depth of Cut (p_adv)
Hard Plastic	8500	0.0254	50%	80%
Soft Plastic	8500	0.0381	60%	70%

The side load and the depth of cut in Table 4 are indicated as a percentage of the mill diameter. The diameter of the selected mill cutter being 3.175 [mm] (1/8 inch), the axial and radial depths of the cut can be computed based on the *depth of cut* and *side load* parameters, respectively. For example, in the case of "hard plastic", the axial depth of cut (i.e., *p_adv* in Figure 12) is 2.54 mm, whereas the radial depth of cut (i.e., *h layer* in Figure 12) is 1.5875 mm.

The *feed per tooth* parameter allows evaluating the TCP speed (i.e., Feed1 in Figure 12), computed as:

$$\text{Feed1 [mm/min]} = \text{feed per tooth [mm]} * \text{teeth number} * \text{Speed (RPM)} \tag{1}$$

In order to contain the temperature during cutting [25] and, therefore, to obtain optimal performance and surface results, a single cutter tool is exploited (i.e., teeth number equal to 1). Hence, in the case of "soft plastic", for instance, the feed1 value is equal to 323.85 mm/min.

5.2. Results

Tests have been conducted using both the specifics for "hard" and "soft" plastic. Better results have been achieved for the "soft plastic" parameters: in this case the residual material remained at the end of working was almost null and the burrs have been easily removed using rotary tools for deburring, leaving the operculum clean.

The final prosthetic device can be found in Figure 13 right after the milling operation, and in Figure 14, after cleaning through a rotary tool (brush with 80 grit size at 50% of the maximum speed of the device, max speed 35 kRPM).

It can be observed as the final surfaces in Figure 14, after deburring, are rather regular and well-finished.

The experimental tests demonstrated the possibility to automatically produce the prosthetic device within the time required by the surgery, and with good finishing of the workpiece.

Moreover, the results achieved through a prototype demo demonstrated the feasibility of having the prosthesis manufactured on-site. The entire process could be automatized by means of an automatic robotic cell, starting from the raw material block and getting to the final sterilized prosthesis. The same robotic system would indeed place the raw workpiece, machining it with the proper tool and, once the machinery is over, grip the workpiece and place it into a sterilizing machine. At the end of the process, the prosthesis could then be directly used by the surgeon.

Figure 13. Mono-cutting bid and "soft plastic": prosthesis removed from the CNC.

Figure 14. Mono-cutting bid and "soft plastic": prosthesis after deburring.

The installation process in an actual hospital would entail the adoption of the robotic cell in a dedicated sterile environment. To this aim, the final design should make use of robots suitably designed for sterile applications, which are already available on the market. They are able to handle decontamination processes through vapor or fluid, and they are designed with special joints and a structure featuring an IP65 degree of protection.

6. Conclusions

The paper presented an algorithm for the generation of an automatic robotic procedure for machining of skull prosthesis. The setup procedure allows to manufacture the prosthesis in real-time during the surgery, based on the actual cut carried out by the surgeon and within the time required to complete the operation.

The input for the algorithm is a digital model of the prosthesis to be realized, which is processed to generate the trajectories to be followed by a robotic system for the manufacturing of the workpiece. A first step of the procedure allows to identify the best placement of the workpiece into the raw material block, to reduce material waste and to identify, at the same time, the dimension of the raw material block needed for machining. This would allow the automatic selection of the raw block between those available, registered in a database, thus reducing the activities to be carried out by medical personnel. The latter should only follow instructions on a monitor to place the proper working piece into the machine.

As a second step, the algorithm automatically generates the toolpath trajectories to be executed on the machine, through an automatic process derived from a slicing procedure. The latter allows to generate zig-zag trajectories to be followed by the milling tool, which is a simple but stable trajectory planning method allowing a straightforward and automatic coding of the toolpath into the program for robot movements. A four d.o.f. at least machine is required, plus one more d.o.f. for tool rotation. To this regards, either six d.o.f manipulators or five-axis Cartesian robot may be exploited for the application.

The results of the prosthesis realized show the good behavior of the automatic procedure, demonstrating the possibility to produce the operculum within the duration of the operation, thus increasing the accuracy of the produced prosthesis and consequently reducing the surgery time. The possibility to realize the prosthesis through a milling machine enables the adoption of plastic material which can undergo radiotherapy without the need to remove the prosthesis from the cranium.

The approach proposed allowed to demonstrate the possibility to automatize the entire production process, thus introducing a fully automated process in a hospital environment. Starting from the digital model of the prosthetic device to be realized, the automation of the orientation of the workpiece within the raw block, the generation of end effector trajectories through an unmanned stable procedure, and finally the automatic programming of the robotic machine, constitute a preparatory work to introduce the automation process in a surgery room. The medical personnel are not expected to operate the machine, except for simple operations like placing the raw block into the machine, authorize the start of the entire procedure, and possibly remove burs using rotary tools for deburring.

Future research should consist in the design of a complete robotic cell, in which the raw block is automatically placed for machining and, once the machinery is over, the prosthesis is gripped and automatically placed into a sterilizing machine. From a procedural point of view, a further future work is related to the development of a specific procedure for the accreditation of the sterile process.

Author Contributions: Conceptualization: H.G., M.C., K.C.; methodology: H.G., K.C.; software: K.C.; validation: K.C.; formal analysis: M.C.; investigation: H.G., K.C.; data curation: K.C.; writing—original draft preparation: M.C.; writing—review and editing: M.C.; visualization: K.C., M.C.; supervision: H.G.; project administration: H.G.; funding acquisition: H.G. All authors have read and agreed to the published version of the manuscript.

Funding: This research received no external funding.

Conflicts of Interest: The authors declare no conflict of interest.

References

1. Peruzzini, M.; Wognum, N.; Bil, C.; Stjepandic, J. Special issue on 'transdisciplinary approaches to digital manufacturing for industry 4.0'. *Int. J. Comput. Integr. Manuf.* **2020**, *33*, 321–324. [CrossRef]

2. Ghosh, U.; Ning, S.; Wang, Y.; Kong, Y.L. Addressing Unmet Clinical Needs with 3D Printing Technologies. *Adv. Healthc. Mater.* **2018**, *7*, e1800417. [CrossRef]

3. Radenkovic, D.; Solouk, A.; Seifalian, A. Personalized development of human organs using 3D printing technology. *Med. Hypotheses* **2016**, *87*, 30–33. [CrossRef]

4. Dobrzański, L.A. Overview and general ideas of the development of constructions, materials, technologies and clinical applications of scaffolds engineering for regenerative medicine. *Arch. Mater. Sci. Eng.* **2014**, *69*, 53–80.

5. Dean, D.; Min, K.-J.; Bond, A. Computer Aided Design of Large-Format Prefabricated Cranial Plates. *J. Craniofacial Surg.* **2003**, *14*, 819–832. [CrossRef]

6. Chrzan, R.; Urbanik, A.; Karbowski, K.; Moskała, M.; Polak, J.; Pyrich, M. Cranioplasty prosthesis manufacturing based on reverse engineering technology. *Med. Sci. Monit.* **2012**, *18*, MT1–MT6. [CrossRef]

7. Guillaume, O.; Geven, M.A.; Varjas, V.; Varga, P.; Gehweiler, D.; Stadelmann, V.A.; Smidt, T.; Zeiter, S.; Sprecher, C.; Bos, R.R.; et al. Orbital floor repair using patient specific osteoinductive implant made by stereolithography. *Biomaterials* **2020**, *233*, 119721. [CrossRef]

8. Gander, T.; Essig, H.; Metzler, P.; Lindhorst, D.; Dubois, L.; Rücker, M.; Schumann, P. Patient specific implants (PSI) in reconstruction of orbital floor and wall fractures. *J. Cranio Maxillofac. Surg.* **2015**, *43*, 126–130. [CrossRef]

9. Stoor, P.; Suomalainen, A.; Lindqvist, C.; Mesimäki, K.; Danielsson, D.; Westermark, A.; Kontio, R.K. Rapid prototyped patient specific implants for reconstruction of orbital wall defects. *J. Cranio Maxillofac. Surg.* **2014**, *42*, 1644–1649. [CrossRef]

10. Schlittler, F.; Vig, N.; Burkhard, J.P.M.; Lieger, O.; Michel, C.; Holmes, S. What are the limitations of the non-patient-specific implant in titanium reconstruction of the orbit? *Br. J. Oral Maxillofac. Surg.* **2020**. [CrossRef]

11. Tarsitano, A.; Battaglia, S.; Crimi, S.; Ciocca, L.; Scotti, R.; Marchetti, C. Is a computer-assisted design and computer-assisted manufacturing method for mandibular reconstruction economically viable? *J. Cranio Maxillofac. Surg.* **2016**, *44*, 795–799. [CrossRef] [PubMed]

12. Kozakiewicz, M.; Wach, T.; Szymor, P.; Zieliński, R. Two different techniques of manufacturing TMJ replacements—A technical report. *J. Cranio Maxillofac. Surg.* **2017**, *45*, 1432–1437. [CrossRef] [PubMed]

13. Huang, G.-Y.; Shan, L.-J. Research on the Digital Design and Manufacture of Titanium Alloy Skull Repair Prosthesis. In Proceedings of the 2011 5th International Conference Bioinformatics Biomedical Engineering, Wuhan, China, 10–12 May 2011. [CrossRef]

14. Keeratihattayakorn, S.; Tangpornprasert, P.; Prasongcharoen, W.; Virulsri, C. Out-of-roundness compensation technique in machining of femoral head prosthesis using conventional CNC machine. *Int. J. Adv. Manuf. Technol.* **2020**, *107*, 2537–2545. [CrossRef]

15. D'Urso, P.S.; Effeney, D.J.; Earwaker, W.J.; Barker, T.M.; Redmond, M.J.; Thompson, R.G.; Tomlinson, F.H. Custom cranioplasty using stereolithography and acrylic. *Br. J. Plast. Surg.* **2000**, *53*, 200–204. [CrossRef] [PubMed]

16. Lee, S.-C.; Wu, C.-T.; Lee, S.-T.; Chen, P.-J. Cranioplasty using polymethyl methacrylate prostheses. *J. Clin. Neurosci.* **2009**, *16*, 56–63. [CrossRef]

17. La Mura, F.; Romanó, P.; Fiore, E.; Giberti, H. Workspace Limiting Strategy for 6 DOF Force Controlled PKMs Manipulating High Inertia Objects. *Robotics* **2018**, *7*, 10. [CrossRef]

18. Beyer, W. *CRC Standard Mathematical Tables and Formulae*; CRC Press: Boca Raton, FL, USA, 1996.

19. Al-Wswasi, M.; Ivanov, A.; Makatsoris, H. A survey on smart automated computer-aided process planning (ACAPP) techniques. *Int. J. Adv. Manuf. Technol.* **2018**, *97*, 809–832. [CrossRef]

20. Lin, A.C.; Gian, R. A Multiple-Tool Approach to Rough Machining of Sculptured Surfaces. *Int. J. Adv. Manuf. Technol.* **1999**, *15*, 387–398. [CrossRef]

21. Bieterman, M.B.; Sandstrom, D.R. A Curvilinear Tool-Path Method for Pocket Machining. *J. Manuf. Sci. Eng.* **2003**, *125*, 709–715. [CrossRef]

22. Carnevale, M.; Castelli, K.; Zaki, A.M.A.; Giberti, H.; Reina, C. Automation of Glue Deposition on Shoe Uppers by Means of Industrial Robots and Force Control. *Mech. Mach. Sci.* **2020**, *91*, 344–352. [CrossRef]

23. Giberti, H.; Sbaglia, L.; Urgo, M. A path planning algorithm for industrial processes under velocity constraints with an application to additive manufacturing. *J. Manuf. Syst.* **2017**, *43*, 160–167. [CrossRef]

24. D'Antona, G.; Davoudi, M.; Ferrero, R.; Giberti, H. A model predictive protection system for actuators placed in hostile environments. In Proceedings of the 2010 IEEE Instrumentation & Measurement Technology Conference, Austin, TX, USA, 3–6 May 2010; pp. 1602–1606. [CrossRef]

25. Alauddin, M.; Choudhury, I.; El Baradie, M.; Hashmi, M. Plastics and their machining: A review. *J. Mater. Process. Technol.* **1995**, *54*, 40–46. [CrossRef]

Publisher's Note: MDPI stays neutral with regard to jurisdictional claims in published maps and institutional affiliations.

 *robotics*

Article

Simulation Assessment of the Performance of a Redundant SCARA

Roberto Bussola [1,†]**, Giovanni Legnani** [1,†]**, Massimo Callegari** [2,†] **and Giacomo Palmieri** [2,†] **and Matteo-Claudio Palpacelli** [2,*,†]

[1] Department of Mechanical & Industrial Engineering, University of Brescia, 25121 Brescia, Italy; roberto.bussola@unibs.it (R.B.); giovanni.legnani@unibs.it (G.L.)

[2] Department of Industrial Engineering & Mathematical Sciences, Polytechnic University of Marche, 60131 Ancona, Italy; m.callegari@univpm.it (M.C.); g.palmieri@univpm.it (G.P.)

* Correspondence: m.palpacelli@univpm.it; Tel.: +39-071-220-4748

† These authors contributed equally to this work.

Received: 13 March 2019; Accepted: 10 June 2019; Published: 12 June 2019

Abstract: The present paper analyses the potential dynamic performance of a novel redundant SCARA robot, currently at the stage of a functional design proposed by a renowned robot manufacturer. The static and dynamic manipulability of the new concept is compared with the conventional model of the same manufacturer by means of computer simulation in typical pick and place tasks arising from industry. The introduction of a further revolute joint in the SCARA robot kinematics leads to some improvements in the kinematic and dynamic behaviour at the expense of a greater complexity. In this paper, the potential of a redundant SCARA architecture in cutting cycle-times is investigated for the first time in performing several tasks. It is shown that, in order to exploit the possible enhancements of the redundant structure, the whole manipulator, mechanics and control must be redesigned according to specific tasks aiming at the optimization of their cycle-time.

Keywords: industrial robot; high speed robot; pick-and-place task; redundant robot; manipulability

---

1. Introduction

The history of assembly technology has a milestone in 1980, when Hiroshi Makino filed the patent of a new concept of robot called SCARA, which stands for Selective Compliance Assembly Robot Arm [1], see Figure 1a; with four axes; it is able to develop a kind of motion, called Schönflies motion, which is very useful in many applications and consists of a linear displacement in three-dimensional space plus one orientation around an axis with fixed direction. The success of the SCARA kinematics is mainly due to the fact that this architecture can realise very high speed movements, unreachable at that time with previous robots like the three-axis R-theta Robot.

Since the 1980s, many other efforts have been made to meet or even overcome the performance of that design, leading in recent years to the completely new concept of parallel kinematics machines (PKMs), like the Delta patented by Clavel [2]. Other parallel kinematic structures that assure fast pick-and-place operation have been derived from the Delta manipulator, e.g., the PKM proposed by Tosi et al. [3] or from other similar concepts [4–6]. The only successful attempt to exploit the SCARA architecture for designing PKMs is probably the so-called dual-arm SCARA robot architecture, which is actually the five-bar linkage showed in Figure 1b.

Robotics **2019**, *8*, 45; doi:10.3390/robotics8020045   www.mdpi.com/journal/robotics

191

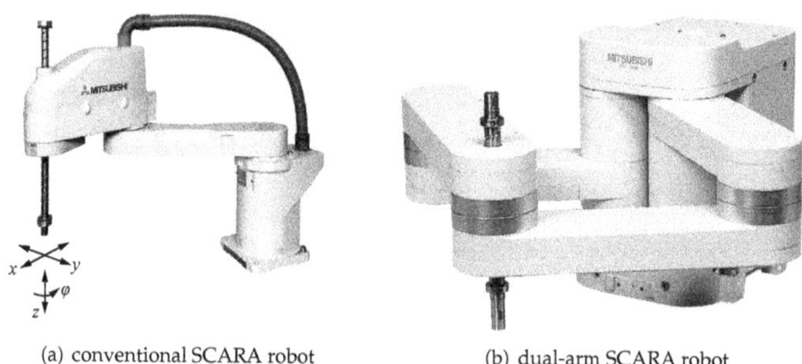

(a) conventional SCARA robot   (b) dual-arm SCARA robot

Figure 1. Different commercial solutions of SCARA robots.

However, the SCARA architecture is still competitive for many applications and, apparently, its structure is simple enough to be considered optimized. On the other hand, for the moment, it does not seem that the possibility to improve performances by using redundant SCARA robots has been considered in order to produce a commercial manipulator: this paper aims at covering this issue. In this case, the position of their gripper on a horizontal plane will depend on three actuators realizing a functional redundancy that can be used, for instance, to increase the robot dexterity and the gripper velocity; in this way, shorter working cycles can be obtained, thus increasing the efficiency of manufacturing operations [7]. Only a few works can be found in the literature dealing with redundant versions of the SCARA robot: some refer to the choice of control schemes and their optimization [8,9], other to dextrous motion control [10] and new redundant architectures for energy savings [11]. On the contrary, many works deal with the exploitation of redundancy to improve the kinematic and dynamic performance of a serial robot [12].

A previous work of the authors Callegari et al. [13] analysed the kinematic and dynamic performance of the concept at a preliminary stage, whereas in this paper the work is significantly extended proposing many dynamic simulations useful to compare the traditional and redundant SCARA designs (indicated as SCARA-Trad and SCARA-Red in the following) in executing typical industrial tasks.

2. Robot Design and Basic Characteristics

A functional design of the proposed redundant manipulator is represented in Figure 2, where a comparison with a traditional SCARA robot of similar size and working area is given. Both robots share the same reach (one meter) and adopt the same end-effector actuators.

Some important technical data provided by an industrial robot manufacturer are shown in Table 1. It can be noted that links are governed by different gear ratios. Motors are hosted in the fixed base and their motion is transmitted to the joints by a system of pulleys and belts, not shown in the figure for simplicity. In this way, each actuator drives the absolute angular position of a single link.

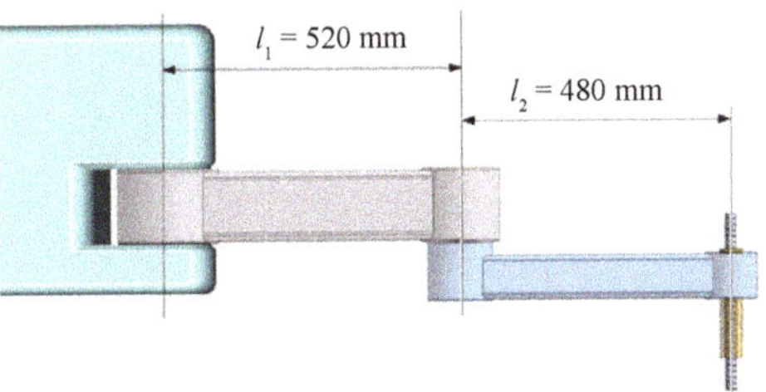

(a) conventional SCARA robot

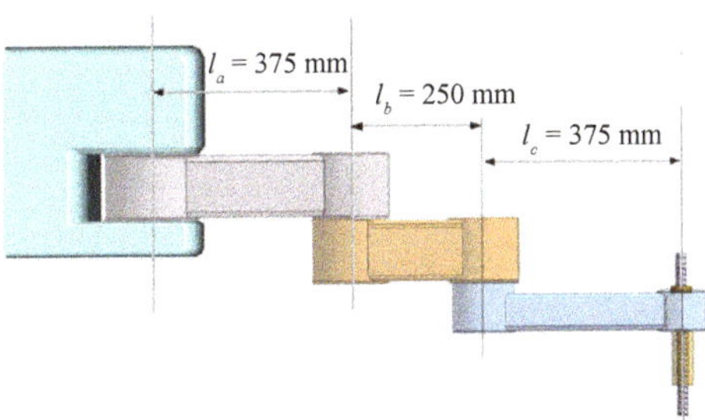

(b) redundant SCARA robot

Figure 2. Technical data of the SCARA robots.

Table 1. Technical data of the conventional and the redundant SCARA robot.

Axis Motor		Rated Torque (Nm)	Max Torque (Nm)	Max Speed (rpm)	Gear Ratios	Links' Length (mm)	Mass (kg)	Inertia (kgm²)
SCARA-Trad	joint 1	9.5	19	5000	30	$l_1 = 520$	12	0.58
	joint 2	(at 2000 rpm)	(at 2000 rpm)		20	$l_2 = 480$	4.8	0.21
SCARA-Red	joint 1	9.5	19	5000	30	$l_a = 375$	11	0.34
	joint 2	(at 2000 rpm)	(at 2000 rpm)		15	$l_b = 250$	6.8	0.15
	joint 3				15	$l_c = 375$	4.4	0.15
end effector (both robots)	spin z-axis	1.59 (at 3000 rpm)	3.18 (at 3000 rpm)	6000	15 40 mm/turn	– –	– –	– –

The kinetostatic properties of the two manipulators can be developed starting from the direct kinematics equations $\mathbf{p} = \mathbf{f}(\mathbf{q})$, which relate the gripper planar position $\mathbf{p} = [x, y]^T$ with the joint coordinate $\mathbf{q}$ represented by the absolute angular position of the links, shown in Figure 3. For the traditional SCARA, it follows:

$$x = l_1 \cos(\varphi_1) + l_2 \cos(\varphi_2),$$
$$y = l_1 \sin(\varphi_1) + l_2 \sin(\varphi_2),$$

(1)

with $\mathbf{q} = [\varphi_1, \varphi_2]^T$, whereas the redundant SCARA equations change as follows:

$$x = l_a \cos(\alpha) + l_b \cos(\beta) + l_c \cos(\gamma),$$
$$y = l_a \sin(\alpha) + l_b \sin(\beta) + l_c \sin(\gamma),$$

(2)

with $\mathbf{q} = [\alpha, \beta, \gamma]^T$.

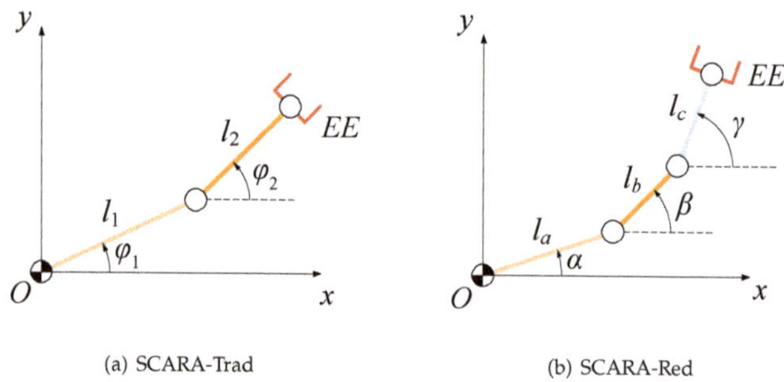

(a) SCARA-Trad   (b) SCARA-Red

Figure 3. Notation: (**a**) traditional SCARA robot, (**b**) redundant SCARA robot.

Cartesian planar velocities and accelerations can be easily developed as:

$$\dot{\mathbf{p}} = \mathbf{J}\dot{\mathbf{q}},$$
$$\ddot{\mathbf{p}} = \mathbf{J}\ddot{\mathbf{q}} + \dot{\mathbf{J}}\dot{\mathbf{q}}, \tag{3}$$

where the dots mark the time derivative, $\mathbf{J}$ is the Jacobian matrix, which is a 2×2 square matrix for the traditional robot and a 2×3 rectangular matrix for the redundant one.

The (theoretical) working areas of the two manipulators, schematically reported in Figure 4, have the shape of a ring with internal radius $r = |l_1 - l_2| = 40\,\text{mm}$ and external radius $R = l_1 + l_2 = 1000\,\text{mm}$ for the traditional SCARA and of a circle with radius $R' = l_a + l_b + l_c = 1000\,\text{mm}$ for the redundant manipulator. The dashed circle with radius $R'' = l_a + l_b - l_c = 250\,\text{mm}$ indicates the portion of the working area that can be reached for any orientation of the third link.

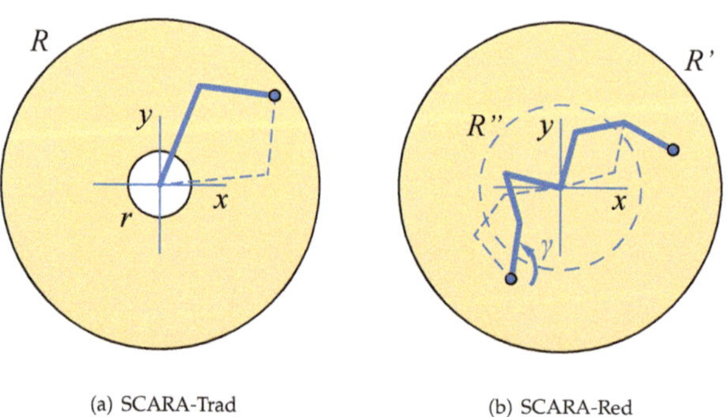

(a) SCARA-Trad   (b) SCARA-Red

Figure 4. Working areas and configurations of the two versions of the robot: (**a**) finite solutions for the traditional SCARA, (**b**) infinite solutions for the redundant SCARA.

It is well known that, for any position $\mathbf{p} = [x, y]^T$ of the gripper, the traditional SCARA robot has two different configurations (Figure 4a), whereas the redundant SCARA robot has infinite solutions (Figure 4b): in this case, for any choice of the angle γ, the robot has two solutions. As discussed by the authors in

Callegari et al. [13], the orientation γ of the SCARA-Red manipulator may be utilised to achieve a better isotropy and higher gripper velocity. This is confirmed by the velocity ellipses defined by Yoshikawa [14]:

$$\dot{\mathbf{p}}^T \left(\mathbf{J}\mathbf{J}^T \right)^{-1} \dot{\mathbf{p}} = k \tag{4}$$

with $k \leq 1$. Some examples of velocity ellipses are shown in Figure 5, where the SCARA robots can be compared in terms of kinematic performance. It can be noted that, due to the radial symmetry around the rotation axis of the first joint, the kinetostatic properties of each manipulator can be studied without considering the whole working area, but it is sufficient to investigate the robot behaviour for gripper positions on a "radial direction", namely the x-axis for convenience. By using the redundancy to optimize the angular position of the last link, as represented in Figure 5c, the velocity ellipses of the SCARA-Red turn out to be bigger than those of the SCARA-Trad.

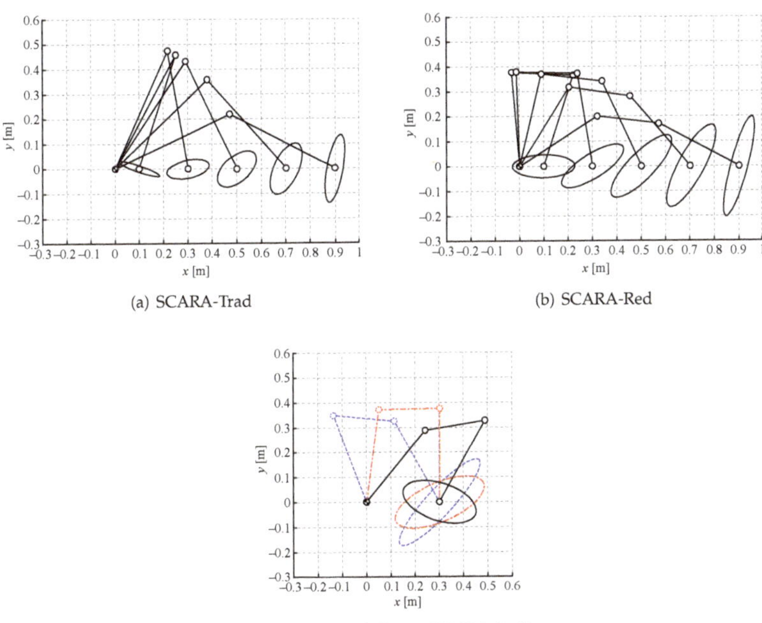

Figure 5. Comparison between velocity ellipses of the SCARA-Trad (a) and SCARA-Red robots (b) for the same gripper position (the SCARA-Red is shown with the same angular position of the last link as the SCARA-Trad). In (c), different ellipses related to different orientations γ of the SCARA-Red at a given end-effector position.

3. Kinematic and Dynamic Characterisation of Robot Performance

As a preliminary phase to the robots' analysis, the kinematic and dynamic models of the two manipulators have been derived in symbolic form and converted in the following canonical form:

$$\boldsymbol{\tau} = \mathbf{M}\ddot{\mathbf{q}} + \tilde{\boldsymbol{\tau}}(\mathbf{q}, \dot{\mathbf{q}}), \tag{5}$$

where $\boldsymbol{\tau} = [\tau_1, ..., \tau_n]^T$ is the vector of joint torques (with $n = 2$ for the SCARA-Trad, $n = 3$ for the SCARA-Red), **M** is the mass matrix and $\tilde{\boldsymbol{\tau}}$ is the term, quadratic in $\dot{\mathbf{q}}$, depending on centripetal and Coriolis accelerations.

A first comparison between the two SCARA robots can be carried out in terms of kinematic and dynamic performance by means of velocity and acceleration polygons, respectively. In fact, at each radial position, the velocity and acceleration of their end-effector along each Cartesian direction allow for drawing a polygon, which is a parallelogram for the SCARA-Trad and a hexagon for the SCARA-Red [13], as shown in Figure 6. A velocity polygon is obtained for each robot by varying the speed of the arm motors in their nominal range (see Table 1). In Figure 6, the orientation γ of the third link of the SCARA-Red is chosen so that the radial velocity v_x is maximized. It is easy to verify from the figure that the SCARA-Red is able to perform a maximum radial velocity higher than the one of the SCARA-Trad, at least under ideal hypotheses. A similar study can be addressed for Cartesian acceleration polygons by imposing motor torques in their operating range (see Table 1) and assuming an initial state with null joint velocities.

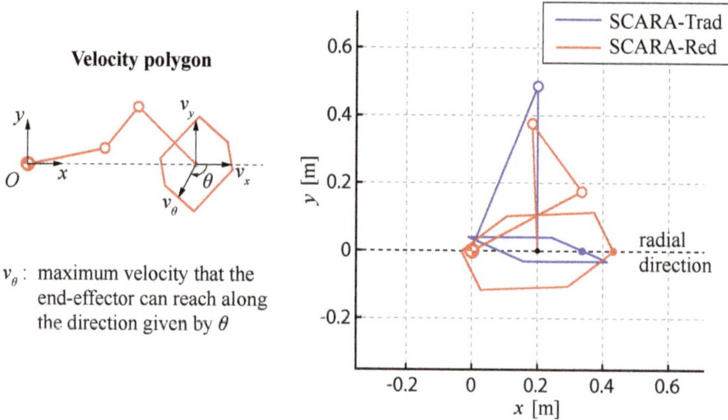

Figure 6. Example of velocity polygons for the SCARA robots drawn at a generic radial position of their end-effector.

In order to highlight the potential of the SCARA-Red with respect to the SCARA-Trad, it is possible to compare the performance of the two robots by graphically superimposing at a common end-effector position the velocity parallelogram of the SCARA-Trad to the envelope of the velocity polygons of the SCARA-Red, obtained by drawing its polygons for different orientations γ, as shown in Figure 7a. In more detail, when a radial position of the end-effector is close to the fixed frame, where the reference system $O - \{x, y\}$ is located, a complete turn of the third link of the SCARA-Red is allowed by the robot kinematics.

The envelope demonstrates that the redundant SCARA has a similar behaviour in terms of velocity in every direction of the horizontal plane. On the contrary, the traditional SCARA can reach the maximum velocity only along the major diagonal of the parallelogram, whereas its performance decreases along a generic angular direction θ, as pointed out by vectors $\mathbf{v}_\theta$. Analogously, in Figure 7b, it is shown how the SCARA-Red performs better than the SCARA-Trad even dynamically, offering higher acceleration in all Cartesian directions.

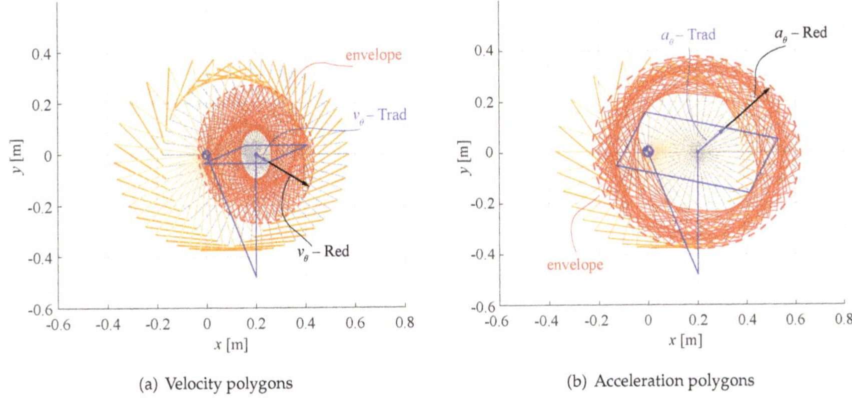

(a) Velocity polygons   (b) Acceleration polygons

Figure 7. Envelope of polygons drawn for different orientation γ.

The manipulability of both the SCARA robots can be deepened by analysing the manipulability indexes I_{MK} and I_{MD}, which respectively provide a measure of kinematic and dynamic isotropy, defined as:

$$I_{MK} = \frac{2\sqrt{\det\left(\mathbf{JJ}^T\right)}}{\mathrm{tr}\left(\mathbf{JJ}^T\right)} \qquad I_{MD} = \frac{2\sqrt{\det\left(\mathbf{JM}^{-1}\mathbf{M}^{-T}\mathbf{J}^T\right)}}{\mathrm{tr}\left(\mathbf{JM}^{-1}\mathbf{M}^{-T}\mathbf{J}^T\right)}. \tag{6}$$

Their expressions result from Yoshikawa, who introduced kinematic and dynamic manipulability measures [14,15], Kim and Khosla, and, more recently, Patel and Sobh [16,17], who gave more insight on dexterity indices. Actually, each index is the ratio between geometric mean and algebraic mean of eigenvalues of each relative matrix ($\mathbf{JJ}^T$ and and $\mathbf{JM}^{-1}\mathbf{M}^{-T}\mathbf{J}^T$ respectively for kinematic and dynamic manipulability), providing an evaluation of how much the eigenvalues deviate from the isotropic condition (equal eigenvalues). As they were defined, indices in Equation (6) are dimensionless and vary in the range $[0, 1]$, with the lower and higher limits related respectively to singular and isotropic configurations. A comparison of the measure of manipulability for the two SCARA robots is shown in Figure 8, where the better kinematic and dynamic behaviour of the redundant SCARA is evident for each radial position. The choice of such indices is due to the demand of better performance in terms of speed and acceleration, in order to have a faster machine in pick-and-place tasks. Other indices could be used to compare the two manipulators when different design specifications are required. For instance, stiffness is an important aspect when interaction forces at the end effector are needed in the machining of workpieces, typically at reduced operating speeds. Energy efficiency could be considered as a secondary purpose, but a more accurate knowledge of actuators and electronic control architecture is essential.

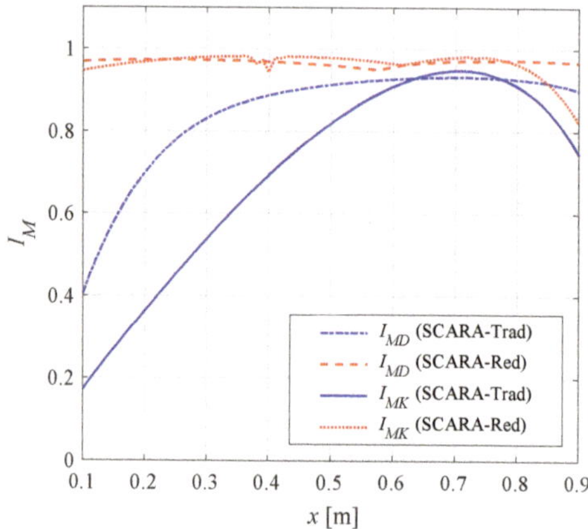

Figure 8. Comparison of kinematic and dynamic manipulability indexes evaluated for the two SCARA robots along the radial direction.

3.1. Kinematic and Dynamic Characterisation of Robot Performance

Preliminary dynamic analysis based on the numerical data of the mass-matrix of the manipulators obtained by the values gathered in Table 1 showed that the dynamic performances of the robots in typical pick-and-place operations are mainly limited by accelerations because motors do not have time to reach their maximum velocity [13]. In these conditions, motors work almost at constant maximum torque and so, to avoid excessive heating, their torque is limited to their rated value. Based on this consideration, a "guaranteed" joint acceleration was defined, which can be achieved in any condition and which is lower than the peak value that can be generated just for short time spans. Moreover, the motion about the vertical direction has not been analysed in this study because it has the same characteristics for the two manipulators. The focus was then kept on the movements in the horizontal plane.

3.2. Dynamic Performance for a Specific Task

To compare the dynamic performance of the two robots, different working cycles are considered to verify the minimum actuation time necessary to perform common tasks usually requested from this class of manipulators. The evaluation of the minimum actuation time is performed considering, for each joint, a maximum absolute velocity $v = 5000\,\text{rpm}$ and a maximum guaranteed acceleration $a = 1628\,\text{rad/s}^2$ for the SCARA-Trad and $a = 1540, 1680, 1260\,\text{rad/s}^2$ for the first, second and third joint of the SCARA-Red, obtaining the following minimum actuation time:

$$
t_{\min} = \begin{cases} \dfrac{\Delta s}{v} + \dfrac{v}{a} & \text{if} \quad \Delta s \geq \dfrac{v^2}{a}, \\[2ex] 2\sqrt{\dfrac{\Delta s}{a}} & \text{otherwise.} \end{cases}
\tag{7}
$$

The two cases in Equation (7) depend on the fact that, for small displacements Δs, the velocity profiles are triangular, whereas, for larger distances, they are trapezoidal. In the present case, triangular profiles are nearly always obtained. The actuation time of each robot is the maximum value among the actuation times of its different joints.

The considered simulation tests could be divided into the following categories:

(a) quick motion from a given *xy* position *A* and a given joint configuration $\mathbf{q}_A$ to a new *xy* position *B* to be reached with the most convenient joint configuration $\mathbf{q}_B$;

(b) repeated motion cycles between two fixed *xy* positions *A* and *B* to be performed with optimal joint configurations $\mathbf{q}_A$ and $\mathbf{q}_B$ (Figure 9);

(c) repeated motion cycles between two *xy* positions to be performed with optimal joint configurations, where the initial (and/or final) position could be given by different locations on a moving conveyor, and the final (and/or initial) position could be a random location in a pallet.

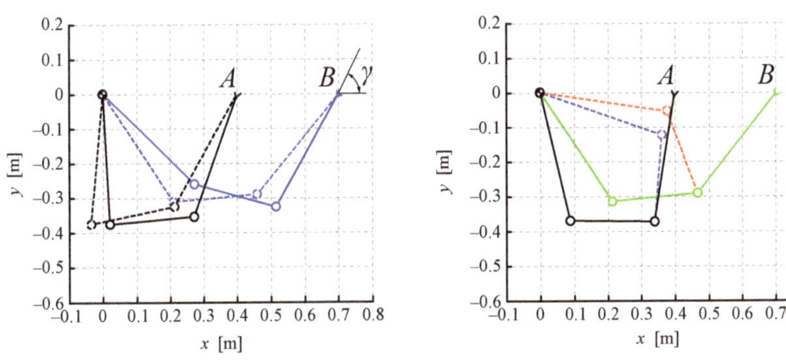

(a) Variable orientation of the end-effector
(b) Assigned orientation of the end-effector

Figure 9. Example of different SCARA-Red configurations considered in the optimization of the pick-and-place cycles: (**a**) the best configuration is obtained assuming a change of the end-effector orientation, (**b**) the best configuration is chosen between the two possible configurations corresponding to an assigned end-effector orientation.

For the case (a), the identification of the optimized motion for the traditional SCARA requires comparing two cases corresponding to the two solutions of the inverse kinematics, whereas, for the case (b), a total of four combinations of the two solutions for initial and final configurations have to be considered. The combination that requires the lowest motion time was considered. For the redundant SCARA, in both cases, an infinite number of combinations occurs. In this case, a finite number of solutions with different values of the angle γ and spaced with a predefined step angle $\Delta\gamma$ was considered. In the present case, it is assumed that $\Delta\gamma = 1°$.

As an example, Figure 10 reports the analysis of the case (a) for the redundant manipulator. The initial position is assigned in the joint space $\mathbf{q}_i = [\alpha_i, \beta_i, \gamma_i]^T$, whereas the final position is assumed in terms of gripper position $\mathbf{p}_f = [x_f, y_f]^T$ and so the angle γ can be freely chosen inside the predefined range. Two solutions are possible for each choice of γ and the configuration with lower motion time is selected. In the example of Figure 10, it is evident that the optimal case corresponds to $\gamma = 53°$ in the second configuration. Generally speaking, the total number of configurations to be considered is $N_a = 2 * 360/\Delta\gamma$.

The case (b) can be analysed in a similar way by considering two configurations for each value of γ in the initial and the final position: the total number of configurations to be compared is $N_b = N_a^2$. The algorithm requires an offline planning phase and cannot be used online due to the high computation times (about $t_a = 0.15$ ms for the first algorithm and about $t_a = 58$ ms for the second algorithm, both implemented in a Delphi XE environment on a PC equipped with Windows 10-64 bit and an Intel(R)

Core(TM) i5-7200U CPU@2.50 GHz processor, (both manufactured by ASUSTek Computer Ink, Peitou, Taipei, Taiwan). These times are reduced to about $t_a = 0.03$ ms and about $t_b = 2.32$ ms for $\Delta\gamma = 5°$. Several initial and final configurations have been considered to cover different pick-and-place tasks, as shown in Figure 11 and detailed in Table 2.

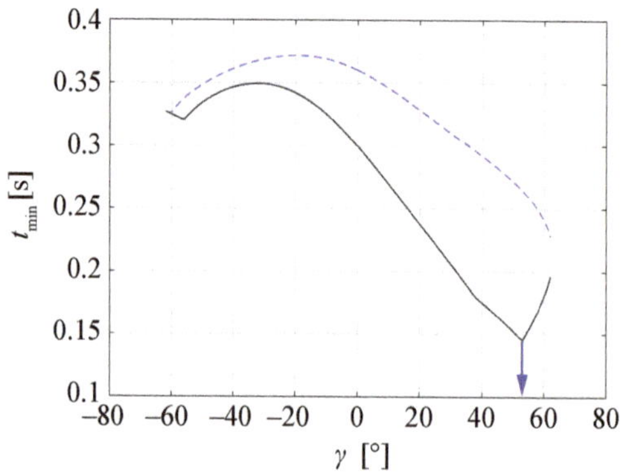

Figure 10. Determination of the optimal cycle for SCARA-Red (cycle a): assigned initial configuration $\mathbf{q}_i = [-18.785, -93.139, 81.000]$, $\mathbf{p}_i = [0.4, 0.0]$ and final assigned configuration $\mathbf{p}_f = [0.7, 0.0]$ (optimized final configuration $\mathbf{q}_{f(t_{\min})} = [-11.278, -64.771, 53.000]$).

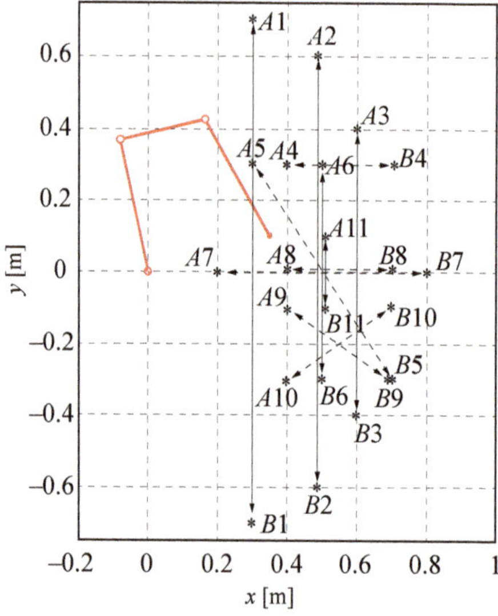

Figure 11. The different pick-and-place tasks considered in the tests (case b).

Table 2. Simulation results of several motion cycles between two fixed xy positions.

Cycle #	p_A	p_B	Length [m]	$\dfrac{t_{AB-red}}{t_{AB-trad}}$	Performance Increment
1 T	A1	B1	1.400	0.899	10%
2 T	A2	B2	1.200	0.741	26%
3 T	A3	B3	0.800	0.848	15%
4 R	A4	B4	0.300	0.809	19%
5 D	A5	B5	0.721	0.719	28%
6 T	A6	B6	0.600	0.639	36%
7 R	A7	B7	0.600	0.973	3%
8 R	A8	B8	0.300	0.948	5%
9 D	A9	B9	0.361	0.858	14%
10 D	A10	B10	0.361	0.723	28%
11 T	A11	B11	0.300	0.631	37%
R = radial direction			maximum	0.973	3%
D = diagonal direction			minimum	0.631	37%
T = tangential direction			average	0.799	20%

The results clearly show that the redundant manipulator performs better than the standard model allowing a reduction of cycle times up to 37%. In the simulations of case (c), five different layouts suggested by the robot manufacturer have been considered: they are characterised by the presence of pallets and conveyors, as shown in Figure 12. In this case, the picking position and the drop point change at each cycle: therefore, due to the variability of cycle times, average values and worst cases must be considered. The presence of conveyors introduces further variability because the objects to be manipulated appear in random positions within a predefined area; they are identified by a vision system and the objects have to be grasped and released "on the fly" during motion. A practical case is illustrated in Figure 13.

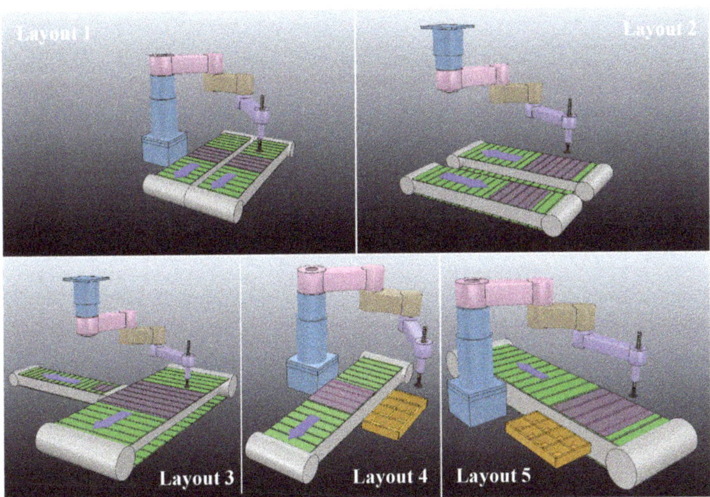

Figure 12. Five different layouts considered in the simulation (case c).

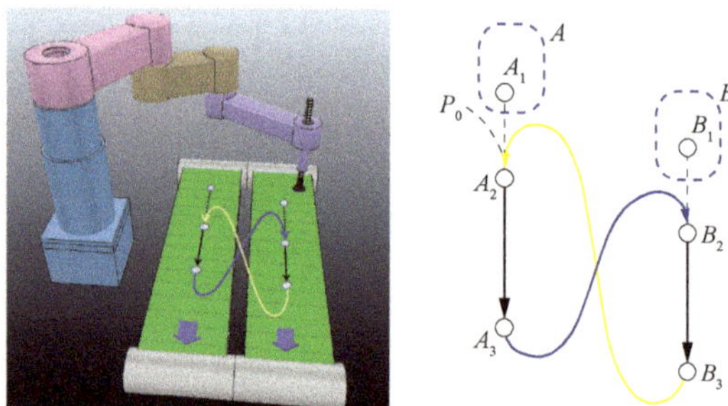

Figure 13. A typical layout used in simulation to test the performance of pick and place cycles on conveyors with random positions of the object to be manipulated.

There are two separated areas indicated by A and B in which some objects may appear, or must be placed, or must be tracked while performing some manipulations or other tasks. A classical motion cycle is composed of two straight segments $A_1 - A_3$ and $B_1 - B_3$ (simulating conveyors) followed at constant velocity plus two curvilinear segments $A_3 - B_1$ and $B_3 - A_1$ followed with minimum actuation time. The generation of the minimum actuation time is easily obtained by generating step-vise set-points for the joints based on the inverse kinematic solutions of the xy point to be reached, the set-points are then filtered using a nonlinear filter based on the concept described in Zanasi et al. [18] and extended in Gerelli et al. [19] (Figure 14). The results produce, for each joint, a trapezoidal (or triangular) velocity profile which guarantees the reaching of the set point in minimum time. This filter is suitable also to track moving set-points. A typical motion generated by this filter is reported in Figure 15. The upper diagram shows the position set-point before and after filtering, whereas the second and the third graph show the filtered velocity and acceleration.

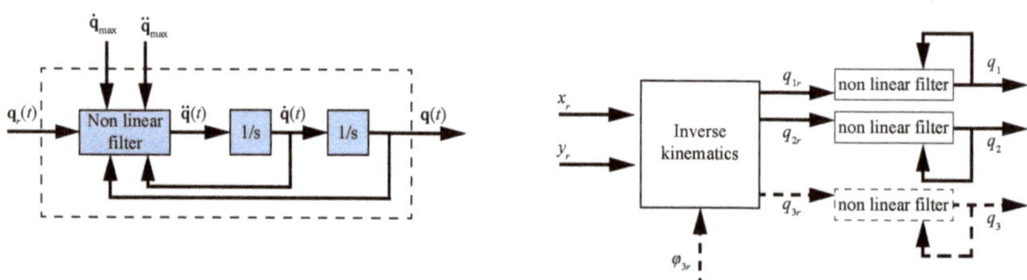

Figure 14. Block diagrams of the nonlinear filter to track a trajectory with minimum actuation time [18,19] on the **left**, and its application to the present task (dashed line just for SCARA-Red) on the **right**.

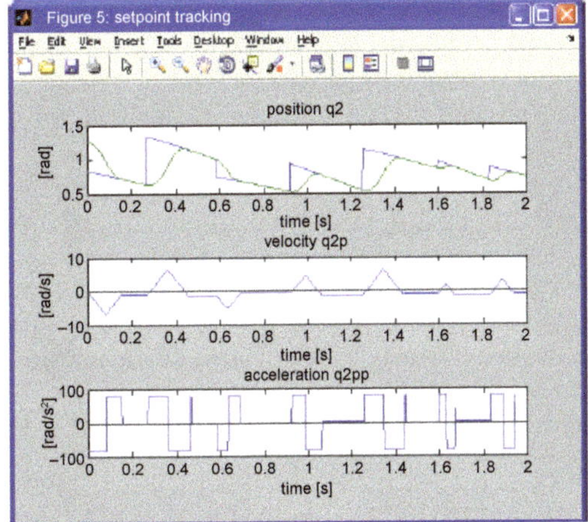

Figure 15. Example of input and output of the nonlinear filter to track a discontinuous set point with a continuous joint motion with bounded velocity and acceleration. From top to bottom: position, velocity and acceleration.

The task cycle is composed of the following steps:

1. the robot is initially at the home position $\mathbf{p}_0 = [x_0, y_0]^T$, with a corresponding joint configuration $\mathbf{q}_0$ and assigned joint velocity $\dot{\mathbf{q}}_0$;
2. a point appears in area A with assigned position $\mathbf{p}_A = [x_A, y_A]^T$ and assigned velocity $\dot{\mathbf{p}}_A$ (position and velocity may change at each cycle);
3. the robot starts moving to reach the moving set points;
4. when the set point is reached (point A_2), it is tracked while performing the requested task (grasping, releasing or other) until point A_3 is reached;
5. a point appears in area B with assigned position $\mathbf{p}_B = [x_B, y_B]^T$ and assigned velocity;
6. the robot starts moving to reach the moving set points;
7. when the set point is reached (point B_2), it is tracked while performing the requested task (grasping, releasing, or other), which ends in B_3;
8. the cycle is repeated from step 2.

Successive cycles are not identical because the points appear in areas A and B in random positions: the motion of the points is due to conveyors that generally move at constant speed, but it cannot be guaranteed. The unpredictable positioning of the target points prevents the use of offline optimization algorithms.

The algorithm that in real time generates the set points of the robot joints is described in the following:

1. at the time t_A, the robot is in point A and a set point appears in coordinate B_1, so the algorithm discussed in the previous section is executed to find the best configuration to reach $\mathbf{p}_B = [x_B, y_B]^T$ assuming null velocity at the end and the corresponding value of the joint coordinates and this value is assumed as reference position $\mathbf{q}_r$ for the joint motion.
2. while the robot is moving toward B_1, its coordinates $\mathbf{p}_B = [x_B, y_B]^T$ change at each instant of time. Data are updated at regular time interval of duration dt. Consequently, at each time t, the reference value for the joint coordinates is updated as $\mathbf{q}(t) = \mathbf{q}(t - \mathrm{d}t) + \mathbf{J}^{\perp}(\mathbf{p}_B - \mathbf{F}(\mathbf{q}_r(t - \mathrm{d}t)))$ to assure

smooth motion during the tracking of the object. $J^{\perp}$ is the Moore–Penrose pseudoinverse of the Jacobian matrix [20], which, in the case of the traditional SCARA, coincides with the inverse of J, whereas it is a 3×2 matrix for the redundant manipulator.

3. the value $\mathbf{q}, \dot{\mathbf{q}}, \ddot{\mathbf{q}}$ for the joint coordinates, velocity and acceleration are evaluated by processing $\mathbf{q}_r$ with the mentioned nonlinear filter.

4. at the time t_{B_1}, the moving point B_1 is reached with a sufficient precision and the next activity is started (moving z axis, object manipulation, ...) while the xy position is continuously tracked.

5. at the time t_{B_2}, the activity is terminated and the robot starts moving to the new moving points appeared in area A using the same procedure used to reach point B (initial evaluation of $\mathbf{q}_r$ and its iterative updating)

6. the moving point A is reached at time t_{A_2} and the manipulating tasks are performed until the time t_{A_3} while continuing to track the moving point

7. operations are cyclically repeated as above.

With the purpose of simulating actual working conditions, the conveyors are moved with different velocity in the different layouts (from $0.05\,\text{m/s}$ up to $0.8\,\text{m/s}$). For each layout shown in Figure 12, some areas where the objects are supposed to be picked or placed are defined. During the simulations, the pick and the place position of the objects are generated randomly inside these areas. Table 3 presents a comparison of the performance of the two manipulators in typical cases. For each layout, the table contains the time t_{AB} (average motion time in xy) for the optimized value of the acceleration (different for each layout) obtained in full pay load conditions and the time t_{AB} for the "guaranteed" value of the acceleration identical for the five layouts. The optimized value of the acceleration is the maximum value that can be achieved in that particular cycle. In the considered cases, the performance of the redundant robot resulted in being better than those of the traditional ones with the exception of the case of the optimized acceleration for Layout 1 in which the traditional SCARA has a slight advantage.

Table 3. Results of simulation runs (times in seconds).

Layout	Optimized Acceleration				Guaranteed Acceleration			
	t_{Trad}	t_{Red}	$\dfrac{t_{Red}}{t_{Trad}}$	Performance Increment	t_{Trad}	t_{Red}	$\dfrac{t_{Red}}{t_{Trad}}$	Performance Increment
1	0.14968	0.15681	1.0476	−5%	0.1863	0.1783	0.9566	4%
2	0.21285	0.20198	0.94893	5%	0.2376	0.2083	0.8768	12%
3	0.23652	0.19305	0.81621	18%	0.2365	0.1934	0.8179	18%
4	0.22855	0.17317	0.75768	24%	0.2438	0.1881	0.7715	23%
5	0.22855	0.17317	0.75768	24%	0.2438	0.1881	0.9019	10%

4. Main Simulation Results

Several issues can be highlighted as a result of the presented work. They are:

(a) The comparison between velocity ellipses of SCARA-Trad and SCARA-Red has shown better kinematic behaviour of the redundant SCARA with respect to the conventional one in terms of higher gripper velocity and with the possibility to use the redundancy to improve its performance.

(b) The redundant robot has shown a better behaviour even for Cartesian accelerations, but the difference decreases when the end-effector moves towards the border of the workspace. Some tests have also shown that Coriolis and Centrifugal terms have a heavier weight for the redundant robot than for the conventional SCARA

(c) Robots' motion is characterized by high speed cycles which do not allow for obtain high velocities; in fact, they are significantly lower than the maximum velocities that the robots can reach. Trapezoidal velocity profiles degenerate in triangular profiles, showing the importance of acceleration abilities with respect to velocity performance. Joint accelerations of the redundant robot, obtained at nominal torque of the motors, has proved to be quite small if compared with joint accelerations of the conventional SCARA, not allowing a full exploitation of the redundant SCARA performances. A better dimensioning of the speed reducers and of the motors can be dealt with in order to improve the redundant robot performance.

(d)   Research about the vertical motion and the rotation of the end-effector has shown some limitations given by motors and speed reducers responsible of such motion. In fact, rated torques associated with such motors are small and they do not allow joint axes to reach high velocities.

(e)   The load at the end-effector has a significant influence on robots' dynamics. The acceleration used in the planning algorithms could be increased according to the load applied at the end-effector.

(f)   Five layouts have been considered as actual work-cells for the definition of tasks. Referring to rated torque values, robots have shown a comparable behaviour in terms of cycle times for different tasks. Therefore, the optimization of the planning algorithm is vital in order to improve the performance of the redundant SCARA. In any case, the redundant robot is better than the conventional robot for specific movements.

(g)   Simulation results are highly dependent on construction parameters, e.g., mass distribution, speed ratios and performance of motors. Therefore, a different mechanical design could lead to different conclusions. Moreover, it is possible that one of the two architectures performs better in a task, whereas the other performs better in another.

(h)   Velocity capabilities of the manipulators deeply depend on the position within the working space.

(i)   Accelerations depend on dynamic parameters of the manipulator and are limited by the available motor torques. It results that, if a single threshold value for the acceleration is selected for each motor, it is necessary to choose a "limited" value that could be "guaranteed" in any situation. Both the limit of the maximum instantaneous torque (intermittent field) and the root mean square (RMS) rate torque (continuous field) have been considered.

(j)   In standard cycles between fixed points, the redundant SCARA manipulator performs much better than the classical one reducing the time in a range between 3% and 35% (20% in average). These results are obtained with the "guaranteed" accelerations.

(k)   In standard cycles with moving points (the five layouts considered), the superiority of the redundant manipulator is less evident and in one case the traditional SCARA performs better. On average, the redundant manipulator performances are better by about 8.5%.

5. Conclusions

The dynamic analysis of both the SCARA robots has been carried out and their performance compared in terms of pick-and-place cycle time. The two architectures have been studied with reference to the physical data supplied by the manufacturer, who also suggested five typical layouts to be considered.

A first analysis of the structures based on the theory of the velocity ellipses shows a theoretical superiority of the redundant SCARA manipulator with respect to the traditional one. Although this superiority cannot be theoretically confirmed due to the limitations in the accelerations for which the trapezoidal velocity profiles degenerates in triangular diagrams, the SCARA-Red higher performances in pick-and-place operation is confirmed by simulation tests over several different working cycles.

The maximum acceleration for the pick-and-place cycles for each different layout has been determined, in order to maximize the performance of the manipulators while respecting the motor limitations (peak and rated torque). Five different values of maximum acceleration were obtained in this way, one for each of the five layouts. Then, by considering all the layouts together, the values of the accelerations have been reduced in order to guarantee their availability in any configuration.

The performances of the traditional and the redundant SCARA were compared in different situations including fast pick-and-place cycle between fixed points, as well as in the five typical pick-and-place operations with moving conveyors to be tracked and random appearance of objects in certain areas. In this case, both optimal and "guaranteed" accelerations were considered. The performance in terms of cycle time highly depends on the strategies adopted to generate the motion of the actuators.

As a conclusion, it can be said that, to have the best performances, the geometrical layout and the acceleration limits must be tuned specifically for each different situation, but reasonable "guaranteed" values of the acceleration may be suggested for preliminary layout installations.

Future studies to further analyse and improve the performance may concern the dynamic optimization of the manipulator (link lengths, masses, gear reduction) and of the layout (position of manipulator and conveyors). This may result in modifying the robot structure, size, some components or developing tools to optimize the layout or the cycle optimization. This analysis may also suggest new concepts in the online planning and control.

Author Contributions: Formal analysis, M.C., G.P. and M.-C.P.; validation, R.B. and G.L.; writing—original draft, M.-C.P. and R.B.; writing—review and editing, M.C. and G.L.

Funding: This research received no external funding.

Conflicts of Interest: The authors declare no conflict of interest.

References

1. Makino, H. Assembly Robot. U.S. Patent 434,150,207, 27 July 1982.
2. Clavel, R. Conception d'un robot parallèle rapide à 4 degrés de liberté. Ph.D Thesis, EPFL (Ecole Polytechnique Fédérale de Lausanne), Lausanne, Switzerland, 1991.
3. Tosi, D.; Legnani, G.; Pedrocchi, N.; Righettini, P.; Giberti, H. Cheope: A new reconfigurable redundant manipulator. *Mech. Mach. Theory* **2010**, *45*, 611–626. [CrossRef]
4. Carricato, M. Fully Isotropic Four-Degrees-of-Freedom Parallel Mechanisms for Schoenflies Motion. *Int. J. Robot. Res.* **2005**, *24*, 397–414. [CrossRef]
5. Altuzarra, O.; Şandru, B.; Pinto, C.; Petuya, V. A symmetric parallel Schönflies-motion manipulator for pick-and-place operations. *Robotica* **2011**, *29*, 853–862. [CrossRef]
6. Xie, F.; Liu, X. Design and Development of a High-Speed and High-Rotation Robot With Four Identical Arms and a Single Platform. *ASME J. Mech. Robot.* **2015**, *7*, 041015. [CrossRef]
7. Reiter, A.; Müller, A.; Gattringer, H. Inverse kinematics in minimum-time trajectory planning for kinematically redundant manipulators. In Proceedings of the IECON 2016—42nd Annual Conference of the IEEE Industrial Electronics Society, Florence, Italy, 23–26 October 2016; pp. 6873–6878.
8. Urrea, C.; Muñoz, R. Joints Position Estimation of a Redundant Scara Robot by Means of the Unscented Kalman Filter and Inertial Sensors. *Asian J. Control* **2016**, *18*, 481–493. [CrossRef]
9. Kumar, A.; Kumar, V.; Gaidhane, P.J. Optimal Design of Fuzzy Fractional Order PI$^\lambda$D$^\mu$ Controller for Redundant Robot. *Procedia Comput. Sci.* **2018**, *125*, 442–448. [CrossRef]
10. Risse, W.; Hiller, M. Dextrous motion control of a redundant SCARA robot. In Proceedings of the 24th Annual Conference of the IEEE Industrial Electronics Society (IECON '98), Aachen, Germany, 31 August–4 September 1998; Cat. No. 98CH36200, Volume 4, pp. 2446–2451.
11. Boscariol, P.; Richiedei, D. Trajectory Design for Energy Savings in Redundant Robotic Cells. *Robotics* **2019**, *8*, 15. [CrossRef]
12. Bowling, A.; Khatib, O. *Modular Decomposition for Optimal Dynamic Design of Redundant Macro/Mini Manipulators*; Springer: Dordrecht, Netherlands, 1996; pp. 29–38.

13. Callegari, M.; Palmieri, G.; Palpacelli, M.C.; Bussola, R.; Legnani, G. Performance Analysis of a High-Speed Redundant Robot. In Proceedings of the MESA 2018—14th ASME/IEEE International Conference on Mechatronic & Embedded Systems & Applications, Oulu, Finland, 2–4 July 2018; pp. 6873–6878.
14. Yoshikawa, T. Manipulability of Robotic Mechanisms. *Int. J. Robot. Res.* **1985**, *4*, 3–9. [CrossRef]
15. Yoshikawa, T. Dynamic manipulability of robot manipulators. In Proceedings of the 1985 IEEE International Conference on Robotics and Automation, St. Louis, MO, USA, 25–28 March 1985; Volume 2, pp. 1033–1038.
16. Kim, J.-O.; Khosla, K. Dexterity measures for design and control of manipulators. In Proceedings of the IROS '91:IEEE/RSJ International Workshop on Intelligent Robots and Systems '91, Osaka, Japan, 3–5 November 1991; Volume 2, pp. 758–763.
17. Patel, S.; Sobh, T. Manipulator Performance Measures—A Comprehensive Literature Survey. *J. Intell. Robot. Syst.* **2015**, *77*, 547–570. [CrossRef]
18. Zanasi, R.; Bianco, C.L.; Tonielli, A. Nonlinear filters for the generation of smooth trajectories. *Automatica* **2000**, *36*, 439–448. [CrossRef]
19. Gerelli, O.; Bianco, C.G.L. A discrete-time filter for the online generation of trajectories with bounded velocity, acceleration, and jerk. In Proceedings of the ICRA 2010—IEEE International Conference on Robotics and Automation, Anchorage, AK, USA, 3–7 May 2010; pp. 3989–3994.
20. Ben-Israel, A.; Greville, T. *Generalized Inverses: Theory and Applications*; Springer: New York, NY, USA, 2003.

Article

Performance Evaluation of a Sensor Concept for Solving the Direct Kinematics Problem of General Planar 3-RPR Parallel Mechanisms by Using Solely the Linear Actuators' Orientations

Stefan Schulz

Workgroup on System Technologies and Engineering Design Methodology, Hamburg University of Technology, 21073 Hamburg, Germany; st.schulz@tuhh.de

Received: 7 June 2019 ; Accepted: 13 August 2019; Published: 16 August 2019

Abstract: In this paper, we experimentally evaluate the performance of a sensor concept for solving the direct kinematics problem of a general planar 3-RPR parallel mechanism by using solely the linear actuators' orientations. At first, we review classical methods for solving the direct kinematics problem of parallel mechanisms and discuss their disadvantages on the example of the general planar 3-RPR parallel mechanism, a planar parallel robot with two translational and one rotational degrees of freedom, where $\underline{P}$ denotes active prismatic joints and R denotes passive revolute joints. In order to avoid these disadvantages, we present a sensor concept together with an analytical formulation for solving the direct kinematics problem of a general planar 3-RPR parallel mechanism where the number of possible assembly modes can be significantly reduced when the linear actuators' orientations are used instead of their lengths. By measuring the orientations of the linear actuators, provided, for example, by inertial measurement units, only two assembly modes exist. Finally, we investigate the accuracy of our direct kinematics solution under static as well as dynamic conditions by performing experiments on a specially designed prototype. We also investigate the solution formulation's amplification of measurement noise on the calculated pose and show that the Cramér-Rao lower bound can be used to estimate the lower bound of the expected variances for a specific pose based exclusively on the variances of the linear actuators' orientations.

Keywords: direct kinematics problem; parallel robots; linear actuators' orientations; assembly modes; general planar 3-RPR parallel mechanism; inertial measurement units; Cramér-Rao lower bound; static and dynamic experiments

---

1. Introduction

The direct kinematics problem is the problem of finding the actual position and orientation, also known as pose, of the moveable manipulator platform with respect to the fixed base platform from the active joints' coordinates. In general, this problem has multiple solutions. For example, for the general planar 3-RPR parallel mechanism, where three linear actuators, that is, active prismatic joints ($\underline{P}$-joints), connect the passive revolute joints (R-joints) of the fixed base platform with those of the moveable manipulator platform, shown in Figure 1, up to six different poses of the manipulator platform are possible for a given set of linear actuators' lengths. These different poses that solve the direct kinematics problem are also known as assembly modes. However, the general questions that have to be answered for solving the direct kinematics problem in terms of control purposes are: (a) how many solutions exist for a given set of active joints' coordinates and (b) which one of them is the actual solution?

Many scientists have focused on answering these questions. The first question is basically solved by reducing the system of kinematic constraint equations to a univariate polynomial equation. Noncomplex solutions of this equation correspond to possible assembly modes of the parallel mechanism, that is, possible ways to assemble it. Among others, Gosselin et al. [1] introduced a polynomial formulation for the direct kinematics problem of the general planar 3-R$\underline{P}$R parallel mechanism and concluded that for a given set of linear actuators' lengths, up to six real solutions can exist. The same result was independently achieved by Peisach, Pennock and Kassner and Wohlhart [2–4] and finally proved by Gosselin and Merlet [5]. Kong and Gosselin [6] even proposed a coordinate-free formulation to avoid dependencies on the chosen reference frame. In contrast to that, Collins [7] used Clifford algebra and Rojas et al. [8] introduced a method based on the bilateration problem to derive the polynomial formulation in a different manner. This distance-based method, however, can even yield two times more solutions compared to classical methods. For the special case where the three revolute base platform joints are aligned, only four solutions exist, see, for example, References [5,6,8].

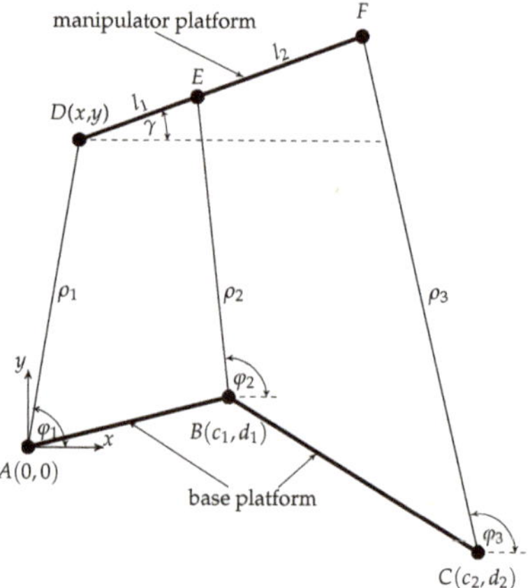

Figure 1. General planar 3-R$\underline{P}$R parallel mechanism with the three base platform joints A, B and C and the three manipulator platform joints D, E and F. The pose of the manipulator platform is given by the position of joint D and the platform's orientation γ with respect to the shown coordinate system.

Finding the univariate polynomial equation makes it possible to calculate all the possible solutions of the direct kinematics problem but it does not identify the actual pose of the manipulator platform. This can be done either by using additional numerical techniques such as Newton-Raphson algorithms with an initial pose estimation [9–15] to transform the system of nonlinear kinematic constraint equations into an explicit or linear problem where a closed-form solution can be found [16–23].

As the linear actuators' lengths are no generalized coordinates, they are only used because they are the active joints' coordinates. Due to the simple inverse kinematics of the general planar 3-R$\underline{P}$R parallel mechanism with a unique solution, the linear actuators' lengths can be directly calculated when the manipulator platform's pose is known. This allows to use other coordinates that are more suitable for solving the direct kinematics problem and, afterwards, calculate the linear actuators' lengths from the obtained manipulator platform's pose. There are several advantages associated with avoiding the linear actuators lengths because (a) reference drives are required to derive the initial lengths, (b) absolute length sensors that do not need reference drives are very expensive and have a limited operation range, (c) possible deformations and backlashes in the linear actuators and joints cannot be determined, and, most importantly, (d) they do not provide a unique solution of the direct kinematics problem.

In order to avoid using the linear actuators' lengths for solving the direct kinematics, we proposed a new sensor concept where the manipulator platform's pose can be uniquely determined from the orientations provided by three inertial measurement units (IMUs) that were placed on top of the manipulator platform as well as on two of the linear actuators [24–26]. For measuring the manipulator platform's orientation, additional wiring effort is required that can cause workspace reductions due to the risk of link-wire interferences. In Reference [27], we therefore suggested using solely the three linear actuators' orientations for solving the direct kinematics problem and derived an analytical formulation that provides the two possible poses of the manipulator platform. Therewith, instead of having up to six assembly modes for the general planar 3-R$\underline{P}$R parallel mechanism when using the linear actuators' lengths that also cannot be found analytically (except for some special cases), we found an analytical expression to calculate them.

As the quality of the formulation's results mainly depends on the quality of the measured linear actuators' orientations, in this paper, we investigate the accuracy of our concept under static as well as dynamic conditions by performing several experiments on a new, specially designed prototype of a general planar 3-R$\underline{P}$R parallel mechanism. For measuring the linear actuators' orientations, we use inertial measurement units that provide linear accelerations and angular velocities of a rigid body in their three axes. Furthermore, we evaluate the maximum achievable accuracy of our formulation and investigate the effect of measurement errors on the calculated manipulator platform's pose by computing the Cramér-Rao lower bound and comparing the results with those of our experiments.

The remainder of this paper is as follows. In Section 2, classical methods for solving the direct kinematics problem of parallel mechanisms and especially the general planar 3-R$\underline{P}$R parallel mechanism are reviewed and their disadvantages are highlighted. In Section 3, we revisit the approach for calculating the number of possible assembly modes when solely the linear actuators' orientations are measured. In Section 4, we then derive the Cramér-Rao lower bound to estimate the variances of the calculated pose of the manipulator platform based on the variances of the linear actuators' orientations. In order to test our concept under static as well as dynamic conditions, in Section 5, we present the experimental results that were performed on a specially designed prototype of a general planar 3-R$\underline{P}$R parallel mechanism. Finally, a conclusion and evaluation is made in Section 6.

Throughout the paper, we use the following notation referring to Figure 1. The three base platform joints are denoted as A, B and C and the three manipulator platform joints as D, E and F. The body-fixed coordinate system of the base platform is located in joint A and the body-fixed coordinate system of the manipulator platform is located in joint D. The position of the manipulator platform with respect to the base platform is given by the coordinates x and y while the orientation of the manipulator platform is given by the angle γ. In the following, the manipulator platform's pose p with respect to the base platform is denoted by the position of the manipulator platform and its orientation:

$$p = \begin{bmatrix} x & y & \gamma \end{bmatrix}^\top . \tag{1}$$

The coordinates of the two remaining base platform joints, B and C, are denoted as c_1 and c_2 in the x-axis and d_1 and d_2 in the y-axis. The three manipulator platform joints D, E and F are aligned and the distance between joint D and joint E is denoted as l_1, whereas the distance between joint E and joint F is denoted as l_2. The linear actuators' lengths are denoted as ρ_1, ρ_2 and ρ_3 and correspond to the distance between the joints A and D, B and E, as well as C and D, respectively. The linear actuators' orientation angles are denoted as φ_1, φ_2 and φ_3 and correspond to the angle between the x-axis and the first, second and third linear actuator, respectively.

2. Review of Classical Solutions for the Direct Kinematics Problem

In the literature, three methods are available for handling the direct kinematics problem of parallel mechanisms. Scientifically, the most interesting method is to derive the echelon form which contains all the solutions of the direct kinematics problem. Here, the system of kinematic constraint equations is reduced to a univariate polynomial equation from which all the possible solutions are then derived. Noncomplex solutions of this equation correspond to possible assembly modes of the parallel mechanism, that is, modes for which the manipulator platform's pose satisfies the requirements of the active joints' coordinates as well as the closure conditions, or, in other words, possible solutions to assemble the parallel mechanism. The echelon form therewith allows to find all the possible solutions of the direct kinematics problem but it does not identify the actual or real pose of the manipulator platform. This problem can be solved by using one of the two following methods.

One possibility of finding the actual solution of the direct kinematics problem is to use iterative techniques such as Newton-Raphson procedures to solve the system of nonlinear kinematic constraint equations. These techniques require a good initial guess of the manipulator platform's pose on the one hand and a determinable pose that is sufficiently far away from a singular configuration on the other hand [28]. In this context, a singularity is a pose where the manipulator platform has at least one uncontrollable instantaneous degree of freedom leading to huge forces in the joints and the linear actuators, see, for example, References [29–31].

As an alternative to additional numerical procedures, in the third method, additional sensor information is used to transform the system of nonlinear kinematic constraint equations into an explicit or linear problem where a closed-form solution can be found. This method allows to find the actual pose of the manipulator platform uniquely and, compared to iterative methods, faster, more accurately and independently from initial pose estimations.

In the following, the three methods are reviewed and their complexity as well as remaining challenges are illustrated on the example of the general planar 3-R$\underline{\text{P}}$R parallel mechanism. As the planar equivalent to the Stewart-Gough platform, this mechanism has been investigated by several scientists in terms of direct kinematics [1–8,32], singularities [29–31,33–35] and control [36–38].

2.1. Analytical Solution

In this section, we review the classical method to derive the assembly modes of the general planar 3-R$\underline{\text{P}}$R parallel mechanism by following the method introduced by Gosselin et al. [1]. In contrast to the classical planar 3-R$\underline{\text{P}}$R parallel mechanism where the manipulator is illustrated as a triangle, we use the mechanism displayed in Figure 1. However, we show that by using the linear actuators' lengths, for this parallel mechanism, up to six solutions for the direct kinematics problem, that is, up to six assembly modes, exist.

The inverse kinematics of the general planar 3-R$\underline{\text{P}}$R parallel mechanism can be written as

$$\rho_1^2 = x^2 + y^2 \,, \tag{2}$$

$$\rho_2^2 = (x + l_1 \cos\gamma - c_1)^2 + (y + l_1 \sin\gamma - d_1)^2 \,, \tag{3}$$

$$\rho_3^2 = \big(x + (l_1 + l_2)\cos\gamma - c_2\big)^2 + \big(y + (l_1 + l_2)\sin\gamma - d_2\big)^2 \,. \tag{4}$$

By subtracting Equation (2) from Equation (3) and Equation (2) from Equation (4), we get

$$\rho_2^2 - \rho_1^2 = Rx + Sy + T \,, \tag{5}$$

$$\rho_3^2 - \rho_1^2 = Ux + Vy + W \tag{6}$$

with

$$
\begin{aligned}
R &= 2l_1 \cos\gamma - 2c_1 \,, \\
S &= 2l_1 \sin\gamma - 2d_1 \,, \\
T &= l_1^2 + c_1^2 + d_1^2 - 2l_1(c_1 \cos\gamma + d_1 \sin\gamma) \,, \\
U &= 2(l_1 + l_2)\cos\gamma - 2c_2 \,, \\
V &= 2(l_1 + l_2)\sin\gamma - 2d_2 \,, \\
W &= (l_1 + l_2)^2 + c_2^2 + d_2^2 - 2(l_1 + l_2)(c_2 \cos\gamma + d_2 \sin\gamma) \,.
\end{aligned}
\tag{7}
$$

From Equation (5), we get

$$x = \frac{\rho_2^2 - \rho_1^2 - Sy - T}{R},$$ (8)

and by inserting this result into Equation (6),

$$y = \frac{R(\rho_3^2 - \rho_1^2 - W) - U(\rho_2^2 - \rho_1^2 - T)}{RV - SU}.$$ (9)

In the same manner, we get

$$x = \frac{V(\rho_2^2 - \rho_1^2 - T) - S(\rho_3^2 - \rho_1^2 - W)}{RV - SU}.$$ (10)

In order to obtain a univariate equation in γ, inserting Equations (9) and (10) into Equation (2) gives us:

$$\rho_1^2 = \frac{\left(V(\rho_2^2 - \rho_1^2 - T) - S(\rho_3^2 - \rho_1^2 - W)\right)^2}{(RV - SU)^2} + \frac{\left(R(\rho_3^2 - \rho_1^2 - W) - U(\rho_2^2 - \rho_1^2 - T)\right)^2}{(RV - SU)^2}.$$ (11)

By applying the Weierstrass substitution

$$X = \tan\frac{\gamma}{2}, \quad \cos\gamma = \frac{1 - X^2}{1 + X^2}, \quad \sin\gamma = \frac{2X}{1 + X^2},$$ (12)

We can get the sixth order polynomial in X, whose six possible solutions can be found numerically. In this context, Wenger et al. [33] investigated the situation where the term $RV - SU$ in Equations (9)–(11) becomes zero. Finally, we can substitute backwards and insert the solutions for γ back into Equations (9) and (10) to obtain the position of the manipulator platform. However, there is no analytical solution for this problem available [8].

As an example, Figure 2 shows the six assembly modes of the general planar 3-R$\underline{P}$R parallel mechanism when using the linear actuators' lengths. Here, we use the following parameters:

$$c_1 = 40\,\text{mm}, \quad d_1 = 10\,\text{mm}, \quad l_1 = 25\,\text{mm}, \quad c_2 = 90\,\text{mm}, \quad d_2 = -20\,\text{mm}, \quad l_2 = 35\,\text{mm}.$$ (13)

With a given set of linear actuators' lengths

$$\rho_1 = 80.6226\,\text{mm}, \quad \rho_2 = 61.7931\,\text{mm}, \quad \rho_3 = 82.9139\,\text{mm},$$ (14)

six assembly modes exist. Due to a faster calculation time, we derive the solution by using the method proposed by Rojas et al. [8]. The coordinates of point D, given by x and y and the orientation of the manipulator platform γ for the six solutions are:

$$\begin{bmatrix} x_{\mathrm{I}} \\ y_{\mathrm{I}} \\ \gamma_{\mathrm{I}} \end{bmatrix} = \begin{bmatrix} 37.3098\,\text{mm} \\ -71.4701\,\text{mm} \\ -59.7539° \end{bmatrix}, \quad \begin{bmatrix} x_{\mathrm{II}} \\ y_{\mathrm{II}} \\ \gamma_{\mathrm{II}} \end{bmatrix} = \begin{bmatrix} -11.5040\,\text{mm} \\ 79.7976\,\text{mm} \\ -50.5183° \end{bmatrix}, \quad \begin{bmatrix} x_{\mathrm{III}} \\ y_{\mathrm{III}} \\ \gamma_{\mathrm{III}} \end{bmatrix} = \begin{bmatrix} 72.6382\,\text{mm} \\ -34.9812\,\text{mm} \\ 38.1265° \end{bmatrix}, \quad (15)$$

$$\begin{bmatrix} x_{\mathrm{IV}} \\ y_{\mathrm{IV}} \\ \gamma_{\mathrm{IV}} \end{bmatrix} = \begin{bmatrix} 10.0000\,\text{mm} \\ 80.0000\,\text{mm} \\ -20.0000° \end{bmatrix}, \quad \begin{bmatrix} x_{\mathrm{V}} \\ y_{\mathrm{V}} \\ \gamma_{\mathrm{V}} \end{bmatrix} = \begin{bmatrix} 36.0067\,\text{mm} \\ 72.1354\,\text{mm} \\ -9.0029° \end{bmatrix}, \quad \begin{bmatrix} x_{\mathrm{VI}} \\ y_{\mathrm{VI}} \\ \gamma_{\mathrm{VI}} \end{bmatrix} = \begin{bmatrix} 79.1195\,\text{mm} \\ 15.4950\,\text{mm} \\ 42.2360° \end{bmatrix}. \quad (16)$$

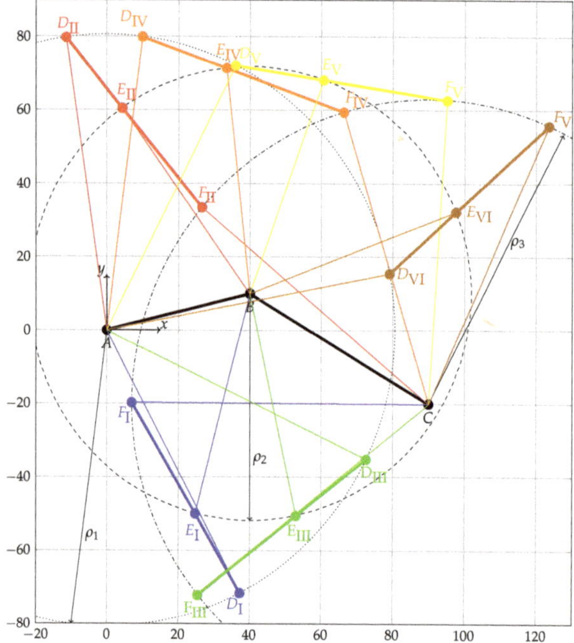

Figure 2. Assembly modes (shown in blue, red, green, orange, yellow and brown) for the manipulator platform of the general planar 3-R$\underline{P}$R parallel mechanism when using the linear actuators' lengths ρ_1, ρ_2 and ρ_3 from Equation (14).

2.2. Numerical Solution

Since we are usually more interested in the actual manipulator platform's pose than in all of the possible poses, it is necessary to distinguish the actual pose from all the others. In the literature, there are numerous methods proposed that aim to find the actual pose of the parallel mechanism. Here, genetic algorithms [39–42], neuronal methods [43,44] and interval analysis methods [15,45] have to be mentioned. In fact, the most common numerical procedures for fast determination of the manipulator platform's pose are iterative techniques such as Newton-Raphson algorithms, see References [11,46–50]. Here, the inverse kinematic equations are used together with a pose estimate for iteratively solving these equations with a multi-dimensional Newton-Raphson algorithm.

All the iterative techniques have in common that they need a pose estimation. With the first guess, they calculate an error between the linear actuators' lengths that correspond to the pose estimate and the measured linear actuators' lengths. By using the measurement model, they vary the pose within several iterations to minimize this error. Different formulations and stop-criteria were proposed to obtain the actual pose (see, for example, Reference [28]) but every iterative method depends on the quality of the first pose estimation. In fact, the pose the algorithm converges to is neither necessarily the actual pose due to the quality of the initial pose estimation nor the closest possible pose next to the initial estimate [28]. The iterative algorithm might also fail to converge in case of singularities [28,51]. In conclusion, the initial pose estimation influences both the pose the algorithm converges to and, not to be neglected, the computation time which corresponds to the number of iterations. For a static case, it is not possible to assure that the actual pose of the manipulator platform can be found at all because, depending on the initial pose estimate, all the solutions are possible. In fact, further information is still required to guarantee that the actual pose can be found.

Fortunately, for dynamic cases such as pose control, the initial guess can be improved during the sampling time and incorrect solutions can be removed. Since the converged pose is available for the previous set of linear actuators' lengths, this pose can be used as an initial guess together with the new set of lengths. Furthermore, the new solution has to be within some boundaries, based on the sampling time, maximum velocities and the latest pose [28].

As an example, we apply the Newton-Raphson algorithm to the general planar 3-R$\underline{P}$R parallel mechanism to compute the actual pose. In case a measurement model h can be found that links the measurements z with the manipulator platform's pose p, this pose can be found iteratively using the Newton-Raphson algorithm:

$$p(i) = p(i-1) + J_h\left(z - h(p(i-1))\right) \tag{17}$$

where i is the iteration step, $p(0)$ the initial pose estimate and J_h the Jacobian of $h(p)$, which is

$$J_h\left(h(p(i))\right) := J_h(i) = \frac{\mathrm{d}h(i)}{\mathrm{d}p(i)}. \tag{18}$$

The algorithm stops when the difference between the measurements z and the results for the measurement model h at the proposed pose $p(i)$ falls below a threshold value Λ, that is,

$$\left\| z - h(p(i)) \right\|_2 < \Lambda. \tag{19}$$

As the measurement model $h(p)$, the inverse kinematic Equations (2)–(4) are used.

The Newton-Raphson algorithm requires exact measurements and a good initial pose estimate. Again, the parameters from Equation (13) are used. As an example, the same linear actuators' lengths from Equation (14) are measured. Now, the Newton-Raphson algorithm is able to compute the pose that meets the conditions given by the measurement model. However, this can be any of the six possible solutions. It therewith depends on the quality of the initial pose estimate. For example, using

$$p(0) = \begin{bmatrix} 10.0000\,\mathrm{mm} & 50.0000\,\mathrm{mm} & 0.0000° \end{bmatrix}^\top \tag{20}$$

as initial pose estimate, after five iterations, the following solution is obtained:

$$\begin{bmatrix} x & y & \gamma \end{bmatrix}^\top = \begin{bmatrix} 10.0000\,\mathrm{mm} & 80.0000\,\mathrm{mm} & -20.0000° \end{bmatrix}^\top, \tag{21}$$

which corresponds to the fourth assembly mode and is shown in blue in Figure 3. But using

$$p(0) = \begin{bmatrix} 50.0000\,\mathrm{mm} & 20.0000\,\mathrm{mm} & 20.0000° \end{bmatrix}^\top, \tag{22}$$

after five iterations, leads to a different solution:

$$\begin{bmatrix} x & y & \gamma \end{bmatrix}^\top = \begin{bmatrix} 79.1195\,\mathrm{mm} & 15.4950\,\mathrm{mm} & 42.2360° \end{bmatrix}^\top, \tag{23}$$

shown in red in Figure 3. This solution corresponds to the fifth assembly mode. Furthermore, for the following initial pose estimate:

$$p(0) = \begin{bmatrix} 0.0000\,\mathrm{mm} & 0.0000\,\mathrm{mm} & 0.0000° \end{bmatrix}^\top, \tag{24}$$

shown in green in Figure 3, the algorithm even fails to converge. Thus, it cannot be guaranteed that the actual pose can be found by using the Newton-Raphson algorithm because an appropriate initial pose estimate has to be provided. Otherwise, the algorithm can converge to the wrong solution or might even fail to converge.

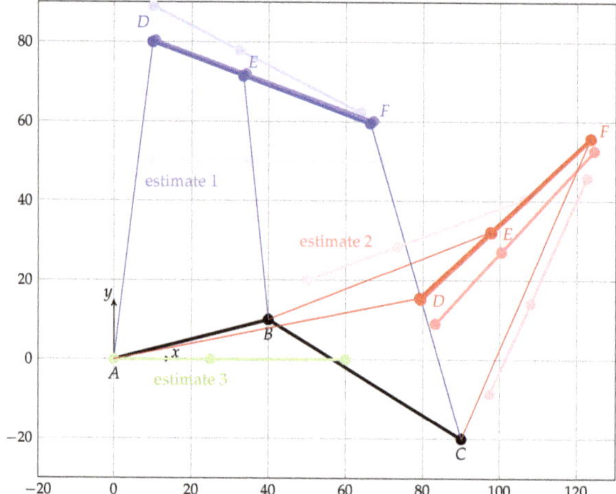

Figure 3. Solutions for the general planar 3-R$\underline{P}$R parallel mechanism when using a Newton-Raphson algorithm with the linear actuators' lengths: solution for $[10\,\text{mm} \quad 50\,\text{mm} \quad 0°]^\top$ (blue), for $[50\,\text{mm} \quad 20\,\text{mm} \quad 20°]^\top$ (red) and for $[0\,\text{mm} \quad 0\,\text{mm} \quad 0°]^\top$ (green) as initial pose estimates.

2.3. Additional Sensor Solution

As a matter of fact, analytical approaches where the inverse kinematic equations are used to obtain a univariate polynomial equation and iterative procedures are both vulnerable to measurement errors, calibration inaccuracies and sensor failure. If only the linear actuators' lengths are used, the manipulator platform's pose cannot be uniquely and unambiguously determined, neither for accurate measurements and optimal calibrated parallel mechanisms nor for perturbed measurements and calibrations. In order to overcome these disadvantages, it is possible to use sensor redundancy. By implementing further sensors, better and more reliable measurement results can be obtained and, in some cases, the actual pose can be determined without additional numerical procedures. In fact, the goal of redundant or auxiliary sensor concepts is to find an explicit or linear formulation for the manipulator platform's pose with the minimum number of sensor information.

The idea of using additional sensors to find the actual pose of the manipulator platform is based on the fact that the linear actuators' lengths are no minimal coordinates and, therewith, are not enough to find a unique solution for the direct kinematics problem. By implementing further sensors, it is possible to get more information about the system's state, reduce the complexity of the constraint equations and therewith, decrease the number of possible assembly modes until only one possible pose of the manipulator platform remains. This allows to solve the direct kinematics equations in considerably less time, only limited by the sampling rate of the sensors but not the calculation time. The introduced information redundancy can later be used to increase the accuracy or to tolerate sensor failure or faulty sensor data, see, for example, Reference [17]. Furthermore, using additional sensors can even enable an auto-calibration of the parallel mechanism [28,52]. However, the type, number and location of the redundant sensors must be chosen very carefully to define a unique solution. Otherwise, it can cause additional problems such as workspace limitations due to the passive legs or joint arrangements, as mentioned in Reference [53]. Furthermore, different sensor types can even reduce the quality of the output by introducing time delays and unwarranted confidences. For example, trusting additional sensors with faulty measurements can lead to incorrect results or even prevent a result from being calculated.

Merlet [28] extensively discussed possible additional sensor concepts and Vertechy et al. [51] presented a very detailed, chronological review. Usually, length sensors and rotary sensors are used as additional sensors to derive the orientations of the linear actuators or passive legs in addition to the linear actuators' lengths, see, for example, References [16–23,51,53–68]. However, several other sensor types were proposed as additional sensors for solving the direct kinematics problem. For example, Baron et al. [69] suggested using a camera in addition to the linear actuators' lengths. For the 6-RUS Hexa-Robot, Hesselbach et al. [70] developed sensors that can be implemented in the passive joints. Inclination sensors can also be used. However, they are more often used for calibration purposes [71]. It can be noticed from the amount of papers dealing with the topic of additional sensor concepts for parallel mechanisms and especially the Stewart-Gough platform that the problem is quite complicated and the proposed solutions are not optimal. In fact, most of the concepts have one or more limitations. One drawback, for example, is the applicability, that is, some additional sensor concepts can only be used for parallel mechanisms with special architecture. The most common limitation is that the base and manipulator platform joints, respectively, should be coplanar, that is, lie on a plane, see, for example, References [18,20,57,58,65,68]. This architecture is often called nearly-general Stewart-Gough platform. Some other concepts require that two or more length sensors are connected to a common point or joint [18,19,63]. It furthermore stands out that all additional sensor concepts use at least one of the linear actuators' lengths and there are only few concepts where less than three linear actuators' lengths are used. In fact, none of the existing concepts completely renounce the lengths of the linear actuators and solely use other sensor information for solving the direct kinematics problem.

As an example for an additional sensor solution, we calculate the actual pose of the general planar 3-RPR parallel mechanism by adding sensor information. One very common possibility to solve the direct kinematics problem is to add supplementary passive linear actuators that are equipped with length sensors. By coinciding the manipulator platform joints of one linear actuator with those of the supplementary linear actuator, see Figure 4, these joints' positions can be uniquely identified and the equations can be simplified to a closed-form solution. Here, for example, two passive linear actuators are added to the parallel mechanism. One is connected to the first and the other one to the third manipulator platform joint.

To the inverse kinematics of the general planar 3-RPR parallel mechanism in Equations (2)–(4), two additional equations for the passive linear actuators can be added:

$$\rho_4^2 = (x - c_4)^2 + (y - d_4)^2 , \tag{25}$$

$$\rho_5^2 = (x + (l_1 + l_2) \cos \gamma - c_5)^2 + (y + (l_1 + l_2) \sin \gamma - d_5)^2 . \tag{26}$$

At first, the intersections of the two circles with the radii ρ_1 starting at point A and ρ_4 starting at point G shall be found, which, in this case, leads to two solutions. By subtracting Equation (2) from Equation (25), an equation for x can be obtained:

$$\rho_4^2 - \rho_1^2 = (x - c_4)^2 + (y - d_4)^2 - x^2 - y^2 = -2c_4 x - 2d_4 y + c_4^2 + d_4^2 , \tag{27}$$

$$\Longleftrightarrow x = \frac{\rho_1^2 - \rho_4^2 - 2d_4 y + c_4^2 + d_4^2}{2c_4} . \tag{28}$$

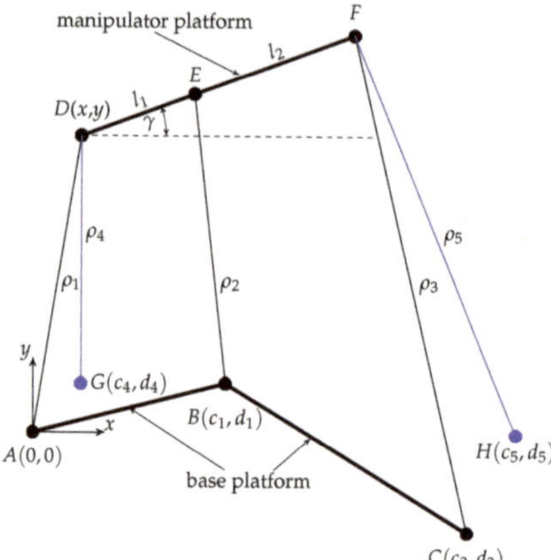

Figure 4. General planar 3-R$\underline{\text{P}}$R parallel mechanism with two additional passive linear actuators with the base platform joints G and H. The active linear actuators are shown in black and the supplementary passive linear actuators are shown in blue.

With Equation (2), the two possible positions $x_{I/II}$ and $y_{I/II}$ of the manipulator platform are derived:

$$\rho_1^2 = \left(\frac{\rho_1^2 - \rho_4^2 - 2d_4 y + c_4^2 + d_4^2}{2c_4} \right)^2 + y^2 \tag{29}$$

$$\Longleftrightarrow y_{I/II} = \frac{\rho_1^2 d_4 - \rho_4^2 d_4 + c_4^2 d_4 + d_4^3}{2(c_4^2 + d_4^2)} \pm \frac{\sqrt{c_4^2(\rho_1^2 + 2\rho_1\rho_4 + \rho_4^2 - c_4^2 - d_4^2)(-\rho_1^2 + 2\rho_1\rho_4 - \rho_4^2 + c_4^2 + d_4^2)}}{2(c_4^2 + d_4^2)} \, , \tag{30}$$

$$x_{I/II} = \frac{\rho_1^2 - \rho_4^2 - 2d_4 y + c_4^2 + d_4^2}{2c_4} = \frac{c_4^4 + c_4^2 d_4^2 \mp d_4\sqrt{c_4^2(c_4^2 + d_4^2)(4\rho_1^2 - c_4^2 - d_4^2)}}{2c_4(c_4^2 + d_4^2)} \, . \tag{31}$$

In a similar way, the two possible positions of point F can be determined. Here, the following substitution is used:

$$\hat{x} = x + (l_1 + l_2)\cos\gamma \, , \qquad \hat{y} = y + (l_1 + l_2)\sin\gamma \, , \tag{32}$$

which reveals two solutions. The angle γ can then be obtained by using the arc tangent with the known horizontal and vertical distances between the points D and F.

As an example, Figure 5 shows the actual solution of the general planar 3-R$\underline{\text{P}}$R parallel mechanism when using the linear actuators' lengths and two additional lengths. Here, the parameters from Equation (13) are used together with

$$c_4 = 10\,\text{mm}\,, \quad d_4 = 10\,\text{mm}\,, \quad c_5 = 100\,\text{mm}\,, \quad d_5 = 0\,\text{mm}\,. \tag{33}$$

With the set of linear actuators' lengths from Equation (14) and

$$\rho_4 = 70.0000\,\text{mm}\,, \quad \rho_5 = 68.3222\,\text{mm}\,, \tag{34}$$

two solutions are obtained. The two possible coordinates of the manipulator platform, given by x and y and the orientation of the manipulator platform γ are:

$$\begin{bmatrix} x_{\mathrm{I}} \\ y_{\mathrm{I}} \\ \gamma_{\mathrm{I}} \end{bmatrix} = \begin{bmatrix} 10.0000\,\mathrm{mm} \\ 80.0000\,\mathrm{mm} \\ -20.0000° \end{bmatrix}, \quad \begin{bmatrix} x_{\mathrm{II}} \\ y_{\mathrm{II}} \\ \gamma_{\mathrm{II}} \end{bmatrix} = \begin{bmatrix} 80.0000\,\mathrm{mm} \\ 10.0000\,\mathrm{mm} \\ -0.7883° \end{bmatrix}, \tag{35}$$

where the second solution is not possible because the distance between the points D_{II} and F_{II} does not satisfy the conditions given by l_1 and l_2. In fact,

$$\left\| D_{\mathrm{II}} F_{\mathrm{II}} \right\|_2 = \sqrt{(80.000\,\mathrm{mm} - 167.7541\,\mathrm{mm})^2 + (10.000\,\mathrm{mm} - 8.7925\,\mathrm{mm})^2}$$

$$= 87.7624\,\mathrm{mm} \neq \sqrt{(l_1 + l_2)^2} = 60\,\mathrm{mm}\,. \tag{36}$$

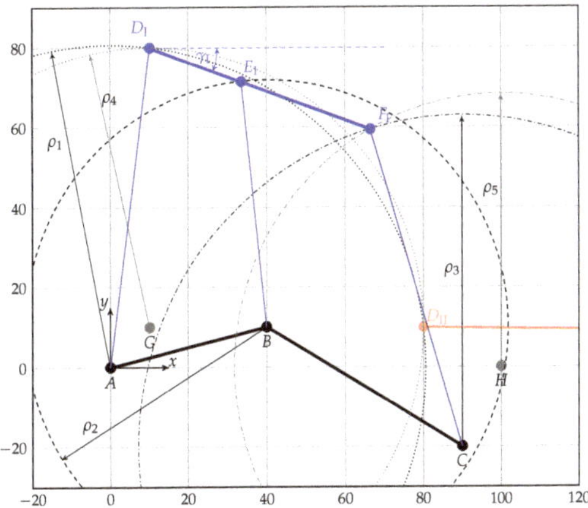

Figure 5. Actual solution (blue) for the general planar 3-R$\underline{P}$R parallel mechanism when using two additional lengths in addition to the linear actuators' lengths. The second solution is shown in red.

3. Assembly Modes when Using the Linear Actuators' Orientations

In the last section, we have seen that all the current concepts for solving the direct kinematics problem have several disadvantages as illustrated on the example of the general planar 3-R$\underline{P}$R parallel mechanism. In this section, we demonstrate that by using the three linear actuators' orientations, the solution of the direct kinematics problem of the general planar 3-R$\underline{P}$R parallel mechanism can be calculated analytically and a maximum of two instead of six assembly modes exist. Here, the elimination method described in Section 2.1 is used where the inverse kinematic equations are used to systematically eliminate unknown variables until a univariate equation is obtained.

For the general planar 3-RPR parallel mechanism shown in Figure 1, the inverse kinematics can be rewritten as

$$\underbrace{\tan \varphi_1}_{A} = \frac{y}{x}\,, \tag{37}$$

$$\underbrace{\tan \varphi_2}_{B} = \frac{y + l_1 \sin \gamma - d_1}{x + l_1 \cos \gamma - c_1}\,, \tag{38}$$

$$\underbrace{\tan \varphi_3}_{C} = \frac{y + (l_1 + l_2) \sin \gamma - d_2}{x + (l_1 + l_2) \cos \gamma - c_2}\,, \tag{39}$$

where we will be using the abbreviations A, B and C in the following. The angles φ_1, φ_2 and φ_3 are the three orientation angles of the linear actuators with respect to the base platform's x-axis and can be obtained, for example, from IMUs that are mounted on the linear actuators. Now, we can rewrite Equation (37):

$$y = Ax\,, \tag{40}$$

and use it in Equation (39):

$$C = \frac{Ax + (l_1 + l_2) \sin \gamma - d_2}{x + (l_1 + l_2) \cos \gamma - c_2}\,. \tag{41}$$

From this, we can derive an expression for x:

$$x = \frac{(l_1 + l_2)(-\sin \gamma + C \cos \gamma) - Cc_2 + d_2}{A - C}\,, \tag{42}$$

and from Equation (40), we can get an expression for y:

$$y = A\frac{(l_1 + l_2)(-\sin \gamma + C \cos \gamma) - Cc_2 + d_2}{A - C}\,. \tag{43}$$

In order to obtain a univariate equation in γ, we use Equations (42) and (43) with the remaining Equation (38) of the inverse kinematics:

$$B = \frac{y + l_1 \sin \gamma - d_1}{x + l_1 \cos \gamma - c_1} = \frac{A\frac{(l_1+l_2)(-\sin\gamma+C\cos\gamma)-Cc_2+d_2}{A-C} + l_1 \sin \gamma - d_1}{\frac{(l_1+l_2)(-\sin\gamma+C\cos\gamma)-Cc_2+d_2}{A-C} + l_1 \cos \gamma - c_1}\,. \tag{44}$$

Now, we have a univariate equation in γ whose two possible solutions are given by

$$\gamma_{\mathrm{I}} = -\operatorname{atan2}\left(\frac{DH - E}{GI}, \frac{F + D}{G}\right)\,, \tag{45}$$

$$\gamma_{\mathrm{II}} = -\operatorname{atan2}\left(\frac{-DH - E}{GI}, \frac{F - D}{G}\right) \tag{46}$$

where

$$D = \sqrt{-I^2\left(-G + \frac{E^2}{I^4}\right)}, \tag{47}$$

$$E = \left(\left((-c_1 + c_2)C + c_1 A - d_2\right)B - (Ac_2 - d_1)C - A(d_1 - d_2)\right)I^2, \tag{48}$$

$$
\begin{aligned}
F = &\left(\left(\left((-c_1 - c_2)l_1 - l_2 c_1\right)B + (d_1 - d_2)l_1 + (d_1 - d_2)l_2\right)C\right.\\
&\left. + c_1 l_1 B^2 - (d_1 - d_2)l_1 B + (c_2 l_1 + c_2 l_2)C^2\right)A^2 + \left((-c_1 + c_2)l_2 B^2 + d_1 l_2 B\right)C^2 - B^2 C d_2 l_2\\
&+ \left(\left(((c_1 - c_2)l_1 + (c_1 - 2c_2)l_2)B - d_1 l_1 - d_1 l_2\right)C^2\right.\\
&\left. + \left(((-c_1 + c_2)l_1 + l_2 c_1)B^2 + ((d_1 + d_2)l_1 - (d_1 - 2d_2)l_2)B\right)C - B^2 d_2 l_1\right)A,
\end{aligned}
\tag{49}
$$

$$G = (C^2 + 1)(A - B)^2 l_2^2 + l_1^2(B - C)^2(A^2 + 1) - 2l_1 l_2(B - C)(AC + 1)(A - B), \tag{50}$$

$$H = A(B - C)l_1 - C(A - B)l_2, \tag{51}$$

$$I = (A - B)l_2 - (B - C)l_1. \tag{52}$$

Finally, we can use the solutions for γ in the Equations (42) and (43) to obtain the position of the manipulator platform. If we are also interested in the linear actuators' lengths, we can calculate them by using inverse kinematics, see Equations (2)–(4).

Figure 6a shows the two assembly modes of the general planar 3-R$\underline{\text{P}}$R parallel mechanism when the linear actuators' orientations are used. We use the same parameters as in Section 2. With a given set of linear actuators' orientation angles

$$\varphi_1 = 82.8750°, \qquad \varphi_2 = 96.0453°, \qquad \varphi_3 = 106.5502°, \tag{53}$$

the two assembly modes can be calculated. Using our method, we can find the following two solutions:

$$\begin{bmatrix} x_I \\ y_I \\ \gamma_I \end{bmatrix} = \begin{bmatrix} 10.0000\,\text{mm} \\ 80.0000\,\text{mm} \\ -20.0000° \end{bmatrix}, \qquad \begin{bmatrix} x_{II} \\ y_{II} \\ \gamma_{II} \end{bmatrix} = \begin{bmatrix} 24.2363\,\text{mm} \\ 193.8902\,\text{mm} \\ 104.5335° \end{bmatrix}. \tag{54}$$

As a second example, Figure 6b shows the two assembly modes of the general planar 3-R$\underline{\text{P}}$R parallel mechanism when the following linear actuators' orientation angles are used:

$$\varphi_1 = 82.8750°, \qquad \varphi_2 = 94.7360°, \qquad \varphi_3 = 101.0877°. \tag{55}$$

In this case, the following two solutions are obtained:

$$\begin{bmatrix} x_I \\ y_I \\ \gamma_I \end{bmatrix} = \begin{bmatrix} 10.0000\,\text{mm} \\ 80.0000\,\text{mm} \\ 20.0000° \end{bmatrix}, \qquad \begin{bmatrix} x_{II} \\ y_{II} \\ \gamma_{II} \end{bmatrix} = \begin{bmatrix} 10.2792\,\text{mm} \\ 82.2340\,\text{mm} \\ 23.6134° \end{bmatrix}. \tag{56}$$

Compared to the first example where the two solutions are far away from each other and the actual solution can be identified easily, in the second example, the two solutions are quite close to each other. This would make the differentiation of the actual solution more difficult, especially when the linear actuators' orientations are perturbed by measurement noise. In fact, when two linear actuators' orientations are identical, the root in Equation (47) becomes negative and no real solution exists. This also corresponds to a direct kinematics singularity.

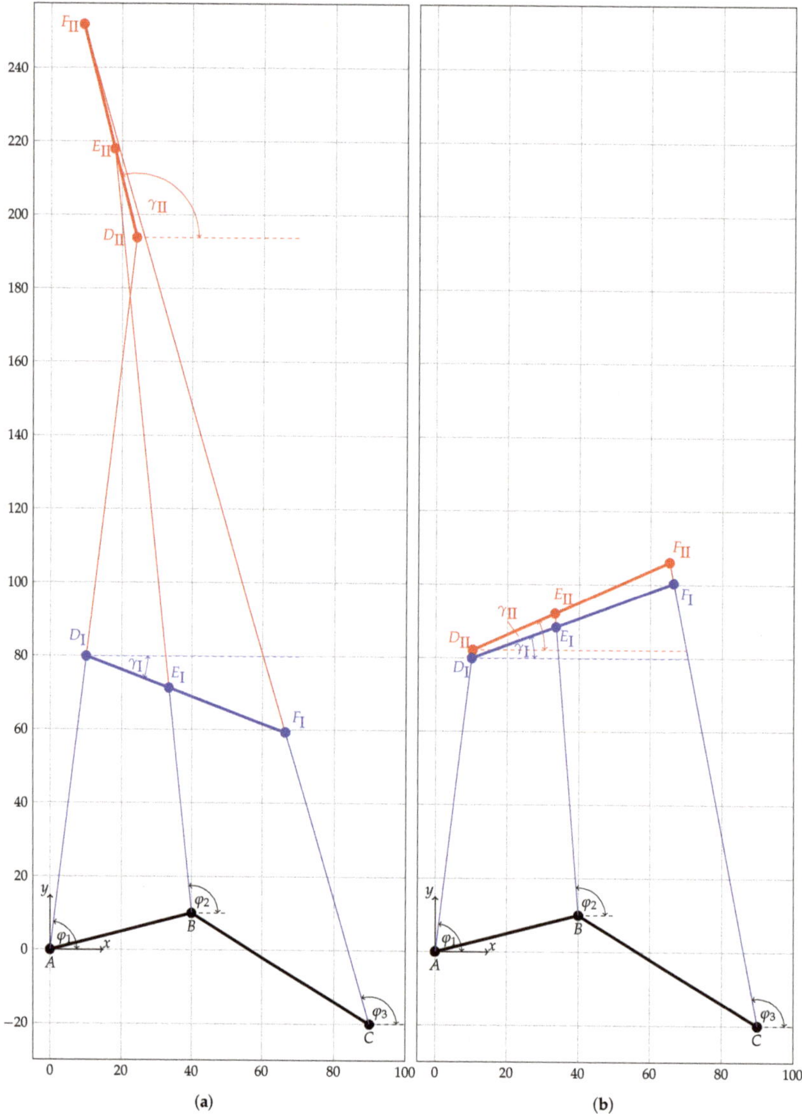

Figure 6. The two assembly modes (shown in blue and red) for the manipulator platform of the general planar 3-RPR parallel mechanism when using the linear actuators' orientations: (**a**) results for $\varphi_1 = 82.8750°$, $\varphi_2 = 96.0453°$ and $\varphi_3 = 106.5502°$ and (**b**) results for $\varphi_1 = 82.8750°$, $\varphi_2 = 94.7360°$ and $\varphi_3 = 101.0877°$.

In general, it can be noticed that, in contrast to the usual six assembly modes, we only have two assembly modes when using the linear actuators' orientations. Furthermore, the assembly modes calculated from the linear actuators' orientations differ from those calculated from the liner actuators' lengths, compare Figures 2 and 6. Finally, the equations that were used for calculating the assembly modes are applicable to every type of general planar 3-RPR parallel mechanisms and can be solved without any numerical methods. In contrast, when using the linear actuators' lengths, there is, in general, no analytical equation available to calculate the assembly modes, see, for example, Reference [8].

4. Cramér-Rao Lower Bound

In order to evaluate the achievable accuracy of the presented approach, based on the expected variances of the linear actuators' orientations, the Cramér-Rao lower bound (CRLB) of the manipulator platform's pose can be computed and compared with the actually measured variances. The CRLB is an estimator that provides the lowest possible mean-squared error among all other estimators. Thus, it can be used to compare existing estimators or algorithms regarding their efficiency on the one hand and to estimate the impact of measurement errors on the calculated pose on the other hand.

By using the inverse kinematic Equations (37)–(39), we can find a relation between the measurement vector z, with

$$z = \begin{bmatrix} \varphi_1 & \varphi_2 & \varphi_3 \end{bmatrix}^\top , \tag{57}$$

and the pose p, with

$$p = \begin{bmatrix} x & y & \gamma \end{bmatrix}^\top , \tag{58}$$

that is given by the measurement model $h(p)$:

$$z = \begin{bmatrix} \operatorname{atan}\left(\frac{y}{x}\right) \\ \operatorname{atan}\left(\frac{y + l_1 \sin\gamma - d_1}{x + l_1 \cos\gamma - c_1}\right) \\ \operatorname{atan}\left(\frac{y + (l_1 + l_2)\sin\gamma - d_2}{x + (l_1 + l_2)\cos\gamma - c_2}\right) \end{bmatrix} =: h(p) . \tag{59}$$

Under the assumption that the measurement vector z is zero-mean Gaussian distributed with its variances $\sigma^2(z_k), k \in \{1, \ldots, 3\}$, that are stored in the covariance matrix C, with

$$C = \operatorname{diag}\left(\sigma^2(z_k)\right) , \tag{60}$$

we can calculate the CRLB as the inverse of the Fisher information matrix F. Its components can be determined as follows:

$$F_{k,l} = \frac{\partial z^\top}{\partial p_k} C^{-1} \frac{\partial z}{\partial p_l} + \frac{1}{2} \operatorname{tr}\left(C^{-1} \frac{\partial C}{\partial p_k} C \frac{\partial C}{\partial p_l}\right) , \tag{61}$$

with $k, l \in \{x, y, \gamma\}$, where $\frac{\partial z}{\partial p_k}$ are the components of the Jacobian J_h of the measurement model $h(p)$:

$$J_h = \frac{dh(p)}{dp} = \begin{bmatrix} \frac{\partial h^\top(p)}{\partial x} & \frac{\partial h^\top(p)}{\partial y} & \frac{\partial h^\top(p)}{\partial \gamma} \end{bmatrix}^\top . \tag{62}$$

In general, the variances $\sigma^2(z_k)$ of the measurement vector z are not constant and the trace in Equation (61) does not vanish so that we have to calculate the derivatives $\frac{\partial C}{\partial p_k}$. Assuming that the variances $\sigma^2(z_k)$, that is, $\sigma^2(\varphi_1)$, $\sigma^2(\varphi_2)$ and $\sigma^2(\varphi_3)$, only depend on the orientation angles φ_1, φ_2 and φ_3, the derivatives $\frac{\partial C}{\partial p_k}$ can be transformed as follows:

$$\frac{\partial C}{\partial p_k} = \frac{\partial C}{\partial z_k} \frac{\partial z_k}{\partial p_k} = \frac{\partial C}{\partial z_k} \frac{\partial h_k(p)}{\partial p_k} , \tag{63}$$

where $\frac{\partial h_k(p)}{\partial p_k}$ is a component of the Jacobian J_h and $\frac{\partial C}{\partial z_k}$ is the derivative of the variance $\sigma^2(z_k)$ for the orientation angle z_k, that is, φ_k:

$$\frac{\partial C}{\partial z_k} = \frac{\partial \operatorname{diag}\left(\sigma^2(z_k)\right)}{\partial z_k} = \operatorname{diag}\left(\frac{\partial \sigma^2(z_k)}{\partial z_k}\right) , \tag{64}$$

and can be derived by experiments.

5. Experiments

5.1. Experimental Device

In order to investigate the accuracy of the direct kinematics solution under static as well as dynamic conditions, we perform experiments on a new, specially designed prototype, see Figure 7. It consists of three identical linear actuators that are connected on the one side to the base platform and on the other side to the manipulator platform. The base and the manipulator platform have integrated revolute joints and, furthermore, the possibility to vary the joints' positions. As linear actuators, we use Actuonix L16-100-35-P with a minimum length of 168 mm, a stroke length of 100 mm and an integrated potentiometer for measuring the current length. The linear actuators are equipped with IMUs to measure their orientation. Here, InvenSense MPU-9250 sensors are chosen as IMUs, where the accelerometer and the gyroscope values are used. An Arduino Mega board with an integrated data acquisition and pose calculation algorithm is mounted inside the experimental device. The Arduino Mega board is furthermore equipped with a display for showing the current pose and a motor shield for controlling the lengths of the linear actuators using a proportional-integral-derivative (PID) controller. For comparing the calculated manipulator platform's pose with the actual pose, we need an independent measurement system. Here, we use image processing to optically analyse the actual manipulator platform's pose, whose joints' positions are equipped with small red dots for optically tracking their position. The positions of the red dots' center points are therefore detected, stored and converted into the positions of the manipulator platform's joints from which the manipulator platform's pose can be calculated. For the experiments on the general planar 3-R$\underline{\text{P}}$R parallel mechanism, we use the following parameters for the base and manipulator platform's joints' coordinates according to Figure 1:

$$c_1 = 170\,\text{mm}\,, \quad d_1 = 10\,\text{mm}\,, \quad l_1 = 70\,\text{mm}\,, \quad c_2 = 280\,\text{mm}\,, \quad d_2 = -20\,\text{mm}\,, \quad l_2 = 30\,\text{mm}\,. \quad (65)$$

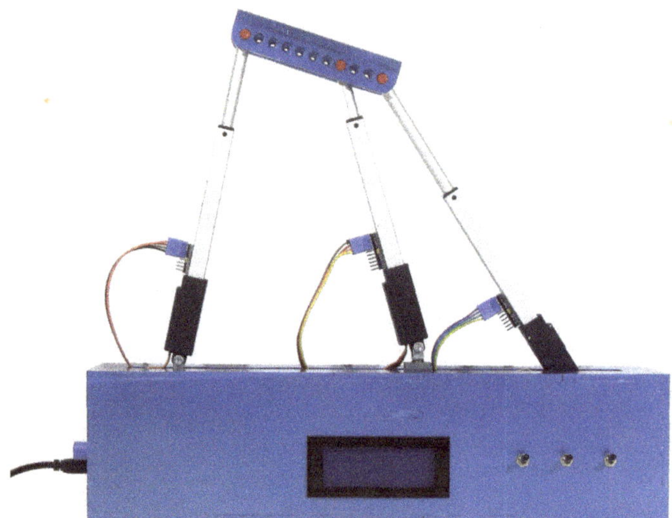

Figure 7. Experimental prototype of the general planar 3-R$\underline{\text{P}}$R parallel mechanism with inertial measurement units (IMUs) mounted on the linear actuators and an Arduino Mega with a display integrated in the base to calculate and show the two assembly modes of the manipulator platform.

5.2. Dynamic Orientation Measurement

The mechanism's y-axis corresponds to the negative gravity vector of the Earth g. The IMUs are mounted on the linear actuators in the way that their x-axes are always parallel to the mechanism's z-axis. For static poses, it is therewith possible to obtain the orientation angles φ_1, φ_2 and φ_3 solely from the accelerometer values of the IMU, $a_{y,k}$ and $a_{z,k}$, where

$$\varphi_{k,\text{acc}} = \text{atan}\, 2\left(a_{k,y}, a_{k,z}\right) . \tag{66}$$

However, when the manipulator platform moves and, therewith, the linear actuators move too, the accelerometer values do not provide accurate results, leading to faulty pose calculations. Robust methods for estimating the actual orientation angles of the linear actuators thus require the IMUs' gyroscope values $\omega_{k,x}$, $\omega_{k,y}$ and $\omega_{k,z}$ in addition to the accelerometer values. The orientation angle φ_k of the kth linear actuator can be obtained, for example, by using a complementary filter with

$$\varphi_{k,\text{com}}(i) = \tau_t \left(\varphi_{k,\text{com}}(i-1) + \omega_{k,x}\Delta t\right) + (1 - \tau_t)\varphi_{k,\text{acc}} , \tag{67}$$

where i is the iteration step, τ_t is the ratio of the gyroscope and accelerometer values and Δt is the time between two measurements. As initial orientation estimates, the results from the accelerometer values, that is, $\varphi_{k,\text{acc}}$, are used.

There are alternatives available for robustly and efficiently estimating the orientations based on IMU measurements including Kalman filtering, nonlinear complementary filters and quaternion based algorithms [72–75]. For the experiments, however, we choose the above introduced complementary filter with $\tau_t = 0.93$ for all the linear actuators. It shows fast responses to changes in the linear actuators' orientations as the gyroscope has a significantly higher impact than the accelerometer. In fact, especially for real-time applications on a low-memory computer, the complementary filter is recommendable because it shows similar accuracy with lower computational complexity compared to other filters. The time between two measurements Δt, which is the inverse of the sampling rate, mainly depends on the computational efficiency of the used algorithms, the programming and the processor. Throughout the experiments, we realized a sampling rate of 53.16 Hz that corresponds to a Δt of 18.81 ms.

5.3. Accuracy of the Orientation Measurements

In order to investigate the dependency of the variances of the orientation angles $\sigma^2(\varphi_k)$ on the orientation angle φ_k itself, we build a test bench, see Figure 8a, where we can mount different IMUs and vary the orientation angle in steps of 5°. Here, we use an InvenSense MPU-9250 sensor which is rotated around its z-axis. For every orientation angle, we take 10,000 measurements with an Arduino Nano and calculate the orientation angle using the accelerometer and gyroscope values. Figure 8b shows the variances of the raw angle φ_{acc}, that is solely calculated from the accelerometer values and the filtered angle φ_{com} for different orientation angles. For the raw orientation angle φ_{acc}, the variances lie between $0.0414°^2$ and $0.1609°^2$. In contrast to that, the filtered angle φ_{com}, shown in Figure 8c, has significantly smaller variances (27 to 38 times smaller) that lie between $0.0015°^2$ and $0.0042°^2$. In order to find a mathematical representation, we added a fifth-order polynomial fit with

$$\sigma^2(\varphi_{\text{acc}}) \approx a_0 + a_1\varphi_{\text{acc}} + a_2\varphi_{\text{acc}}^2 + a_3\varphi_{\text{acc}}^3 + a_4\varphi_{\text{acc}}^4 + a_5\varphi_{\text{acc}}^5 , \tag{68}$$

$$\sigma^2(\varphi_{\text{com}}) \approx a_0 + a_1\varphi_{\text{com}} + a_2\varphi_{\text{com}}^2 + a_3\varphi_{\text{com}}^3 + a_4\varphi_{\text{com}}^4 + a_5\varphi_{\text{com}}^5 , \tag{69}$$

where the constants a_0–a_5 are given in Table 1.

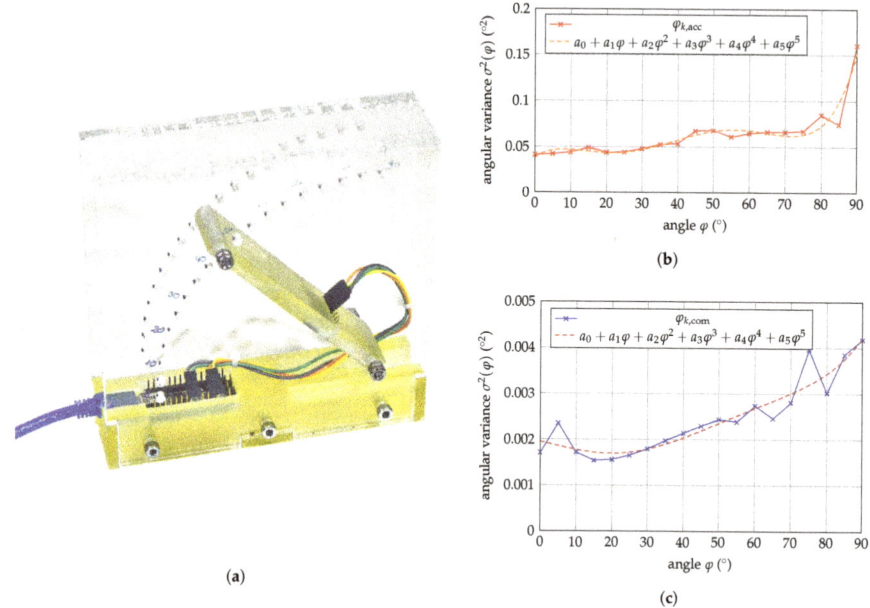

(a)

(b)

(c)

Figure 8. Experimental test bench to investigate IMUs on the dependency of the orientation angles' variances on the orientation angle (**a**). Experimental results of the InvenSense MPU-9250: (**b**) variances of the raw angle and suitable fifth-order polynomial fit and (**c**) variances of the filtered orientation angle and suitable fifth-order polynomial fit.

Table 1. Constants of the fifth-order polynomial for describing the orientation angle's variances of the raw and the filtered orientation angle as a function of the orientation angle itself.

	a_0	a_1	a_2	a_3	a_4	a_5
φ_{acc}	3.8553×10^{-2}	2.7241×10^{-1}	-2.7631×10^{-2}	1.0431×10^{-3}	-1.5467×10^{-5}	7.8730×10^{-8}
φ_{com}	1.9579×10^{-3}	-1.6773×10^{-5}	-5.1628×10^{-7}	5.0914×10^{-8}	-8.2501×10^{-10}	4.2277×10^{-12}

5.4. Accuracy of Static Pose Detections

As a first experiment on our general planar 3-R$\underline{P}$R parallel mechanism, we investigate how accurate the assembly modes can be obtained under static conditions when solely the linear actuators' orientations are used. We therefore investigate ten randomly chosen static poses of the manipulator platform where we take 500 measurements and calculate the two assembly modes from the measured linear actuators' orientation angles. In this context, we compare the accuracy for the assembly modes that can be obtained when raw orientation angles φ_{acc} and filtered orientation angles φ_{com} are used. In addition to this, we compare our experimental results with those provided by the CRLB. As the ground truth, we use the actual manipulator platform's pose whose joints' positions are optically analyzed by using image processing.

Table 2 shows the ten investigated, randomly chosen static poses. We choose the coordinates for the manipulator platform poses between 64.62 mm and 155.25 mm in the x-axis, between 157.14 mm and 216.06 mm in the y-axis and between $-20.45°$ and $15.84°$ for the platform orientation. Furthermore, Table 2 shows the mean values of the calculated poses after 500 measurements calculated from the raw orientation angles. First of all, it can be noticed that solution I, that is calculated from Equation (45), always corresponds to the actual pose while solution II, that is calculated from Equation (46), always corresponds to the second assembly mode with higher y-coordinates. Second of all, it can be noticed that solution I and solution II are sufficiently far away from each other so that they can be distinguished unambiguously. Finally, when comparing the actual pose with the mean value of the calculated poses, it can be noticed that solution I has an offset error that varies from pose to pose between -5.16 mm and 0.11 mm in the x-axis, between -12.22 mm and 1.03 mm in the y-axis and between $-1.48°$ and $6.42°$ for the platform orientation. The offset errors do not seem to have any dependencies.

Table 2. Investigated static poses and mean values of the calculated poses (solution I and solution II) after 500 measurements obtained from the raw orientation angles. Dimensions are in mm and $°$.

Pose	Actual Pose $\begin{bmatrix} x & y & \gamma \end{bmatrix}^{\mathsf{T}}$	Solution I $\begin{bmatrix} x_{\mathrm{I}} & y_{\mathrm{I}} & \gamma_{\mathrm{I}} \end{bmatrix}^{\mathsf{T}}$	Solution II $\begin{bmatrix} x_{\mathrm{II}} & y_{\mathrm{II}} & \gamma_{\mathrm{II}} \end{bmatrix}^{\mathsf{T}}$
1	$\begin{bmatrix} 146.76 & 190.46 & 14.01 \end{bmatrix}^{\mathsf{T}}$	$\begin{bmatrix} 148.10 & 190.89 & 14.63 \end{bmatrix}^{\mathsf{T}}$	$\begin{bmatrix} 309.13 & 398.44 & -146.85 \end{bmatrix}^{\mathsf{T}}$
2	$\begin{bmatrix} 90.71 & 212.00 & -20.38 \end{bmatrix}^{\mathsf{T}}$	$\begin{bmatrix} 94.96 & 220.66 & -24.55 \end{bmatrix}^{\mathsf{T}}$	$\begin{bmatrix} 146.43 & 340.08 & -84.27 \end{bmatrix}^{\mathsf{T}}$
3	$\begin{bmatrix} 137.55 & 206.21 & -7.71 \end{bmatrix}^{\mathsf{T}}$	$\begin{bmatrix} 137.68 & 206.42 & -6.67 \end{bmatrix}^{\mathsf{T}}$	$\begin{bmatrix} 250.39 & 375.34 & -107.89 \end{bmatrix}^{\mathsf{T}}$
4	$\begin{bmatrix} 155.25 & 191.61 & 15.72 \end{bmatrix}^{\mathsf{T}}$	$\begin{bmatrix} 155.41 & 190.95 & 17.04 \end{bmatrix}^{\mathsf{T}}$	$\begin{bmatrix} 322.82 & 396.65 & -151.26 \end{bmatrix}^{\mathsf{T}}$
5	$\begin{bmatrix} 123.65 & 211.69 & -11.96 \end{bmatrix}^{\mathsf{T}}$	$\begin{bmatrix} 124.50 & 211.93 & -11.72 \end{bmatrix}^{\mathsf{T}}$	$\begin{bmatrix} 217.26 & 369.73 & -99.65 \end{bmatrix}^{\mathsf{T}}$
6	$\begin{bmatrix} 69.22 & 215.68 & -12.16 \end{bmatrix}^{\mathsf{T}}$	$\begin{bmatrix} 74.53 & 228.82 & -18.83 \end{bmatrix}^{\mathsf{T}}$	$\begin{bmatrix} 124.70 & 382.65 & -90.18 \end{bmatrix}^{\mathsf{T}}$
7	$\begin{bmatrix} 107.01 & 190.51 & 0.71 \end{bmatrix}^{\mathsf{T}}$	$\begin{bmatrix} 107.98 & 192.76 & -0.73 \end{bmatrix}^{\mathsf{T}}$	$\begin{bmatrix} 219.71 & 392.17 & -120.73 \end{bmatrix}^{\mathsf{T}}$
8	$\begin{bmatrix} 64.62 & 186.05 & 15.84 \end{bmatrix}^{\mathsf{T}}$	$\begin{bmatrix} 66.82 & 191.85 & 10.75 \end{bmatrix}^{\mathsf{T}}$	$\begin{bmatrix} 156.08 & 448.07 & -144.02 \end{bmatrix}^{\mathsf{T}}$
9	$\begin{bmatrix} 125.21 & 161.73 & 13.32 \end{bmatrix}^{\mathsf{T}}$	$\begin{bmatrix} 125.31 & 162.46 & 13.50 \end{bmatrix}^{\mathsf{T}}$	$\begin{bmatrix} 276.07 & 357.91 & -153.10 \end{bmatrix}^{\mathsf{T}}$
10	$\begin{bmatrix} 132.37 & 157.14 & 8.30 \end{bmatrix}^{\mathsf{T}}$	$\begin{bmatrix} 132.71 & 158.64 & 7.45 \end{bmatrix}^{\mathsf{T}}$	$\begin{bmatrix} 284.36 & 339.89 & -142.95 \end{bmatrix}^{\mathsf{T}}$

Figure 9 shows the position and orientation errors between the investigated static poses and solution I that was obtained experimentally from the IMUs' values with 500 repetitions as boxplots. The results from the raw accelerometer values are shown in red and the results from the complementary filtered orientation angles are shown in blue. Comparing the results from the raw accelerometer values with those obtained from the filtered orientation angles, it can be noticed that both show similar offset errors but, most importantly, the results for the filtered orientation angles have significantly lower variances. In fact, throughout the ten investigated poses, the variances of the position and orientation errors obtained with the raw accelerometer values are approximately 27 times higher than those obtained with the filtered orientation angles. This applies for all axes (8.3 to 56.1 times higher for the x-axis, 14.2 to 37.5 times higher for the y-axis and 14.2 to 33.3 times higher for the platform orientation). From the results shown in Figure 9, it can also be noticed that the variances in the axes are not constant and show dependencies on the position and orientation of the manipulator platform. In fact, the best results for the raw accelerometer values were obtained for the poses 9 and 10 where the position and orientation errors show variances of only 0.19 mm^2 to 0.34 mm^2 in the x-axis, 1.60 mm^2 to 1.63 mm^2 in the y-axis and $2.43°^2$ to $2.96°^2$ for the platform orientation. In contrast to that, poses 2 and 6 show the highest variances for the raw accelerometer values with 5.07 mm^2 to 7.33 mm^2 in the x-axis, 65.15 mm^2 to 74.55 mm^2 in the y-axis and $20.36°^2$ to $22.71°^2$ for the platform orientation. The same applies for the filtered orientation angles.

In conclusion, the manipulator platform's pose can be obtained quite accurately with only small offset errors. Nevertheless, the variances obtained for the raw accelerometer values are very high but can be significantly improved when using the filtered orientation angles instead. Here, variances between 0.006 mm^2 and 0.155 mm^2 for the x-axis, between 0.051 mm^2 and 2.450 mm^2 for the y-axis and between $0.073°^2$ and $0.743°^2$ can be obtained. In comparison to the other poses, the variances for poses 2 and 6 are comparably higher. One possible reason may be that only poses 2, 6 and 8, the orientation angle φ_2 is above $90°$, whereas φ_2 is below $90°$ for the other poses. However, pose 8 does not show the high variances that would be expected.

The variances of the position and orientation errors depend on the variances of the measurements and, more importantly, the amplification of the solution formulation. The CRLB allows to calculate the lower bound of the variances that we can expect for a specific pose based on the variances of the measurements, that is, the orientation angles. Therewith, we can compare our experimental results for each pose with those calculated by the CRLB. Figure 10 shows the position and orientation errors obtained experimentally from the filtered orientation angles in blue and the simulated position and orientation errors calculated with the CRLB in purple. The CRLB only requires the actual pose, the measurement model and the variances of the measurements. Here, we use the polynomial in Equation (69) to estimate the variances of the filtered orientation angles.

Table 3 shows the variances and CRLB's results for the first five static poses when using raw orientation angles and when using filtered orientation angles. From the first five investigated static poses, it can already be noticed that the variances of the position and orientation errors obtained from experiments correspond with those calculated by the CRLB. The same applies for the poses six to ten (not displayed). For all the poses, the difference between the experimental results from the filtered orientation angles and the CRLB's results are very small and do not exceed $0.15\,\text{mm}^2$ in the x-axis, $0.28\,\text{mm}^2$ in the y-axis and $0.32^{\circ 2}$ for the platform orientation. When using the CRLB together with the polynomial in Equation (68) as the variances of the measurements, we can estimate the variances of the position and orientation errors of the results obtained with the raw accelerometer values similarly accurate (not displayed in Figure 10). Here, the difference between the raw experimental results and the CRLB's results are not higher than $1.42\,\text{mm}^2$ in the x-axis, $8.00\,\text{mm}^2$ in the y-axis and $-5.82^{\circ 2}$ for the platform orientation.

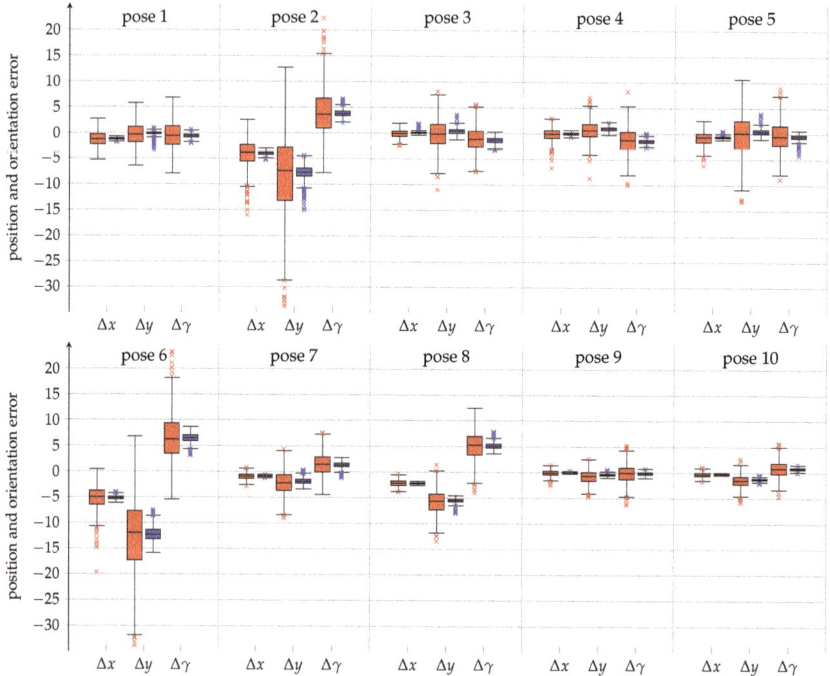

Figure 9. Results for the ten investigated static poses with 500 repetitions obtained experimentally from the raw accelerometer values (red) and the filtered orientation angles (blue). The errors in each axis, Δx, Δy and $\Delta \gamma$, are displayed in a boxplot. Dimensions are in mm and °. The box corresponds to the area in which the middle 50% of the errors lie while the whiskers indicate the area in which the middle 99.3% of the errors lie.

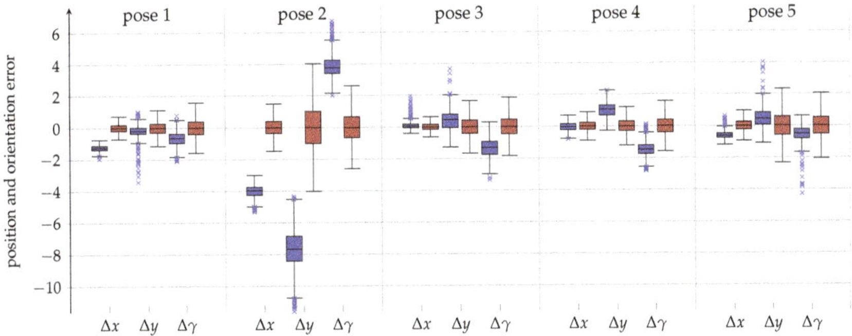

Figure 10. Results for the first five investigated static poses with 500 repetitions obtained experimentally from the filtered orientation angles (blue) and by simulation using the corresponding Cramér-Rao lower bound (CRLB) (purple). The errors in each axis, Δx, Δy and $\Delta \gamma$, are displayed in a boxplot. Dimensions are in mm and $°$. The box corresponds to the area in which the middle 50% of the errors lie while the whiskers indicate the area in which the middle 99.3% of the errors lie.

Table 3. Variances and results for the Cramér-Rao lower bound for the first five static poses when using raw orientation angles and when using filtered orientation angles. The variances are displayed as $\left[\sigma^2(x)\ \sigma^2(y)\ \sigma^2(\gamma)\right]^{\top}$. Dimensions are in mm^2 and $°^2$.

Pose	Variances for Raw Orientation Angles		Variances for Filtered Orientation Angles	
	Experiments	**CRLB**	**Experiments**	**CRLB**
1	$\begin{bmatrix} 1.8895 \\ 4.4329 \\ 6.9274 \end{bmatrix}$	$\begin{bmatrix} 1.6150 \\ 4.1013 \\ 7.6160 \end{bmatrix}$	$\begin{bmatrix} 0.0337 \\ 0.2244 \\ 0.2335 \end{bmatrix}$	$\begin{bmatrix} 0.0715 \\ 0.1697 \\ 0.3470 \end{bmatrix}$
2	$\begin{bmatrix} 7.3331 \\ 65.1472 \\ 20.3617 \end{bmatrix}$	$\begin{bmatrix} 7.7918 \\ 57.1507 \\ 23.6220 \end{bmatrix}$	$\begin{bmatrix} 0.1493 \\ 2.4503 \\ 0.6109 \end{bmatrix}$	$\begin{bmatrix} 0.2968 \\ 2.2853 \\ 0.9335 \end{bmatrix}$
3	$\begin{bmatrix} 0.6376 \\ 7.9631 \\ 6.0536 \end{bmatrix}$	$\begin{bmatrix} 1.2437 \\ 9.4596 \\ 9.7948 \end{bmatrix}$	$\begin{bmatrix} 0.0767 \\ 0.4116 \\ 0.3189 \end{bmatrix}$	$\begin{bmatrix} 0.0582 \\ 0.3787 \\ 0.4445 \end{bmatrix}$
4	$\begin{bmatrix} 1.5456 \\ 3.5633 \\ 5.6761 \end{bmatrix}$	$\begin{bmatrix} 2.5668 \\ 4.9758 \\ 7.7996 \end{bmatrix}$	$\begin{bmatrix} 0.0594 \\ 0.1653 \\ 0.2352 \end{bmatrix}$	$\begin{bmatrix} 0.1070 \\ 0.2006 \\ 0.3483 \end{bmatrix}$
5	$\begin{bmatrix} 1.6415 \\ 14.4296 \\ 8.1143 \end{bmatrix}$	$\begin{bmatrix} 3.0573 \\ 18.2156 \\ 13.9312 \end{bmatrix}$	$\begin{bmatrix} 0.0593 \\ 0.4560 \\ 0.3463 \end{bmatrix}$	$\begin{bmatrix} 0.1263 \\ 0.7334 \\ 0.5923 \end{bmatrix}$

In conclusion, by only knowing the measurement variances of the IMUs, it is possible to predict the manipulator platform's variances very accurately for any pose in the workspace without experiments at all. As the experimental results match the CRLB's results, we can furthermore conclude that the solution formulation proposed in Section 4 is the optimal estimator with the lowest amplification of measurement variances on the position and orientation variances. In the measurement model for the CRLB, we assumed the measurement error to be zero-mean Gaussian. The experiments indicate that this is not true. By also including these offset errors and the nonlinearity of the IMUs into the measurement model, it would be possible to predict the offset error of the manipulator platform's pose in addition to its variances. As an alternative, in order to realize zero-mean Gaussian measurement errors, it is also be possible to eliminate the offset errors and the nonlinearity of the IMUs by doing further calibrations.

5.5. Comparing Analytic Orientation-Based Results with Iterative Length-Based Results for Static Pose Detections

In Section 2, we reviewed classical methods for solving the direct kinematics problem of the general planar 3-R$\underline{P}$R parallel mechanism and mentioned that iterative methods like the Newton-Raphson algorithm are most often used for finding the actual pose of the manipulator platform from the linear actuators' lengths. The accuracy mainly depends on the initial estimate and the accuracy of the measured linear actuators' lengths. The linear actuators that are used in our prototype of the general planar 3-R$\underline{P}$R parallel mechanism have integrated potentiometers. Consequently, the lengths are measured indirectly with the problem that the actual lengths of the linear actuators are not measured and rely on the linearities of the potentiometers. In addition, the analog inputs of the Arduino Mega board have a limited resolution of 10 bit leading to a maximum resolution of 0.0977 mm/bit for the linear actuators' lengths (stroke length of the linear actuators divided by the resolution of Arduino Mega). Nevertheless, the obtained linear actuators' lengths can be used together with an initial estimate and the Newton-Raphson algorithm to calculate the actual pose of the manipulator platform. In order to evaluate the quality of the measured linear actuators' lengths and to guarantee convergence of the Newton-Raphson algorithm, we used the actual pose as initial estimate. Table 4 shows the mean offset errors of the Newton-Raphson algorithm for the measured linear actuators' lengths. Furthermore, it shows the mean offset errors for the ten static poses obtained with the analytic formulation proposed in Section 3 and the raw orientation angles.

Table 4. Investigated static poses and mean offset errors Δx, Δy and $\Delta \gamma$ of the analytic, orientation-based formulation and the iterative length-based solution (Newton-Raphson algorithm). Dimensions are in mm and $°$. Poses, where the algorithm fails to converge are indicated by a $--$.

Pose	Actual Pose			Offset Error Solution I			Offset Error Newton-Raphson Algorithm		
	$\begin{bmatrix} x & y & \gamma \end{bmatrix}^{\mathsf{T}}$			$\begin{bmatrix} \Delta x & \Delta y & \Delta \gamma \end{bmatrix}^{\mathsf{T}}$			$\begin{bmatrix} \Delta x & \Delta y & \Delta \gamma \end{bmatrix}^{\mathsf{T}}$		
1	146.76	190.46	14.01	−1.25	−0.23	−0.66	8.18	−9.22	4.09
2	90.71	212.00	−20.38	−3.99	−7.83	3.88	−1.85	−2.77	1.58
3	137.55	206.21	−7.71	0.11	0.39	−1.35	−0.42	−3.79	1.38
4	155.25	191.61	15.72	−0.05	1.03	−1.48	2.77	−4.41	0.78
5	123.65	211.69	−11.96	−0.59	0.51	−0.60	−0.82	−2.92	1.36
6	69.22	215.68	−12.16	−5.16	−12.22	6.42	1.32	−3.69	2.25
7	107.01	190.51	0.71	−0.90	−1.85	1.26	−3.12	0.21	−1.08
8	64.62	186.05	15.84	−2.21	−5.55	5.16	−32.48	5.86	−11.14
9	125.21	161.73	13.32	−0.02	−0.41	−0.15	−−	−−	−−
10	132.37	157.14	8.30	−0.25	−1.26	0.82	−−	−−	−−

From Table 4, it can be observed that, except for the poses 2 and 6, the mean error of the Newton-Raphson algorithm is significantly higher than the mean error of the analytic formulation. In fact, the mean errors of the Newton-Raphson algorithm are 1.5 to even 20 times higher than the mean errors of the analytic formulation. The mean errors spread between −32.48 mm and 8.18 mm in the x-axis, between −9.22 mm and 5.86 mm in the y-axis and between −11.14$°$ and 4.09$°$ for the platform orientation. For the poses 9 and 10, the Newton-Raphson algorithm even converged to a completely wrong solution although the actual pose is used as the initial pose estimate. In contrast, for poses 2 and 6, the Newton-Raphson algorithm shows more accurate results than the analytic formulation. The mean errors in the calculated poses indicate that there is an offset between the actual and the measured linear actuators' lengths that needs to be removed by calibration. By comparing the actual lengths with the measured lengths, however, only small errors in the lengths measurements were recognized (±1.5 mm). Other than for the linear actuators' orientations and consequently the results for the analytic formulation, the variances for the Newton-Raphson algorithm are nearly zero (0.21 mm^2 in the x-axis, 0.09 mm^2 in the y-axis and 0.06$^{\circ 2}$ for the platform orientation) since the lengths do not change under static conditions and the potentiometers' readings only differ by ±1 bit. This indicates that if the lengths are measured correctly and with a sufficiently high resolution, the manipulator platforms' pose can be found with the Newton-Raphson algorithm more robustly than from the unfiltered orientation angles. In the current form, that is, using the linear actuators' potentiometer values, only slightly lower variances as for the filtered orientation angles can be obtained. However, the Newton-Raphson algorithm requires at least three to five iterations to converge, whereas the analytic formulation provides an explicit formulation without any iteration steps. Furthermore, if the initial pose estimate is changed away from the actual pose, the required number of iterations increase and we cannot guarantee that the Newton-Raphson algorithm will always converge to the correct solution.

In conclusion, the Newton-Raphson algorithm together with the linear actuators' lengths shows higher offset errors but lower variances than the analytic formulation where solely the linear actuators' orientations are used. However, for pose detections where no accurate initial pose estimate can be provided, for example, in the beginning of an experiment or after restarting the system, the convergence of the Newton-Raphson algorithm cannot be guaranteed.

5.6. Accuracy of Dynamic Pose Detections

As a second experiment on our general planar 3-R$\underline{\text{P}}$R parallel mechanism, we investigate how accurate the manipulator platform's pose can be obtained under dynamic conditions when solely the linear actuators' orientations are used. Therefore, we continuously move the manipulator platform dynamically by changing the linear actuators' lengths adequately using a PID controller that minimizes the differences between the target and the actual lengths. We let the linear actuators run with 12 V leading to higher velocities (±40 mm/s and $\pm15^\circ$/s). During the experiment, we measure and filter the linear actuators' orientations with the maximum possible sampling rate, that still is 53.16 Hz and calculate the two assembly modes using the formulation proposed in Section 3. As the ground truth, we again use image processing to optically analyse the actual manipulator platform's pose, whose joints' positions are equipped with small red dots for optically tracking their position (the images are recorded with 30 fps). Figure 11 shows the trajectories of the manipulator platform's joints in blue, red and green, respectively, during the dynamic experiment. The entire dynamic experiment is also shown in the video of the Supplementary Material.

Figure 12 shows the manipulator platform's pose during the dynamic experiment calculated from the raw (red) and the filtered (blue) linear actuators' orientations. As reference (black), the positions and orientations calculated from the optically analysed manipulator platform joints are displayed. During the experiment, the manipulator platform's pose ranges between 97.6 mm and 154.5 mm in the x-axis, between 177.1 mm and 219.1 mm in the y-axis and between 11.6° and 24.7° for the platform orientation. Here, we only use solution I of the proposed formulation calculated from Equation (45). Solution II range between 165.9 mm and 333.7 mm in the x-axis, between 332.2 mm and 492.6 mm in the y-axis and between -179.9° and 179.8° for the platform orientation and is therewith sufficiently far away from solution I.

The poses calculated from the raw accelerometer values are significantly noisier than the poses calculated with the complementary filtered orientation angles. In fact, the complementary filter's results are at least two times more accurate than the unfiltered results and match the actual manipulator platform's pose quite well. Especially in the x-axis, the complementary filter's results are comparatively accurate and do not exceed a position error of ±5 mm. For the platform orientation, the complementary filter's results show errors mainly between -10° and 5°. Only between second 22 and 25 of the experiment, the complementary filter shows strangely big position and orientation errors. The same but with a smaller impact, happens at second 3 of the experiment. In both cases, the calculated orientation angle of the manipulator platform drifts away by $10°$ (at 3 s) and even $30°$ (at 23 s). The reason for this is probably that, due to the high velocity and the sampling rate, the Arduino Mega looses some measuring information leading to inaccurate linear actuators' orientations.

Figure 13 summarises the position and orientation errors of the raw and the filtered orientation angles in boxplots. Both the poses calculated from the raw accelerometer values and the poses calculated with the complementary filter do not show any offset errors. Due to the high measurement noise, the poses calculated from the raw accelerometer values are very noisy and show huge errors. In total, only 50% of the errors range between -15.3 mm and 11.9 mm for the x-axis, between -23.2 mm and 22.3 mm for the y-axis and between -21.9 mm and 19.3 mm for the platform orientation. Furthermore, the the proposed formulation often fails to solve Equations (45) and (46) from the raw accelerometer values. Apparently, the root in Equation (47) becomes negative. This can be traced back to the noisy IMUs' measurements under fast motions. Hence, from the raw accelerometer values, the pose cannot be obtained sufficiently accurate. In contrast to that, the results calculated from the complementary filter are significantly more accurate and robust. Here, lower variances of the position and orientation errors can be obtained.

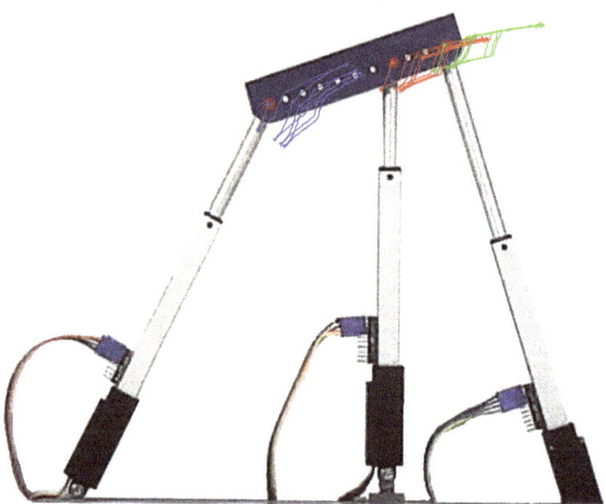

Figure 11. Trajectories of the first (blue), second (red) and third (green) manipulator platform joint during the dynamic experiment. The trajectories were recorded by a camera with 30 fps and the joints' positions were analysed using image processing.

For comparison, we additionally used the linear actuators' lengths and the Newton-Raphson algorithm to calculate the manipulator platform's pose iteratively. These results, however, are not calculated on the Arduino Mega due to the required initial estimate in the beginning of the experiment and, more importantly, the significantly longer calculation time. In fact, using the linear actuators' lengths and the Newton-Raphson algorithm is at least ten times slower than using the proposed analytic algorithm together with the filtered orientations. However, even though the actual pose of the manipulator platform was given as an initial pose estimate, for this experiment, the Newton-Raphson algorithm converged to a completely wrong pose in the beginning, that is,

$$\begin{bmatrix} x & y & \gamma \end{bmatrix}^\top = \begin{bmatrix} 172.0\,\text{mm} & -142.4\,\text{mm} & 233.3° \end{bmatrix}^\top ,\tag{70}$$

and did not return from there. Hence, the results of the Newton-Raphson algorithm does not match the actual pose at all. Possible reason are the small errors in the linear actuators' lengths and the missing robustness of the Newton-Raphson algorithm.

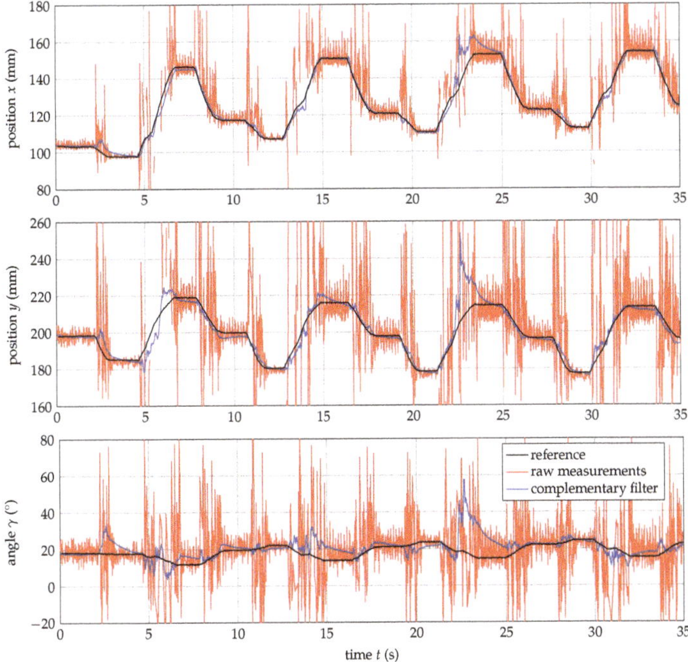

Figure 12. Pose of the manipulator platform during the dynamic experiment calculated from the raw (red) and the filtered (blue) linear actuators' orientations: (**a**) x-position, (**b**) y-position and (**c**) orientation angle γ. As reference (black), the positions and orientations calculated from the optically analysed manipulator platform joints are used.

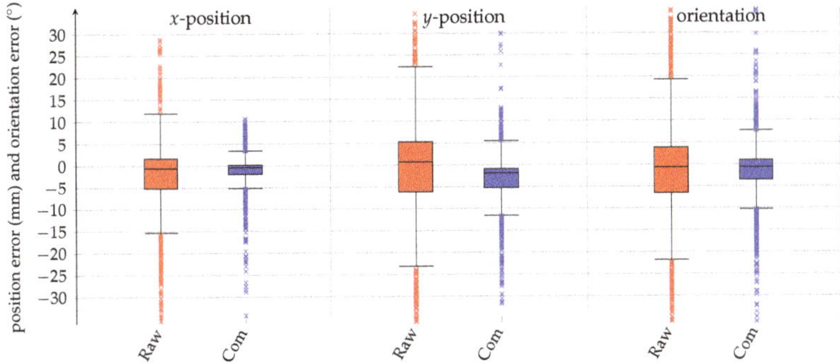

Figure 13. Boxplots of the position and orientation errors of the manipulator platform's pose during the experiment calculated with the raw orientation angles (red) and the complementary filtered orientation angles (blue). The box corresponds to the area in which the middle 50% of the errors lie while the whiskers indicate the area in which the middle 99.3% of the errors lie.

In conclusion, the results obtained from the filtered linear actuators' orientation angles together with our formulation proposed in Section 3 are capable of calculating the actual manipulator platform's pose even under dynamic conditions. Comparably small offset errors and variances can be obtained throughout the dynamic experiment. It therewith is significantly more accurate and robust than the raw orientation angles. Nevertheless, the variances obtained with the complementary filter are still too high for an accurate pose control. However, it outperforms the Newton-Raphson algorithm in terms of accuracy, robustness and computational efficiency.

6. Conclusions

In this paper, we first reviewed classical methods for solving the direct kinematics problem of parallel mechanisms and discussed their disadvantages on the example of the general planar 3-R$\underline{\text{P}}$R parallel mechanism. In order to avoid these disadvantages, we proposed a sensor concept together with an analytical formulation for solving the direct kinematics problem of a general planar 3-R$\underline{\text{P}}$R parallel mechanism. By measuring the orientations of the linear actuators, provided, for example, by inertial measurement units, the number of possible assembly modes can be reduced down to two when using the linear actuators' orientations instead of their lengths. Finally, we experimentally evaluated the accuracy of our direct kinematics solution under static as well as dynamic conditions by performing experiments on a specially designed prototype.

The static experiments prove that it is possible to calculate the two possible assembly modes of the manipulator platform from the linear actuators' orientations. For the investigated general planar 3-R$\underline{\text{P}}$R parallel mechanism, the two solutions of the direct kinematics problem are sufficiently far away from each other to distinguish between them. By using the raw accelerometer values to calculate the linear actuators' orientation angles, the variances in the orientation angles are quite high leading to huge variances in the calculated poses of the manipulator platform. The mean results, however, are quite precise. By using a complementary filter instead, where the linear actuators' orientation angles are calculated from the IMUs' accelerometer and gyroscope values, the variances in the orientation angles are significantly smaller (27 to 38 times) leading also to smaller variances in the calculated poses of the manipulator platform. Here, variances between $0.006\,\text{mm}^2$ and $0.155\,\text{mm}^2$ for the x-axis, between $0.051\,\text{mm}^2$ and $2.450\,\text{mm}^2$ for the y-axis and between $0.073°^2$ and $0.743°^2$ can be obtained.

By using the Cramér-Rao lower bound (CRLB) together with the known variances of the linear actuators' orientation angles, it is possible to estimate the variances of the calculated manipulator platform's pose in each axis for every pose in the workspace. For the static measurements, the experimental results match the CRLB's results so that we can conclude that the proposed solution formulation is the optimal estimator with the lowest amplification of measurement variances on the position and orientation variances.

The dynamic experiment also indicates that the raw accelerometer values are too noisy to be used for accurately and robustly calculating the manipulator platform's pose. Throughout the experiment, the results show huge variances. Furthermore, the proposed formulation furthermore fails to solve Equations (45) and (46) that, can also be traced back to the noisy raw measurements. Much more accurate and robust results can be obtained from the filtered orientations angles. The complementary filter shows significantly lower variances of the position and orientation errors and no offset error. Therewith, the proposed analytic algorithm enables to actually calculate the manipulator platform's pose even under dynamic conditions. The risk of confusion between the two assembly modes never existed during the experiments since solution I, provided by Equation (45), always corresponds to the actual pose.

The analytic formulation for calculating the two assembly modes of the manipulator platform from the linear actuators' orientations presented in Section 3 can be further generalized. In fact, in the model of the general planar 3-R$\underline{\text{P}}$R parallel mechanism, we assumed that the three manipulator platform joints D, E and F are aligned. This model is sufficiently general to show that the planar 3-R$\underline{\text{P}}$R parallel mechanism can have up to six assembly modes. However, it does not correspond to the most general case where no constraints are given for the base and the manipulator platform joints. In future, we will focus on finding an analytic formulation for calculating the assembly modes even for this case.

By obtaining a unique solution of the direct kinematics problem without requiring the linear actuators' lengths, in future, it is possible to actually benefit especially in the control of parallel mechanisms. Usually, the measured linear actuators' lengths are compared with the target lengths that are provided by inverse kinematics for a given target pose, see Figure 14a. Due to the direct kinematics problem and the existence of singularities in the workspace of parallel mechanisms, we cannot guarantee that the linear actuators' lengths correspond to only one pose and it is possible that the manipulator platform is in (or moves to) a different pose than expected. This problem can be solved by using more suitable coordinates, for example, the linear actuators' orientations. When the direct kinematics problem provides a unique solution or, in this case, the two solutions are far away from each other, we can ensure that the manipulator platform always moves to the target pose. Figure 14b shows a pose control concept where the linear linear actuators' orientation angles φ are used. For a given target pose p_{target}, the target orientation angles φ_{target} can be calculated from inverse kinematics. They can be compared with the measured orientation angles φ_{is} and the required deviation of the orientation angles $\Delta\varphi$ can be calculated and given to the controller, for example, a PID controller. The controller then calculates an appropriate output u for the system that, in turn, produces the system output. Using the proposed sensor concept, the system output can be measured, for example, with IMUs mounted on the linear actuators. These measurements are filtered and finally compared with the new target orientation angles. In contrast to usual control concepts where we cannot guarantee that the pose that belongs to the measurements, in general, the linear actuators' lengths, is actually the target pose (indicated by the dashed line in Figure 14a. However, by using the proposed control concept shown in Figure 14b, we actually can. In this context, controllability of the robot is essential. Briot et al. [76] proposed an interesting approach to the analysis of the controllability of parallel mechanisms.

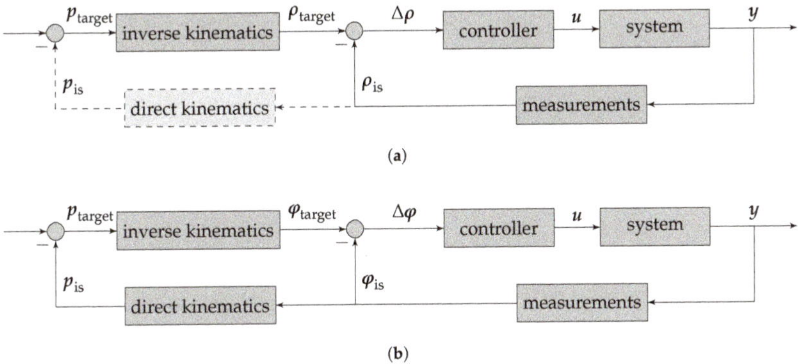

Figure 14. Conventional (**a**) and proposed control concept (**b**) for controlling the manipulator platform's pose of a parallel mechanism. The conventional control concept uses the linear actuators' lengths, whereas the proposed control concept uses the linear actuators' orientations. In contrast to the conventional control concept, the proposed control concept can guarantee an analytic solution of the direct kinematics problem.

Supplementary Materials: The following are available online at http://www.mdpi.com/2218-6581/8/3/72/s1, Video S1: Demonstration_Video_3RPR. A video including the dynamic experiment, the Matlab code and other information are available online at https://github.com/stefanschulz85/Assembly-Modes-of-a-3-RPR-parallel-Mechanism-when-Using-the-Linear-Actuators-Orientations (doi:10.5281/zenodo.3240459).

Funding: This research received no external funding.

Acknowledgments: Stefan Schulz would like to thank Arthur Seibel for his valuable suggestions and Aniruidha Nagaraj Vyasamudra for his commitment during his project work. The publication of this article was supported by the Deutsche Forschungsgemeinschaft (DFG, German Research Foundation)—Projektnummer 392323616 and Hamburg University of Technology (TUHH) in the funding programme "Open Access Publishing".

Conflicts of Interest: The author declare no conflict of interest.

References

1. Gosselin, C.M.; Sefrioui, J.; Richard, M.J. Solutions polynomiales au problème de la cinématique directe des manipulateurs parallèles plans à trois degrés de liberté. *Mech. Mach. Theory* **1992**, *27*, 107–119. [CrossRef]
2. Peisach, E.E. Determination of the position of the member of three-joint and two-joint four member Assur groups with rotational pairs. *Machinowedenie* **1985**, *5*, 55–61. (In Russian)
3. Pennock, G.R.; Kassner, D.J. Kinematic analysis of a planar eight-bar linkage: Application to a platform-type robot. *J. Mech. Des.* **1992**, *114*, 87–95. [CrossRef]
4. Wohlhart, K. Direct kinematic solution of the general planar Stewart platform. In Proceedings of the 3rd International Conference on Computer Integrated Manufacturing, Troy, NY, USA, 20–22 May 1992; pp. 403–411.
5. Gosselin, C.; Merlet, J.P. The direct kinematics of planar parallel manipulators: Special architectures and number of solutions. *Mech. Mach. Theory* **1994**, *29*, 1083–1097. [CrossRef]
6. Kong, X.; Gosselin, C. Forward displacement analysis of third-class analytic 3-R$\underline{P}$R planar parallel manipulators. *Mech. Mach. Theory* **2001**, *39*, 1009–1018. [CrossRef]
7. Collins, C.L. Forward kinematics of planar parallel manipulators in the Clifford algebra of $\mathbb{P}^2$. *Mech. Mach. Theory* **2002**, *37*, 799–813. [CrossRef]
8. Rojas, N.; Thomas, F. The forward kinematics of 3-R$\underline{P}$R planar robots: A review and a distance-based formulation. *IEEE Trans. Robot.* **2011**, *27*, 143–150. [CrossRef]
9. Mimura, N.; Funahashi, Y. A new analytical system applying 6 DOF parallel link manipulator for evaluating motion sensation. In Proceedings of 1995 IEEE International Conference on Robotics and Automation, Nagoya, Japan, 21–27 May 1995; pp. 227–233. [CrossRef]
10. Gosselin, C.M. Parallel computational algorithms for the kinematics and dynamics of planar and spatial parallel manipulators. *ASME J. Dyn. Syst. Meas. Control* **1996**, *118*, 22–28. [CrossRef]
11. McAree, P.R.; Daniel, R.W. A fast, robust solution to the Stewart platform forward kinematics. *J. Robot. Syst.* **1996**, *13*, 407–427. [CrossRef]
12. Šika, Z.; Kočandrle, V.; Stejskal, V. An investigation of properties of the forward displacement analysis of the generalized Stewart platform by means of general optimization methods. *Mech. Mach. Theory* **1998**, *33*, 245–253. [CrossRef]
13. Der-Ming, K. Direct displacement analysis of a Stewart platform mechanism. *Mech. Mach. Theory* **1999**, *34*, 453–465. [CrossRef]
14. Dhingra, A.K.; Almadi, A.N.; Kohli, D. A Groebner-Sylvester hybrid method for closed-form displacement analysis of mechanisms. *ASME J. Mech. Des.* **2000**, *122*, 431–438. [CrossRef]
15. Merlet, J.P. Solving the forward kinematics of a Gough-type parallel manipulator with interval analysis. *Int. J. Robot. Res.* **2004**, *23*, 221–235. [CrossRef]
16. Shi, X.; Fenton, R.G. Forward kinematic solution of a general 6 DOF Stewart platform based on three point position data. In Proceedings of the Eighth World Congress on the Theory of Machines and Mechanism, Prague, Czechoslovakia, 26–31 August 1991; pp. 1015–1018.
17. Stoughton, R.; Arai, T. Optimal sensor placement for forward kinematics evaluation of a 6-DOF parallel link manipulator. In Proceedings of the IEEE/RSJ International Workshop on Intelligent Robots and Systems (IROS), Osaka, Japan, 3–5 November 1991; pp. 785–790. [CrossRef]
18. Cheok, K.C.; Overholt, J.L.; Beck, R.R. Exact methods for determining the kinematics of a Stewart platform using additional displacement sensors. *J. Robot. Syst.* **1993**, *10*, 689–707. [CrossRef]
19. Merlet, J.P. Closed-form resolution of the direct kinematics of parallel manipulators using extra sensors data. In Proceedings of the IEEE International Conference on Robotics and Automation (ICRA), Atlanta, GA, USA, 2–6 May 1993; pp. 200–204. [CrossRef]
20. Han, K.; Chung, W.; Youm, Y. New resolution scheme of the forward kinematics of parallel manipulators using extra sensor data. *ASME J. Mech. Des.* **1996**, *118*, 214–219. [CrossRef]
21. Parenti-Castelli, V.; Gregorio, R.D. Real-time computation of the actual posture of the general geometry 6-6 fully-parallel mechanism using two extra rotary sensors. *J. Mech. Des.* **1998**, *120*, 549–554. [CrossRef]
22. Bonev, I.A.; Ryu, J. A new method for solving the direct kinematics of general 6-6 Stewart platforms using three linear extra sensors. *Mech. Mach. Theory* **2000**, *35*, 423–436. [CrossRef]

23. Vertechy, R.; Parenti-Castelli, V. Accurate and fast body pose estimation by three point position data. *Mech. Mach. Theory* **2007**, *42*, 1170–1183. [CrossRef]
24. Schulz, S.; Seibel, A.; Schreiber, D.; Schlattmann, J. Sensor concept for solving the direct kinematics problem of the Stewart-Gough platform. In Proceedings of the IEEE/RSJ International Conference on Intelligent Robots and Systems (IROS), Vancouver, BC, Canada, 24–28 September 2017; pp. 1959–1964. [CrossRef]
25. Schulz, S.; Seibel, A.; Schlattmann, J. Closed-form solution for the direct kinematics problem of planar 3-RPR parallel mechanisms. In Proceedings of the IEEE International Conference on Robotics and Automation (ICRA), Brisbane, Australia, 21–25 May 2018; pp. 968–973. [CrossRef]
26. Seibel, A.; Schulz, S.; Schlattmann, J. On the direct kinematics problem of parallel mechanisms. *J. Robot.* **2018**, *2018*, 2412608. [CrossRef]
27. Schulz, S.; Seibel, A.; Schlattmann, J. Assembly modes of general planar 3-RPR parallel mechanisms when using the linear actuators' orientations. In *Advances in Mechanism and Machine Science, Proceedings of the 15th IFToMM World Congress on Mechanism and Machine Science, Krakow, Poland, 30 June–4 July 2019*; Uhl, T., Ed.; Springer: Cham, Switzerland, 2019; Volume 73, pp. 279–288. [CrossRef]
28. Merlet, J.P. *Parallel Robots*; Springer: Dordrecht, The Netherlands, 2006. [CrossRef]
29. Merlet, J.P. Singular configurations of parallel manipulators and Grassmann geometry. *Int. J. Robot. Res.* **1992**, *8*, 45–56. [CrossRef]
30. Li, H.; Gosselin, C.M.; Richard, M.J. Determination of maximal singularity-free zones in the workspace of planar three-degree-of-freedom parallel mechanisms. *Mech. Mach. Theory* **2006**, *41*, 1157–1167. [CrossRef]
31. Zein, M.; Wenger, P.; Chablat, D. Non-singular assembly-mode changing motions for 3-RPR parallel manipulators. *Mech. Mach. Theory* **2008**, *43*, 480–490. [CrossRef]
32. Wenger, P.; Chablat, D. Kinematic analysis of a class of analytic planar 3-RPR parallel manipulators. In *Computational Kinematics*; Kecskeméthy, A., Müller, A., Eds.; Springer: Berlin, Germany, 2009; pp. 43–50. [CrossRef]
33. Wenger, P.; Chablat, D.; Zein, M. Degeneracy study of the forward kinematics of planar 3-RPR parallel manipulators. *ASME J. Mech. Des.* **2006**, *129*, 1265–1268. [CrossRef]
34. Briot, S.; Arakelian, V.; Bonev, I.A.; Chablat, D.; Wenger, P. Self-motions of general 3-RPR planar parallel robots. *Int. J. Robot. Res.* **2008**, *27*, 855–866. [CrossRef]
35. Caro, S.; Binaud, N.; Wenger, P. Sensitivity analysis of 3-RPR planar parallel manipulators. *ASME J. Mech. Des.* **2009**, *129*, 121005. [CrossRef]
36. Staicu, S. Power requirement comparison in the 3-RPR planar parallel robot dynamics. *Mech. Mach. Theory* **2009**, *44*, 1045–1057. [CrossRef]
37. Chablat, D.; Jha, R.; Caro, S. A framework for the control of a parallel manipulator with several actuation modes. In Proceedings of the IEEE International Conference on Industrial Informatics (INDIN), Poitiers, France, 19–21 July 2016; pp. 190–195. [CrossRef]
38. Moezi, S.A.; Rafeeyan, M.; Zakeri, E.; Zare, A. Simulation and experimental control of a 3-RPR parallel robot using optimal fuzzy controller and fast on/off solenoid valves based on the PWM wave. *ISA Trans.* **2016**, *61*, 265–286. [CrossRef]
39. Boudreau, R.; Turkkan, N. Solving the forward kinematics of parallel manipulators with a genetic algorithm. *J. Robot. Syst.* **1996**, *13*, 111–125. [CrossRef]
40. Sheng, L.; L. Wan-Long, D.Y.C.; Liang, F. Forward kinematics of the Stewart platform using hybrid immune genetic algorithm. In Proceedings of the IEEE International Conference on Mechatronics and Automation, Luoyang, China, 25–28 June 2006; pp. 2330–2335. [CrossRef]
41. Rolland, L.; Chandra, R. Forward kinematics of the 6-6 general parallel manipulator using real coded genetic algorithms. In Proceedings of the IEEE/ASME Conference on Advanced Intelligent Mechatronics (AIM), Singapore, 14–17 July 2009; pp. 1637–1642. [CrossRef]
42. Rolland, L.; Chandra, R. The forward kinematics of the 6-6 parallel manipulator using an evolutionary algorithm based on generalized generation gap with parent-centric crossover. *Robotica* **2016**, *34*, 1–22. [CrossRef]
43. Yee, C.S.; Lim, K.B. Forward kinematics solution of Stewart platform using neural networks. *Neurocomputing* **1997**, *16*, 333–349. [CrossRef]
44. Parikh, P.J.; Lam, S.S.Y. A hybrid strategy to solve the forward kinematics problem in parallel manipulators. *IEEE Trans. Robot.* **2005**, *21*, 18–25. [CrossRef]

45. Didrit, O.; Petitot, M.; Walter, E. Guaranteed solution of direct kinematic problems for general configurations of parallel manipulator. *IEEE Trans. Robot. Autom.* **1998**, *14*, 259–266. [CrossRef]

46. Dieudonne, J.E.; Parrish, R.V.; Bardusch, R.E. *An Actuator Extension Transformation for a Motion Simulator and Inverse Transformation Applying Newton-Raphson's Method*; NASA Technical Report TN D-7067; NASA Langley Research Center: Hampton, VA, USA, 1972.

47. Nguyen, C.C.; Zhou, Z.L.; Antrazi, S.S.; Campbell, C.E. Efficient computation of forward kinematics and Jacobian matrix of a Stewart platform-based manipulator. In Proceedings of the IEEE Proceedings of the SOUTHEASTCON '91, Williamsburg, VA, USA, 7–10 April 1991; pp. 869–874. [CrossRef]

48. Merlet, J.P. Direct kinematics of parallel manipulator. *IEEE Trans. Robot. Autom.* **1993**, *9*, 842–846. [CrossRef]

49. Liu, K.; Fitzgerald, J.M.; Lewis, F.L. Kinematic analysis of a Stewart platform manipulator. *IEEE Trans. Ind. Electron.* **1993**, *40*, 282–293. [CrossRef]

50. Yang, C.; Zheng, S.; Jin, J.; Zhu, S.; Han, J. Forward kinematics analysis of parallel manipulator using modified global Newton-Raphson method. *J. Cent. South Univ. Technol.* **1996**, *17*, 1264–1270. [CrossRef]

51. Vertechy, R.; Parenti-Castelli, V. Robust, fast and accurate solution of the direct position analysis of parallel manipulators by using extra-sensors. In *Parallel Manipulators, towards New Applications*; Wu, H., Ed.; I-Tech Education and Publishing: Vienna, Austria, 2008; pp. 133–154.

52. Zhuang, H. Self calibration of parallel mechanisms with a case study on Stewart platform. *IEEE Trans. Robot. Autom.* **1997**, *13*. [CrossRef]

53. Chiu, Y.J. Forward kinematics of a general fully parallel manipulator with auxiliary sensors. *Int. J. Robot. Res.* **2001**, *20*, 401–414. [CrossRef]

54. Arai, T.; Cleary, K.; Nakamura, T. Design, Analysis and Construction of a Prototype Parallel Link Manipulator. In Proceedings of the IEEE/RSJ International Workshop on Intelligent Robots and Systems (IROS), Ibaraki, Japan, 3–6 July 1990; pp. 205–212. [CrossRef]

55. Baron, L.; Angeles, J. The direct kinematics of parallel manipulators under redundant sensors. *IEEE Trans. Robot. Autom.* **2000**, *16*, 12–19. [CrossRef]

56. Baron, L.; Angeles, J. The kinematic decoupling of parallel manipulators using joint-sensor redundancy. *IEEE Trans. Robot. Autom.* **2000**, *16*, 644–651. [CrossRef]

57. Bonev, I.A.; Ryu, J.; Kim, N.J.; Lee, S.K. A simple new closed-form solution of the direct kinematics of parallel manipulators using three linear extra sensors. In Proceedings of the IEEE/ASME International Conference on Advanced Intelligent Mechatronics (AIM), Atlanta, GA, USA, 19–23 September 1999; pp. 526–530. [CrossRef]

58. Bonev, I.A.; Ryu, J.; Kim, N.J.; Lee, S.K. A closed-form solution to the direct kinematics of nearly general parallel manipulators with optimally located three linear extra sensors. *IEEE Trans. Robot. Autom.* **2001**, *17*, 148–156. [CrossRef]

59. Etemadi-Zanganeh, K.; Angeles, J. Real-time direct kinematics of general six-degree-of-freedom parallel manipulators with minimum-sensor data. *J. Robot. Syst.* **1995**, *12*, 833–844. [CrossRef]

60. Han, K.; Chung, W.; Youm, Y. Local Structurization for the Forward Kinematics of Parallel Manipulators Using Extra Sensor Data. In Proceedings of the IEEE International Conference on Robotics and Automation (ICRA), Nagoya, Japan, 21–27 May 1995; pp. 514–520. [CrossRef]

61. Nair, R.; Maddocks, J.H. On the Forward Kinematics of Parallel Manipulators. *Int. J. Robot. Res.* **1994**, *13*, 171–188. [CrossRef]

62. Innocenti, C. Closed-Form Determination of the Location of a Rigid Body by Seven In-Parallel Linear Transducers. *J. Mech. Des.* **1998**, *120*, 293–298. [CrossRef]

63. Parenti-Castelli, V.; Gregorio, R.D. Determination of the Actual Configuration of the General Stewart Platform Using Only One Additional Sensor. *J. Mech. Des.* **1999**, *121*, 21–25. [CrossRef]

64. Parenti-Castelli, V.; Gregorio, R.D. A New Algorithm Based on Two Extra-Sensors for Real-Time Computation of the Actual Configuration of the Generalized Stewart-Gough Manipulator. *J. Mech. Des.* **2000**, *122*, 294–298. [CrossRef]

65. Tancredi, L.; Merlet, J.P. Extra sensors data for solving the forward kinematics problem of parallel manipulators. In Proceedings of the 9th World Congress on the Theory of Machines and Mechanisms, Milan, Italy, 29 August–2 September 1995; pp. 2122–2126.

66. Tancredi, L.; Teillaud, M.; Merlet, J.P. Forward Kinematics of a Parallel Manipulator with Additional Rotary Sensors Measuring the Position of Platform Joints. In *Computational Kinematics*; Merlet, J.P., Ravani, B., Eds.; Solid Mechanics and Its Applications; Kluwer Academic Publishers: Dordrecht, The Netherlands, 1995; Volume 40, pp. 261–270._27. [CrossRef]

67. Vertechy, R.; Dunlop, G.R.; Parenti-Castelli, V. An accurate algorithm for the real-time solution of the direct kinematics of 6-3 Stewart platform manipulators. In *Advances in Robot Kinematics*; Lenarčič, J., Thomas, F., Eds.; Solid Mechanics and Its Applications; Kluwer Academic Publishers: Dordrecht, The Netherlands, 2002; Volume 40, pp. 369–378. [CrossRef]

68. Jin, Y. Exact solution for the forward kinematics of the general Stewart platform using two additional displacement sensors. In Proceedings of the 23th ASME Biennial Mechanism Conference, Minneapolis, MN, USA, 11–14 September 1994; pp. 491–945.

69. Baron, L.; Angeles, J. A linear algebraic solution of the direct kinematics of parallel manipulators using a camera. In Proceedings of the 9th World Congress on the Theory of Machines and Mechanisms, Milano, Italy, 29 August–2 September 1995; pp. 1925–1929.

70. Hesselbach, J.; Bier, C.; Pietsch, I.; Plitea, N.; Buttenbach, S.; Wogersien, A.; Guttler, J. Passive joint-sensor applications for parallel robots. In Proceedings of the IEEE/RJS International Conference on Intelligent Robots and Systems (IROS), Sendai, Japan, 28 September–2 October 2004; pp. 3507–3512. [CrossRef]

71. Besnard, S.; Khalil, W. Calibration of parallel robot using two inclinometers. In Proceedings of the IEEE International Conference on Robotics and Automation (ICRA), Detroit, MI, USA, 10–15 May 1999; pp. 1758–1763. [CrossRef]

72. Yun, X.; Lizarraga, M.; Bachmann, E.R.; McGhee, R.B. An improved quaternion-based Kalman filter for real-time tracking of rigid body orientation. In Proceedings of the IEEE/RSJ International Conference on Intelligent Robots and Systems (IROS), Las Vegas, NV, USA, 27–31 October 2003; pp. 1074–1079. [CrossRef]

73. Mahony, R.; Hamel, T.; Pflimlin, J.M. Nonlinear complementary filters on the special orthogonal group. *IEEE Trans. Autom. Control* **2008**, *53*, 1203–1218. [CrossRef]

74. Madgwick, S.O.H.; Harrison, A.J.L.; Vaidyanathan, R. Estimation of IMU and MARG orientation using a gradient descent algorithm. In Proceedings of the IEEE International Conference on Rehabilitation Robotics (ICORR), Zurich, Switzerland, 29 June–1 July 2011; pp. 179–185. [CrossRef]

75. Valenti, R.G.; Dryanovski, I.; Xiao, J. Keeping a good attitude: A quaternion-based orientation filter for IMUs and MARGs. *Sensors* **2015**, *15*, 19302–19330. [CrossRef]

76. Briot, S.; Martinet, P.; Rosenzveig, V. The hidden robot: An efficient concept contributing to the analysis of the controllability of parallel robots in advanced visual servoing techniques *IEEE Trans. Robot.* **2015**, *31*, 1337–1352. [CrossRef]

 *robotics*

Article

Additive Manufacturing as an Essential Element in the Teaching of Robotics

Kevin Castelli [1,†] **and Hermes Giberti** [2,*,†]

1   Dipartimento di Ingegneria Meccanica e Industriale, Università degli Studi di Brescia, Via Branze 38, 25123 Brescia, Italy

2   Dipartimento di Ingegneria Industriale e dell'Informazione, Università degli Studi di Pavia, Via A. Ferrata 5, 27100 Pavia, Italy

*   Correspondence: hermes.giberti@unipv.it

†   These authors contributed equally to this work.

Received: 29 June 2019; Accepted: 16 August 2019; Published: 20 August 2019

Abstract: This paper aims to describe how additive manufacturing can be useful in enhancing a robotic course, allowing students to focus on all aspects of the multidisciplinary components of this subject. A three-year experience of the course of "robotic system design" is presented to support the validity of the use of this technology in teaching. This course is specifically aimed at Master of Science (MSc) Mechanical Engineering students and therefore requires one to view the subject in all its aspects including those which are not conventionally taken into consideration such as mechanical design, prototyping and the final realization.

Keywords: robotics teaching; 3D-printing; low-cost robots; education; Arduino

---

1. Introduction

Modern technology requires an ever-increasing awareness and knowledge of multidisciplinary subjects. In order to reach this objective, Universities offer study paths that combine a number of different disciplines. In particular, traditional engineering courses during the last decade have tended to mix topics related to mechanics, information technologies, electronics and the like to give rise to better prepared professional figures [1]. In order to perfect this approach, it is necessary to combine all the above elements in a unique experience capable of letting students understand the interaction between them. A robot is, by definition, the perfect example of a multidisciplinary device [2]. It combines the knowledge and science of mechanical engineering, electronics, computer science and automation. This, however, must of necessity be related to the practical application in order to give the students a complete teaching experience. Whilst the theoretical aspect should never be underplayed, it is equally important that the student does not restrict him/herself to mere workshop experiences but is also encouraged to express their creativity in the solving of practical problems. Therefore, hands-on lessons, lab sessions and flipped classroom philosophies have been preferred. These three teaching techniques can be enhanced and made more effective through the use of suitable technological tools such as additive manufacturing devices. For this reason, there are several examples of courses that use additive manufacturing (AM) solutions as help in teaching. An example of how AM has already been blended into University courses as well as how AM has become a subject in itself is presented in the work of Jamison Go and A. John Hart [3]. An extensive overview of the use of AM in teaching and education is also shown in the paper of Simon Ford and Tim Minshall [4].

The present article focuses on how Additive Manufacturing can be an essential element in the teaching of robotics, comparing the experience available in the scientific literature and presenting the approach used to teach mechanical students at the course of "Robotic System Design" at the Politecnico di Milano (previouly called "Robot Mechanics" [5]). Several examples of robotics lab sessions can be found around the world. For instance: [6] describes three robotic courses held at the United States Naval Academy; Ritsumeikan University of Kyoto has also developed a practical robotics education program [7]; whilst some laboratory exercises are presented in [8] to give students some practical experience; [9] reports some web-based programming tools to generate three dimensional models of robots using Denavit–Hartenberg parameters. Nevertheless, all these courses mainly focus on programming aspects in that they rely on ready-made kits or open-source architectures that can easily be found on the web. Papers [10,11] are examples of how these kits have been developed for a robotic course. Whilst [12,13] indicate how low-cost robots are helpful devices for teaching purposes (robots non only for the sake of robotics but also for the sake of STEM). Kits or pre-built solutions may well be compelling tools to increase the student's passion in the subject, but they lack creativity and aspects related to the solving of problems. Moreover, students tend to assemble mechanical, electronic and IT parts to make the devices present in the kit operative, but do not really grasp how the different technologies and disciplines interact between them to give rise to a robotics system.

In order to overcome these limitations, besides using a mix of innovative teaching techniques such as hands-on lessons, lab sessions, and the flipped classroom, students were requested to project, build and operate a robot starting only from the functional specifications of the machine. In particular, the kinematic chain and the task to be performed were assigned. Thus the students are exposed to an all-encompassing teaching experience, highly multidisciplinary with theoretical aspects supported by practical implementation. Solely due to the diffusion and use of AM techniques which simplify the production of components and giving the creativity of students free rein allowing them to verify, by experiment, the efficacy of their projects. In addition, the 3D printer itself being a robotic system, students come to use, control and master these types of devices so as to build suitable parts in the construction of their robots. This paper not only describes this new teaching experience but also how AM affects the outline of the course itself. Robotics courses, in fact, generally focus on kinematic, trajectory planning and control, without actually forcing the students to concretely experience how design and realization, in fact, affect real systems.

From the 2015/2016 academic year, a project was assigned to the students enrolled at the "Robotic System Design" course proposed in the final year of the MSc in Mechanical Engineering at Politecnico di Milano. Besides programming tasks, also available in the previously mentioned courses, students are required to design the mechanical components of their robotic system, which will be 3D printed entirely from scratch. Fused deposition modeling (FDM) printers have been selected because they are safe, easy to operate and above all affordable. Printers used are "Sharebot 42" and the plastic material employed is polylactic acid (PLA). The final project basically consists of a three-degree of freedom (DOF) robotic configuration that students, dived into groups, have to develop in its mechanical, electrical and control aspects. In order to stimulate the students' creativity, they are required to perform an additional task of their choice. A challenge is scheduled for the end of the course where all the groups are required to give a practical demonstration. As a matter of fact, as highlighted in [14], the competition encourages increased effort on the part of the students. As an added incentive, the winning team gets to keep their creation while the losing teams must return the hardware given to them. This innovative course has two main advantages: firstly the students are faced with actual implementation problems and to make their own practical choices, thus helping them to develop a critical sense; simultaneously they can unleash their imagination. These hands-on classes, where students design the robot, are supported by theoretical lectures aimed at providing them a solid background to deal with problems related to kinematics, dynamics, and motion planning. One of the main objectives of the hands-on classes is to help students acquire practical experience and to put into practice the concepts studied during theoretical classes. As stated in [15], providing just merely theoretical knowledge is not sufficient and in some cases might well be counter-productive since it could confuse the student or make him/her accustomed to approaching actual problems with a limited perspective.

The following are set out as follows: Section 2 reports how AM reshaped the course; Section 3 deals with the hardware and software required to actuate the robots; Section 4 describes the course projects, while Section 5 shows their developments and evaluation; finally Section 6 reports the feedback from students.

2. AM Changes a Course Syllabus

The exploitation of AM as a didactic tool enforces to redefine the course outline and schedule. Students are required to learn in a short time different basics knowledge to start dimensioning their robots to be able to complete their project by the end of the semester.

As concerns theoretical concepts, a selective compliance assembly robot arm (SCARA) robot is used as an example, also throughout the course, to teach the simple models that will be used to start outlining the machine. A brief introduction on mechanical components (for example reducers and transmission units) gives an overview of what can be used for the final realization. The last initial knowledge package deals on AM and rapid electronic prototyping (i.e., Arduino) to start familiarizing with these tools.

With these basic knowledge available students can start and continue developing their machines while the theoretical classes deepen and generalize the topics already discussed for 6 dof system and move on to new topics such as: parallel kinematic machine, trajectory planning, control algorithm and calibration.

Here the topics of the "Robotics System Design" course are discussed.

- Planar robots: forward and inverse kinematics. The next topic is represented by velocity and kineto–static analyses, the Jacobian matrix is introduced as well as the concepts of velocity and manipulability ellipsoids. During this first stage, great attention is set on the identification of the workspace of the machine. Once presented the notions of kinematics, dynamic models of the example robot are introduced. Even if the course is mainly focused on serial kinematic robots, some examples of parallel kinematic architectures are presented highlighting the pros and cons of the two families of robots.
- Coordinate transform and 3D kinematics: the concept of coordinate transformation is introduced adopting the pose matrix approach. Equipped with these notions the kinematic analysis is extended to three-dimensional robots. DH parameters are presented and differential kinematics is introduced. At this point, student are advised of the difficulty of solving the inverse kinematic problem for a generic serial manipulator.
- Dynamics: as for the kinematic analysis, dynamics is extended to three dimensional systems. Action, inertial and momentum matrices are introduced, and equations of motion for a generic serial manipulator are presented both using Newton equations and Lagrange formulation.
- Trajectory planning: various kinds of motion laws are presented analyzing in which conditions is more convenient to use a specific one with respect to another. Among the algorithms presented it is worth to mention: point to point motion laws (e.g., TVP, cycloidal), splines, linear polynomial interpolation with parabolic blends.
- Control algorithms: different control algorithms and architectures are presented, centralized and decentralized ones, in joint space and workspace. Examples of force control architectures are also provided.
- Mechanical components: dedicated mechanical components for robotics are presented, such as harmonic and cycloidal reducers, rotary ball spline, roller pinion systems, grippers and the like.

3. Rapid Prototyping in Electronics

In order to actuate the 3D printed robots, motors are required. Thanks to low-cost easy to assemble hardware, students can animate their creations without any significant prior knowledge on electronics or industrial robot programming. Thus, a box containing some electronic hardware is delivered to each group. A box features an Arduino UNO board (Figure 1a) as controller; a TinkerKit SensorShield

(Figure 1b) to facilitate wiring I/O; a joystick (Figure 1c), a potentiometer (Figure 1d) and a button (Figure 1e) as input devices; and four Servo motors (Figure 1f). The additional servo motor can be used to open and close a gripper in correspondence of the end-effector, according to the specific task assigned. The servos have a nominal operating range from 0 to 180 degrees, speed of 10 rad/s, stall torque of 1.8 Kgf · cm at 5 V.

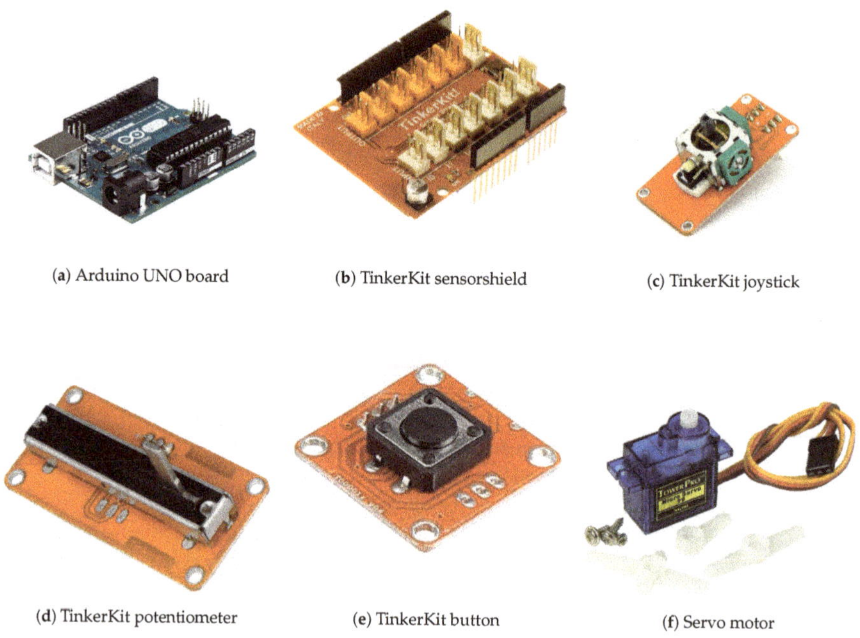

(**a**) Arduino UNO board	(**b**) TinkerKit sensorshield	(**c**) TinkerKit joystick
(**d**) TinkerKit potentiometer	(**e**) TinkerKit button	(**f**) Servo motor

Figure 1. Components provided to the students.

The members of the groups can start building some simple circuit and understand how electrical circuit works following the examples present in the Arduino IDE. Low voltage I/O allows them to make mistake without damaging the hardware or harm themselves. In this initial phase, they should also assess the capability of the hardware and material that are available to them, to determine the optimal region where their servos can and should be operated. This is accomplished by starting with labs on programming and control. Figure 2 shows some examples of the final electric circuits built by the students.

Programming and Control:

In order to meet such a requirement, students need to be introduced to programming in the Arduino environment. The first examples deal with how to read analog and digital inputs coming respectively from the joystick or the potentiometer and the pushbutton; on how to use the servo library to control the servomotors. Then, they have to combine these sketches to generate an appropriate command for the servos to move the robot to the desired pose from the read input. Once the three analog signals are acquired (two for the joystick and one for the potentiometer), the information has to be translated into a movement of the end-effector, thus an additional step is required, that is the solution of the inverse kinematic problem. Once the corresponding joint coordinates are found it is possible to send the command signal to the servos. Of course, both the signals read from the joystick and the potentiometer present some offsets so it is necessary to tune the program.

Once the basic knowledge has been assimilated, they are asked to establish a serial communication between Matlab and the Arduino board. This step shows the problematic behind communication among the several units of a control system of an industrial robot [16] (Figure 3).

Figure 2. Examples of the physical implementation of the electric system.

```
if (Serial.available() > 0) {
   if (flag_state==0){
      count=Serial.parseFloat();
      flag_state=1;
   }
   for ( int k = 0; k < count; k++) {
      buf[k] = Serial.parseFloat();
   }
```

```
for i=1:size(Q,2)
   if i==1
      fprintf(s, '%1.f',size(Q,1));
   end
   fprintf(s, flag_vect,Q(:,i));
end
```

Figure 3. Example of the serial communication code for both Arduino and Matlab.

4. Project Description and Assessment

During the first class, the students are informed of the robotic configuration and the task that they must realize and solve. With some differences from one year to the other, especially related to the robotic configuration and evaluated task, the main points that students need to carry out are:

- Study of the kinematic configuration assigned;
- Numerical analysis to determine working space and motor torque required;
- Manual control (joystick);
- Mandatory mission to be assessed for the challenge at the end of the course;
- Task of choice.

These points assure the final requirement that is to design, build and control a three-DOF serial robot for a specific task.

In Figures 4 and 5 show the kinematic scheme of two machines assigned and the built versions by students. Note that with simple kinematic architectures students are able to analyze the design process of the robot in all its aspects without encountered mathematical problems that can divert attention from the goal. Due to the limited time (48 h between labs and hands-on classes) and the number of students (ranging from 50 to 70 people), they are divided into groups of 4–5 each. This allows them to split the required work freely within the group, thus relieving students, also attending other courses, of overburden and assuring the completion of the project by the end of the course.

During the day of the competition, groups present their robots attaching a short report outlining the steps and assessments carried out during the design phase and the performances of every single machine is recorded. The ranking is obtained weighting objective parameters such as time and precision for the assigned task. Let us consider the third year case as a practical example. Students were asked to present a drawing SCARA able to trace a 6×6 square on a piece of paper; the drawing was then analyzed by means of image processing to extrapolate precise information on the performance.

Students are then required to take an oral exam to attest both their theoretical knowledge and their involvement in each phase of the design. Groups are free to also develop their own task. The majority opts for vision-based tasks such as object location and placing by means of image processing algorithms developed in Matlab.

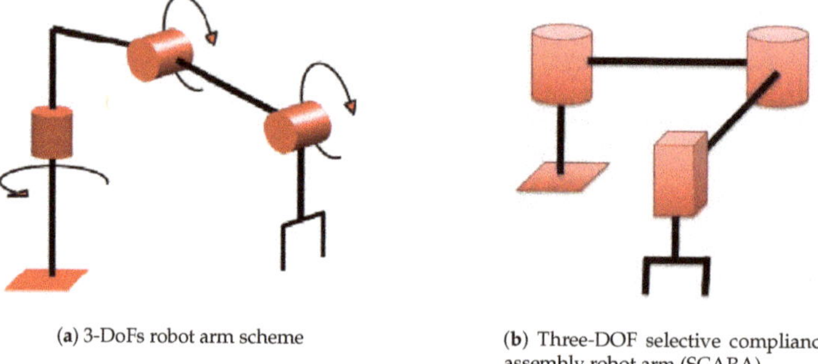

(**a**) 3-DoFs robot arm scheme

(**b**) Three-DOF selective compliance assembly robot arm (SCARA)

Figure 4. Three-degree of freedom (DOF) robotic configurations.

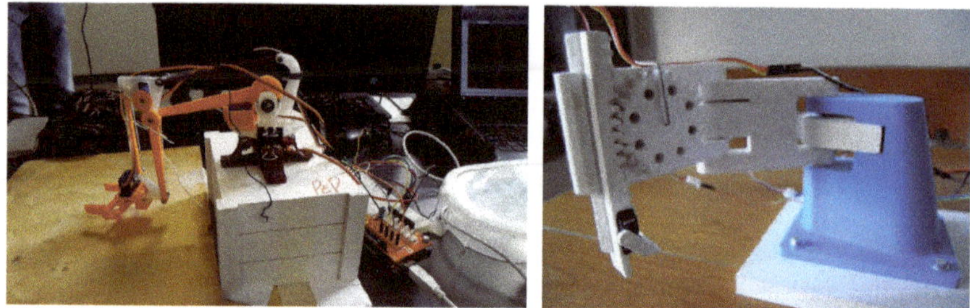

Figure 5. Three-DOF robotic configurations: examples.

5. Course Development and Outcomes

In this section the main steps of the project are discussed especially underlining the aspects related to AM. Some robots build in the last three years will also be shown as practical examples.

5.1. Mechanical Design

Students, knowing the characteristics of their hardware and the task that they have to do, can start dimensioning their robots. They are expected to design structural components from scratch; they are left free to choose the dimensions of the links and of all other components. The reason behind this choice is to help them strengthen their design capabilities; they have a to find a good compromise in the total dimensions of the robot. As a matter of fact, very short links will result in a reduced workspace. On the other hand, very long and bulky elements will result in the servo motors reaching their limits in terms of deliverable torque. Servo motors are quite delicate components, so proper solutions have to be found in order to avoid axial and radial overloading of the output shaft. 3D printing offers an interesting tool for students to test and verify their calculation, ingenuity and theoretical knowledge against real system. The mechanical design is done take into consideration:

- Task constraints;
- Dependency on the kinematic parameters;
- Servo minimum displacements and range;
- Power requirements.

Students can start dimensioning their machines using a simplified approach to solve the inverse dynamic problem [17]. Once written the kinematic equation (if S are the coordinates in the work space,

Q in the joint space and L the kinematic parameters, then $S = F(Q, L)$), it is possible to obtain the Jacobian matrix (defined as $J = \partial F / \partial Q$). If the virtual work principle is applied, this can be obtained:

$$F_q = -J_d^T (F_{si} + F_{se})$$

where,

$$F_{si} = M\ddot{S}_d \qquad \ddot{S}_d = J_d\ddot{Q} + \dot{J}_d\dot{Q}.$$

It has been indicated with F_{si} the vector containing the inertial forces, F_{se} the applied forces, with S_d the position vector in the work space of all the points where forces and torques are applied (augmented case), J_d is the Jacobian matrix obtained for the augmented case and M is a diagonal matrix containing the inertia terms. By assigning a simple trajectory (constant acceleration profile) to the end effector, it is possible to obtain an estimation of the torques. Students are then faced with another designing decision: how to couple the servos to the links. In Figure 6, some coupling cases can be found. Some groups designed their project directly connecting the motor to the joint axis other by means of a transmission: gears are used with a 1:1 ratio just to realize more compact solution and evenly distributing the weight along the link; speed reducers ($\tau < 1$) to increase torque at the base joint to actuate the moving part of the robot; belt drives and four-bar mechanism to drive the joints while fixing the motor on the ground, thus reducing the weight on the moving parts of the robot. All of the transmission components are 3D printed. Since the structure will be 3D printed, it is necessary to take into account that the adopted technology prevents to realize some components: even if it is possible to use support material, thin and embossed components are very hard to realize; if the printed components are too long they might bend due to thermal stresses related to different cooling; if support material is employed, in general, the surface finishing of the face in direct contact with the support material is quite poor, leading to an increase of the friction coefficient, thus more power required at the motor axis. Students are therefore encouraged not only to design for assembly but also for 3D printing.

5.2. 3D-Printing

The 3D CAD (Computer-aided design) models for the base, links and end effector (Figure 7) need then to be manufactured. Students are briefly introduced therefore to the world of FDM technology discussing the overall printing cycle from CAD to post-processing. The attention is focused on practical aspects to show how the printing parameters affect the final build: significance of the temperature at the nozzle and at the plate, as well as nozzle to layer height ratio, printing velocity, infill (throughout the build), when to use support material (supports and rafts) and why use skirts and brims. With such knowledge, they are able to understand the philosophy behind a slicing software and why input variables are threated as such (Figure 8a).

In Figure 8b it is possible to see some components imported in Slic3r to generate the Gcode that will be fed to the 3D printer. Students are taught how to operate the assigned printer in order to leave them the freedom to correct and adjust their initial design according to the obtained results or simply to correct some printing defects.

Once the robot has been printed and assembled (usually the 3D-printed components require also post-processing), groups experience the difference from design and real system. Due to building defects, excessive and incorrect material removal (during post-processing), total center point (TCP) positioning precision reduces. Thus groups are spontaneously encouraged to perform a simple calibration to obtain a least-square estimation of the real kinematic parameters in order to increase their winning chances.

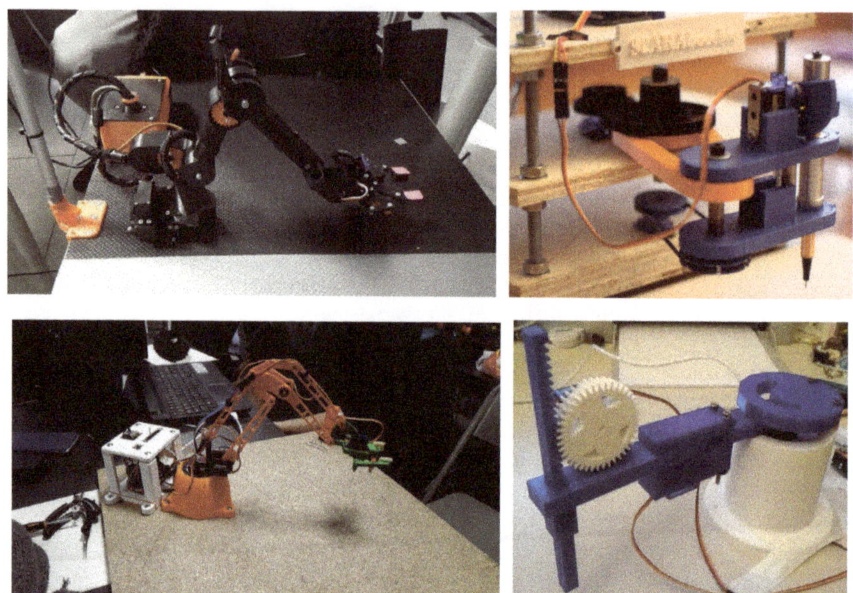

Figure 6. Servos-links coupling examples.

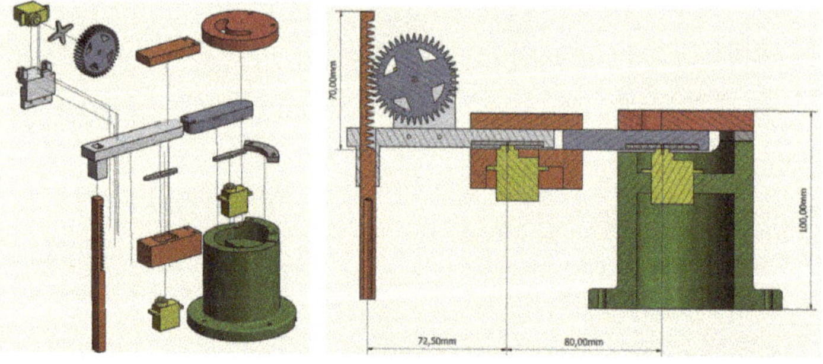

Figure 7. Selective compliance assembly robot arm (SCARA) solution example.

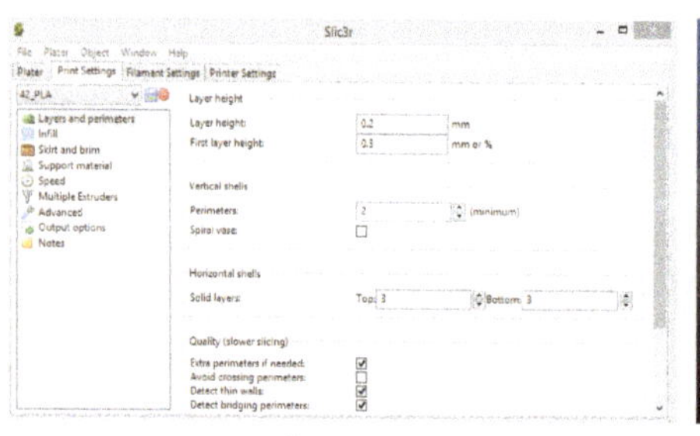

(**a**) Slic3r settings

(**b**) Robot components imported in Slic3r

Figure 8. Slicing.

5.3. Multibody Simulation:

The multibody simulation is aimed at assessing the goodness of the overall project. Indeed, once the structural components are designed, it is possible to have an estimate of their inertial properties, necessary to build the multibody model. Some students determined approximately the density of their pieces or by weighting them and dividing by the volume or by multiplying the density of the PLA by the infill percentage ($\rho = \rho_{PLA} \times \%_{infill}$). The software used to perform the multibody simulation is Simulink, in particular a package belonging to the Simscape library called Simmechanics. Given the trajectory planned it is possible to perform a simulation in order to assess that the motor torque required do not exceed the limits. If this is the case, it is necessary to modify the trajectory maybe reducing speeds and accelerations, or, in the worst case, to redesign some of the components to reduce inertial forces.

5.4. Trajectory Planning:

In order to carry out the task assigned, students must develop a feasible trajectory according to the project requirements and hardware capabilities. Usually a basic operation is also required to familiarize groups with industrial jobs such as pick and place with obstacle avoidance. The main task usually requires to combine and adjust different path planning algorithms presented during the course.

During the third academic year, groups were asked to: starting from the home position, reach a point P1 (the center of the square) with an overturned "*U*" shape trajectory and mark the spot on a piece of paper; with the same trajectory, position the drawing tool in a vertex and from there draw a 6 cm square adopting the linear polynomial interpolation with parabolic blends. Upon request, the robot should also be able to repeat the square five times as well as drawing 20 parallel lines spaced 3, 2 or 1 millimeter.

All the algorithms used to define the motion laws are written in Matlab; once established the serial communication, the servos are directly controlled from Matlab passing the joint coordinate to the Arduino board with a fixed interval of time. It has been decided to opt for this solution instead of introducing ROS (Robot Operating System), especially its message-passing framework, in order to comply to the background of the students (it would require knowledge on Linux, Python and C++, topics that are not dealt with in previous courses).

5.5. Competition and Project Evaluation:

The 2017–2018 academic year will be addressed as explanatory example.

To assure an objective analysis of the results, the final list was obtained equally weighting speed and precision. The speed was recorded by means of Matlab functions (tic and toc) between the first and last package send via serial communication and checked, during the competition, with a chronometer.

The precision rank was the result of the combination of 12 different parameter. The values were extrapolated from the drawings by means of a Matlab script developed for the occasion to perform the image processing; the picture saw the square and a reference system to easily infer pixel to centimeter conversion (Figure 9a). The algorithm converts the image into binary image (Figure 9b); once filtered and reconstructed, the square is described in the x-y plane (Figure 9c). At the end, the sides are detected an compared with the ideal ones (Figure 9c) to beget the results.

The top ranking group for speed received one point, the second two points and the last fourteen; the same for the precision. At the end the scores were summed and the group with the least points was declared the winner.

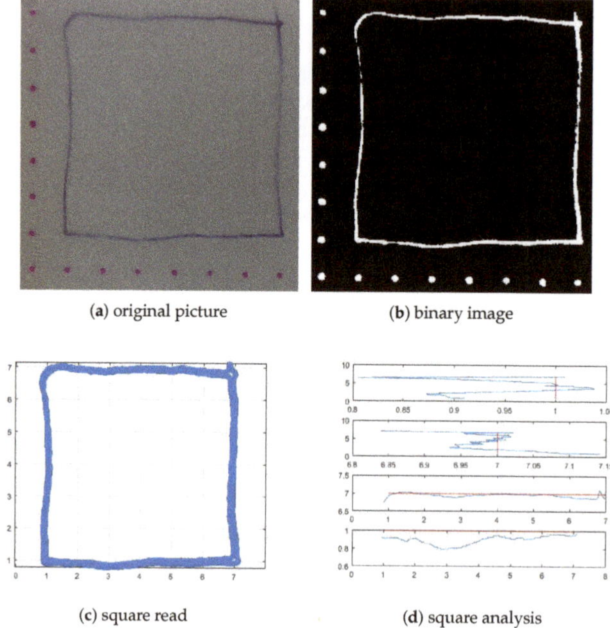

(a) original picture

(b) binary image

(c) square read

(d) square analysis

Figure 9. The 6 × 6 cm square reading example

6. Can AM Enhance Didactics?

At the end of the course students were asked to fill an anonymous survey in order to understand if this new technology was able to improve the course didactics. It was also conducted to have a feedback of their experience highlighting in which way the course helped them improving their skills and which elements should be improved for future years.

The survey is divided into three sections: the first part is aimed at verifying which percentage of the total work each student spent in each of the sub-areas; the second set of questions is focused on students self-evaluation, in particular on assessing which is their feeling about the impact on their skills after attending the course; the last part wants to identify which aspects of the course could be changed in order to provide a better experience especially on the involvement of AM in the course. The data reported in the next sections refer to the classes starting from the 2016/2017 academic year.

6.1. First Section

Students were asked to indicate which percentage of their work was spent in each area of the hands-on classes. Figure 10a shows the averaged percentage value spent in each area.The mechanical design required more time with respect to other areas, in particular in relation to Multibody Simulation. This result puts in evidence the lack of experience in designing mechanical components before attending the course. Figure 10b shows that the vast majority of the students thinks that the total effort of their group-mates is excellent. This result highlights the enthusiasm of students in attending the course.

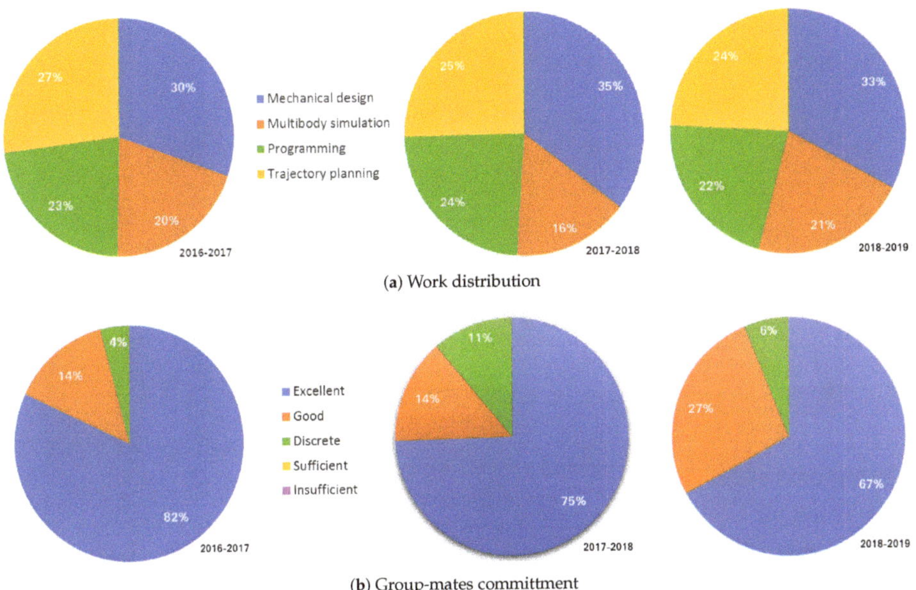

(**a**) Work distribution

(**b**) Group-mates committment

Figure 10. Survey results: first section.

6.2. Second Section

Students are asked to indicate which benefits they acquired after attending the course. Looking at Figure 11, it is possible to notice that less then half of the students thinks that the course helped improving their capabilities of designing a generic automatic system, while around the 70% thinks that the course contributed to improve their design skills related to a robotic system. Even if the 50% thinks that the course made them passionate about robotic subjects, only the 30% of them is fond of programming. This last result seems to be in line with the motivations that brought the authors to organize the course as described above as it highlights that some of them is more interested in the mechatronics design of robotic systems than programming.

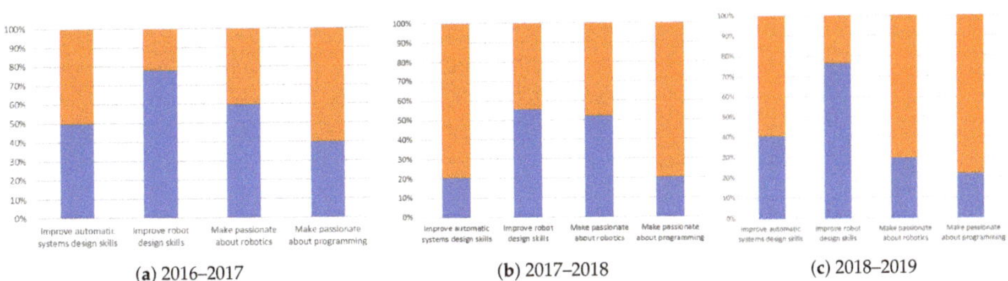

(**a**) 2016–2017

(**b**) 2017–2018

(**c**) 2018–2019

Figure 11. Survey results: second section.

6.3. Third Section

The last section of the survey wants to investigate how the experience provided during the course could be improved, but primarily how AM was perceived. The idea of realizing the robot components using a 3D-printer was welcomed with a great enthusiasm since, as can be noticed from Figure 12a, the 80% asserts that the 3D-printing technology is stimulating and funny. As could be expected not everyone was pleased with this activity, but only few students thought that was a boring activities.

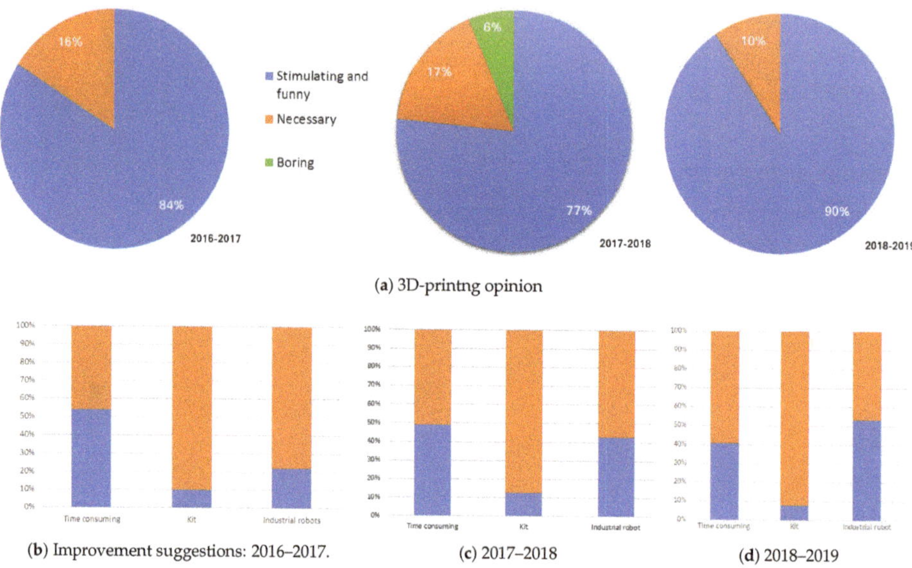

(a) 3D-printng opinion

(b) Improvement suggestions: 2016–2017.

(c) 2017–2018

(d) 2018–2019

Figure 12. Survey results: third section.

In the last questions students are asked to provide their opinion on the overall time spent for the project, on the possibility of using a ready-made kit to assemble and on the possibility of using industrial robots instead of a hand made robot. The answers to these questions seems to reflect the previous considerations; as a matter of fact, even if the 50% states that the project is very time-consuming compared to the CFU offered by the course, only the 10% thinks that the introduction of a ready-made kit would be beneficial, and the 40% believes that using industrial robot would be useful. These results underline that students rather spend more time engaging themselves in design activities then assembling robot kits. It's understandable their desire to also work on industrial robot to grasp, while attending university, the working reality they see themselves employed in.

7. Conclusions

This paper describes the authors' experience in adopting FDM 3D printing for an hands-on robotic course at the Politecnico di Milano. The experience, gained in the course of three years of work, shows that AM can change and improve the approach to the teaching of robotics, as shown by the positive feedback from students. The developed course is discussed highlighting the opportunities offered by AM and electronics prototyping. 3D printing was found to be a valid tool to engage students not only in the programming and control aspects, but also that of mechanical design; this is done by assigning the development of a 3 dof serial kinematics robot capable of performing a given task with the help of only some low-cost electrical hardware handed out to each group. Taking into account the positive results obtained with the use of 3D printing, for future classes the 4D printing technique [18] can be considered in order to enhance the designing experience of the students.

Author Contributions: Conceptualization, H.G. and K.C.; methodology, H.G. and K.C.; software, K.C.; validation, H.G. and K.C.; formal analysis, K.C.; writing–original draft preparation, K.C.; writing–review and editing, H.G. and K.C.; supervision, H.G.

Funding: This research received no external funding

Conflicts of Interest: The authors declare no conflict of interest.

Abbreviations

The following abbreviations are used in this manuscript:

AM	Additive manufacturing
FDM	Fused deposition modeling
SCARA	Selective compliance assembly robot arm
DH	Denavit–Hartenberg
TCP	Tool center point

References

1. Giberti, H.; Cinquemani, S. Motor-reducer sizing through a MATLAB-based graphical technique. *IEEE Trans. Educ.* **2012**, *55*, 552–558. [CrossRef]
2. Giberti, H.; Cinquemani, S.; Ambrosetti, S. 5R 2dof parallel kinematic manipulator—A multidisciplinary test case in mechatronics. *Mechatronics* **2013**, *23*, 949–959. [CrossRef]
3. Go, J.; Hart, A.J. A framework for teaching the fundamentals of additive manufacturing and enabling rapid innovation. *Addit. Manuf.* **2016**, *10*, 76–87. [CrossRef]
4. Ford, S.; Minshall, T. Invited review article: Where and how 3D printing is used in teaching and education. *Addit. Manuf.* **2019**, *25*, 131–150. [CrossRef]
5. Giberti, H.; Fiore, E. The "robot mechanics" course experience at Politecnico di Milano. *Mech. Mach. Sci.* **2018**, *49*, 583–590._61. [CrossRef]
6. Piepmeier, J.A.; Bishop, B.E.; Knowles, K.A. Modern robotics engineering instruction. *IEEE Robot. Autom. Mag.* **2003**, *10*, 33–37. [CrossRef]
7. Nagai, K. Learning while doing: Practical robotics education. *IEEE Robot. Autom. Mag.* **2001**, *8*, 39–43. [CrossRef]
8. Krotkov, E. Robotics laboratory exercises. *IEEE Trans. Educ.* **1996**, *39*, 94–97. [CrossRef]
9. Robinette, M.F.; Manseur, R. ROBOT-DRAW, an internet-based visualization tool for robotics education. *IEEE Trans. Educ.* **2001**, *44*, 29–34. [CrossRef]
10. Vandevelde, C.; Wyffels, F.; Ciocci, M.C.; Vanderborght, B.; Saldien, J. Design and evaluation of a DIY construction system for educational robot kits. *Int. J. Technol. Des. Educ.* **2016**, *26*, 521–540. [CrossRef]
11. Wong, N.; Cheng, H.H. CPSBot: A Low-Cost Reconfigurable and 3D-Printable Robotics Kit for Education and Research on Cyber-Physical Systems. In Proceedings of the 12th IEEE/ASME International Conference on Mechatronic and Embedded Systems and Applications (MESA) Auckland, New Zealand, 29–31 August 2016; pp. 1–6.
12. Ceccarelli, M. Robotic teachers' assistants. *IEEE Robot. Autom. Mag.* **2003**, *10*, 37–45. [CrossRef]
13. Armesto, L.; Fuentes-Durá, P.; Perry, D. Low-cost Printable Robots in Education. *J. Intell. Robot. Syst.* **2016**, *81*, 1. [CrossRef]
14. Murphy, R.R. 'Competing' for a robotics education. *IEEE Robot. Autom. Mag.* **2001**, *8*, 44–55. [CrossRef]
15. Jung, S. Experiences in Developing an Experimental Robotics Course Program for Undergraduate Education. *IEEE Trans. Educ.* **2013**, *56*, 129–136. [CrossRef]
16. Castelli, K.; Giberti, H. A preliminary 6 Dofs robot based setup for fused deposition modeling. *Mech. Mach. Sci.* **2019**, *68*, 249–257._27 [CrossRef]
17. Legnani, G.; Fassi, I. *Robotica Industriale*; EAN: Città Studi, Torino, Italy, 2019.
18. Khoo, Z.X.; Teoh, J.E.M.; Liu, Y.; Chua, C.K.; Yang, S.; An, J.; Leong, K.F.;Yeong, W.Y. 3D printing of smart materials: A review on recent progresses in 4D printing. *Virtual Phys. Prototyp.* **2015**, *10*, 103–122. [CrossRef]

Article

A Novel 3-URU Architecture with Actuators on the Base: Kinematics and Singularity Analysis

Raffaele Di Gregorio

Department of Engineering, University of Ferrara, 44122 Ferrara, Italy; raffaele.digregorio@unife.it;
Tel.: +39-0532974828

Received: 30 June 2020; Accepted: 28 July 2020; Published: 31 July 2020

Abstract: Translational parallel manipulators (TPMs) with DELTA-like architectures are the most known and affirmed ones, even though many other TPM architectures have been proposed and studied in the literature. In a recent patent application, this author has presented a TPM with three equal limbs of Universal-Revolute-Universal (URU) type, with only one actuated joint per limb, which has overall size and characteristics similar to DELTA robots. The presented translational 3-URU architecture is different from other 3-URUs, proposed in the literature, since it has the actuators on the frame (base) even though the actuated joints are not on the base, and it features a particular geometry. Choosing the geometry and the actuated joints highly affects 3-URU's behavior. Moreover, putting the actuators on the base allows a substantial reduction of the mobile masses, thus promising good dynamic performances, and makes the remaining part of the limb a simple chain constituted by only passive R-pairs. The paper addresses the kinematics and the singularity analysis of this novel TPM and proves the effectiveness of the new design choices. The results presented here form the technical basis for the above-mentioned patent application.

Keywords: lower-mobility manipulator; translational parallel manipulator; kinematics; mobility analysis; singularity analysis

---

1. Introduction

Parallel manipulators (PMs) feature two rigid bodies, one fixed (base) and the other mobile (platform), connected to one another through a number of kinematic chains (limbs). Translational PMs (TPMs) are 3-degrees-of-freedom (DOF) PMs whose platform can perform only spatial translations. TPMs are a particular family of lower-mobility PMs. DELTA-like architectures [1,2] are the most known and affirmed [3] TPM architectures, even though many (see [4–8] for instances and for further Refs.) other TPM architectures have been proposed and studied in the literature.

Lower-mobility PMs must be preferred to 6-degrees-of-freedom (6-DOF) PMs in all the industrial manipulation tasks that do not require a general spatial motion since they have simpler and faster architectures. Unfortunately, among the usual PM singularities [9,10] that fall inside the operational space, lower-mobility PMs may have particular singularities, named "constraint singularities" [11], where they can change their operating mode. Thus, the identification of architectures with wide regions of the operational space that are free from singularities, which is central for PMs, becomes somehow more complex and critical in the design of lover-mobility PMs.

TPMs with 3-URU[1] architectures [8] have been studied by many researchers. Such architectures feature three limbs of Universal-Revolute-Universal (URU) type that simultaneously connect the

---

[1]   Hereafter, P, R, S, and U stand for prismatic pair, revolute pair, spherical pair and universal joint. Additionally, the serial kinematic chains constituting the PM limbs are indicated by a string of such capital letters that give the sequence of joint types encountered by moving from the base to the platform along the considered limb.

platform to the base. The ones proposed in the literature [12–14] have the R-pairs, adjacent to the base, as actuated joints or, when presented as a spatial mechanism without actuated joints [11], have the axes of the three R-pairs adjacent to the base (to the platform) that are coplanar and with a common intersection. Changing the actuated joints and/or modifying the base (the platform) geometry affect the behavior of the machine in a substantial manner as regard both to the load redistribution among the links and to the functional aspects (e.g., useful workspace sizes and location).

The novelty of the translational 3-URU proposed in this paper, hereafter named LaMaViP 3-URU, stands in the fact that:

(i)   the actuators are on the base even though the actuated joints are not on the base,

(ii)   in each URU limb, the actuated R-pair is the one not adjacent to the base in the U-joint adjacent to the base, and

(iii) it has a particular base (platform) geometry where the axes of the three R-pairs adjacent to the base (to the platform) share a common intersection point but are not coplanar.

Putting the actuators on the base allows a significant reduction of the mobile masses, thus promising good dynamic performances, and makes the remaining part of the limb a simple chain constituted by only passive R-pairs.

This paper addresses the kinematics and the singularity analysis of the LaMaViP 3-URU and proves the effectiveness of the new design choices by demonstrating that the adopted design choices provide wide free-from-singularity regions of the operational space. The results presented here form the technical basis for a patent application of the author.

The organization of the paper is as follows. Section 2 presents the LaMaViP 3-URU together with some background concepts and the adopted notations. Section 3 analyzes the instantaneous kinematics and identifies the singularity loci. Then, Section 4 discusses the results and draws the conclusions.

2. The Novel Translational 3-URU

Out of constraint singularities [8,11], a 3-URU architecture is a TPM if it is manufactured and assembled so that in each URU limb the axes of the two ending R-pairs are parallel to one another and the axes of the three intermediate R-pairs are all parallel [8].

Figure 1 shows the reference geometry for a LaMaViP 3-URU. The geometry of Figure 1 has the axes of the three R-pairs adjacent to the base (to the platform) that are mutually orthogonal and share a common intersection point.

With reference to Figure 1,

- $Ox_by_bz_b$ and $Px_py_pz_p$ are two Cartesian references fixed to the base and to the platform, respectively; without losing generality, these two references have been chosen with the homologous coordinate axes that are parallel to one another[2];

- A_i (B_i) for i = 1,2,3 are the centers of the U joints adjacent to the base (to the platform);

- without losing generality [15], in the i-th limb, i=1,2,3, the points A_i and B_i are assumed to lie on the same plane perpendicular to the axes of the three intermediate R-pairs; such plane intersects at C_i the axis of the R-pair between the two U-joints;

- e_1, e_2, and e_3 are unit vectors of the coordinate axes x_b, y_b, and z_b (x_p, y_p, and z_p), respectively, and, at the same time, unit vectors of the three R-pair axes fixed to the base (to the platform);

- g_i, i = 1, 2, 3, is the unit vector parallel to the axes of the three intermediate R-pairs of the i-th limb.

---

[2]   It is worth noting that the parallelism of the coordinate axes is kept during the motion since the analyzed 3-URU is translational.

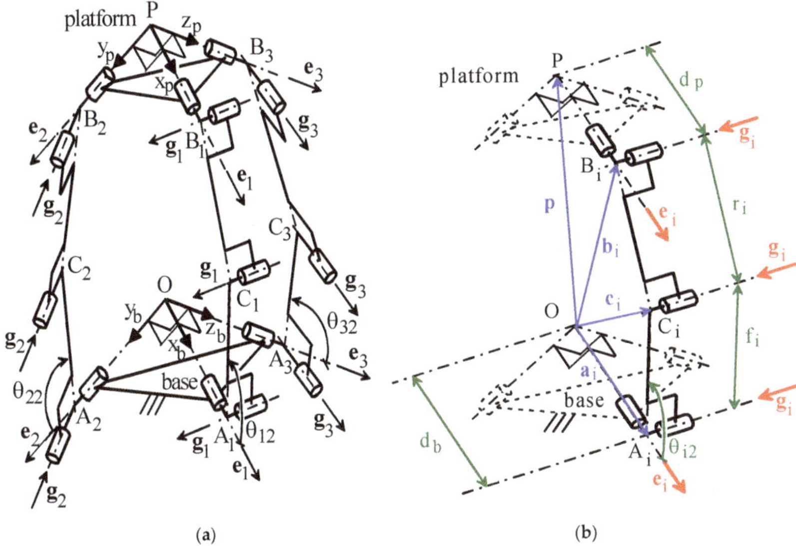

Figure 1. LaMaViP 3-URU with the R-pair axes that are fixed in the base (platform) mutually orthogonal: (**a**) overall scheme and notations, (**b**) detailed scheme of the i-th limb.

Moreover, the following definition/choices are introduced:

- $d_p = B_1P = B_2P = B_3P$;
- $d_b = A_1O = A_2O = A_3O$;
- in each URU limb, the five R-pairs are numbered with an index, j, that increases by moving from the base toward the platform; the actuated joint is the second R-pair;
- the angle θ_{ij}, for i = 1, 2, 3, and j = 1, ... , 5, is the joint variable of the j-th R-pair of the i-th limb; the actuated-joint variables are the angles θ_{i2}, i = 1, 2, 3 (see Figure 1); also, the phase reference of the angles θ_{i1}, i = 1, 2, 3, are given by the relationships (see Figure 1):
- $g_1 = \cos\theta_{11}\, e_2 + \sin\theta_{11}\, e_3,\ g_2 = -\cos\theta_{21}\, e_1 + \sin\theta_{21}\, e_3,\ g_3 = \cos\theta_{31}\, e_1 + \sin\theta_{31}\, e_2$;
- θ_{iM}, for i = 1, 2, 3, is the rotation angle of the motor shaft (see Figure 2) of the actuator of the i-th limb;
- $f_i = A_iC_i$, for i = 1, 2, 3; $r_i = B_iC_i$, for i = 1, 2, 3;
- $h_i = g_i \times e_i$, for i = 1, 2, 3;
- $u_i = (C_i - A_i)/f_i = \cos\theta_{i2}\, e_i + \sin\theta_{i2}\, h_i$, for i = 1, 2, 3;
- $v_i = (B_i - C_i)/r_i = \cos\theta_{i3}\, u_i + \sin\theta_{i3}\, (\cos\theta_{i2}\, h_i - \sin\theta_{i2}\, e_i)$ for i = 1, 2, 3, which also defines the phase reference of the angle θ_{i3};
- $p = (P - O) = xe_1 + ye_2 + ze_3$, where $(x, y, z)^T$ collects the coordinates of point P in $Ox_by_bz_b$; such coordinates also identify the platform pose during motion since the studied 3-URU is translational;
- $a_i = (A_i - O) = d_be_i$, for i = 1, 2, 3;
- $b_i = (B_i - O) = p + d_pe_i$, for i = 1, 2, 3;
- $c_i = (C_i - O) = a_i + f_iu_i$, for i = 1, 2, 3.

Figure 2 shows a possible mechanical transmission, based on a bevel gearbox that actuates the second R-pair of the i-th limb by keeping the actuator on the base. Figure 3 shows a constructive scheme of the i-th URU limb. Figures 2 and 3 highlight that the actual construction of the proposed type of URU limb is quite simple.

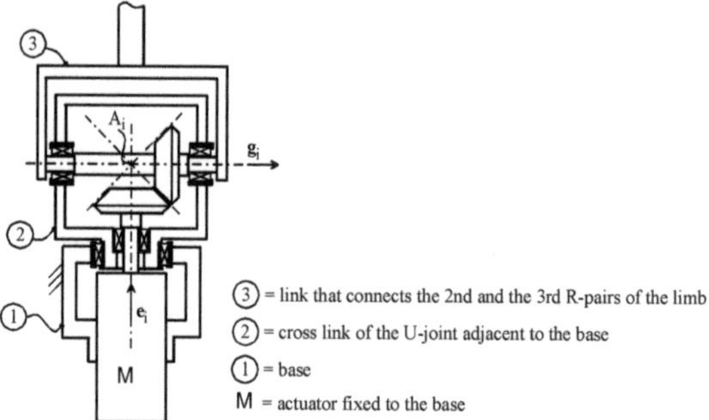

Figure 2. A possible mechanical transmission, based on a bevel gearbox, for actuating the 2nd R-pair of the i-th limb by keeping the actuator fixed to the base.

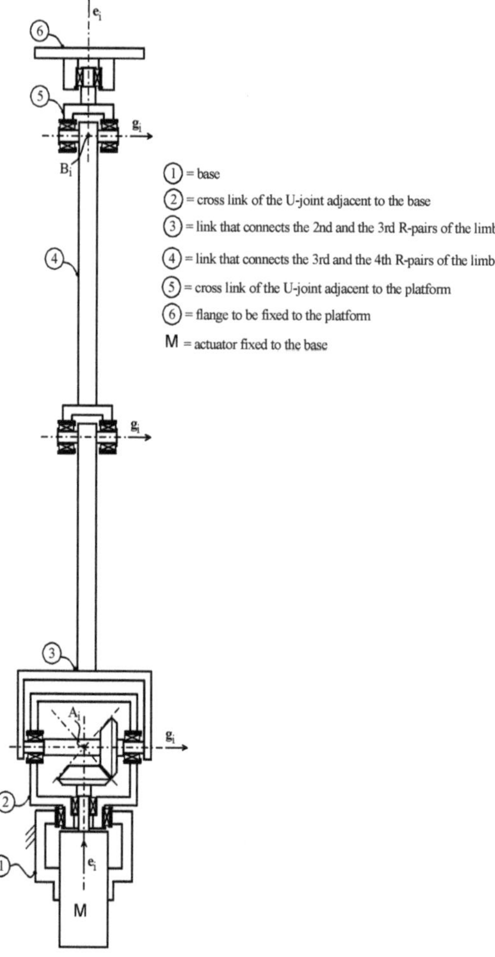

Figure 3. Constructive scheme of the i-th Universal-Revolute-Universal (URU) limb.

3. Mobility Analysis

In this section, the instantaneous input–output relationship of the LaMaViP 3-URU is deduced, and then, it is used for determining its singularity loci. The instantaneous input–output relationship is a linear mapping that relates the actuated-joint rates (instantaneous inputs), which, in the studied 3-URU, are $\dot{\theta}_{i2}$, $i = 1, 2, 3$, and the platform twist (instantaneous outputs), that is, $\hat{\$} = \left(\dot{\mathbf{p}}^T, \boldsymbol{\omega}^T \right)^T$ where $\boldsymbol{\omega}$ is the angular velocity of the platform.

In the case under study, the three URU limbs allow the platform angular velocity to be expressed in the following three different ways

$$\boldsymbol{\omega} = \left(\dot{\theta}_{i1} + \dot{\theta}_{i5} \right) \mathbf{e}_i + \left(\dot{\theta}_{i2} + \dot{\theta}_{i3} + \dot{\theta}_{i4} \right) \mathbf{g}_i \qquad\qquad i = 1, 2, 3 \tag{1}$$

whose dot product by $\mathbf{h}_i$ ($= \mathbf{g}_i \times \mathbf{e}_i$) yields

$$\mathbf{h}_i \cdot \boldsymbol{\omega} = 0 \qquad\qquad i = 1,2,3 \tag{2}$$

Moreover, $\dot{\mathbf{p}}$ enters into the following kinematic relationships[3]

$$\dot{\mathbf{p}} - \boldsymbol{\omega} \times (\mathbf{p} - \mathbf{b}_i) = \dot{\mathbf{b}}_i = \left(\dot{\theta}_{i1}\, \mathbf{e}_i + \dot{\theta}_{i2}\, \mathbf{g}_i \right) \times (\mathbf{b}_i - \mathbf{a}_i) + \dot{\theta}_{i3}\, \mathbf{g}_i \times (\mathbf{b}_i - \mathbf{c}_i) \qquad i = 1, 2, 3 \tag{3}$$

whose dot product by $(\mathbf{b}_i - \mathbf{c}_i) = r_i\, \mathbf{v}_i$ yields

$$\mathbf{v}_i \cdot \dot{\mathbf{p}} + [\mathbf{v}_i \times (\mathbf{p} - \mathbf{b}_i)] \cdot \boldsymbol{\omega} = \dot{\theta}_{i2}\, [\mathbf{g}_i \times (\mathbf{b}_i - \mathbf{a}_i)] \cdot \mathbf{v}_i \qquad i = 1, 2, 3 \tag{4}$$

Equations (2) and (4) provide the following instantaneous input–output relationship for the LaMaViP 3-URU

$$\begin{bmatrix} \mathbf{V} & \mathbf{T} \\ \mathbf{0}_{3\times3} & \mathbf{H} \end{bmatrix} \begin{pmatrix} \dot{\mathbf{p}} \\ \boldsymbol{\omega} \end{pmatrix} = \begin{bmatrix} \mathbf{G} \\ \mathbf{0}_{3\times3} \end{bmatrix} \begin{pmatrix} \dot{\theta}_{12} \\ \dot{\theta}_{22} \\ \dot{\theta}_{32} \end{pmatrix} \tag{5}$$

where $\mathbf{0}_{3 \times 3}$ is the 3×3 null matrix,

$$\mathbf{V} = \begin{bmatrix} \mathbf{v}_1^T \\ \mathbf{v}_2^T \\ \mathbf{v}_3^T \end{bmatrix}, \; \mathbf{T} = \begin{bmatrix} [\mathbf{v}_1 \times (\mathbf{p} - \mathbf{b}_1)]^T \\ [\mathbf{v}_2 \times (\mathbf{p} - \mathbf{b}_2)]^T \\ [\mathbf{v}_3 \times (\mathbf{p} - \mathbf{b}_3)]^T \end{bmatrix}, \; \mathbf{H} = \begin{bmatrix} \mathbf{h}_1^T \\ \mathbf{h}_2^T \\ \mathbf{h}_3^T \end{bmatrix} \tag{6a}$$

and

$$\mathbf{G} = \begin{bmatrix} [\mathbf{g}_1 \times (\mathbf{b}_1 - \mathbf{a}_1)] \cdot \mathbf{v}_1 & 0 & 0 \\ 0 & [\mathbf{g}_2 \times (\mathbf{b}_2 - \mathbf{a}_2)] \cdot \mathbf{v}_2 & 0 \\ 0 & 0 & [\mathbf{g}_3 \times (\mathbf{b}_3 - \mathbf{a}_3)] \cdot \mathbf{v}_3 \end{bmatrix} \tag{6b}$$

Since the actuators are not directly mounted on the actuated joints, Equation (5) has to be accompanied by additional equations coming from the kinematic analysis of the actuation device (Figure 2) in order to implement control algorithms. Such equations can be deduced as follows. With reference to Figures 2 and 3, the following formulas can be stated

$$^i\boldsymbol{\omega}_{21} = \dot{\theta}_{i1} \mathbf{e}_i, \; {}^i\boldsymbol{\omega}_{32} = \dot{\theta}_{i2} \mathbf{g}_i, \; {}^i\boldsymbol{\omega}_{M1} = \dot{\theta}_{iM} \mathbf{e}_i \qquad i = 1, 2, 3 \tag{7}$$

---

[3] In Equation (3), the first equality is obtained by rearranging the kinematic relationship $\dot{\mathbf{p}} = \dot{\mathbf{b}}_i + \boldsymbol{\omega} \times (\mathbf{p} - \mathbf{b}_i)$ whereas, the last equality is deduced by introducing the kinematic relationship $\dot{\mathbf{c}} = \left(\dot{\theta}_{i1} \mathbf{e}_i + \dot{\theta}_{i2} \mathbf{g}_i \right) \times (\mathbf{c}_i - \mathbf{a}_i)$ into the expression of the velocity of Bi when considered a point of the link CiBi (see Figure 1b), that is, $\dot{\mathbf{b}} = \dot{\mathbf{c}}_i + \left[\dot{\theta}_{i1} \mathbf{e}_i + \left(\dot{\theta}_{i2} + \dot{\theta}_{i3} \right) \mathbf{g}_i \right] \times (\mathbf{b}_i - \mathbf{c}_i)$.

where $^i \omega_{pq}$ denotes the angular velocity of link p with respect to link q in the i-th limb, and the index M denotes the motor shaft. In addition, the relative motion theorems [16] states that

$$^i \omega_{M2} = {}^i \omega_{M1} - {}^i \omega_{21} = \left(\dot{\theta}_{iM} - \dot{\theta}_{i1}\right) e_i \qquad i = 1, 2, 3 \tag{8}$$

Eventually, let k_i be the speed ratio of the bevel gearbox of the i-th limb, the following relationship must hold:

$$k_i = \frac{^i \omega_{32} \cdot g_i}{^i \omega_{M2} \cdot e_i} = \frac{\dot{\theta}_{i2}}{\dot{\theta}_{iM} - \dot{\theta}_{i1}} \qquad i = 1, 2, 3 \tag{9}$$

which yields

$$\dot{\theta}_{i2} = k_i (\dot{\theta}_{iM} - \dot{\theta}_{i1}) \qquad i = 1, 2, 3 \tag{10}$$

whose integration gives

$$\theta_{i2} = k_i [(\theta_{iM} - \theta_{i1}) - (\theta_{iM|0} - \theta_{i1|0})] \qquad i = 1, 2, 3 \tag{11}$$

where $\theta_{iM|0}$ and $\theta_{i1|0}$ are the values of θ_{iM} and θ_{i1}, respectively, when θ_{i2} is equal to zero.

Equation (10) relates the actuated-joint rates to the angular velocities of the motor shafts and involves the non-actuated joint rates $\dot{\theta}_{i1}$, for i = 1, 2, 3. The dot product of Equation (3) by g_i, after some algebraic manipulations, relates the joint rates $\dot{\theta}_{i1}$, for i = 1, 2, 3, to the platform twist as follows:

$$\dot{\theta}_{i1} = \frac{g_i \cdot \dot{p} + [g_i \times (p - b_i)] \cdot \omega}{h_i \cdot (b_i - a_i)} \qquad i = 1, 2, 3 \tag{12}$$

The introduction of $\dot{\theta}_{i1}$'s expressions given by Equation (12) into Equation (10) and, then, of the resulting expressions of $\dot{\theta}_{i2}$ into Equation (4) yields

$$\left(v_i + k_i \frac{[g_i \times (b_i - a_i)] \cdot v_i}{h_i \cdot (b_i - a_i)} g_i\right) \cdot \dot{p} + \left(v_i \times (p - b_i) + k_i \frac{[g_i \times (b_i - a_i)] \cdot v_i}{h_i \cdot (b_i - a_i)} g_i \times (p - b_i)\right) \cdot \omega = \dot{\theta}_{iM} k_i [g_i \times (b_i - a_i)] \cdot v_i \quad i = 1, 2, 3 \tag{13}$$

System (13) is the direct relationship between the angular velocities of the motor shafts, $\dot{\theta}_{iM}$, for i = 1, 2, 3, and the platform twist, that is, it is the instantaneous-kinematics model necessary to the control system of the machine which replaces the first three equations of system (5).

3.1. Singularity Analysis

The availability of the instantaneous input–output relationship allows the solution of two instantaneous-kinematics' problems [10]: the forward instantaneous-kinematics (FIK) problem and the inverse instantaneous-kinematics (IIK) problem. The FIK problem is the determination of the platform twist for assigned values of the actuated-joint rates; vice versa, the IIK problem is the determination of the actuated joint rates for an assigned value of the platform twist.

Singular configurations (singularities) are the PM configurations where one or the other or both of the two above-mentioned problems are indeterminate [9,10]. In particular [9], type-I singularities refer to the indetermination of the IIK problem, type-II singularities refer to the indetermination of the FIK problem, and type-III singularities refer to the indetermination of both the two problems. From a kinematic point of view, type-I singularities correspond to limitations of the instantaneous mobility of the platform and are located at the workspace boundary; they are present in all the manipulators and are sometimes called "serial singularities". Differently, type-II singularities are mainly inside the workspace and correspond either (a) to a local increase of platform's instantaneous DOFs[4] or (b), without any local variation of platform's instantaneous DOFs, to some platform DOFs that locally become non-controllable through the actuated joints (i.e., the physical constraints locally become no longer independent). They are present only in closed kinematic chains (i.e., in PMs) and are sometimes called "parallel singularities".

Type-II(b) singularities may occur in any PM; whereas, type-II(a) singularities may occur only in lower-mobility PMs, whose limb connectivity[5] is higher than the PM DOFs. Type-II(a) singularities are named "constraint singularities" [11] since the additional platform DOFs acquired at such singularities may make the platform change its type of motion (operating mode). In particular, in a TPM, such additional DOFs can only be instantaneous rotations which may make the platform exit from the pure-translation operating mode; that is why TPMs' constraint singularities are also named "rotation singularities" and TPMs' type-II(b) singularities are also named "translation singularities" [8].

3.1.1. Rotation (Constraint) Singularities of LaMaViP 3-URU

The platform translation is guaranteed if and only if the constraints applied to the platform by the three URU limbs make the platform angular velocity, $\boldsymbol{\omega}$, equal to zero. The last three equations of system (5) are able to impose $\boldsymbol{\omega} = 0$, if the determinant of the coefficient matrix, $\mathbf{H}$, is different from zero. Therefore, the constraint singularities are the configurations that satisfy the geometric condition[6]

$$\det(\mathbf{H}) = \mathbf{h}_1 \cdot (\mathbf{h}_2 \times \mathbf{h}_3) = 0 \tag{14}$$

Equation (14) is satisfied when the unit vectors $\mathbf{h}_i$, for i = 1, 2, 3, are coplanar. Since the i-th unit vector $\mathbf{h}_i$ is perpendicular to the plane passing through the coordinate axis of $Ox_b y_b z_b$ with the direction of $\mathbf{e}_i$ where the unit vector $\mathbf{g}_i$ lies on (that is, to the plane where the cross link of the i-th U-joint lies on (see Figure 1)) and the three so-identified planes always share point O as common intersection, such a geometric condition occurs when these three planes simultaneously intersect themselves in a common line passing through point O (see Figure 4).

From an analytic point of view, the notations introduced in Section 2 make it possible to write

$$\mathbf{g}_i = \frac{\mathbf{e}_i \times (\mathbf{b}_i - \mathbf{a}_i)}{\left|\mathbf{e}_i \times (\mathbf{b}_i - \mathbf{a}_i)\right|} = \frac{\mathbf{e}_i \times [\mathbf{p} + (d_p - d_b)\mathbf{e}_i]}{\left|\mathbf{e}_i \times [\mathbf{p} + (d_p - d_b)\mathbf{e}_i]\right|} = \frac{\mathbf{e}_i \times \mathbf{p}}{\left|\mathbf{e}_i \times \mathbf{p}\right|} \qquad i = 1,2,3 \tag{15a}$$

and

$$\mathbf{h}_i = \mathbf{g}_i \times \mathbf{e}_i = \frac{(\mathbf{e}_i \times \mathbf{p}) \times \mathbf{e}_i}{\left|\mathbf{e}_i \times \mathbf{p}\right|} = \frac{\mathbf{p} - (\mathbf{e}_i \cdot \mathbf{p})\mathbf{e}_i}{\left|\mathbf{e}_i \times \mathbf{p}\right|} \qquad i = 1,2,3 \tag{15b}$$

Then, the introduction of the analytic expression of $\mathbf{p}$ (i.e., $\mathbf{p} = x\mathbf{e}_1 + y\mathbf{e}_2 + z\mathbf{e}_3$) into Equation (15b) yields

$$\mathbf{h}_1 = \frac{y\mathbf{e}_2 + z\mathbf{e}_3}{\sqrt{y^2 + z^2}}; \; \mathbf{h}_2 = \frac{x\mathbf{e}_1 + z\mathbf{e}_3}{\sqrt{x^2 + z^2}}; \; \mathbf{h}_3 = \frac{x\mathbf{e}_1 + y\mathbf{e}_2}{\sqrt{x^2 + y^2}} \tag{16}$$

---

[4]   It is worth stressing that platform's instantaneous DOFs may be different from the mechanism instantaneous DOFs since they depend on how effective are the mechanism constraints on the platform instantaneous motion and that they cannot exceed the DOF number of a free rigid body.

[5]   According to [17], here, the term "limb connectivity" denotes the DOF number the platform would have if it were connected to the base only through that limb.

[6]   It is worth reminding that the determinant of a 3×3 matrix is the mixed product of its three rows (or column) vectors.

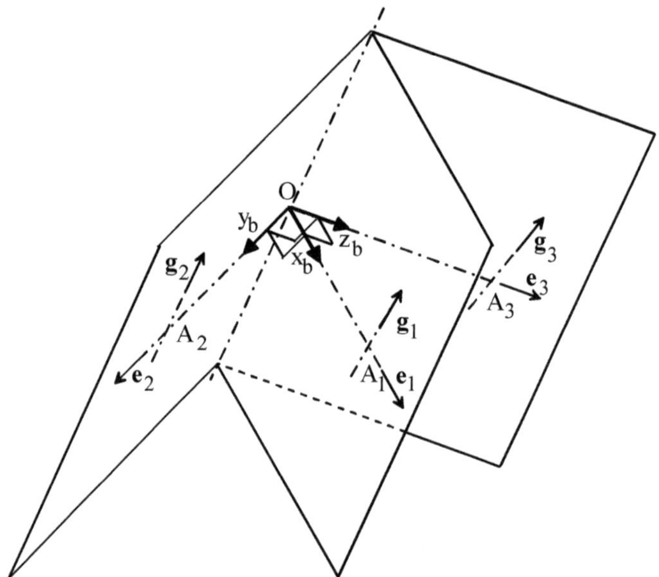

Figure 4. Geometric condition that identifies a rotation (constraint) singularity.

Eventually, the introduction of the explicit expressions given by Equation (16) into the singularity condition (14) provides the following analytic equation of the geometric locus of the rotation (constraint) singularities

$$\mathbf{h}_1 \cdot (\mathbf{h}_2 \times \mathbf{h}_3) = \frac{2xyz}{\sqrt{(x^2 + z^2)(x^2 + y^2)(y^2 + z^2)}} = 0 \tag{17}$$

The analysis of Equation (17) reveals that the rotation singularity locus is constituted by the 3 coordinate planes $x = 0$, $y = 0$, and $z = 0$ (Figure 5). Additionally, the analysis of Figure 1, of Formula (16) and Equation (2) reveals that

- when point P lies on the $y_b z_b$ coordinate plane (i.e., $x = 0$), the three unit vectors $\mathbf{h}_i$, for $i = 1, 2, 3$, (see Formulas (16)) are all parallel to the $y_b z_b$ coordinate plane; therefore, the component of ω along $\mathbf{e}_1$ is not locked (see Equations (2)) and the platform can perform rotations around axes parallel to the x_b axis;
- when point P lies on the $x_b z_b$ coordinate plane (i.e., $y = 0$), the three unit vectors $\mathbf{h}_i$, for $i = 1, 2, 3$, (see Formulas (16)) are all parallel to the $x_b z_b$ coordinate plane; therefore, the component of ω along $\mathbf{e}_2$ is not locked (see Equations (2)) and the platform can perform rotations around axes parallel to the y_b axis;
- when point P lies on the $x_b y_b$ coordinate plane (i.e., $z = 0$), the three unit vectors $\mathbf{h}_i$, for $i = 1, 2, 3$, (see Formulas (16)) are all parallel to the $x_b y_b$ coordinate plane; therefore, the component of ω along $\mathbf{e}_3$ is not locked (see Equations (2)) and the platform can perform rotations around axes parallel to the z_b axis.

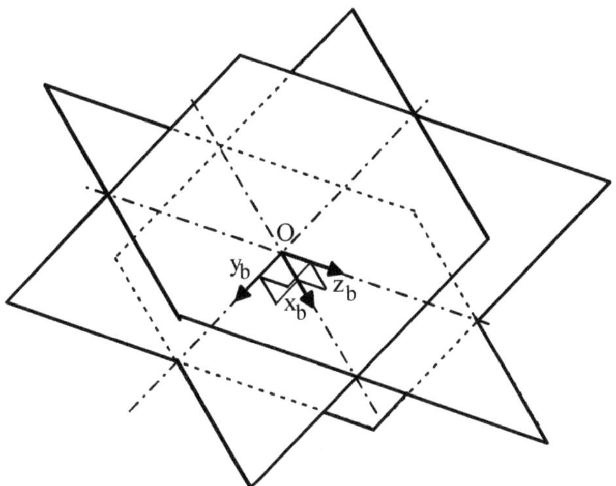

Figure 5. The rotation (constraint) singularity locus.

As a consequence, when P lies on a coordinate axis the platform locally acquires 2 rotational DOFs; whereas, when P coincides with O (i.e., x = y = z = 0) the platform locally acquires 3 rotational DOFs, even though the expression at the left-hand side of Equation (17) becomes indeterminate in all these cases.

In short, the rotation-singularity locus is constituted by three mutually orthogonal planes (i.e., the three coordinate planes of $Ox_by_bz_b$). Such a locus leaves eight wide simply-connected convex regions (i.e., the eight octants of $Ox_by_bz_b$) of the operational space, where the platform is constrained to translate. Inside any of these regions, the useful workspace of the studied 3-URU can be safely located. Moreover, since $\omega = 0$ in them, the instantaneous input–output relationship (i.e., system (5)) simplifies itself as follows

$$\mathbf{V}\dot{\mathbf{p}} = \mathbf{G}\,\dot{\theta}_2 \tag{18}$$

where $\dot{\theta}_2 = \left(\dot{\theta}_{12}, \dot{\theta}_{22}, \dot{\theta}_{32}\right)^T$; whereas, the instantaneous-kinematics model necessary to the machine control (i.e., system (14)) simplifies itself as follows

$$\left(\mathbf{v}_i + k_i \frac{[\mathbf{g}_i \times (\mathbf{b}_i - \mathbf{a}_i)] \cdot \mathbf{v}_i}{\mathbf{h}_i \cdot (\mathbf{b}_i - \mathbf{a}_i)} \mathbf{g}_i\right) \cdot \dot{\mathbf{p}} = \dot{\theta}_{iM}\, k_i [\mathbf{g}_i \times (\mathbf{b}_i - \mathbf{a}_i)] \cdot \mathbf{v}_i \qquad i = 1, 2, 3 \tag{19}$$

3.1.2. Translation (Type-II(b)) Singularities of LaMaViP 3-URU

Out of constraint singularities, system (18) is the instantaneous input–output relationship to consider. With reference to system (18), the FIK is the determination of $\dot{\mathbf{p}}$ for an assigned $\dot{\theta}_2$. This problem has a unique solution if and only if the determinant of the coefficient matrix, $\mathbf{V}$, is different from zero. Therefore, the translation singularities are the configurations that satisfy the geometric condition

$$\det(\mathbf{V}) = \mathbf{v}_1 \cdot (\mathbf{v}_2 \times \mathbf{v}_3) = 0 \tag{20}$$

Equation (20) is satisfied when the unit vectors $\mathbf{v}_i$, for i = 1, 2, 3, are coplanar. This geometric condition occurs when the three segments B_iC_i, i = 1, 2, 3, (see Figure 1) are all parallel to a unique plane (see Figure 6). From an analytic point of view, the adopted notations (see Section 2 and Figure 1) bring to light the following relationships

$$(\mathbf{b}_i - \mathbf{c}_i) = r_i\,\mathbf{v}_i = \mathbf{p} + (d_p - d_b)\,\mathbf{e}_i - f_i\,(\cos\theta_{i2}\,\mathbf{e}_i + \sin\theta_{i2}\,\mathbf{h}_i) \qquad i = 1, 2, 3 \tag{21}$$

which, after the introduction of the analytic expressions of $\mathbf{p}$ (i.e., $\mathbf{p} = x\mathbf{e}_1 + y\mathbf{e}_2 + z\mathbf{e}_3$) and of $\mathbf{h}_i$ (i.e., Equations (16)), become

$$\mathbf{b}_1 - \mathbf{c}_1 = [x + (d_p - d_b) - f_1 \cos\theta_{12}]\mathbf{e}_1 + [1 - f_1\, m_1\, \sin\theta_{12}]\, y\mathbf{e}_2 + [1 - f_1\, m_1\, \sin\theta_{12}]\, z\mathbf{e}_3 \tag{22a}$$

$$\mathbf{b}_2 - \mathbf{c}_2 = [1 - f_2\, m_2\, \sin\theta_{22}]\, x\mathbf{e}_1 + [y + (d_p - d_b) - f_2 \cos\theta_{22}]\mathbf{e}_2 + [1 - f_2\, m_2\, \sin\theta_{22}]\, z\mathbf{e}_3 \tag{22b}$$

$$\mathbf{b}_3 - \mathbf{c}_3 = [1 - f_3\, m_3\, \sin\theta_{32}]\, x\mathbf{e}_1 + [1 - f_3\, m_3\, \sin\theta_{32}]\, y\mathbf{e}_2 + [z + (d_p - d_b) - f_3 \cos\theta_{32}]\mathbf{e}_3 \tag{22c}$$

with

$$m_1 = \frac{1}{\sqrt{y^2 + z^2}},\, m_2 = \frac{1}{\sqrt{x^2 + z^2}},\, m_3 = \frac{1}{\sqrt{x^2 + y^2}}, \tag{23}$$

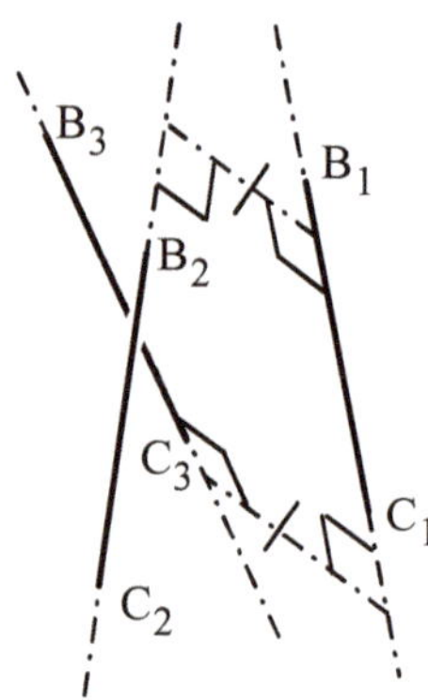

Figure 6. Geometric condition that identifies a translation singularity.

Eventually, the product of Equation (20) by the non-null constant $r_1\ r_2\ r_3$ yields the equivalent equation

$$(\mathbf{b}_1 - \mathbf{c}_1) \cdot [(\mathbf{b}_2 - \mathbf{c}_2) \times (\mathbf{b}_3 - \mathbf{c}_3)] = 0 \tag{24}$$

which, after the introduction of Formulas (22a), (22b) and (22c), becomes the following analytic expression of the translation-singularity locus

$$xyz\, [1 - n_2 n_3 - n_1 n_2 - n_1 n_3 + 2\, n_1 n_2 n_3] + xy\, q_3(1 - n_1 n_2) + xz\, q_2(1 - n_1 n_3) + yz\, q_1(1 - n_2 n_3) +$$
$$x\, q_2 q_3 + y\, q_1 q_3 + z\, q_1 q_2 + q_1 q_2 q_3 = 0$$

$$\tag{25}$$

where

$$n_1 = [1 - f_1\, m_1\, \sin\theta_{12}];\ n_2 = [1 - f_2\, m_2\, \sin\theta_{22}];\ n_3 = [1 - f_3\, m_3\, \sin\theta_{32}] \tag{26a}$$

$$q_1 = (d_p - d_b) - f_1 \cos\theta_{12};\ q_2 = (d_p - d_b) - f_2 \cos\theta_{22};\ q_3 = (d_p - d_b) - f_3 \cos\theta_{32} \tag{26b}$$

The actuated-joint variables, θ_{12}, θ_{22}, and θ_{32}, can be eliminated from Equation (25) by using the solution formulas of the inverse position analysis [18] reported in Appendix A. In doing so, Equation (25) becomes an equation that contains only the geometric constants of the machine and the platform pose coordinates, x, y, and z. Such equation, which is the analytic expression of a surface (the translation-singularity surface) in $Ox_b y_b z_b$, can be exploited, during design, to determine the optimal values of the geometric constants of the machine that move the translation singularities into regions of the operational space which are far from the useful workspace.

3.1.3. Serial (Type-I) Singularities of LaMaViP 3-URU

The solution of the IIK problem involves only the first three equations of system (5). The analysis of these three equations reveals that they can be separately solved with respect to θ_{i2}, i = 1, 2, 3, since matrix **G** is diagonal, and that the solution is indeterminate when at least one of the following geometric condition is satisfied (see Figure 1):

$$\mathbf{g}_i \cdot [(\mathbf{b}_i - \mathbf{a}_i) \times (\mathbf{b}_i - \mathbf{c}_i)] = \mathbf{g}_i \cdot [(\mathbf{c}_i - \mathbf{a}_i) \times (\mathbf{b}_i - \mathbf{c}_i)] = f_i\, r_i\, \sin\theta_{i3} = 0 \qquad i = 1,2,3 \qquad (27)$$

The i-th Equation (27) is satisfied when the i-th limb is fully extended ($\theta_{i3} = 0$) or folded ($\theta_{i3} = \pi$). These two geometric conditions identify two concentric spherical surfaces with point A_i as center, which point B_i must lie on. From an analytic point of view, since $\mathbf{b}_i = \mathbf{p} + d_p \mathbf{e}_i$ and $\mathbf{a}_i = d_b \mathbf{e}_i$, the equations of these two spherical surfaces in $Ox_b y_b z_b$ can be written as follows (here, the square of a vector denotes the dot product of the vector by itself)

$$(\mathbf{b}_i - \mathbf{a}_i)^2 = [\mathbf{p} + (d_p - d_b)\mathbf{e}_i]^2 = \mathbf{p}^2 + (d_p - d_b)^2 + 2\,(d_p - d_b)\,\mathbf{p} \cdot \mathbf{e}_i = (f_i + r_i)^2 \qquad i = 1,2,3 \quad (28a)$$

$$(\mathbf{b}_i - \mathbf{a}_i)^2 = [\mathbf{p} + (d_p - d_b)\mathbf{e}_i]^2 = \mathbf{p}^2 + (d_p - d_b)^2 + 2\,(d_p - d_b)\,\mathbf{p} \cdot \mathbf{e}_i = (f_i - r_i)^2 \qquad i = 1,2,3 \quad (28b)$$

Equation (28) are also the equations of the reachable-workspace boundaries. Therefore, the reachable workspace of the LaMaViP 3-URU can be analytically defined by the following system of inequalities

$$(f_1 - r_1)^2 \le x^2 + y^2 + z^2 + (d_p - d_b)^2 + 2\,(d_p - d_b)\,x \le (f_1 + r_1)^2 \qquad (29a)$$

$$(f_2 - r_2)^2 \le x^2 + y^2 + z^2 + (d_p - d_b)^2 + 2\,(d_p - d_b)\,y \le (f_2 + r_2)^2 \qquad (29b)$$

$$(f_3 - r_3)^2 \le x^2 + y^2 + z^2 + (d_p - d_b)^2 + 2\,(d_p - d_b)\,z \le (f_3 + r_3)^2 \qquad (29c)$$

In the case $d_b = d_p$ and $f_i = r_i = R$ for i = 1, 2, 3, inequalities (29) give a sphere with center O and radius 2R as reachable workspace (see Figure 7).

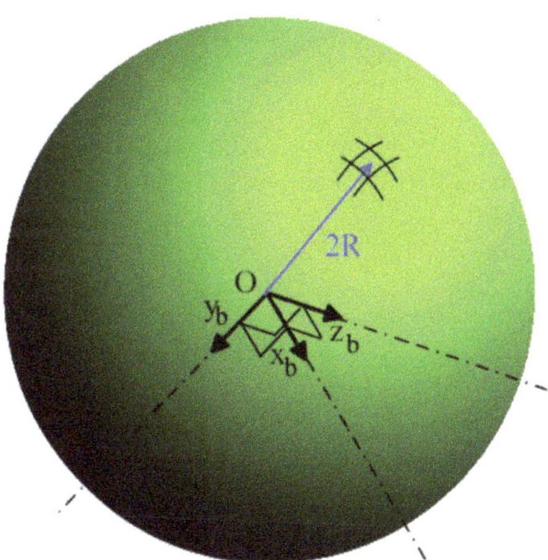

Figure 7. Reachable workspace in the case $d_b = d_p$ and $f_i = r_i = R$ for i = 1,2, 3.

3.2. Singularity Analysis of the Actuation Device

Since the actuators are not directly mounted on the actuated joint in the LaMaViP 3-URU, the motion transmission must be analyzed to check whether there are configurations (hereafter called "actuation singularities") in which the relationship (i.e., Equations (10)) between the actuated-joint rates, $\dot{\theta}_{i2}$, $i = 1, 2, 3$, and the angular velocities of the motor shafts, $\dot{\theta}_{iM}$, $i = 1, 2, 3$, is indeterminate. In this subsection, such relationship is deduced and analyzed.

The introduction of $\boldsymbol{\omega} = \mathbf{0}$ and of $\dot{\mathbf{p}} = \mathbf{V}^{-1}\mathbf{G}\,\dot{\boldsymbol{\theta}}_2$ (see Equation (18)) into Equation (12) yields

$$\dot{\boldsymbol{\theta}}_1 = \mathbf{N}\,\mathbf{M}\mathbf{V}^{-1}\mathbf{G}\dot{\boldsymbol{\theta}}_2 \tag{30}$$

with

$$\mathbf{M} = \begin{bmatrix} \mathbf{g}_1^T \\ \mathbf{g}_2^T \\ \mathbf{g}_3^T \end{bmatrix},\ \mathbf{N} = \begin{bmatrix} [\mathbf{h}_1 \cdot (\mathbf{b}_1 - \mathbf{a}_1)]^{-1} & 0 & 0 \\ 0 & [\mathbf{h}_2 \cdot (\mathbf{b}_2 - \mathbf{a}_2)]^{-1} & 0 \\ 0 & 0 & [\mathbf{h}_3 \cdot (\mathbf{b}_3 - \mathbf{a}_3)]^{-1} \end{bmatrix} \tag{31}$$

Then, the introduction of Equation (30) into Equation (10), after some rearrangements, gives the sought-after relationship between the actuated-joint rates, and the angular velocities of the motor shafts, that is,

$$\mathbf{S}\,\dot{\boldsymbol{\theta}}_2 = \mathbf{K}\,\dot{\boldsymbol{\theta}}_M \tag{32}$$

with $\dot{\boldsymbol{\theta}}_M = (\dot{\theta}_{1M}, \dot{\theta}_{2M}, \dot{\theta}_{3M})^T$, $\mathbf{S} = \mathbf{I}_{3\times3} + \mathbf{K}\mathbf{N}\mathbf{M}\mathbf{V}^{-1}\mathbf{G}$ where, $\mathbf{I}_{3\times3}$ is the 3×3 identity matrix, and

$$\mathbf{K} = \begin{bmatrix} k_1 & 0 & 0 \\ 0 & k_2 & 0 \\ 0 & 0 & k_3 \end{bmatrix};\ \mathbf{V}^{-1} = \frac{1}{\mathbf{v}_1 \cdot (\mathbf{v}_2 \times \mathbf{v}_3)}[\mathbf{v}_2 \times \mathbf{v}_3, \mathbf{v}_3 \times \mathbf{v}_1, \mathbf{v}_1 \times \mathbf{v}_2] \tag{33}$$

The expansion of the above expression of matrix $\mathbf{S} = [s_{ij}]$ gives the following explicit expression of its ij-th entry, s_{ij} for $i,j = 1, 2, 3$,

$$s_{ij} = \delta_{ij} + \frac{k_i[\mathbf{g}_i \cdot (\mathbf{v}_{(i+1)\bmod 3} \times \mathbf{v}_{(i+2)\bmod 3})][\mathbf{g}_j \times (\mathbf{b}_j - \mathbf{a}_j)] \cdot \mathbf{v}_j}{\mathbf{v}_1 \cdot (\mathbf{v}_2 \times \mathbf{v}_3)[\mathbf{h}_i \cdot (\mathbf{b}_i - \mathbf{a}_i)]} \qquad i,j = 1,2,3 \tag{34}$$

where δ_{ij} denotes the Kronecker delta and the subscript "(n+m) mod 3" denotes the sum with modulus 3 of the two integers n and m as defined in modular arithmetic [19].

The analysis of matrix $\mathbf{S}$ immediately reveals that, when matrix $\mathbf{V}$ is not invertible (i.e., when Equation (20) is satisfied), relationship (32) is indeterminate. Such a condition does not provide further reductions of the regions where the useful workspace can be located since it coincides with the translation-singularity locus (i.e., with Equation (20)) analyzed in Section 3.1.2. Over this condition, Equation (32) fails to give unique values of the actuated-joint rates, $\dot{\theta}_{i2}$, $i = 1, 2, 3$, for assigned values of the angular velocities of the motor shafts, $\dot{\theta}_{iM}$, $i = 1, 2, 3$, when the determinant of matrix $\mathbf{S}$ is equal to zero, that is, when the following geometric condition is satisfied

$$\det(\mathbf{S}) = \mathbf{s}_1 \cdot (\mathbf{s}_2 \times \mathbf{s}_3) = 0 \tag{35}$$

where $\mathbf{s}_i$, for $i = 1, 2, 3$, are the column vectors of matrix $\mathbf{S}$. Therefore, an actuation singularity occurs when the three vectors $\mathbf{s}_i$, for $i = 1, 2, 3$, are coplanar. From an analytic point of view, Equation (35) is the equation of a surface in $Ox_by_bz_b$, which corresponds to the actuation-singularity locus. Such equation can be put in the form $f(x, y, z) = 0$ by exploiting the above-reported expressions of the terms appearing in Equation (34) and can be used to size the geometric constants and the speed ratios k_i, $i = 1, 2, 3$, so that the actuation singularity locus is far from the useful workspace.

From the point of view of the platform control, the presence of the actuation singularities justifies the difference between System (18) and System (19). In particular, unlike System (18), System (19) yields the following geometric expression of the translation-singularity locus

$$\left(\mathbf{v}_1 + k_1\frac{[\mathbf{g}_1 \times (\mathbf{b}_1 - \mathbf{a}_1)] \cdot \mathbf{v}_1}{\mathbf{h}_1 \cdot (\mathbf{b}_1 - \mathbf{a}_1)}\mathbf{g}_1\right) \cdot \left[\left(\mathbf{v}_2 + k_2\frac{[\mathbf{g}_2 \times (\mathbf{b}_2 - \mathbf{a}_2)] \cdot \mathbf{v}_2}{\mathbf{h}_2 \cdot (\mathbf{b}_2 - \mathbf{a}_2)}\mathbf{g}_2\right) \times \left(\mathbf{v}_3 + k_3\frac{[\mathbf{g}_3 \times (\mathbf{b}_3 - \mathbf{a}_3)] \cdot \mathbf{v}_3}{\mathbf{h}_3 \cdot (\mathbf{b}_3 - \mathbf{a}_3)}\mathbf{g}_3\right)\right] = 0 \tag{36}$$

which imposes the zeroing of the mixed product of the three vectors that dot multiply $\dot{p}$ in the three equations of System (19). The i-th vector, for i = 1, 2, 3, of this vector triplet is associated to the i-th limb and lies on a plane spanned by the two unit vectors v_i and g_i. Differently from Equation (20), which is satisfied by the coplanarity of the three unit vectors v_i, i = 1, 2, 3, Equation (36) is satisfied by the coplanarity of these other three vectors that are not aligned with the unit vectors v_i, i = 1, 2, 3, any longer. Equation (36) can be put in the form f(x, y, z) = 0 by exploiting the above-reported expressions of the terms appearing in it and can be used as an alternative to Equations (20) and (35) to size the geometric constants and the speed ratios k_i, i = 1, 2, 3, so that both the translation and the actuation singularity loci are far from the useful workspace.

4. Conclusions

The kinematics and the singularity analysis of a novel translational architecture of 3-URU type, named LaMaViP 3-URU, have been addressed. With respect to other translational 3-URU, the novelty of the LaMaViP 3-URU stands on the fact that i) it has the actuators on the base even though the actuated joints are not on the base, ii) in each URU limb, the actuated R-pair is the one not adjacent to the base in the U-joint adjacent to the base, and (iii) it has a particular base (platform) geometry where the axes of the three R-pairs adjacent to the base (to the platform) share a common intersection point, but are not coplanar. These features are the premises to have a translational 3-URU with overall sizes and performances similar to the ones of the DELTA robot.

Here, the instantaneous input–output relationship of the LaMaViP 3-URU has been deduced together with the instantaneous relationship that directly relates the platform twist to the angular velocities of the 3 motor shafts. Then, the singularity analysis has been addressed. Both the geometric and the analytic conditions that identify all the singularities of the LaMaViP 3-URU have been determined.

The results of this study prove that there are eight wide simply-connected convex regions of the operational space where the platform is constrained to translate and the useful workspace can be safely located, which makes the proposed architecture a viable alternative to other translational PMs. Additionally, the reachable-workspace boundaries equations, the translation, and the actuation singularity loci equations as a function of the geometric constants and of the transmission constants have been provided. Such equations are all the necessary tools for the dimensional synthesis of the LaMaViP 3-URU. These results form the technical basis of a patent application of the author.

Future works on the LaMaViP 3-URU will address the dimensional synthesis of the LaMaViP 3-URU together with the kinematic and dynamic performance analyses.

5. Patents

The results of this work form the basis for the following Italy patent application:

Di Gregorio, R. Meccanismo Parallelo Traslazionale. Patent No. 102020000006100, 23 March 2020.

Funding: This work has been developed at the Laboratory of Mechatronics and Virtual Prototyping (LaMaViP) of Ferrara Technopole, supported by FAR2019 UNIFE funds.

Conflicts of Interest: The author declares that he has no conflict of interest and that the funders had no role in the design of the study, in the collection, analyses, or interpretation of data, in the writing of the manuscript, or in the decision to publish the results.

Appendix A. Inverse Position Analysis

The inverse position analysis (IPA) of the LaMaViP 3-URU consists of the determination of the actuated-joint variables (i.e., the angle θ_{12}, θ_{22}, and θ_{32}) for assigned values of the platform pose parameters (i.e., point P's coordinates x, y, and z). This problem has been solved in [18]. In this appendix the solution illustrated in [18] is briefly summarized.

By using the introduced notations, the following relationships can be deduced (see [18] for details):

$$\alpha_i^2 + \beta_i^2 + f_i^2 - r_i^2 - 2 f_i (\alpha_i \cos \theta_{i2} + \beta_i \sin \theta_{i2}) = 0 \qquad i = 1,2,3 \qquad (A1)$$

where $\alpha_1 = x + d_p - d_b$, $\alpha_2 = y + d_p - d_b$, $\alpha_3 = z + d_p - d_b$, $\beta_1 = \sqrt{y^2 + z^2}$, $\beta_2 = \sqrt{x^2 + z^2}$, $\beta_3 = \sqrt{x^2 + y^2}$.

The introduction of the trigonometric identities $\cos\theta_{i2} = (1 - t_i^2)/(1 + t_i^2)$ and $\sin\theta_{i2} = 2t_i/(1 + t_i^2)$, where $t_i = \tan(\theta_{i2}/2)$, into Equations (A1) transforms them into quadratic equations whose solutions are

$$t_i = \frac{2\,f_i\beta_i \mp \sqrt{4 f_i^2(\alpha_i^2 + \beta_i^2) - \left(\alpha_i^2 + \beta_i^2 + f_i^2 - r_i^2\right)^2}}{\left(\alpha_i + f_i\right)^2 + \beta_i^2 - r_i^2} \qquad i = 1, 2, 3 \qquad (A2)$$

Formulas (A2) provide up to two values of θ_{i2}. From a geometric point of view, these two solutions per limb correspond to the up to two intersections of two circumferences that lie on the plane perpendicular to the unit vector $\mathbf{g}_i$ and passing through A_i and B_i, one with center at A_i and radius f_i and the other with center at B_i and radius r_i. These intersections are the possible positions of point C_i (see Figure 1) compatible with an assigned platform pose.

References

1. Clavel, R. Delta, a fast robot with parallel geometry. In Proceedings of the 18th International Symposium on Industrial Robots, Sydney, Australia, 26–28 April 1988; pp. 91–100, ISBN 0-948507-97-7.
2. Clavel, R. Device for the Movement and Positioning of an Element in Space. Patent No. 4976582, 11 December 1990.
3. Brinker, J.; Corves, B. A Survey on Parallel Robots with Delta-like Architecture. In Proceedings of the 14th IFToMM World Congress, Taipei, Taiwan, 25–30 October 2015. [CrossRef]
4. Hervè, J.M.; Sparacino, F. Structural synthesis of "parallel" robots generating spatial translation. In Proceedings of the 5th International Conference on Advanced Robotics—ICAR1991, Pisa, Italy, 19–22 June 1991; pp. 808–813.
5. Tsai, L.W. Kinematics of a Three-dof Platform with Three Extensible Limbs. In *Recent Advances in Robot Kinematics*; Lenarcic, J., Parenti-Castelli, V., Eds.; Kluwer Academic Publishers: Dordrecht, The Netherlands, 1996; pp. 401–410, ISBN 978-94-010-7269-4.
6. Gogu, G. *Structural Synthesis of Parallel Robots—Part 2: Translational Topologies with Two and Three Degrees of Freedom*; Springer: Heidelberg, Germany, 2009; ISBN 978-90-481-8202-2.
7. Kong, X.; Gosselin, C.M. *Type Synthesis of Parallel Mechanisms*; Springer: Heidelberg, Germany, 2007; ISBN 978-3-642-09118-6.
8. Di Gregorio, R. A Review of the Literature on the Lower-Mobility Parallel Manipulators of 3-UPU or 3-URU Type. *Robotics* **2020**, *9*, 5. [CrossRef]
9. Gosselin, C.M.; Angeles, J. Singularity analysis of closed-loop kinematic chains. *IEEE Trans. Robot. Automat.* **1990**, *6*, 281–290. [CrossRef]
10. Zlatanov, D.; Fenton, R.G.; Benhabib, B. A unifying framework for classification and interpretation of mechanism singularities. *ASME J. Mech. Des.* **1995**, *117*, 566–572. [CrossRef]
11. Zlatanov, D.; Bonev, I.A.; Gosselin, C.M. Constraint Singularities as C-Space Singularities. In *Advances in Robot. Kinematics: Theory and Applications*; Lenarčič, J., Thomas, F., Eds.; Springer: Dordrecht, Germany, 2002; pp. 183–192.
12. Huda, S.; Takeda, Y. Kinematic analysis and synthesis of a 3-URU pure rotational parallel mechanism with respect to singularity and workspace. *J. Adv. Mech. Des. Syst. Manuf.* **2007**, *1*, 81–92. [CrossRef]
13. Huda, S.; Takeda, Y. Kinematic Design of 3-URU Pure Rotational Parallel Mechanism with Consideration of Uncompensatable Error. *J. Adv. Mech. Des. Syst. Manuf.* **2008**, *2*, 874–886. [CrossRef]
14. Carbonari, L.; Corinaldi, D.; Palpacelli, M.; Palmieri, G.; Callegari, M. A Novel Reconfigurable 3-URU Parallel Platform. In *Advances in Service and Industrial Robotics*; Ferraresi, C., Quaglia, G., Eds.; Springer: Dordrecht, Germany, 2018; pp. 63–73.
15. Di Gregorio, R.; Parenti-Castelli, V. A Translational 3-DOF Parallel Manipulator. In *Advances in Robot. Kinematics: Analysis and Control*; Lenarcic, J., Husty, M.L., Eds.; Kluwer: Norwell, MA, USA, 1998; pp. 49–58.
16. Ardema, M.D. *Newton-Euler Dynamics*; Springer: New York, NY, USA, 2005.
17. Hunt, K.H. *Kinematic Geometry of Mechanisms*; Clarendon Press: Oxford, UK, 1990.

18. Di Gregorio, R. Position Analysis of a Novel Translational 3-URU with Actuators on the Base. In *New Advances in Mechanisms, Mechanical Transmissions and Robotics. MTM & Robotics 2020*; Corves, B., Lovasz, E., Hüsing, M., Maniu, I., Gruescu, C., Eds.; Springer: Dordrecht, Germany, 2020. (in press)
19. Davenport, H. *The Higher Arithmetic*, 8th ed.; Cambridge University Press: New York, NY, USA, 2008; pp. 31–33.

 *robotics*

Article

Functional Design of a 6-DOF Platform for Micro-Positioning

Matteo-Claudio Palpacelli [1,*,†], **Luca Carbonari** [2,†], **Giacomo Palmieri** [1], **Fabio D'Anca** [3], **Ettore Landini** [4] and **Guido Giorgi** [5]

1   Dipartimento DIISM, Università Politecnica delle Marche, 60131 Ancona, Italy; g.palmieri@univpm.it
2   Dipartimento DIMEAS, Politecnico di Torino, 10129 Torino, Italy; luca.carbonari@polito.it
3   Istituto Nazionale di AstroFisica (INAF)—Osservatorio Astronomico, 90134 Palermo, Italy; fabio.danca@gmail.com
4   Fab Crea Srl, 16149 Genova, Italy; ettore.landini@fabcrea.it
5   Vacuum Fab Srl, 20010 Cornaredo, Italy; guido.giorgi@vacuumfab.it
*   Correspondence: m.palpacelli@univpm.it; Tel.: +39-071-220-4748
†   These authors contributed equally to this work.

Received: 14 October 2020; Accepted: 18 November 2020; Published: 23 November 2020

Abstract: Parallel kinematic machines (PKMs) have demonstrated their potential in many applications when high stiffness and accuracy are needed, even at micro- and nanoscales. The present paper is focused on the functional design of a parallel platform providing high accuracy and repeatability in full spatial motion. The hexaglide architecture with 6-$\underline{P}$SS kinematics was demonstrated as the best solution according to the specifications provided by an important Italian company active in the field of micro-positioning, particularly in vacuum applications. All the steps needed to prove the applicability of such kinematics at the microscale and their inherent advantages are presented. First, the kinematic model of the manipulator based on the study's parametrization is provided. A global conditioning index (GCI) is proposed in order to optimize the kinetostatic performance of the robot, so that precise positioning in the required platform workspace is guaranteed avoiding singular configurations. Some numerical simulations demonstrate the effectiveness of the study. Finally, some details about the realization of a physical prototype are given.

Keywords: kinematic optimization; vacuum applications; micro-positioning; parallel kinematics machines

---

1. Introduction

The ability to change the orientation and position of a sample with high precision has become an indispensable requisite in many applications, ranging from testing of microelectronic components, to assembly of optoelectronic components, to measurements in extreme environments such as ultra-high vacuums, to exposure to cryogenic radiation and temperature typical of synchrotrons, and more. These functions are normally performed by robotic platforms with limited workspace and a small range of rotation. Compliant devices often fulfill such requirements allowing precise movements only in a subdomain of the typically required six degrees of freedom, as is well documented in [1–7]. Parallel kinematic machines (PKMs) are usually preferred to serial devices because of their intrinsic features, such as stiffness, positioning accuracy, and repeatability. Many examples of commercial devices based on parallel kinematics architectures can be found in industry, especially in the measurement sector.

The most widely used PKM in such fields is the Gough–Stewart platform [8], which has been commercially proposed in many sizes in order to respond to different application domains. Many examples

Robotics 2020, 9, 99; doi:10.3390/robotics9040099   www.mdpi.com/journal/robotics

of hexapods, whose operating principle is used to orient and manipulate microelectronic components, can be found in the scientific literature [9–11]. The main drawback of these manipulators, shared by most of the PKMs, is the limited workspace caused by the fixed location of both active and passive joints at the base frame, with which the legs of the manipulator are connected.

This limit can be overcome by a well-known mechanism from École Polytechnique Fédérale of Zürich, which takes the name of Hexaglide for its particular architecture. It belongs to the family of Hexapods, with six legs and six actuated rails at the fixed base [12]. In fact, the mechanical solution of having six parallel rails at the base can significantly extend the manipulator workspace, at least in one direction, providing a sort of extruded volume. One of the benefits could be the possibility of combining classic operations of pick and place, exploiting the whole length of the rails and covering a large workspace, to specific alignment and assembly tasks, for instance those typical of the microelectronics industry, in reduced sections of the rails by repeating them in several parallel workstations.

There are many examples in the literature of Hexaglide PKMs, but some of them are conceived with a different arrangement of the actuated rails at the base frame, for instance along the edges of an equilateral triangle [13] or in a star configuration [14], or even in more complex configurations [15,16], losing the ability to translate the entire mechanics in a prevalent direction. On the contrary, other studies are focused on their parallelism [17,18], but the proposed mechanical designs are oriented towards large machines where accuracy and repeatability are not the main objectives.

The present work proposes the functional design of a Hexaglide manipulator aimed at precision applications in the electronics and measurement sectors, with the aim of revealing a new potential use by showing its kinematic performance when designed for carrying out microscale tasks. Therefore, the first part of the paper is focused on the kinematics modeling of the manipulator, whereas the rest of the paper proposes an optimization procedure that, starting from one of its multiple inverse kinematics solutions, results in a functional design that fulfills the specifications set by an industrial manufacturer. Finally, a prototype designed with the aforementioned procedure is shown. Such prototype is based on a different inverse kinematics solution that is subject to a confidentiality agreement and currently cannot be disclosed.

2. Design Specifications

The design specifications of the Italian manufacturer concern the kinematic and static performance required by the manipulator. They result from a market analysis in the microelectronics sector, as previously indicated. These can be summarized as follows:

- overall dimensions of $350 \times 350 \times 225$ mm (XYZ),
- translation of $125 \times 50 \times 25$ mm along the x-, y- and z-directions, respectively,
- translation resolution of about 0.5 μm, repeatability of ± 1 μm, maximum speed of 5 mm/s,
- rotation range of $\pm 5°$ about x-, y- and z-axes,
- rotation resolution of about $2 \times 10^{-4°}$, repeatability of $\pm 2 \times 10^{-4°}$,
- payload of 1.5 kg.

3. The Hexaglide Kinematics

As already mentioned in the previous sections, one of the most suitable kinematic architectures that can meet such desired specifications is the Hexaglide. Six actuated sliders, equally oriented in a given direction of the fixed base, provide the manipulator with a main direction of translation. A typical configuration of the manipulator is obtained with six rigid legs of constant length, and a hexagonal or triangular moving platform according to the connection of the legs in a 6-6 [19] or 6-3 scheme [20] by means of spherical joints, as shown in Figure 1. Six of the twelve spherical joints, the ones at the base or the others at the moving platform, can be replaced by universal joints for a non-under-constrained version of the manipulator, avoiding the spin of the legs about their axis. It is often preferred that there is a reduction of the rails at the frame with three units, where two sliders move in pairs [21–23].

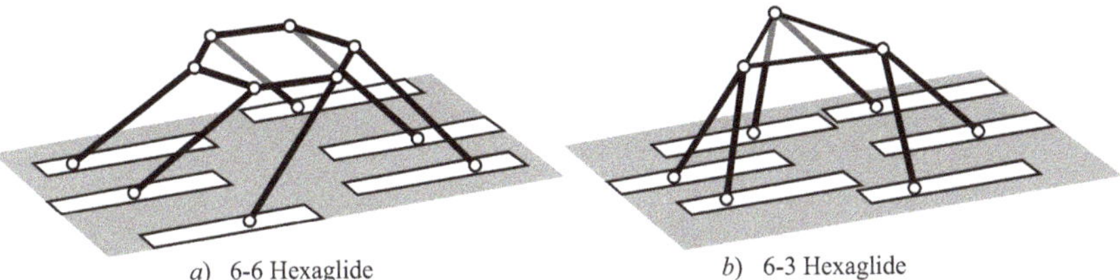

Figure 1. Functional solutions for the Hexaglide manipulator.

3.1. Model Parametrization

The pose of the moving platform of a manipulator is typically defined by means of a homogeneous transformation expressed in terms of a 4×4 matrix, called $^0_1\mathbf{T}$ in the following. It involves the rotation matrix $^0_1\mathbf{R}$ between the mobile reference system $\{1\}$, placed on the moving platform, and the fixed reference system $\{0\}$, located at the base frame, and the relative position vector $^0\mathbf{p}_1$ amongst their origins. The two reference systems and other important geometric data related to the Hexaglide are shown in Figure 2. The rotation matrix $^0_1\mathbf{R}$ is also interpreted in this work as a sequence of *roll*, *pitch*, and *yaw* elemental rotations, namely a matrix multiplication of the form $^0_1\mathbf{R} = \mathbf{R}_z(\theta_z)\mathbf{R}_y(\theta_y)\mathbf{R}_x(\theta_x)$, with obvious meanings of terms. The well-established study's parametrization can be conveniently used in the implementation of calculation routines in order to avoid representation singularities for $^0_1\mathbf{R}$:

$$
^0_1\mathbf{T} = \begin{bmatrix} ^0_1\mathbf{R} & ^0\mathbf{p}_1 \\ \mathbf{0}_{1\times3} & 1 \end{bmatrix} =
$$

$$
= \begin{bmatrix}
x_0^2 - x_1^2 - x_2^2 + x_3^2 & 2(x_0 x_1 - x_2 x_3) & 2(x_0 x_2 + x_1 x_3) & 2(x_0 y_3 - x_1 y_2 + x_2 y_1 - x_3 y_0) \\
2(x_0 x_1 + x_2 x_3) & -x_0^2 + x_1^2 - x_2^2 + x_3^2 & 2(x_1 x_2 - x_0 x_3) & 2(x_0 y_2 + x_1 y_3 - x_2 y_0 - x_3 y_1) \\
2(x_0 x_2 - x_1 x_3) & 2(x_1 x_2 + x_0 x_3) & -x_0^2 - x_1^2 + x_2^2 + x_3^2 & 2(x_1 y_0 - x_0 y_1 + x_2 y_3 - x_3 y_2) \\
0 & 0 & 0 & x_0^2 + x_1^2 + x_2^2 + x_3^2
\end{bmatrix}. \tag{1}
$$

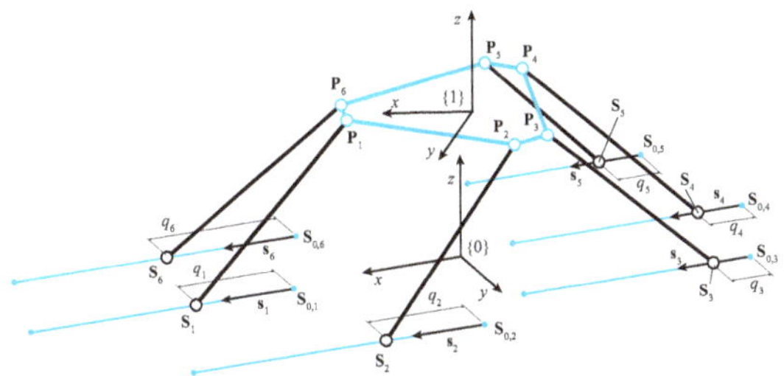

Figure 2. Kinematic scheme of the 6-P̱SS Hexaglide.

The eight study parameters, here gathered in the vector $\mathbf{x} = [x_0, x_1, x_2, x_3, y_0, y_1, y_2, y_3]^T$, are constrained by the study quadric:

$$S: \quad s = 0, \tag{2}$$

with

$$s = x_0 y_0 + x_1 y_1 + x_2 y_2 + x_3 y_3. \tag{3}$$

Moreover, the following constraint normalization equation must hold in order to avoid a projective transformation:

$$N: \quad n = 0, \tag{4}$$

with

$$n = x_0^2 + x_1^2 + x_2^2 + x_3^2 - 1. \tag{5}$$

The full inverse transformations are also given for the sake of completeness:

$$
\begin{aligned}
x_3 &= \frac{1}{2}\sqrt{{}_1^0 T_{1,1} + {}_1^0 T_{2,2} + {}_1^0 T_{3,3} + 1} & \quad y_0 &= -\frac{1}{2}\left({}_1^0 T_{1,4} x_3 + {}_1^0 T_{2,4} x_2 - {}_1^0 T_{3,4} x_1\right) \\
x_0 &= \frac{{}_1^0 T_{3,2} - {}_1^0 T_{2,3}}{4 x_3} & \quad y_1 &= -\frac{1}{2}\left(-{}_1^0 T_{1,4} x_2 + {}_1^0 T_{2,4} x_3 + {}_1^0 T_{3,4} x_0\right) \\
x_1 &= \frac{{}_1^0 T_{1,3} - {}_1^0 T_{3,1}}{4 x_3} & \quad y_2 &= -\frac{1}{2}\left({}_1^0 T_{1,4} x_1 - {}_1^0 T_{2,4} x_0 + {}_1^0 T_{3,4} x_3\right) \\
x_2 &= \frac{{}_1^0 T_{2,1} - {}_1^0 T_{1,2}}{4 x_3} & \quad y_3 &= -\frac{1}{2}\left({}_1^0 T_{1,4} x_0 + {}_1^0 T_{2,4} x_1 + {}_1^0 T_{3,4} x_2\right)
\end{aligned}
\tag{6}
$$

where ${}_1^0 T_{i,j}$ stands for the element of matrix ${}_1^0 \mathbf{T}$ at the ith row and jth column.

The position of the lower spherical joints is parametrized as follows:

$$\mathbf{S}_i = \mathbf{S}_{0,i} + q_i \mathbf{s}_i, \tag{7}$$

where $\mathbf{S}_i$ is the variable position of the attachment point of the ith leg on the frame, corresponding to the ith spherical joint center, driven by the actuated prismatic joints at the base, $\mathbf{S}_{0,i}$ is the fixed point on each slider that corresponds to $\mathbf{S}_i$ when in the home position, $\mathbf{s}_i$ is the unit vector giving the direction of the linear sliding, and q_i is the ith actuated joint variable. All the q_i can be gathered in the actuated joint vector $\mathbf{q} = [q_1, q_2, q_3, q_4, q_5, q_6]^T$. The generality of the expression can be noted, where $\mathbf{s}_i$ can still be different for each prismatic joint.

The ith leg can be represented by vector $\mathbf{L}_i$ whose expression is given by

$$\mathbf{L}_i = \mathbf{P}_i - \mathbf{S}_i = L_i \mathbf{l}_i, \tag{8}$$

where L_i is its constant length and $\mathbf{l}_i$ is the unit vector giving its direction from $\mathbf{S}_i$ to $\mathbf{P}_i$.

Finally, the moving platform, whose geometric scheme is shown in Figure 3, is parametrized by means of two parameters: r is the radius of a circle, on which the six centers of the upper spherical joints lie, and φ is the semi-central angle of the circular sector between pairs of $\mathbf{P}_i$ points, namely $\{\mathbf{P}_6, \mathbf{P}_1\}, \{\mathbf{P}_2, \mathbf{P}_3\}, \{\mathbf{P}_4, \mathbf{P}_5\}$. Such arrangement is motivated by the need to avoid the perfect hexagonal symmetry of singular configurations inside the manipulator workspace, while maintaining a certain symmetry. Matrix ${}_1^0 \mathbf{T}$ in (1) can be used to find the absolute coordinates of each point $\mathbf{P}_i$ in the fixed reference frame $\{0\}$ when the coordinates in the moving reference frame $\{1\}$ are known:

$$\begin{bmatrix} {}^{0}\mathbf{P}_i \\ 1 \end{bmatrix} = {}^{0}_{1}\mathbf{T} \begin{bmatrix} {}^{1}\mathbf{P}_i \\ 1 \end{bmatrix}. \tag{9}$$

From here on, all vectors are considered written in the fixed frame, unless explicitly stated.

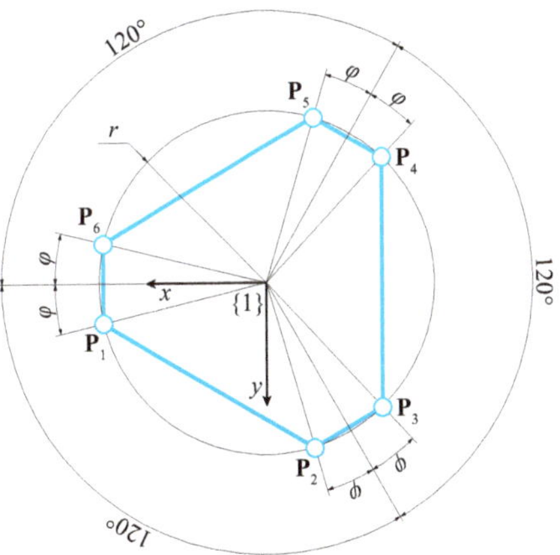

Figure 3. Geometric scheme of the moving platform.

3.2. Inverse Position Kinematics (IPK)

The mapping from the pose coordinates of the moving platform to the actuated joint displacements along the six linear axes can easily be obtained if the rigid body constraint is imposed to the length of the legs of the manipulator. It consists of the following expression:

$$E_i: \quad e_i = 0, \tag{10}$$

with

$$e_i: \quad (\mathbf{P}_i - \mathbf{S}_i)^T (\mathbf{P}_i - \mathbf{S}_i) - L_i^2.$$

The subscript i is related to the ith leg according to the notation shown in Figure 4. The resulting system of equations is given by as many constraints as the number of legs, with the addition of the study quadric (S) and the normalization equation (N):

$$\{E_1 : E_2 : E_3 : E_4 : E_5 : E_6 : S : N\}. \tag{11}$$

The expressions of points $\mathbf{S}_i$ and $\mathbf{P}_i$ are given by Equations (7) and (9), respectively. As expected, the system provides two inverse kinematics solutions for each leg, whose expression in compact form is given by:

$$q_{i_{1,2}} = \mathbf{s}_i^T \mathbf{a}_i \pm \sqrt{\left(\mathbf{s}_i^T \mathbf{a}_i\right)^2 - \mathbf{a}_i^T \mathbf{a}_i + L_i^2}, \tag{12}$$

where $\mathbf{a}_i$ is the ith vector resulting from the following expression:

$$\mathbf{a}_i = {}^{0}\mathbf{p}_1 + {}^{0}_{1}\mathbf{R}\,{}^{1}\mathbf{P}_i - \mathbf{S}_{0,i}. \tag{13}$$

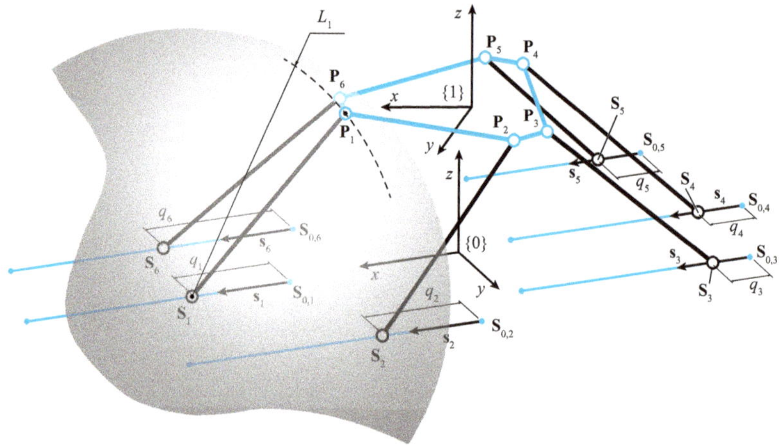

Figure 4. Constraint manifold of the *i*th leg.

Both solutions in Equation (12) are acceptable and provide 2^6 theoretical inverse kinematic solutions for the Hexaglide, if all combinations are taken into account. Therefore, a selection routine for the best solution that meets the design specifications, together with the best kinematic performance, is required. Some geometrical and physical assumptions can be made: the zero configuration for each slider, namely the absolute position of each point $\mathbf{S}_{0,i}$ at the fixed base, can be associated to one end of the relative actuated rail, whereas the other end corresponds to the maximum stroke q_{max}, providing an always positive or at least null q_i. Moreover, the legs of the manipulator must never reach a vertical configuration during its operation. Such condition is required to avoid singularities, as shown in the following section about differential kinematics. The foregoing assumptions are summarized as follows:

$$0 \leq q_i \leq q_{max}\ (i = 1, \ldots, 6) \quad \wedge \quad \begin{array}{l} \mathbf{s}_i^T \mathbf{S}_i > \mathbf{s}_i^T \mathbf{P}_i + \varepsilon\ (i = 1, 2, 6) \\ \mathbf{s}_i^T \mathbf{S}_i < \mathbf{s}_i^T \mathbf{P}_i - \varepsilon\ (i = 3, 4, 5) \end{array}, \tag{14}$$

with ε as large as desired and $\mathbf{s}_i$ directed along the x-axis of the fixed frame $\{0\}$.

3.3. Direct Position Kinematics (DPK)

The mapping from the actuated joint vector $\mathbf{q}$ to the pose coordinates of the moving platform expressed by vector $\mathbf{x}$, or analogously given by matrix $^0_1\mathbf{T}$, is more complex than the inverse problem, as expected for PKMs. It can be demonstrated that in this case such problem has 40 solutions and numerical approaches are often preferred to closed-form models because of their easier software implementation. Numerical solutions depend on initial conditions and it is not generally easy to converge to the right solution among the many possibilities. As an initial condition, a matrix $^0_1\mathbf{T}_0$ based on an identity matrix $\mathbf{I}_{3\times3}$ and an arbitrary position vector $^0\mathbf{p}_{1,0}$ inside the manipulator workspace, or alternatively a starting Cartesian vector $\mathbf{x}_0$, is chosen. The algorithm conceived to solve the problem consists of the following steps:

1. Input: vector $\mathbf{q}$;
2. a discrete sequence of actuated joint vectors, regulated by the index k, is obtained by means of a linear interpolation ranging from $\mathbf{q}_{in}$ ($= \mathbf{q}_0$) to $\mathbf{q}_{fin}$ ($= \mathbf{q}$), with $\mathbf{q}_0$ resulting from the *IPK* of $\mathbf{x}_0$;

3. an iterative Newton–Raphson algorithm, this time regulated by the index i, is progressively used for each $\mathbf{q}_k$ of the sequence to evaluate the vector $\mathbf{x}_k$ that verifies the constraint manifold in (11) with a desired level of accuracy, each iteration starting from the previous solution $\mathbf{x}_{k-1}$;

4. Output: vector $\mathbf{x}$ of study parameters or directly $^0_1\mathbf{T}$ if a matrix form is preferred.

In more detail, at each iteration of step 3, a new estimate of the solution $\mathbf{x}_k$ is sought by means of the aforementioned Newton–Raphson algorithm:

$$\mathbf{x}_{i+1} = \mathbf{x}_i - \mathbf{J}_i^{-1}\mathbf{v}_i,\tag{15}$$

where $\mathbf{v}_i$ and $\mathbf{J}_i$ are:

$$\mathbf{v}_i = \begin{bmatrix} e_1: & |\mathbf{P}_1(\mathbf{x}_i) - \mathbf{S}_1(\mathbf{q_k})|^2 - L_1^2 \\ \vdots & \vdots \\ e_6: & |\mathbf{P}_6(\mathbf{x}_i) - \mathbf{S}_6(\mathbf{q_k})|^2 - L_6^2 \\ s: & s(\mathbf{x}_i) \\ n: & n(\mathbf{x}_i) \end{bmatrix}, \quad \mathbf{J}_i = \frac{\partial}{\partial \mathbf{x}_i^T}\begin{bmatrix} e_1 \\ \vdots \\ e_6 \\ s \\ n \end{bmatrix}_{8\times8}.\tag{16}$$

3.4. Differential Kinematics

The Jacobian matrix of the manipulator, which maps the Cartesian velocity vector $\dot{\mathbf{x}}$ of the moving platform from the actuated joint velocity vector $\dot{\mathbf{q}}$, is presented in the classic partitioned form:

$$\mathbf{J}_x\dot{\mathbf{x}} = \mathbf{J}_q\dot{\mathbf{q}},\tag{17}$$

where $\dot{\mathbf{x}} = [\boldsymbol{\omega}^T, \mathbf{v}^T]^T$ with $\boldsymbol{\omega}$ the angular velocity and $\mathbf{v}$ the linear velocity of the moving platform, and $\dot{\mathbf{q}} = [\dot{q}_1, \dot{q}_2, \dot{q}_3, \dot{q}_4, \dot{q}_5, \dot{q}_6]^T$. The expression of the matrices $\mathbf{J}_x$ and $\mathbf{J}_q$, called respectively the geometric and analytical Jacobians of the manipulator, are obtained by making use of kinematic torsors:

$$\mathbf{J}_x = \begin{bmatrix} \$_{r,1}^T \\ \vdots \\ \$_{r,6}^T \end{bmatrix} \quad \text{and} \quad \mathbf{J}_q = \begin{bmatrix} \$_{r,1}^T\$_1 & \cdots & 0 \\ \vdots & \ddots & \vdots \\ 0 & \cdots & \$_{r,6}^T\$_6 \end{bmatrix},\tag{18}$$

where $\$_i = [\mathbf{0}^T, \mathbf{s}_i^T]^T$ is the six-dimensional kinematic torsor related to the ith actuated prismatic joint and $\$_{r,i}^T = [(\mathbf{P}_i \times \mathbf{l}_i)^T, \mathbf{l}_i^T]$ is the dual torsor to passive joints of each ith leg. When the square matrix $\mathbf{J}_x$ has full rank, the full Jacobian matrix can be found as:

$$\mathbf{J} = \mathbf{J}_x^{-1}\mathbf{J}_q.\tag{19}$$

A look at the components of the matrices $\mathbf{J}_x$ and $\mathbf{J}_q$ in (18) allows identification of the singular configurations of the manipulator. The two matrices lose rank according to the following conditions:

- $\det(\mathbf{J}_x) = 0$ when any pair of reciprocal torsors $\$_{r,i}$ and $\$_{r,j}$ (with $i \neq j$) satisfies the relation $\$_{r,i} \parallel \$_{r,j}$;
- $\det(\mathbf{J}_q) = 0$ when $\mathbf{l}_i \perp \mathbf{s}_i$ is verified at least for one of the ith legs.

4. Optimization Problem

An objective function Φ is defined with the aim of maximizing the kinematic performance of the manipulator, laying the foundation for a geometrical optimization. One of the most frequently used tools in kinematic optimization problems is the condition number of the Jacobian matrix. However, besides not being a finite index, it merges the kinematic characteristics of different natures when applied to full motion manipulators. It follows that two sub-matrices extracted from matrix **J** related to angular and linear velocities independently seem more suitable for a guided optimization, instead of using the full Jacobian matrix [24]. Two indexes result from this partitioning:

$$cn_\omega = \sqrt{\frac{\sigma_{\min}\left(\mathbf{J}_\omega\mathbf{J}_\omega^T\right)}{\sigma_{\max}\left(\mathbf{J}_\omega\mathbf{J}_\omega^T\right)}} \quad \text{and} \quad cn_v = \sqrt{\frac{\sigma_{\min}\left(\mathbf{J}_v\mathbf{J}_v^T\right)}{\sigma_{\max}\left(\mathbf{J}_v\mathbf{J}_v^T\right)}} \,, \tag{20}$$

where $\mathbf{J}_\omega$ is the 3×6 matrix given by the first three rows of **J** and analogously $\mathbf{J}_v$ by the second three. The terms $\sigma_{\min}$ and $\sigma_{\max}$ represent the minimum and maximum eigenvalues of the matrices inside the round brackets. The two condition numbers in (20) are local indexes and have the property of being confined in the range $0 \leq cn \leq 1$, with isotropic and singular conditions revealed by the upper and lower extremes respectively. Further useful indexes with global properties can be defined if the condition numbers in (20) are integrated on the manipulator workspace W [25], where the latter in the present case refers to the desired workspace defined in Section 2:

$$GCI_\omega = \frac{\int_W cn_\omega dW}{\int_W dW} \quad \text{and} \quad GCI_v = \frac{\int_W cn_v dW}{\int_W dW} \,. \tag{21}$$

Finally, an objective function Φ based on (21) and tailored to meet the design specifications can be defined as:

$$\Phi\left(\mathbf{p}_h, \boldsymbol{\lambda}\right) = 1 - \left(k_\omega GCI_\omega + k_v GCI_v\right), \tag{22}$$

where the weights k_ω and k_v are in the range $[0, 1]$ and can be adjusted according to kinematic needs and respecting the relationship $k_\omega + k_v = 1$, favoring translation over rotation performance if required, or other combinations. As highlighted in (22), function Φ depends on $\mathbf{p}_h = [y_h, z_h, \theta_z]^T$, namely the y and z coordinates of the relative position vector between frames $\{1\}$ and $\{0\}$ in the home configuration of the manipulator, and the yaw rotation about the vertical z axis. The home configuration is defined as the central configuration of the manipulator, from which the required translations and rotations can be performed without incurring singularities, with a horizontal top surface of its moving platform. This latter condition is assumed for practical reasons, in fact the top surface can be used as a horizontal support plane, narrowing the study only to rotations about the z axis. At this stage of the study, the x coordinate between frames $\{1\}$ and $\{0\}$ is arbitrary, therefore, a variability of x_h is not taken into account. Finally, the vector $\boldsymbol{\lambda}$ in (22) collects all the geometric parameters of the manipulator that a user wants to optimize. In the present case, with reference to Section 3, it is defined as:

$$\boldsymbol{\lambda} = \left[\varphi, r, L, S_{1,y}, S_{3,y}, S_{4,y}, S_{6,y}\right], \tag{23}$$

with the y-position of the ith slider with respect to the fixed frame $\{0\}$ represented by $S_{i,y}$ and all the legs considered with equal length $L_i = L$. It can be noted that $S_{2,y}$ and $S_{5,y}$ do not participate in the optimization process because the respective linear guides are the most external along the y-direction and are confined within the maximum range of 350 mm already mentioned in Section 2.

Geometric Optimization

A preliminary study about the expected overall size of the manipulator was carried out. The design specifications of Section 2 allow limitation of the range of variability of some geometrical parameters, so that optimization results are effectively constrained around an arbitrary initial configuration taken as a first guess. Eventually, they are checked again after all parameters are optimized.

The maximum height of the moving reference frame {1} with respect to the fixed frame {0} is limited to 200 mm, leaving the remaining millimeters to mechanically design the moving platform. According to the results of the kinematic study in Section 3.4, a maximum tilt of 75° for each leg from the horizontal plane is considered in order to be sufficiently far from singular configurations. It follows that a hypothesized length for the legs is 210 mm, as shown in Figure 5a. In addition, the unit vectors $\mathbf{l}_i$ of the legs must be as far as possible from being parallel, suggesting a staggered disposition of the linear guides at the base. Their first attempt arrangement is shown in Figure 5b, with a distance of 300 mm between the external guides in order to take into account their physical dimensions.

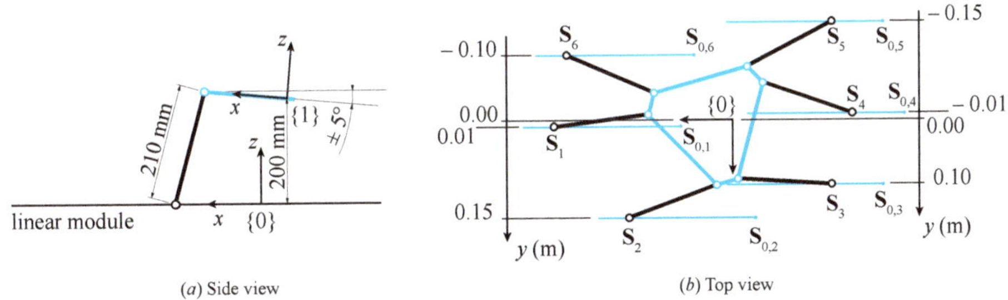

Figure 5. Preliminary design: (**a**) side view, (**b**) top view.

The choice of a first guess value for the geometric parameters of the moving platform is due to technical reasons: on the one hand, the maximization of the free surface where a load can be placed, and, on the other hand, the need for enough space to realize a physical connection of the legs with the platform. Their values are: $\varphi_0 = 10°$ and $r_0 = 75$ mm, where the subscript 0 has the meaning of the initial value of the optimization procedure.

As already mentioned, the home position of the sliders along the *x*-direction, corresponding to the home configuration of the moving platform, is defined only after the optimization of the other geometric parameters. In fact, the zero of each slider can be placed so that the actual home position falls in the center of the available stroke of the sliders.

The objective function in (22) collects information from the robot about its whole required workspace volume. However, as expected by virtue of the particular features of the Hexaglide along a specific direction, the study of Φ is narrowed to a single cross-section of the workspace: without loss of generality the condition $x = 0$ is chosen. This assumption is justified by the behavior of the manipulator that does not depend on the translation in the *x*-direction, but only on the maximum dimension of the sliders and their initial position. A range of ±25 mm and ±12.5 mm in the *y*- and *z*-directions, respectively, and rotations of ±5° about all Cartesian axes are considered (see Section 2).

The pre-sizing values of the optimization parameters, together with their assumed lower and upper range limits, are shown in Table 1. Their values result from a preliminary analysis (a sort of Monte Carlo method) of the objective function subject to discrete variations of the parameters in their admissible range. Values of the objective function above 0.7 allow a large number of combinations of parameters to be excluded from the study. A further refinement around the most promising combinations allows the choosing of the initial values gathered in Table 1, approximated to whole numbers, with a corresponding Φ of about 0.51. Rotations greater than five degrees for θ_z in Table 1 are justified by the search of its home configuration value, around which the required range of ±5° can be obtained with the best kinematic performance. A similar approach is followed for the other design parameters.

Table 1. Initial values, lower and upper limits of the optimization parameters.

	φ	θ_z	r	L	$S_{1,y}$	$S_{3,y}$	$S_{4,y}$	$S_{6,y}$	y_h	z_h
	[°]					[mm]				
Init	10	0	75	210	10	100	−10	−100	0	70
Inf	5	−20	50	150	0	0	−60	−150	−100	50
Sup	15	20	100	250	60	150	0	0	100	200

The optimization procedure is repeated 10 times iteratively, each time providing the result obtained in the previous step as starting point, in order to improve the local minimum point of the objective function. This method is necessary because of the sensitivity of the problem to initial conditions.

5. Optimization Results

The geometrical parameters and the home position coordinates summarized in Table 1 are processed by means of the Matlab Optimization Toolbox by MathWorks. A constrained minimization problem is tackled in order to find the required optimal values according to the minimization of the function Φ proposed in (22), with equal weights $k_\omega = k_v = 0.5$. Hard constraints are imposed on Equation (11), while soft constraints are imposed on the variability of vector **x**, ensuring that Cartesian y and z translations together with x and y rotations in their required ranges are satisfied. Another important soft constraint ensures that none of the six legs exceeds the maximum tilt of 80°, a less stringent condition with respect to the one mentioned in the previous sections, in order to give more flexibility to the optimum search. Table 2 shows each calculation step progressively as rows increase. It is easy to verify that, after the first few steps, the variability of parameters tends to decrease and results stabilize around some reference values, with quite a small deviation. Finally, a Φ of about 0.48 is obtained, derived from a slow progressive increase at each step of the optimization process.

Table 2. Optimized values of geometric parameters and home configuration.

	φ	θ_z	r	L	$S_{1,y}$	$S_{3,y}$	$S_{4,y}$	$S_{6,y}$	y_h	z_h	Φ
Step	[°]						[mm]				−
0	10	0	75	210	10	100	−10	−100	0	70	0.512
1	9.99	−8.63	56.63	208.66	49.97	138.07	−3.28	−118.84	4.69	140.95	0.498
2	10.02	−9.41	59.90	239.33	57.48	135.60	−3.70	−116.95	46.54	146.15	0.497
3	10.12	−10.92	59.58	245.18	57.62	135.78	−3.65	−117.13	50.27	150.48	0.494
4	9.98	−11.68	57.82	249.86	59.89	148.93	−0.14	−118.10	50.83	155.30	0.493
5	10.05	−9.65	60.49	248.47	57.76	136.49	−3.72	−116.62	53.29	152.75	0.490
6	10.05	−9.92	57.80	249.97	59.97	149.72	−0.03	−118.08	50.92	155.44	0.490
7	9.98	−8.93	60.59	248.57	57.76	136.54	−3.72	−116.57	53.43	152.80	0.487
8	10.08	−9.88	58.13	249.32	59.49	145.60	−0.73	−117.93	50.99	154.60	0.486
9	10.04	−9.92	57.80	249.96	59.97	149.72	−0.03	−118.11	50.85	155.47	0.482
10	10.06	−10.24	61.14	248.53	57.79	136.80	−3.79	−116.27	54.00	152.58	0.478
Result	10	−10	60	250	60	140	0	−115	50	155	0.482

Resulting design values are highlighted in the last row of Table 2, where an approximation of figures to average and more practical values is proposed, remembering that better conditions are associated with smaller values of Φ. Some important information about the home configuration is already available from the optimization procedure, for instance vector $\mathbf{p}_h = [y_h, z_h, \theta_z]^T$ in (22) is now completely known. However, the absolute position of the actuated carriages in the home configuration is only partially defined, knowing their y and z components (see Figure 5, Table 2 with $S_{i,z} = 0$ for $i = 1, ..., 6$). Their x components can be obtained from the application of the inverse kinematics algorithm proposed in (12), where the Cartesian pose of the moving platform is assigned: $x_h = 0$, $\theta_x = \theta_y = 0$, together with the remaining parameters in Table 2. A null value can be also assigned to the components of $S_{0,i}$. Such choice allows the absolute x position of the carriages to be found with respect to the fixed reference frame $\{0\}$ when all the linear guides at the platform base are perfectly side by side. Ultimately, the absolute coordinates of the S_i points, for $i = 1, ..., 6$, in the home configuration become:

$$S_1 = \begin{bmatrix} 0.2553 \\ 0.06 \\ 0 \end{bmatrix} ; \quad S_2 = \begin{bmatrix} 0.1761 \\ 0.15 \\ 0 \end{bmatrix} ; \quad S_3 = \begin{bmatrix} -0.2182 \\ 0.14 \\ 0 \end{bmatrix} ;$$

$$S_4 = \begin{bmatrix} -0.2384 \\ 0 \\ 0 \end{bmatrix} ; \quad S_5 = \begin{bmatrix} -0.1598 \\ -0.15 \\ 0 \end{bmatrix} ; \quad S_6 = \begin{bmatrix} 0.1850 \\ -0.115 \\ 0 \end{bmatrix} ,$$

$$(24)$$

with values in meters. In (24) only the x coordinates result from a computation, and this justifies the figures down to a tenth of a millimeter. In order to locate the points $S_{0,i}$ to their actual position and to maximize the admissible travel in the x-direction, their zero position must be placed so that:

$$S_{0,i} = S_i - \frac{q_{max}}{2} s_i. \tag{25}$$

6. Kinematic Performance

Once the optimal geometric dimensions are determined, a further step towards the definition of the mechanical elements can be done by defining the linear modules at the fixed base. The Hercules Stages HLS-M-185-DC-Arom1Vpp by Vacuum Fab Srl with a coreless DC motor, a linear optical scale with a ± 50 nm resolution, a stroke of 185 mm, and an actuation force of 212 N are used. If an ultra-high vacuum application is the target, RodRail eXtreme linear translation stages, produced by the same manufacturer, with a stroke of 190 mm and a repeatability of 1 µm in a vacuum, can be used instead.

A map of the Cartesian positioning errors obtained by virtue of the nameplate data of the linear stages is shown in Figure 6 on a vertical yz-plane ($x_h = 0$), covering the manipulator workspace. The white dot represents the Cartesian home configuration of the platform inside its workspace. A fixed orientation is considered in all Cartesian positions, the same as the home configuration, namely $\theta_x = 0$, $\theta_y = 0$ and $\theta_z = -10°$. The map in Figure 6 is coherent with the one of the determinants of the Jacobian matrix shown in Figure 7, obtained by analyzing the expression in (19) inside the manipulator workspace. In fact, where the value of the determinant is small, the manipulator in the relative Cartesian position is less prone to errors due to uncertainties in the displacement of the linear modules.

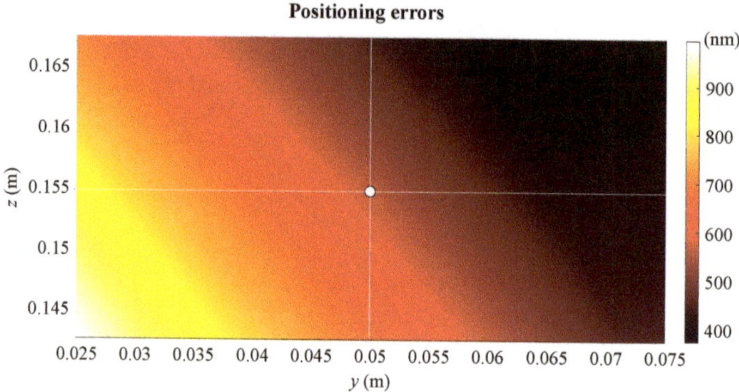

Figure 6. Map of the Cartesian positioning errors (fixed orientation: $\theta_x = 0$, $\theta_y = 0$, $\theta_z = -10°$).

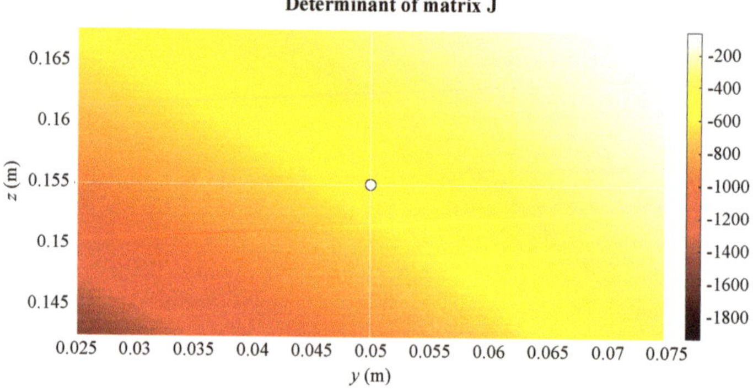

Figure 7. Map of the determinant of **J** (fixed orientation: $\theta_x = 0$, $\theta_y = 0$, $\theta_z = -10°$).

The maximum absolute value of positioning error obtained in the center of the workspace is about 0.6 μm. Such value is the worst possible considering every direction and any combination of errors for the actuated carriages, but it significantly reduces according to the direction chosen, even by an order of magnitude. If the first three rows of the Jacobian matrix are considered instead of the second three, orientation errors can be investigated. Figure 8 shows a trend analogous to the other map, where an error in the center of the workspace of about 3×10^{-4} degrees can be read, related to the worst condition. Other studies could be proposed taking into account a change of orientation of the moving platform with a fixed Cartesian position, but results are quite similar to the ones already proposed and do not really add more insight. Finally, a force polyhedron can be evaluated in static conditions to show if the manipulator meets the requirement of a 1.5 kg payload. Figure 9 refers to the static performance in terms of Cartesian forces in every direction when the manipulator is in its home configuration and no torques are provided at the end effector. The maximum force magnitude is about 800 N, whereas the minimum is about 80 N, but it is easy to verify that the highest performance is related to almost vertical directions, allowing the weight of the mechanical parts of the manipulator to be borne as well as the required payload. Eventually, it can be concluded that the obtained results almost entirely meet the specifications of Section 2.

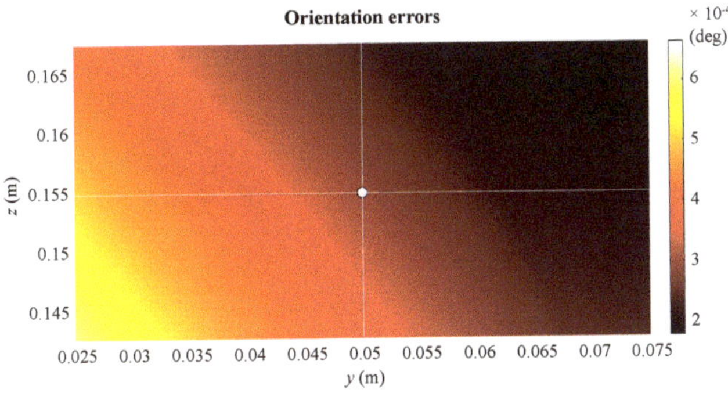

Figure 8. Map of the orientation errors (around the given orientation $\theta_x = 0$, $\theta_y = 0$, $\theta_z = -10°$).

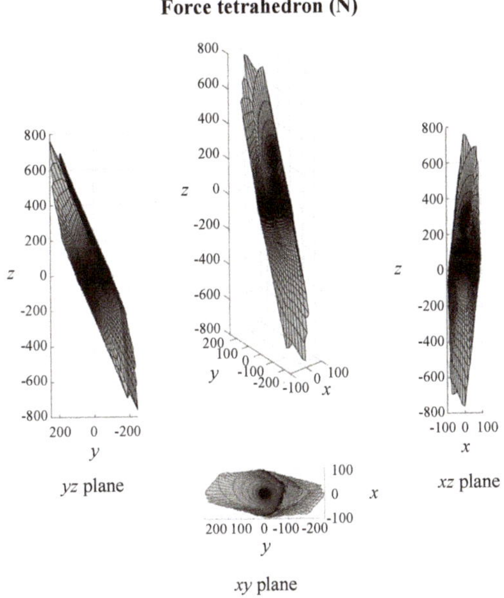

Figure 9. Static force performance in home configuration.

7. Conclusions

In this study we proposed a compact version of a Hexaglide parallel manipulator, commissioned by an Italian company operating in the micro-positioning sector. The main feature of the manipulator is its ability to exploit a Cartesian direction along which the entire mechanism can move without inner relative motions, with the aim of serving a few stations arranged for different mechatronic tasks. All the steps needed to functionally design a high-precision manipulator were addressed. The effect of commercial linear modules on the Cartesian positioning errors of the manipulator were investigated. The resolution of their sensors allows a high repeatability to be achieved along the sliding direction of the prismatic rails and high performance along the other directions, whose worst condition is still less than a micrometer for translations and less than a thousandth of a degree for rotations. Tailored linear modules with a longer stroke can be used to increase the workspace of the manipulator, while preserving the other kinetostatic features. Finally, all key functional design information of the manipulator has been given to prototype a compact Hexaglide.

An even more compact design can be obtained with a different leg tilt and similar stages with a shorter stroke of 150 mm, as shown in Figure 10 for a prototype of another confidential version of the Hexaglide by Vacuum Fab Srl. Figure 10 also shows the mechanical solution that can be used for the twelve spherical joints, based on a magnetic principle with self-centering. Preliminary tests on such a prototype have demonstrated a repeatability for linear and angular movements of about 0.2 μm and better than 0.001°, respectively.

Figure 10. Prototype of a more compact version of the Hexaglide manipulator with similar performance by Vacuum Fab Srl.

Author Contributions: Conceptualization, L.C. and M.-C.P.; methodology, L.C. and M.-C.P.; software, E.L. and F.D.; formal analysis, L.C. and G.P.; investigation, L.C. and G.P.; resources, M.-C.P.; data curation, G.P. and M.-C.P.; writing—original draft preparation, M.-C.P.; writing—review and editing, G.P.; supervision, M.-C.P. and G.G.; project administration, G.G.; funding acquisition, G.G. All authors have read and agreed to the published version of the manuscript.

Funding: This research received no external funding.

Conflicts of Interest: The authors declare no conflict of interest.

References

1. Ahn, C.; Seo, T.; Kim, J.; Kim, T.-W. High-tilt parallel positioning mechanism development and cutter path simulation for laser micro-machining. *Comput. Aided Des.* **2007**, *39*, 218–228.
2. Shimizu, Y.; Peng, Y.; Kaneko, J.; Azuma, T.; Ito, S.; Gao, W.; Lu, T.-F. Design and construction of the motion mechanism of an XY micro-stage for precision positioning. *Sens. Actuators A Phys.* **2013**, *201*, 395–406.
3. Zhang, D.; Li, X.; Ni, M.; Li, P.; Dong, L.; Rui, D.; Zhang, J. Design of 6 degree-of-freedom micro-positioning mechanism for optical elements. In Proceedings of the IEEE 13th International Conference on Electronic Measurement & Instruments (ICEMI), Yangzhou, China, 20–22 October 2017.
4. Xiao, R.; Shao, S.; Xu, M.; Jing, Z. Design and Analysis of a Novel Piezo-Actuated $XY\theta_z$ Micropositioning Mechanism with Large Travel and Kinematic Decoupling. *Adv. Mater. Sci. Eng. Hindawi* **2019**, *2019*, 5461725.
5. Tian, Y.; Shirinzadeh, B.; Zhang, D.; Zhong, Y. Modelling and analysis of a three-revolute parallel micro-positioning mechanism. *Proc. Inst. Mech. Eng. Part C J. Mech. Eng. Sci.* **2011**, *225*, 1273–1286.
6. Palpacelli, M.-C.; Palmieri, G.; Callegari, M. A Redundantly Actuated 2-Degrees-of-Freedom Mini Pointing Device. *J. Mech. Robot.* **2012**, *4*, 031012.

7. Palmieri, G.; Callegari, M.; Carbonari, L.; Palpacelli, M.C. Mechanical design of a mini pointing device for a robotic assembly cell. *Meccanica* **2015**, *50*, 1895–1908.
8. Lin, L.-C.; Tsay, M.-U. Modeling and control of micropositioning systems using Stewart platforms. *J. Robot. Syst.* **2000**, *17*, 17–52.
9. Liu, Z.; Cai, C.; Yang, M.; Zhang, Y. Testing of a MEMS Dynamic Inclinometer Using the Stewart Platform. *Sensors* **2019**, *19*, 4233.
10. Mura, A. Multi-dofs MEMS displacement sensors based on the Stewart platform theory. *Microsyst. Technol.* **2012**, *18*, 575–579.
11. Shi, H.; Duan, X.; Su, H. Optimization of the workspace of a MEMS hexapod nanopositioner using an adaptive genetic algorithm. In Proceedings of the IEEE International Conference on Robotics and Automation (ICRA), Hong Kong, China, 31 May–7 June 2014.
12. Rehsteiner, F.; Neugebauer, R.; Spiewak, S.; Wieland, F. Putting Parallel Kinematics Machines (PKM) to Productive Work. *Ann. CIRP* **1999**, *48*, 345–350.
13. Bernal, J.; Campa, R.; Soto, I. Kinematics and dynamics modeling of the 6-3-PUS-type Hexapod parallel mechanism. *J. Mech. Sci. Tech.* **2018**, *32*, 4555–4570.
14. Giberti, H.; La Mura, F.; Resmini, G.; Parmeggiani, M. Fully Mechatronical Design of an HIL System for Floating Devices. *Robotics* **2018**, *7*, 39.
15. Nabavi, S.N.; Shariatee, M.; Enferadi, J.; Akbarzadeh, A. Parametric design and multi-objective optimization of a general 6-PUS parallel manipulator. *Mech. Mach. Theory* **2020**, *152*, 103913.
16. Guo, J.; Wang, D.; Fan, R.; Chen, W.; Zhao, G. Kinematic calibration and error compensation of a hexaglide parallel manipulator. *Proc. Inst. Mech. Eng. Part B J. Eng. Manuf.* **2019**, *233*, 215–225.
17. Dong, W.; Du, Z.; Xiao, Y.; Chen, X. Development of a parallel kinematic motion simulator platform. *Mechatronics* **2013**, *23*, 154–161.
18. Spiewak, S.A.; Ludwick, S.J.; Hauer, G. A Test Setup for Evaluation of Harmonic Distorsions in Precision Inertial Sensors. *J. Manuf. Sci. Eng.* **2013**, *135*, 021015.
19. Gao, F.; Li, W.; Zhao, X.; Jin, Z.; Zhao, H. New kinematic structures for 2-, 3-, 4-, and 5-DOF parallel manipulator designs. *Mech. Mach. Theory* **2002**, *37*, 1395–1411.
20. Liu, C.H.; Huang, K.-C.; Wang, Y.-T. Forward position analysis of 6-3 Linapod parallel manipulators. *Meccanica* **2012**, *47*, 1271–1282.
21. Abtahi, M.; Pendar, H.; Alasty, A.; Vossoughi, G. Kinematics and Singularity Analysis of the Hexaglide Parallel Robot. In Proceedings of the ASME International Mechanical Engineering Congress and Exposition, Boston, MA, USA, 31 October–6 November 2008.
22. Yen, P.-L.; Ke, Z.-W.; Lu, T.-S.; Lu, C.-W. Development of a New Safety-Enhanced Surgical Robot Using the Hexaglide Structure. In Proceedings of the IEEE International Conference on Systems, Man and Cybernetics, The Hague, The Netherlands, 10–13 October 2004.
23. Kong, M.; Zhang, Y.; Sun, L.; Du, Z. Analysis of a New Workspace of the Hexaglide as a Motion Simulator for Fuel Tanker Trucks. In Proceedings of the International Conference on Mechatronics, Kumamoto, Japan, 8–10 May 2007.
24. Merlet, J.P. Jacobian, Manipulability, Condition Number, and Accuracy of Parallel Robots. *ASME J. Mech. Des.* **2006**, *128*, 199–206.
25. Gosselin, C.; Angeles, J. A Global Performance Index for the Kinematic Optimization of Robotic Manipulators. *ASME J. Mech. Des.* **1991**, *113*, 220–226.

Publisher's Note: MDPI stays neutral with regard to jurisdictional claims in published maps and institutional affiliations.

Article

Design of Soft Grippers with Modular Actuated Embedded Constraints

Gabriele Maria Achilli [1], Maria Cristina Valigi [2], Gionata Salvietti [3] and Monica Malvezzi [3,*]

[1]   Department of Engineering, University of Perugia, 05100 Terni, Italy; gabrielemaria.achilli@studenti.unipg.it
[2]   Department of Engineering, University of Perugia, 06125 Perugia, Italy; mariacristina.valigi@unipg.it
[3]   Department of Information Engineering and Mathematics, University of Siena, 53100 Siena, Italy; gionata.salvietti@unisi.it
*   Correspondence: monica.malvezzi@unisi.it

Received: 31 October 2020; Accepted: 4 December 2020; Published: 6 December 2020

Abstract: Underactuated, modular and compliant hands and grippers are interesting solutions in grasping and manipulation tasks due to their robustness, versatility, and adaptability to uncertainties. However, this type of robotic hand does not usually have enough dexterity in grasping. The implementation of some specific features that can be represented as "embedded constraints" allows to reduce uncertainty and to exploit the role of the environment during the grasp. An example that has these characteristics is the Soft ScoopGripper a gripper that has a rigid flat surface in addition to a pair of modular fingers. In this paper, we propose an upgraded version of the Soft ScoopGripper, developed starting from the limits shown by the starting device. The new design exploits a modular structure to increase the adaptability to the shape of the objects that have to be grasped. In the proposed device the embedded constraint is no rigid neither unactuated and is composed of an alternation of rigid and soft modules, which increase versatility. Moreover, the use of soft material such as thermoplastic polyurethane (TPU) reduces the risk of damage to the object being grasped. In the paper, the main design choices have been exploited and a finite element method (FEM) analysis through static simulation supports a characterization of the proposed solution. A complete prototype and some preliminary tests have been presented.

Keywords: wearable robots; underactuated robots; robotic manipulation

1. Introduction

1.1. Robotic Hands

The ability of grasping and manipulating objects with robotic systems and devices in a safe and robust way has been represented for at least three decades [1] as an important and challenging research topic in robotics, especially when high performance is required and/or in uncertain and unstructured environments, due to the complexity of grasping and manipulation tasks and device limits [2]. Robotic grippers are composed of an assembly of rigid joints and links [3]. The actuators can directly be applied to the links or the joints. Indeed, the actuators can be placed at the gripper base using cables or tendon-like structures. Robotic grippers can use different types of sensors, such as encoders, torque sensors, or accelerometers, to obtain information about position, velocity, and distance from the object [4]. A robotic gripper is normally used to grasp object placed in a definite space; research in this area of robotics led to the development of devices able to recognize targets as an object, or an obstacle and create the best path planning strategy to complete assigned tasks [5–7]. Although several interesting solutions have been presented in the literature, the design and control of robotic hands and grippers still represent an open and challenging problem, and several different solutions have been proposed. Some implementations resemble an anthropomorphic structure [8,9] while other ones exploit underactuation and parallel kinematic structures [10]. Also, the applications cover a wide range of possibilities: from humanoids [8,11], to prostheses [12], to space applications [10]. A wide variety of solutions have been proposed in terms of mechanical structure and actuation systems, not limited to electro-mechanic transmissions, such as pneumatic system [13] or shape memory alloy [14]. One interesting research branch regards the development of solutions exploiting underactuation [15–17] and modularity [18], to reduce the complexity of the hand by maintaining a suitable level of performance. Another aspect to be considered in the mechanical design of robotic hands is represented by transmission systems [19]. In particular, in the literature, tendon-driven mechanisms have been widely used in articulated-finger robotic hands. Several interesting issues on tendon-driven mechanisms regarding tendon redundancy and joint stiffness adjustability for a robotic mechanism driven with redundant tendons are discussed in [20].

Given the high number of degrees of freedom (DoFs) needed by a robotic hand and finger, underactuation is an important aspect to be considered to simplify the mechanical structure, reduce the mass, and improve robustness, in designing safe and robust robotic hands and fingers. The term underactuated refers to mechanisms that have fewer actuators than DoFs: it is evident that reducing the number of actuated variables in robotic hands could decrease the overall manipulability properties and the capability to adapt to different shapes and dimensions of grasped objects, on the other hand it has been demonstrated that underactuation associated with compliance can provide interesting properties to the device, like for instance self-adaptability [21,22]. Underactuated mechanisms help to achieve an adaptive grasp similar to human grasping. Also, human hands present some coupling between joints, e.g., the distal interphalange joints of the fingers are not independently controllable [23–26].

1.2. Differential Mechanisms

When a few motors are employed to activate many joints, driving simultaneously the opening and closing motion of the fingers, a differential mechanism is necessary. For instance in [27–29] a simple differential mechanism was used to decouple finger motions when one of them was constrained, for example when one of the fingers contacts an object or an external surface. In the literature there are many applications of differential mechanisms for robotic fingers and hands [30]. In [31], a differential system based on gears is used for a novel architecture of robotic hand and the properties of differential mechanisms arranged in cascade via parallel or serial connections is studied. In [32], a planetary gear solution and a fluid T-pipe scheme are described. In [33] a moving pulley differential mechanism was used, while in [13], a differential with a T-shape fluid mechanism and the connected seesaw circuit is presented. In [34], an underactuated anthropomorphic gripper for prosthetic applications was presented, in which a mechanical lever inside the palm allowed to extend the grasping capabilities and improve the force transmission ratio of the gripper. This mechanism was further developed in [35], whereby the differential mechanism included a set of locking buttons that allow the user to stop the motion of each finger.

1.3. The Scoop Hand

One of the strategies that can be employed with underactuated compliant hands to compensate the lack of control is represented by environmental constraint exploitation [36]. This exploitation has been inspired by human behaviours, that often use environmental features (e.g., flat surfaces, edges, corners, etc.) in an active way to support the hand in reaching the object or to reposition it [37]. In [38], a flat active surface was added to an underactuated modular gripper to embed on the hand a constraint surface that could help in grasping and manipulating objects. In the solution presented in [38] the flat surface was approximately rigid and had only one actuated DoF at its base allowing a limited bending.

1.4. Paper Contribution

In this paper, we present an improvement of the idea proposed in [38], guided by: (i) a further exploitation of system compliance by means of flexible materials and (ii) providing a closure motion also to the flat embedded constraint. Different solutions were evaluated and compared, in terms of complexity, payload, versatility. The solution that has been identified as optimal has a mobile and compliant proximal part and a rigid distal element shaped to easy the approach to the object to be grasped. This solution has been prototyped and some preliminary tests were performed.

1.5. Paper Organization

In the first part of the paper, we explain the context of grasping with modular robotic hands and the principal devices used for this purpose. We analyze and compare some proposals to improve a device already tested and working, to characterize some specific functions. Some developed prototypes, the design phases and the studies of the functions which have in the field of grasping are shown. Finite element method (FEM) simulations are carried out to evaluate behavior in working conditions. Finally, the proposed versions are compared with the starting model to evaluate the improvements.

2. Motivation and Starting Point

2.1. Modular Hands

The field of robotic anthropomorphic grippers has been very thorough in recent years, through the introduction of modular structures [39]. The modular approach increases the possible configurations of the device and improve its versatility. Using modules with the same geometrical dimensions, the gripper can be disassembled and reassembled to adapt it to new and different tasks [40]. Another important point of modular structures concerns the low-cost of their production: it will be necessary to produce and assemble copies of a few types of module for realizing different configurated devices [39].

Grippers are usually designed to grasp single objects with a definite shape: this characteristic leads to an optimization of the device design. The modular structure can improve this limit by assembling the device differently [39]. The soft modular hand grippers can adapt to many different shapes of various grasped objects. Moreover, mechanical elements, such as a differential system [38], allow an asymmetrical grasp, allowing them to take even more objects, if properly identified.

In soft manipulation many grippers exploit the external environment to reduce the uncertainties in grasping. In this case, the external environment can be a constraint able to improve the capability to grasp [36].

The addition of embedding constraint into a gripper allows to manipulate a specific object that have an unknown shape, maybe even inserted within a group of other different objects. An underactuated soft hand is an example of this kind of device.

2.2. The Starting Point Solution, Main Features, and Limits

This paper proposes an upgraded version of "The Soft ScoopGripper" [38], a modular, underactuated soft gripper with embodied constraints (Figure 1a). This device is composed of two modular fingers actuated by a single tendon through a differential system and a rigid scoop, which represents the constraint. The rigid modular parts of the fingers are connected together using flexible joints: this configuration allows to have a deformable structure able to adapt to the shape of objects. The rigid scoop, that is connected to the base of the gripper with a flexible prismatic hinge, increases the certainty of the grasp (Figure 1b). An important feature of the gripper is the capability of grasping object not necessary placed on a regular surface, as a plane. Again, the large scoop increases the ability to bear weighty object put on it.

However, using the Soft ScoopGripper, some attention must be paid to its limitations. The grasped objects have to be adequate in term of stiffness to avoid their damage. Use depends on the shape of the objects, and for this reason a flat surface is required. For the gasp of small objects, the user needs a good dexterity to prevent them from falling; in the case of targets with different shapes, versatility and a better reconfiguration is required. According to these issues, this paper proposes many improvements and skills, such as a new design and different movement structure.

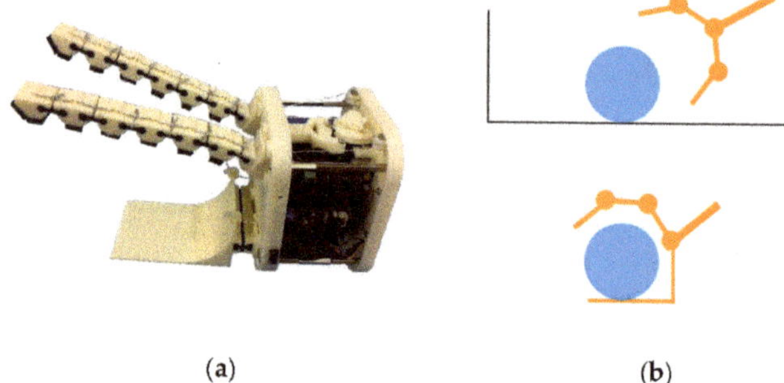

(a)   (b)

Figure 1. The *Soft ScoopGripper* prototype presented in [38] (**a**) and the main idea of soft hands with embodied constraints (**b**). (© 2019 IEEE)

3. Design Improvements

3.1. Exploiting Soft Materials

The first improvement proposed in this paper consists in making flexible and active the embedded constraint, that in the first prototype was realized with a rather stiff material (acrylonitrile styrene acrylate, ASA) and could only rotate with respect to the palm, thanks to an elastic connection in the proximal part. This almost rigid realization makes the structure not enough adaptable to the environment uncertainties and to the object. A rigid and thin structure could also provide high contact pressures during the grasp and damage fragile objects (e.g., vegetables, fruits, etc.). As a first improvement of the device, in order to guarantee a higher adaptability, the rigid link has been designed to be realized with a flexible material, in particular TPU was chosen since it is easy to manufacture with standard FDM technologies. TPU is a highly elastic polymer, used for applications that require a good flexibility; this material has a low Young's modulus value compared to a rigid plastic, such as ASA [41]. Two holes were designed into the single body scoop for the routing of two artificial tendons, from the top to the bottom, needed to actuate the soft hand. Moreover, a pair of holes are planned for inserting cylindrical steel pins, which allow for the recovery of a suitable level of mechanical stiffness in the central part of the structure and allow to lift heavier objects.

A FEM analysis was realised on the CAD model (Figure 2a,b). This process was used to show the distribution of displacement and stress over the soft structure. The test was realized simulating the grasp and lift of an object with a mass of 2 kg, so a weight force of 19.61 N was considered, which was placed on the end of the scoop. The results (Figure 3a,b) show that there are no high stress areas, with the exception of the contact region between the soft scoop and the support. However, at the distal part of the scoop, a displacement of 79.01 mm was shown (Figure 3c). It is evident that the high flexibility of the employed material would lead to deformations unsuitable for a stable and robust grasp of objects with masses higher than 1–2 kg, which can be assimilated to a weight force of 10–20 N. Four prototypes of the model were 3D printed in TPU material, using different infill percentages. The infill density percentage has an influence on the material mechanical properties, e.g., on the Young modulus (see Table 1) [42].

Table 1. Four different infill density percentages applied during the 3D printing process and the corresponding variation of Young's modulus E, that affect the stiffness and the flexibility of the scoop.

Prototype	Infill Percentage %	E [MPa]
#1	30	1.38
#2	50	2.07
#3	70	6.53
#4	90	9.45

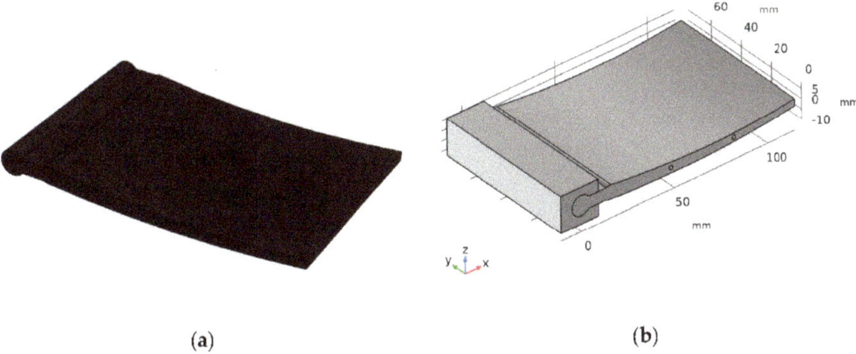

(a)

(b)

Figure 2. (**a**) A 3D view of a scoop made of soft material. (**b**) The FEM model for the simulation of the static stress over the scoop.

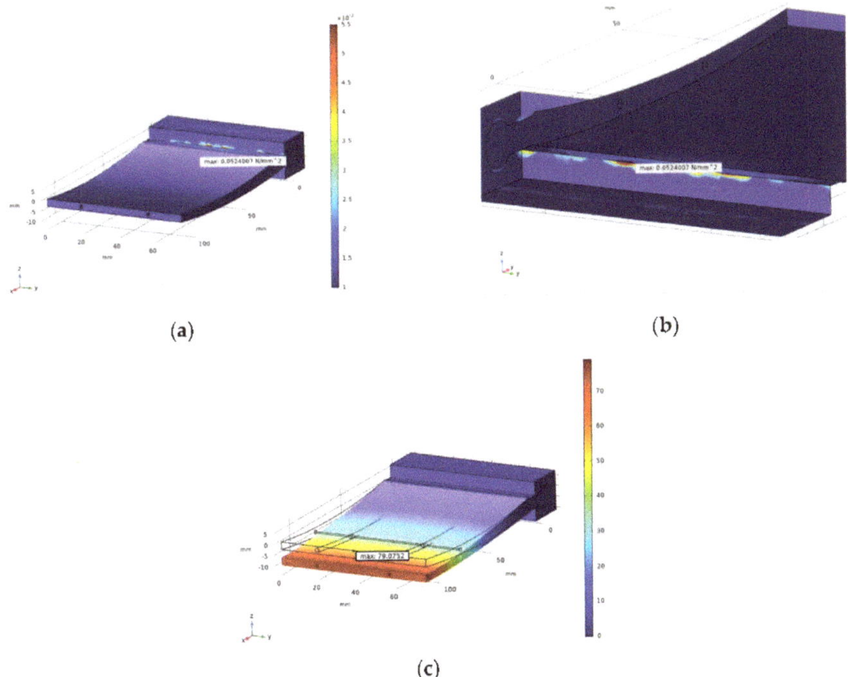

(a)

(b)

(c)

Figure 3. Results of the FEM analysis on the flexible version, a load of 19.61 N on the tip edge was applied. (**a**) equivalent Von Mises stress distribution and (**b**) detail area; (**c**) overall displacement.

Therefore, the prototypes were tested on an experimental setup, fixing the support with the scoop to a surface, inserting the artificial tendons and cylindrical pins inside the holes. The test was divided into two parts: analysing the flexion of the scoop by applying a force to the tendons, to simulate the action of the differential mechanism [28], and applying a test weight to the surface to verify the results obtained from the FEM. The results of this test show a larger value of the deformation with respect to the FEM evaluations in the deformation, when a 19.61 N load was applied, the deformation was not measured due to its irregularities and not repeatability. Experimental results also showed some issues with the device actuation. Both the numerical evaluations and the experimental measures showed that this solution presents several limitations: using a part entirely in TPU makes the scoop excessively deformable and unable to stably grasp objects with mass higher than 1–2 kg, and furthermore the actuation system present on the device is not optimal for this design. To proceed with this solution, it would be necessary to review the entire structure, totally overturning the pre-existing system, starting from the positioning of the motors to the entire kinematic chain that is the basis of the robot hand. For this reason, it was decided not to continue with the development of this version and to change the entire structure of the component.

3.2. Modular Elements/1

Starting from the Soft ScoopGripper finger structure, the design of the phalanges was changed to make them as much similar as possible to the scoop (see Figure 1a), with a different articulation system. This idea was an extension of the rigid modules realized in ASA, which replace the couple of fingers with a single articulated platform, as shown in (Figure 4a).

This version retains the advantage of the actuation mechanism already present on the original device, redesigning the new paired modules. During grasping, the modules get close each other, forming a continuous surface (Figure 4b), where the object can be leaned. This feature would allow to grasp even small objects. The new modules are 3D printed in ASA and are connected by flexible passive elements realized with TPU. To avoid damage to the objects, TPU strips are placed on the inner face of the modules.

This design allows to overcome some limits of the previous design, but during the development of the prototype, some problems were highlighted, such as the assembly time, that was excessive and not suitable for a device that is intended to be easily customizable and adaptable to task variations. Furthermore, the installation of the TPU profiles on the internal surface of the modules would not guarantee stability, as the glue present appears to have difficulties in adhesion in the long term. These problems encountered in the preliminary prototyping stages have required to develop a new solution which makes the device even more versatile and easier to assemble.

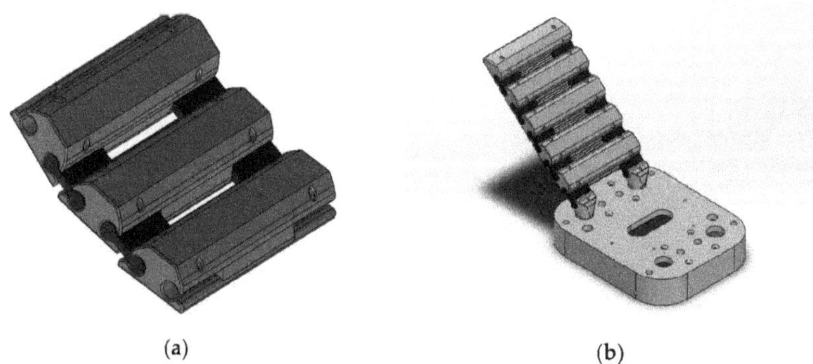

(a) (b)

Figure 4. Modular coupled fingers: (**a**) The scoop is substituted by an articulated platform with the same modular structure of the finger, and larger rigid elements; (**b**) The prototype CAD design.

3.3. Modular Elements/2

The structure of the third solution was inspired by the fingers of the initial Soft ScoopGripper. In particular a motor fixed on the base of the device rotates a pulley that is connected to an artificial tendon fixed to a slide; the slide is connected to a separate pair of fingers through other two tendons, composed of rigid modules and flexible joints, and fixed on the last module of each single finger. Through the analysis of the Soft ScoopGripper, the proposed solution includes TPU modules inside the new rigid modules, while not compromising the characteristics and the mobility of the device (see Figure 5g).

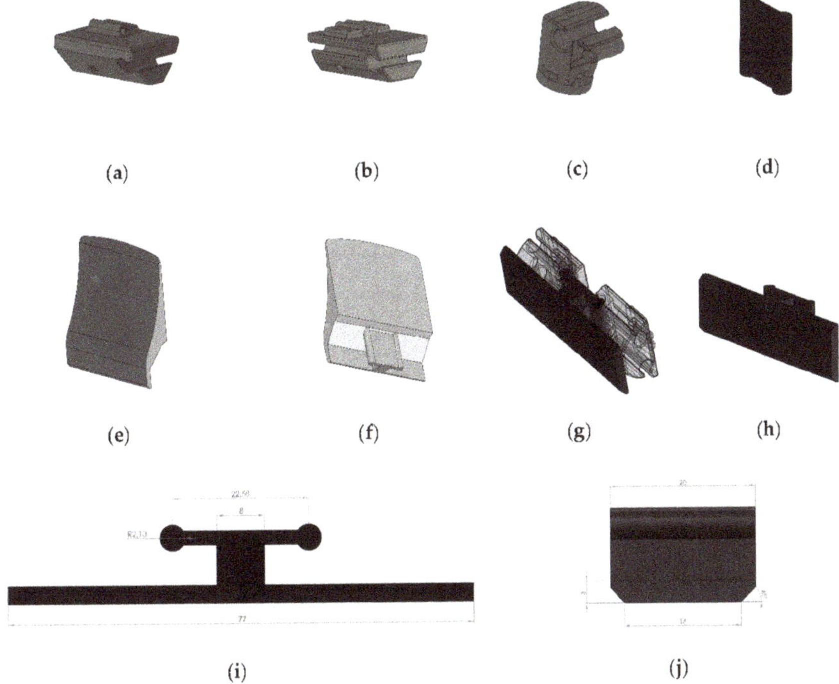

Figure 5. Modular elements that compose the gripper: (**a**) Ending module, (**b**) middle module and (**c**) starting module of the robotic finger of the gripper. (**d**) Flexible joint placed between a pair of modules. (**e**) Module B, front and (**f**) back view. (**g**) Two middle modules joined module A. (**h**) Module A, (**i**) top and (**j**) side view.

A flap with a through hole was added on the module to allow the inserting of a TPU module (Figure 5a,b). To ensure stability and at the same time ease of assembly, different types of modules have been provided. All these modules, made of ASA, and are connected each other through elastic TPU joints (Figure 5d). The same profile of the TPU joints was used, thus keeping the hole diameter unchanged; this choice improves further modularity and takes advantage of an already tested and functional design. Once the flap was developed, we proceeded to the design of the TPU module. A classification was performed, dividing them into two types, module A (Figure 5h) and module B (Figure 5e,f). They differ in the function that they will have to perform. Module A will be used on each pair of phalanges, while the end module will be used only on the top of the scoop, where this must have a shape that allows it to be positioned below the object to be manipulated. After some preliminary tests on 3D printed versions of module B, it was decided to use ASA as material for these components.

The handle of Soft ScoopGripper can be modelled and customized according to the user's size hand for a better grip and a further development of this gripper will be the use of 3D scanner techniques [43,44] to customize the design of the support handle.

4. Analysis and Comparison

Numerical Evaluations with FEM Analysis

The next phase of this work was the study of static stresses applied to the scoop, using FEM tools. The first step was the definition of a simplified parametric CAD structure similar to the device (see Figure 6). Some details on the model that were not necessary for the analysis were removed to simplify the analysis without influencing the accuracy of the results. The simplifications were introduced for device shape and the elements that compose it, the structure, the constraints and forces used for the kinematic analysis. The upper part of the prototype was examined, holding the two base phalanges using fixed constraints, and applying a roller on the end phalanges: in this way a realistic scoop behavior was emulated. Two different system configurations have been developed: the first, which a constraint was applied to the top of the scoop, to reproduce the behavior when the force was applied by the artificial tendons; the other configuration assesses the resistance of the device when there is no such force, and was obtained by removing the constraint. For both configurations, four tests were carried out with different conditions. A force equal to 19.61 N was applied, equivalent to lifting an object with a mass of 2 kg; this value was chosen to compare the results obtained with the FEM analysis with the ones obtained with the real prototype, which were carried out with objects having a maximum mass of 2 kg. The force was applied to different modules with the same constraint conditions (Table 2), to evaluate the behavior of the device at the different positions that can be assumed by the object taken during grasping. A summary of the results is shown in (Table 3).

Table 2. Configuration of force application.

Case	Modules to which Force is Applied
1	Modules 2 and 3
2	Module 3
3	Modules 3 and 4
4	Module 4

Table 3. Results of FEM analysis performed in the two different configurations.

Case	Gripper with No Constraint			Gripper with Constraint		
	Analysis	Min Value	Max Value	Analysis	Min Value	Max Value
1	Displacement	0 mm	22.3 mm	Displacement	0 mm	0.507 mm
	Von Mises Stress	5.50×10^{-4} N/mm^2	22.7 N/mm^2	Von Mises Stress	6.49×10^{-5} N/mm^2	3.2 N/mm^2
2	Displacement	0 mm	16.6 mm	Displacement	0 mm	0.604 mm
	Von Mises Stress	3.48×10^{-4} N/mm^2	15.3 N/mm^2	Von Mises Stress	6.66×10^{-5} N/mm^2	2.82 N/mm^2
3	Displacement	0 mm	22.3 mm	Displacement	0 mm	0.419 mm
	Von Mises Stress	8.13×10^{-4} N/mm^2	27.4 N/mm^2	Von Mises Stress	2.19×10^{-5} N/mm^2	1.76 N/mm^2
4	Displacement	0 mm	33.7 mm	Displacement	0 mm	22.3 mm
	Von Mises Stress	5.95×10^{-4} N/mm^2	18.9 N/mm^2	Von Mises Stress	5.90×10^{-5} N/mm^2	1.96 N/mm^2

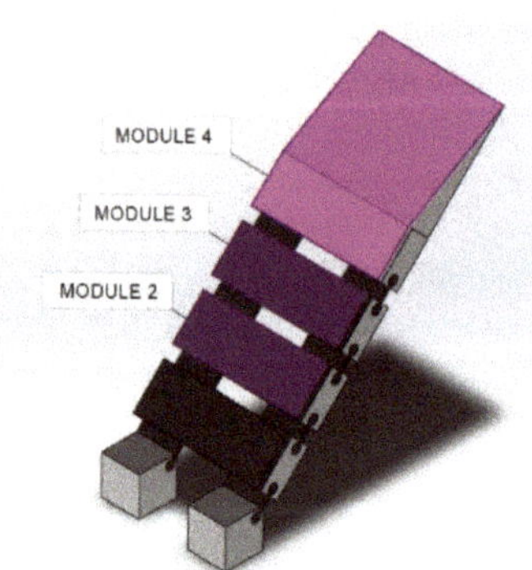

Figure 6. Simplified 3D model for FEM analysis.

5. Prototyping and Testing

FEM analysis results showed some issues concerning the excessive deformation of the device in action. The choice of using TPU as material for the realization has been identified as the main cause of these issues. The elastic modulus E of this material combined with its difficult printability for FDM technology [45] limits its applicability for the parts that require a high accuracy. Because of these problems, ASA was used to construct the end part of the scoop. This material has different properties with respect to TPU (summarized in Table 4) in particular it is much stiffer and therefore it allows to realize components with a lower flexibility. In addition, ASA is a material that is easier to be printed with respect to TPU, allowing for better resolution and thinner thicknesses. The prototype made with 3D printing additive manufacturing technology is visible in (Figure 7a,b).

Table 4. Mechanical properties of the materials under study.

Material Properties	ASA	TPU
Elastic modulus (E)	29 N/mm^2	15.2 N/mm^2
Poisson ratio	1.03	1.29
Density	1070 kg/m^3	1200 kg/m^3

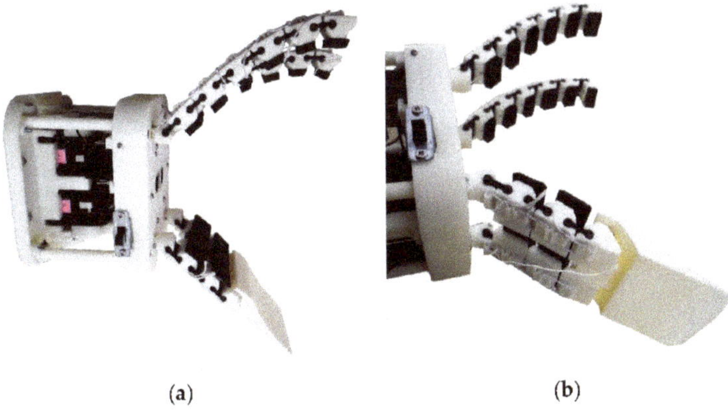

(a)   (b)

Figure 7. The soft gripper with modular embedded constraints prototype (**a**) and a detail of the modular scoop (**b**).

Closure motion: The assembled device was subject to a series of tests to allow the tracking of the closure movements and end-tip trajectories followed by the scoop in a real set up. Three different configurations were considered, which allowed to evaluate the overall behavior of the device: grasping without object (Figure 8a); grasping of an object placed in the center of the scoop (Figure 8b); grasping of an object placed on the outer edge of the scoop (Figure 8c). Comparing the data extracted from the measurements of the different configurations (Table 5), it was possible to describe how the scoop grasps an object, according to its position. The maximum displacement and rotation were found in case 1, where no object was present; this can be explained due to the absence of an obstacle that slows down the movement of the scoop. In case 2, however, the minimum values conducted in this test were found: this could be caused by the higher torque required by the actuator to move the object, which has a lower distance between the point of application of the force and the rotation center, compared to case 3.

Gripper Control. In addition to the FEM analysis and the study of the trajectory, it was necessary to test the behavior in a real environment to characterize the device. A structure designed for the Soft ScoopGripper [38] was used to analyze all the capabilities of the gripper. This structure is composed of two parts: a handle, molded in ASA, to allow the operator to grab the device easily, and an interface with two buttons, which allow the opening (extension) and closing (flexing) of the scoop, inserted inside the handle itself. Figure 9 reports the finite state machine (FSM) developed to control the gripper, valid for both the finger part and the scoop. The single pressure of the closing button activates the e1 event, which triggers the closing action (flexing) of the fingers and the blade, until contact with an object is detected; upon detection of the presence of the object, by monitoring the torque applied to the electric motor by the microcontroller, the device stops the bending, and passes to the next state, contact/torque mode. In this phase, it is possible to adjust the force applied to the object being gripped by consecutively pressing the relative button, which starts a new state of flexion. This sequence is interrupted when a predetermined threshold is exceeded, which defines the maximum allowable flexion of both the scoop and the finger, called fully flexed. At this stage, the object is completely grasped by the device. The second button of the control interface allows you to activate the extension function, which reduces the moment applied by the electric motor and releases the object. This function can be activated at any time during the process.

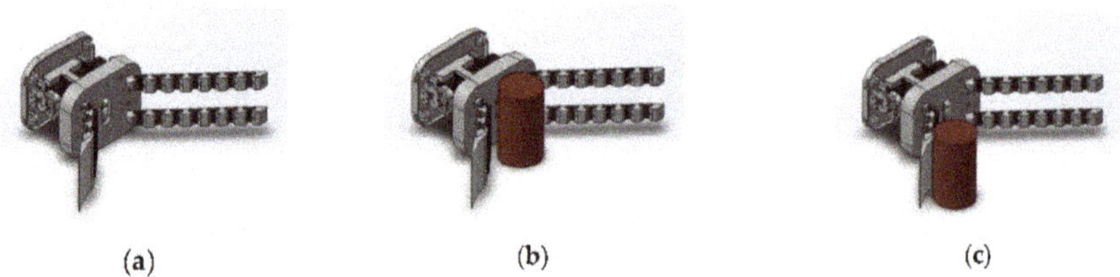

(a) (b) (c)

Figure 8. Three different configurations employed to study the movements of the device: (**a**) grasping without object. (**b**) Grasping of an object placed in the center of the scoop. (**c**) Grasping of an object placed on the outer edge of the scoop.

Table 5. Data obtained from the analysis of the rotation and translation when the scoop is grasping.

	Configuration	Rotations [deg]	Translations [mm]
		Z	Y
1	No object	73.5	64.3
2	Center	53.9	49.3
3	Outer edge	72.2	59.7

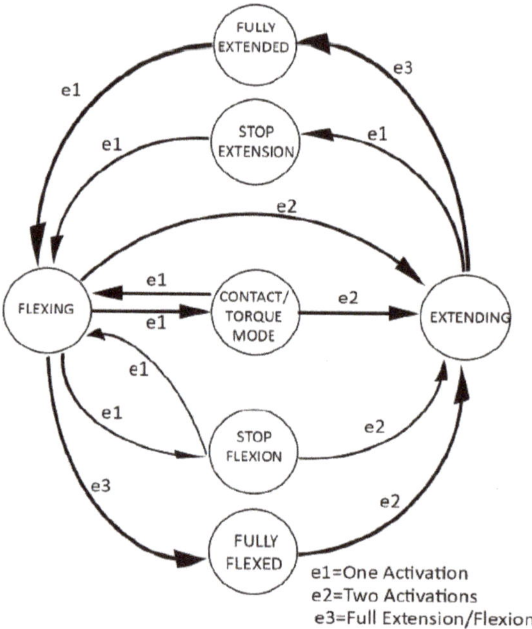

Figure 9. Finite State Machine for gripper control.

Tests with the YCB object set: Following what has already been done for the Soft ScoopGripper [38], tests based on the so-called YCB object and model set and benchmarking were performed [46]. The YCB

Object and model set have been designed for facilitating benchmarking in robotic manipulation and they are quite widely adopted in the robotics community to test and compare the grasp and manipulation capabilities of robotic hands and grippers. The availability of an object set for benchmarking allows to compare different types of end-effector, evaluating their grasping characteristics on a heterogeneous group of objects, which differ in shape, weight and stiffness. All the experimentation phases were carried out both for the new version of the scoop and for the previous one, so to have a comparison criterion in the same conditions and with the same samples, to note the differences and the capabilities of the different solutions. To grip different types of objects, it was necessary to apply a reconfiguration of the fingers: the mechanical system designed allows to rotate the components at the base of the two fingers to allow grasping objects having a cylindrical symmetry, without having to rotate the device (Table 6, case 3). Using a scoop reduces the force required on the fingers to hold the object; the scoop will take most of the weight, while the fingers will keep the object stable, to not lose the grasp. The comparison table between the two devices shows how the new version, despite a greater compliance given by a less rigid structure, allows to support heavy objects (Table 6, case 2), also adapting to the shapes of objects with non-flat surfaces (Table 6, cases 1 and 3). This feature has shown greater safety in the grip compared to the previous version. In Table 6 (case 4), it is shown that the contact surface formed by the TPU modules, with greater flexibility than the rigid ASA scoop, limits the risk of damage.

Table 6. Comparison between different object grasped with the gripper.

	Case 1	Case 2	Case 3	Case 4
A				
B				

6. Conclusions and Future Work

6.1. Conclusions

Table 7 summarizes a qualitative comparison relating to the limits of the Soft ScoopGripper previously reported, analyzing which of these had been overcome by the proposed prototypes. The prototype 3 can overcome all the limits encountered in the original project. Indeed, in prototype 1, the main problems have not been sufficiently resolved, due to the excessive deformability of the chosen material (TPU) and the implementation system, with little improvement possibilities, due to the limits shown. Prototype 2 was used as a basic idea for the development of the prototype 3, as it provided the opportunity to showcase the potential of the modular design, despite not having accomplished a real experimentation phase.

In summary, prototype 3, that uses a modular system having a high ability to adapt to the object, shows good versatility and the possibility of reconfiguration. This allows to move from a structure made up of two pairs of fingers to one having a couple of fingers and a scoop in an assembly time of a few minutes, thanks to the developed interchangeable modules. Furthermore, as visible from the comparative scheme (Table 6), despite the use of a yielding structure and with greater deformability, the grasping and sealing capacity of the gripping objects is unchanged compared to the Soft ScoopGripper.

6.2. Future Work

The studies carried out so far have mainly focused on the design and choice of the most suitable device. Future studies will be used to perform an experimental quantitative characterization of the gripper. Furthermore, tests will be carried out according to ISO 14539 standard to validate the work.

Currently, we are also identifying the suitable application fields for this type of gripper. The ability to manipulate objects by simulating the action exerted by a human hand is required in those areas where the integrity of the product itself and that of the operator performing the task prevail. Two fields have been identified, namely agri-food and waste industry. In both areas, grasping ability and not damaging the object are required, and this device is suitable for this task. Therefore, we are working to replicate a convey-belt system on a laboratory scale to verify grasping capabilities in a more complex environment. Another part of the research will be the study of other types of materials and manufacturing for the realization of gripper components.

Table 7. Summary comparison between three different prototypes.

Soft Scoop Gripper Properties	Prototypes		
	1	2	3
Capability to adapting to non-flat rigid surfaces, where the stiffness of the material and the scoop shape make it difficult to insert it under an object.	No	Yes	Yes
Avoiding damage of grabbed objects, always due to the stiffness of the material of which the scoop is made.	Yes	Yes	Yes
Enough adaptability to the shape of objects; if the shape of the object does not have a flat surface, the grip is almost exclusively performed by the fingers, and the scoop works as a constraint only.	Yes	Yes	Yes
Good mobility of the scoop; no limited movement, which does allow a secure grip if the device is trying to grasp small objects.	No	Yes	Yes
Capability to selecting a particular target within a heterogeneous mix of different shape objects, as extracting a ball from inside a basket of toys.	No	Yes	Yes
Versatile and easily reconfigurable.	No	No	Yes

Author Contributions: G.M.A. contributed to investigation, conceptualization, data curation and writing (original draft, review & editing); M.C.V. contributed to supervision and writing (original draft); G.S. contributed to conceptualization and writing (review & editing); M.M. contributed to methodology and supervision. All authors have read and agreed to the published version of the manuscript.

Funding: The work was partially funded by "Fondo Ricerca di Base 2019, dell'Università degli Studi di Perugia" for the project:" Studio di base per tecnologia aptica: modellazione teorica e validazione sperimentale del toccare umano in 3D" and by project INBOTS CSA, Inclusive Robotics for a better Society, Program H2020-ICT-2017-1; grant agreement n. 780073.

Conflicts of Interest: The authors declare no conflict of interest.

References

1. Mason, M.T.; Salisbury, J.K., Jr. *Robot Hands and the Mechanics of Manipulation*; The MIT Press: Cambridge, MA, USA, 1985.

2. Carbone, G. *Grasping in Robotics*; Springer: Berlin/Heidelberg, Germany, 2012; Volume 10.

3. Melchiorri, C.; Kaneko, M. Robot Hands. In *Springer Handbook of Robotics*; Siciliano, B., Khatib, O., Eds.; Springer: Berlin/Heidelberg, Germany, 2008; pp. 345–360. [CrossRef]

4. Shintake, J.; Cacucciolo, V.; Floreano, D.; Shea, H. Soft robotic grippers. *Adv. Mater.* **2018**, *30*, 1707035. [CrossRef] [PubMed]

5. Haidegger, T.; Galambos, P.; Rudas, I.J. Robotics 4.0–Are we there yet? In Proceedings of the 2019 IEEE 23rd International Conference on Intelligent Engineering Systems (INES), Gödöllő, Hungary, 25–27 April 2019; pp. 000117–000124.

6. Kehoe, B.; Berenson, D.; Goldberg, K. Toward cloud-based grasping with uncertainty in shape: Estimating lower bounds on achieving force closure with zero-slip push grasps. In Proceedings of the 2012 IEEE International Conference on Robotics and Automation, St Paul, MN, USA, 14–18 May 2012; pp. 576–583.

7. Takács, K.; Mason, A.; Christensen, L.B.; Haidegger, T. Robotic Grippers for Large and Soft Object Manipulation. In Proceedings of the IEEE 20th International Symposium on Computational Intelligence and Informatics (CINTI 2020), Budapest, Hungary, 5–7 November 2020.

8. Fukaya, N.; Toyama, S.; Asfour, T.; Dillmann, R. Design of the TUAT/Karlsruhe humanoid hand. In Proceedings of the 2000 IEEE/RSJ International Conference on Intelligent Robots and Systems (IROS 2000) (Cat. No. 00CH37113), Takamatsu, Japan, 31 October–5 November 2000; pp. 1754–1759.

9. Butterfaß, J.; Grebenstein, M.; Liu, H.; Hirzinger, G. DLR-Hand II: Next generation of a dextrous robot hand. In Proceedings of the 2001 ICRA. IEEE International Conference on Robotics and Automation (Cat. No. 01CH37164), Seoul, Korea, 21–26 May 2001; pp. 109–114.

10. Martin, E.; Desbiens, A.L.; Laliberté, T.; Gosselin, C. SARAH hand used for space operation on STVF robot. In Proceedings of the International Conference on Intelligent Manipulation and Grasping, Genova, Italy, 1–2 July 2004; pp. 279–284.

11. Roccella, S.; Carrozza, M.C.; Cappiello, G.; Dario, P.; Cabibihan, J.-J.; Zecca, M.; Miwa, H.; Itoh, K.; Marsumoto, M. Design, fabrication and preliminary results of a novel anthropomorphic hand for humanoid robotics: RCH-1. In Proceedings of the 2004 IEEE/RSJ International Conference on Intelligent Robots and Systems (IROS)(IEEE Cat. No. 04CH37566), Sendai, Japan, 28 September–2 October 2004; pp. 266–271.

12. Dechev, N.; Cleghorn, W.; Naumann, S. Multiple finger, passive adaptive grasp prosthetic hand. *Mech. Mach. Theory* **2001**, *36*, 1157–1173. [CrossRef]

13. Raparelli, T.; Mattiazzo, G.; Mauro, S.; Velardocchia, M. Design and development of a pneumatic anthropomorphic hand. *J. Robot. Syst.* **2000**, *17*, 1–15. [CrossRef]

14. Maffiodo, D.; Raparelli, T. Three-fingered gripper with flexure hinges actuated by shape memory alloy wires. *Int. J. Autom. Technol.* **2017**, *11*, 355–360. [CrossRef]

15. Niola, V.; Penta, F.; Rossi, C.; Savino, S. An underactuated mechanical hand: Theoretical studies and prototyping. *Int. J. Mech. Control* **2015**, *16*, 11–19.

16. Cosenza, C.; Niola, V.; Savino, S. Analytical study for the capability implementation of an underactuated three-finger hand. In *New Trends in Medical and Service Robotics*; Springer: Berlin/Heidelberg, Germany, 2019; pp. 161–168.

17. Niola, V.; Rossi, C.; Savino, S.; Timpone, F. Study of an underactuated mechanical finger driven by tendons. *Int. J. Autom. Technol.* **2017**, *11*, 344–354. [CrossRef]

18. Maffiodo, D.; Raparelli, T. Comparison among different modular SMA actuated flexible fingers. In *Advances in Italian Mechanism Science*; Springer: Cham, Switzerland, 2018; pp. 324–331.

19. Carbone, G.; Rossi, C.; Savino, S. Performance comparison between FEDERICA hand and LARM hand. *Int. J. Adv. Robot. Syst.* **2015**, *12*, 90. [CrossRef]

20. Kobayashi, H.; Hyodo, K.; Ogane, D. On tendon-driven robotic mechanisms with redundant tendons. *Int. J. Robot. Res.* **1998**, *17*, 561–571. [CrossRef]

21. Laliberté, T.; Gosselin, C.M. Underactuation in space robotic hands. In Proceedings of the Sixth International Symposium on Artificial Intelligence, Robotics and Automation in Space ISAIRAS: A New Space Odyssey, St-Hubert, QC, Canada, 18–22 June 2001.

22. Hussain, I.; Salvietti, G.; Spagnoletti, G.; Prattichizzo, D. The soft-sixthfinger: A wearable emg controlled robotic extra-finger for grasp compensation in chronic stroke patients. *IEEE Robot. Autom. Lett.* **2016**, *1*, 1000–1006. [CrossRef]

23. Laliberte, T.; Birglen, L.; Gosselin, C. Underactuation in robotic grasping hands. *Mach. Intell. Robot. Control* **2002**, *4*, 1–11.

24. Ceccarelli, M.; Tavolieri, C.; Lu, Z. Design considerations for underactuated grasp with a one DOF anthropomorphic finger mechanism. In Proceedings of the 2006 IEEE/RSJ International Conference on Intelligent Robots and Systems, Beijing, China, 9–15 October 2006; pp. 1611–1616.

25. Wu, L.; Carbone, G.; Ceccarelli, M. Designing an underactuated mechanism for a 1 active DOF finger operation. *Mech. Mach. Theory* **2009**, *44*, 336–348. [CrossRef]

26. Carbone, G.; Yao, S.; Ceccarelli, M.; Lu, Z. Design and simulation of a new underactuated mechanism for LARM hand. *Robotica* **2017**, *35*, 483–497.

27. Hussain, I.; Malvezzi, M.; Gan, D.; Iqbal, Z.; Seneviratne, L.; Prattichizzo, D.; Renda, F. Compliant gripper design, prototyping, and modeling using screw theory formulation. *Int. J. Robot. Res.* **2020**, 0278364920947818.

28. Malvezzi, M.; Iqbal, Z.; Valigi, M.C.; Pozzi, M.; Prattichizzo, D.; Salvietti, G. Design of Multiple Wearable Robotic Extra Fingers for Human Hand Augmentation. *Robotics* **2019**, *8*, 102. [CrossRef]

29. Malvezzi, M.; Valigi, M.C.; Salvietti, G.; Iqbal, Z.; Hussain, I.; Prattichizzo, D. Design criteria for wearable robotic extra–fingers with underactuated modular structure. In Proceedings of the International Conference of IFToMM ITALY, Naples, Italy, 29–30 November 2018; pp. 509–517.

30. Birglen, L.; Gosselin, C.M. Force analysis of connected differential mechanisms: Application to grasping. *Int. J. Robot. Res.* **2006**, *25*, 1033–1046. [CrossRef]

31. Zappatore, G.A.; Reina, G.; Messina, A. Analysis of a highly underactuated robotic hand. *Int. J. Mech. Control* **2017**, *18*, 17–24.

32. Birglen, L.; Gosselin, C.M. On the force capability of underactuated fingers. In Proceedings of the 2003 IEEE International Conference on Robotics and Automation (Cat. No. 03CH37422), Taipei, Taiwen, 14–19 October 2003; pp. 1139–1145.

33. Massa, B.; Roccella, S.; Carrozza, M.C.; Dario, P. Design and development of an underactuated prosthetic hand. In Proceedings of the 2002 IEEE international conference on robotics and automation (Cat. No. 02CH37292), Washington, DC, USA, 11–15 May 2002; pp. 3374–3379.

34. Baril, M.; Laliberte, T.; Gosselin, C.; Routhier, F. On the design of a mechanically programmable underactuated anthropomorphic prosthetic gripper. *J. Mech. Des.* **2013**, *135*. [CrossRef]

35. Kontoudis, G.P.; Liarokapis, M.V.; Zisimatos, A.G.; Mavrogiannis, C.I.; Kyriakopoulos, K.J. Open-source, anthropomorphic, underactuated robot hands with a selectively lockable differential mechanism: Towards affordable prostheses. In Proceedings of the 2015 IEEE/RSJ International Conference on Intelligent Robots and Systems (IROS), Hamburg, Germany, 28 September–2 October 2015; pp. 5857–5862.

36. Eppner, C.; Deimel, R.; Alvarez-Ruiz, J.; Maertens, M.; Brock, O. Exploitation of environmental constraints in human and robotic grasping. *Int. J. Robot. Res.* **2015**, *34*, 1021–1038. [CrossRef]

37. Heinemann, F.; Puhlmann, S.; Eppner, C.; Élvarez-Ruiz, J.; Maertens, M.; Brock, O. A taxonomy of human grasping behavior suitable for transfer to robotic hands. In Proceedings of the 2015 IEEE International Conference on Robotics and Automation (ICRA), Seattle, WA, USA, 25–30 May 2015.

38. Salvietti, G.; Iqbal, Z.; Malvezzi, M.; Eslami, T.; Prattichizzo, D. Soft hands with embodied constraints: The soft scoopgripper. In Proceedings of the 2019 International Conference on Robotics and Automation (ICRA), Montreal, QC, Canada, 20–24 May 2019; pp. 2758–2764.

39. Shen, Y.; Salemi, W.; Rus, B.; Moll, D.; Lipson, M. Modular self-reconfigurable robot systems. *IEEE Robot. Autom. Mag.* **2007**, *14*, 43–52.

40. Prattichizzo, D.; Malvezzi, M.; Hussain, I.; Salvietti, G. The sixth-finger: A modular extra-finger to enhance human hand capabilities. In Proceedings of the 23rd IEEE International Symposium on Robot and Human Interactive Communication, Edinburgh, UK, 25–29 August 2014; pp. 993–998.

41. Reppel, T.; Weinberg, K. Experimental determination of elastic and rupture properties of printed ninjaflex. *Tech. Mech. Sci. J. Fundam. Appl. Eng. Mech.* **2018**, *38*, 104–112.

42. Hussain, I.; Salvietti, G.; Malvezzi, M.; Prattichizzo, D. On the role of stiffness design for fingertip trajectories of underactuated modular soft hands. In Proceedings of the 2017 IEEE International Conference on Robotics and Automation (ICRA), Singapore, 29 May–3 June 2017; pp. 3096–3101.

43. Logozzo, S.; Valigi, M.C.; Canella, G. Advances in optomechatronics: An automated tilt-rotational 3D scanner for high-quality reconstructions. *Photonics* **2018**, *5*, 42. [CrossRef]

44. Valigi, M.C.; Logozzo, S.; Canella, G. A new automated 2 DOFs 3D desktop optical scanner. In *Advances in Italian Mechanism Science*; Springer: Berlin/Heidelberg, Germany, 2017; pp. 231–238.

45. Harris, C.G.; Jursik, N.J.; Rochefort, W.E.; Walker, T.W. Additive Manufacturing with Soft TPU—Adhesion Strength in Multimaterial Flexible Joints. *Front. Mech. Eng.* **2019**, *5*, 37. [CrossRef]

46. Calli, B.; Walsman, A.; Singh, A.; Srinivasa, S.; Abbeel, P.; Dollar, A.M. Benchmarking in manipulation research: The YCB object and model set and benchmarking protocols. *arXiv* **2015**, arXiv:1502.03143.

Publisher's Note: MDPI stays neutral with regard to jurisdictional claims in published maps and institutional affiliations.

 **robotics**

Article

A Feasibility Study of a Robotic Approach for the Gluing Process in the Footwear Industry †

Kevin Castelli, Ahmed Magdy Ahmed Zaki, Yevheniy Dmytriyev, Marco Carnevale * and Hermes Giberti *

Dipartimento di Ingegneria Industriale e dell'Informazione, Università di Pavia, via Ferrata 5, 27100 Pavia, Italy;
kevin.castelli@unipv.it (K.C.); ahmedmagdyahme.zaki01@universitadipavia.it (A.M.A.Z.);
yevheniy.dmytriyev01@universitadipavia.it (Y.D.)
* Correspondence: marco.carnevale@unipv.it (M.C.); hermes.giberti@unipv.it (H.G.)
† This paper is an extended version of our paper by Carnevale, M.; Castelli, K.; Zaki, A.M.A.; Giberti, H.; Reina, C. Automation of Glue Deposition on Shoe Uppers by Means of Industrial Robots and Force Control. *Mech. Mach. Sci.* **2021**, *91*, 344–352.

Abstract: Manufacturing processes in the shoe industry are still characterized to a large extent by human labour, especially in small and medium craft enterprises. Even when machinery is adopted to support manufacturing operations, in most cases an operator has to supervise or carry out the task. On the other hand, craft footwear industries are called to respond to continuous challenges to face the globalization effects, so that a rapid adaptability to customer needs is required. The industry 4.0 paradigms, which are taking place in the industrial environments, represent an excellent opportunity to improve the efficiency and quality of production, and a way to face international competitors. This paper analyses and proposes a robotic cell to automatize the process of glue deposition on shoe upper, which exploits a new means of depositing the glue compared to State-of-Art applications. While the latter mainly adopt glue gun spraying systems or pneumatic syringes, the proposed robotic cell is based on an extrusion system for the deposition of molten material originally in the form of a filament, similar to all extent to those adopted for Fused Deposition Modeling (FDM). Two cell solutions are designed and tested. In the former the extruder is the robot end effector and the shoe upper is grounded to the cell frame. In the second, being the reciprocal, the shoe last is clamped to the robot wrist and the extruder is fixed to the cell frame. The peculiarities of the two solutions are pointed out and compared in terms of cell layout, hardware, programming software and possibility to develop collaborative applications. A self developed slicing software allows designing the trajectories for glue deposition based on the CAD model of the shoe upper, also allowing driving the inclination of the extruder nozzle with respect to the vectors normal to the upper surface. Both the proposed cell layouts permit to achieve good quality and production times. The solution with the mobile extruder is able to deposit glue at highest end-effector speed (up to 200 mm/s). On the other hand, the solution with the mobile shoe upper and fixed extruder seems to be more appropriate to enhance collaborative applications.

Keywords: industrial automation; footwear manufacturing; shoe upper bonding; slicing software; robotics

Citation: Castelli, K.; Zaki, A.M.A.; Dmytriyev, Y.; Carnevale, M.; Giberti, H. A Feasibility Study of a Robotic Approach for the Gluing Process in the Footwear Industry. *Robotics* **2021**, *10*, 6. https://doi.org/10.3390/robotics10010006

Received: 30 October 2020
Accepted: 26 December 2020
Published: 31 December 2020

1. Introduction

The shoe industry is characterized by both functional and fashion goals, the main functions a shoe must perform being protection, comfort, and style. A large variety of materials can be used to satisfy both the requirements: despite leather being the primary raw material in earlier styles, it has remained at present a standard mainly for expensive dress shoes, whereas many other types of materials can be exploited to enhance any shoe's specific aim.

To cover a large variety of types and styles, while offering a desired-by-the-market product, a growing number of big brands and startups rely on the potential of footwear mass customization strategies [1]. This means that the flexibility of the production has to be guaranteed, usually with the side effect of rising the complexity and the cost of the production plants. To allow the manufacturing and assembling process to be modified and adapted quickly and in an easy way to any modifications to the shoe design, it is necessary to adopt re-configurable machines which can be reprogrammed without major changes. However, most of shoe production processes currently require the intervention of manpower to be executed [2], with strong implications on production times and safety. The introduction of automation would imply a reduction of the number of human operators required, thus reducing production costs [2].

In the last two decades, many authors introduced solutions for the automation of parts of the entire process of shoe manufacturing. Some example can be found in [3] for lasting operation, in [4] for grinding, in [3,5,6] for roughing, in [7–9] for sole bonding, in [10] for sole grasping, in [11] for finishing operations. In [12] the authors propose an automated production plant for mass customized production, and in [2] the advantages of using a new integrated design and production process are outlined.

The coupling between the shoe upper and the sole is one of the latest operations to be performed, and it is often identified as the bottleneck of the production [13]. The lower and upper parts are united using different techniques, which depend on the type of shoe to be produced, as well as on the machinery and technology available to carry out the job. There are three major assembling techniques that are used in the shoe industry [14]:

- cementing: the upper body and the lower part are assembled together using adhesives;
- injection: the sole is injected into a mold, in direct contact with the upper part;
- stitching: the two parts are assembled using threads.

Among the above mentioned procedures, the cementing process offers several advantages of a more flexible and homogeneous joint, better aesthetic properties and the feasibility of the automation process [15]. However, the cementing activity is recognized as one of the most critical operations in the shoe manufacturing [15]. It includes the preparations of the materials, the cleaning and treating of the surfaces to be assembled, and the preparation of the chosen adhesive solutions. The upper is usually stretched over a last (i.e., a fixture representing the shape of the foot) and attached to the bottom through an assembling process [16].

Most of the state-of-art papers dealing with the issue of automation of glue deposition exploit a non-contact deposition system based on a spraying gun moved by robot, with the workpiece placed on a worktop [7,8,17]. On the other hand, in fields of application other than the footwear industry, gluing systems with a fixed syringe and the workpiece moved by the robot were proposed too [18].

The most recent and advanced solutions in shoe manufacturing [7,8] consist of depositing glue on the sole, so that the problem can often be assimilated to a 2D problem. In case of soles for shoes with heels a 3D analysis is required, and the third dimension is only exploited to keep the glue gun at a constant distance from the sole surface.

A very relevant enhancement of this research also consists of the possibility to define automatically the robot path based on the outcome of visual techniques [7,8,17,19], also able to identify the sole shape and position.

This paper focuses on the design of a robotic cell for the automation of the process of glue deposition on the shoe upper, which exploits a new means for glue deposition, based on an extrusion system to melt and deposit the adhesive material on the shoe upper through a nozzle. The extruded glue, initially in the form of a solid wire, is a prototype product produced by a third-party company. This technology allows a more localized deposition of material, compared to standard methods. It is derived from the field of additive manufacturing [20], where advanced robotic solutions were proposed to overcome the limits of traditional cartesian robot. The proposed system is therefore completely representative and assimilable to an Additive Manufacturing application, carried out by exploiting a 6 d.o.f. robot [21].

During the glue deposition process, a proper contact force must be maintained between the extruder nozzle and the upper surface to get proper glue deposition, so that the proposed robotic cell exploits a force feedback control in parallel to the position control. Compared to previous robotic solutions for the shoe industry [6], this paper also presents an enhancement for the tool path generation: in the proposed application the orientation of the extruder nozzle can be indeed continuously oriented with respect to the vector normal to the shoe upper surface, as an outcome of a dedicated slicing software based on the digital model of the shoe upper.

Two cell solutions are discussed in the paper:

- Mobile extruder cell type: the last is anchored to the ground and the robot holds a mobile extruder for glue deposition;
- Fixed extruder cell type: the extruder for glue deposition is grounded to a fixed frame, and the shoe last is anchored to the robot and moved to reach the desired poses with respect to the fixed glue extruder.

Both the solutions are investigated and compared, with the aim of assessing the most feasible one. The first discriminant factor is the feasibility to realize the related robotic cell layout, which involves the issue of manipulating the hot-extruder through a robot in the former case, or using the robot to grip the shoe last in the second case. The two layouts involve two different ways of using the robotic resources: in case of mobile extruder the robot is necessarily dedicated to a single process and must be retooled to execute different operations. On the other hand, when the robot holds the shoe last, several processes can be executed with different fixed tools (e.g., cementing, roughing, power pressing) without the need to dismount the shoe last from the robot wrist.

The paper is organized as follows: Section 2 describes and compares the layouts of the two proposed solutions for the robotic cell, focusing on the extrusion system, on the exploitation of the force control to guarantee a proper contact between the extruder nozzle and the deposition surface, and on the electronic control system for synchronizing the robot movements and the glue flow from the extruder. Section 3 describes the slicing software developed to define the path for glue deposition on the upper surfaces and discusses the main differences in the software required for the two alternative solutions, with moving and fixed extruders. Section 4 reports the experiments carried out and the achieved results. Finally, conclusions are drawn in Section 5, outlining the pros and cons of the two proposed layouts.

2. Robotic Cell Layouts

The robotic cells proposed for the automated deposition process are composed of two main subsystems:

- A system for deposition of molten material, consisting of an extrusion system to be commanded synchronously with the robot's movement, so as to provide the exact quantity of adhesive.
- A 6 d.o.f. manipulator, which fulfills the task of positioning the system for material deposition relatively to the shoe upper.

Figure 1 represents the two alternative configurations proposed for the usage of the robot manipulator, both of them aimed at governing the relative position and orientation between the glue supply system and the 3D-shaped surface of the shoe. A former configuration (Figure 1a) relies on a mobile extruder, and consequently on an orientable nozzle, directly connected to the robot wrist, with the shoe being steadily grounded into the operational space. Conversely, the second configuration is based on a fixed nozzle, grounded into the operational space, with the robot wrist holding and handling the last and the shoe upper worn onto it. This layout is represented in Figure 1b. The extruder is held by a supporting structure, whereas the last is clamped by a fastener linked to the robot wrist. During the adhesive deposition the tool supplies a suitable flow of glue at a chosen rate, in order to deposit the required adhesive film over the curved surface. The material has to be deposited on both the bottom and the side part of the shoe upper.

Figure 1 represents the cell layouts exploiting the collaborative Techman Robot TM5-700, even if also the industrial robot Mitsubishi RV-2F-Q has been considered in the present analysis.

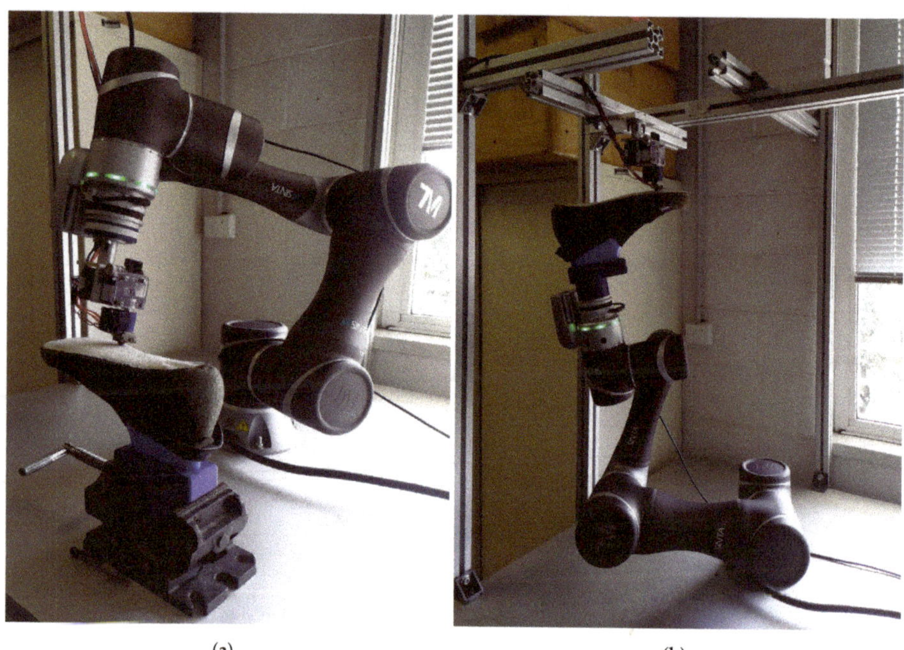

(a)   (b)

Figure 1. Experimental test layouts: (**a**) Mobile extruder; (**b**) Mobile shoe last.

The extrusion system, specifically designed to facilitate the filament provision, can be exploited in both the configurations proposed, being either mounted on the wrist of the robot or on a fixed frame which would keep it in a set location. It was specifically designed and assembled with customized components, to enable the possibility to interchange any item, code and robot during the tuning and the experimental phases. It is composed of a stepper motor (NEMA 17, 400 Steps per Revolution, 0.34 Nm stall torque) to control the filament feeding to an hot-end, at the desired flow rate. The adhesive adopted is a prototype product produced by an third-party Company. It is in the form of a filament with diameter equal to 3 mm, and has to wet the shoe upper surface in order to guarantee a stable and durable bond. The extruder motor is coupled with a speed reducer (ratio 1/3) that directly drives the filament to the hot-end at a feeding speed of around 2 mm/s. The hot end is able to guarantee the temperatures required by the specific glue adopted, in the range 285–295 °C. The nozzle diameter is equal to 0.8 mm.

The control of the extrusion system is achieved by an Arduino Mega-based board combined with a shield featuring as stepper driver (DRV8825 up to 1/32 microstepping), with a MOSFET transistor commanding a heat cartridge (12 V, 40 W) and with a temperature sensor (PT1000) placed in proximity of the extrusion nozzle. Based on this measure and on the reference temperature, the specifically Arduino firmware closes a PID feed-back loop to command the MOSFET and the heat cartridge.

The extrusion system can be set to act as a server or client in a LAN, to which the robot is connected too. While the connection is on, the Arduino firmware awaits incoming messages to operate the extruder. The possible commands are related to the step speed, the temperature set and read, fan speed and software reset.

The value of material flow rate as a function of the robot speed was preliminary defined based on the standard equation governing the deposition process in Additive Manufacturing processes, i.e., the continuity equation between the mass of filament pushed forward by the stepper motor and the mass deposited on the substrate.

$$A_{wire} V_{robot} = \frac{\pi d_n^2}{4} V_{extrusion} = \frac{\pi d_f^2}{4} V_{feed} \tag{1}$$

Equation (1) represents the relationship between the flow rates at the printing bed (i.e., shoe upper surface), at the nozzle output, and at the feeding gear respectively, under the assumption that the material density ρ is not significantly affected by the extrusion process. The cross sections at the nozzle outlet and at the feeding gear can be considered circular in shape, with diameters d_n and d_f respectively, which allows a straightforward computation of the corresponding areas. In standard additive manufacturing applications a gap h is kept between the nozzle and the deposition surface during the deposition process. In such a case, when the extruded material hits the deposition surface, the resulting geometry can be expected to resemble a rectangle with rounded sides, providing that the width w of such rectangle is minor than the outer diameter of the nozzle D_n [22].

The area of the strand on the substrate can be therefore written as [23]:

$$A_{wire} = wh - \left(1 - \frac{\pi}{4}\right)h^2 \tag{2}$$

The selection of the parameters w and h allows evaluating the area A_{wire} in Equation (1). The value of the deposition speed V_{robot} (i.e., relative speed between the nozzle and the shoe upper surface along the deposition path) allows evaluating the actual speed of the extruded filament V_{feed}.

The above described formulas can be applied in the case of an actual additive manufacturing process. However, when the deposited material is an adhesive in the form of a filament, which was the case considered in the present paper, the nozzle has to get in contact with a proper force with the deposition surface, as to enable the adhesion of the molten material. In such a case, the suitable relationship between the material flow rate and the robot speed has to be calibrated experimentally.

The design of the path that the manipulator has to follow to generate the needed trajectories between the nozzle and the upper surfaces is one of the most relevant issues of the proposed application. The trajectories are firstly evaluated based on a CAD model of the shoe upper, and on a slicing software specifically developed for this purpose. The software outputs a six dimensional path (i.e., positions and angular orientations) to be followed during glue deposition on both the bottom and side part of the shoe upper. Once this trajectories are defined off-line with respect to the upper geometry, the robot motion has to be designed. Two different procedures can be followed according to the setup being either with mobile extruder or mobile last.

For the former case, the extruder can be considered as the end-effector of the robot, once its dimensions are fully defined, and the robot can consequently be easily programmed to make the nozzle get in contact with the upper surface with the desired poses. On the other hand, an alternative algorithm is adopted in the case of the movable last. In this latter case, the surfaces of the shoe upper, on which the fixed nozzle has to draw the path, are moved in three dimensions together with the last. The poses output by the slicing software are firstly referred to a single point of the last, whose distance with respect to the clamp, and consequently to the robot wrist, is known. The details of the slicing software, and the way it is used to generate the end-effector path, are described in the next paragraph Section 3.

During the glue deposition process, a proper contact force must be maintained between the extruder nozzle and the upper surface. Inaccuracies in the shoe upper CAD model or misalignment in the last clamp can lead to detachments or sudden increase of the contact force, the latter resulting in a nozzle closure. For these reasons, a mere geometrical approach might require a calibration after each change of setup, which would definitely be rather time consuming and unsuitable for customized mass production. On the contrary, an effective solution would require all the production phases to be prepared in advance, during the design phase of the specific shoe model. For this reason, the propose robotic cell exploits a force feedback control, in parallel to the control position: the tentative trajectories previously defined off-line are adjusted online through a force control with constant reference force along the direction of the nozzle axis. An external load cell is attached between the robot wrist and the extruder in the setup with movable extruder (see Figure 1a), and between the robot wrist and the shoe upper in the setup with fixed extruder (Figure 1b). The target maximum linear speed for glue deposition is 200 mm/s, which should be reached guaranteeing a proper contact force between the extruder tool and the shoe upper to be glued, as discussed in Section 4.

3. Path Generation Algorithm

The algorithm to generate the robot poses can be divided into three parts: a first one defining the path to be followed on the shoe upper surfaces during the glue deposition process; a second one exploiting this path to effectively command the robot in the joint space; a third part for singularity and collision checks.

As for the first step, there are, in the literature, several methods exploitable to extract the path from the CAD model of a shoe, mainly based on the computation of the intersection between the object and a reference surface. As an example, in case of a STEP model the intersection can be found analytically as in [24,25]. In case of a STL model, some examples of the slicing process are [26], which exploits a method based on neighbouring points to get the surface variation, and [27], tailored to spherical cutting surfaces.

The method proposed for the present application is instead based on an algorithm normally used to evaluate overlapping objects in collision detection theory [28]. The slicing procedure is carried out as a procedure for identification of collisions between the shoe object and a selected surface, then identifying the locus of the intersecting points. This method allows exploiting the same algorithm libraries for both the path planning and collision checks, thus reducing the time required to update the software in case of any modifications to the robotic cell required to accomplish with a change of production. The solution proposed offers to the end users a series of libraries and combinable functions to ease the path extraction from the CAD model, by queuing perimeters and infills.

The software, described in Section 3.1, is able to output a 6 dimensional path characterized by the position coordinates (x, y, z) and by the components $(cos(\xi), cos(\psi), cos(\phi))$ of the unit vector normal to the surface in each point, defined in an absolute reference frame. When generating the deposition path, therefore, the software allows considering not only the positions where the shoe upper and the extruder nozzle have to get in contact, but also the orientation of the nozzle axis with respect to the surface normal direction.

The second part of the algorithm consists of the programming of the robot poses and of joint positions to generate the desired trajectories on the shoe upper. As already mentioned, this step is rather straightforward in the case of the mobile extruder solution, whereas some additional geometrical transformation must be applied in the case with fixed extruder and mobile shoe upper. In the former case, indeed, the extruder can be defined as a tool in the robot controller, by specifying its constant dimensions and the constant distances between the tool center point and the robot wrist. Once the position of the shoe upper, clamped and grounded to the frame of the robotic cell is set, and once the deposition path with respect to the shoe upper is fully defined as an output of the slicing software, the robot can be easily programmed. An example of such an application is presented in [29], where a standard six DOF anthropomorphic manipulator (ABB IRB2400/16) is equipped with a force controlled head to accomplish roughing task on the side and bottom surface of a shoe upper.

On the other hand, in the case of mobile shoe with the last clamped to the robot wrist, the path output by the slicing software must be referenced to a unique frame belonging to the shoe last, whose origin's position and angular position with respect to the robot wrist are known. The mathematical steps of this transformation are described in Section 3.2.

The singularity and collision check algorithms adopted in this work are carried out with the self-implemented code described in Section 3.3, allowing firstly to solve the inverse kinematic and to identify the presence of any singularity in the solutions, and then to check for self-collisions between the robot arm and the tools, or collision between the robot arms and any other object in the robotic cell.

3.1. Slicing Software

The slicing software performs its calculation on an input CAD model of the shoe upper geometry, in the format of an STL file, which must be available from the shoe manufacturer. Both ASCII or binary STL files can be directly loaded in a Matlab procedure (*stlread* function), or Python procedure (*numpy-stl* library, *PyPi* repository), which return the mesh in terms of connectivity list and vertices. Normal vectors can be also obtained from the built-in functions, or directly computed from the definition of each triangle of the mesh. The loaded data then undergo a pre-processing routine to repair possible defects in the model (e.g., gaps, mixed normal, overlaps), and to reduce the size of the stored variables by eliminating repetitions of the same points (i.e., shared ordinates) defining the mesh.

As an example, for a starting geometry of 104,338 triangles defining a mesh of a shoe upper (corresponding to 313,014 points), only 52,169 points unequivocally describe the related geometry. As a rule of thumb, the point to describe the desired geometry are reduced by one sixth.

A further viable mean to reduce the computational costs of the path generation consists of cutting the geometry under analysis. In the case of the shoe upper, only the portion of the upper where the glue is to be deposited can be selected. This region is identified as the area of interaction between the upper and the sole in the final shoe.

A graphical representation of the input and output of this step can be observed in Figure 2, representing the original picture, the model of the shoe upper and the selected portion where the glue has to be placed. The figure also represent the position of the seam.

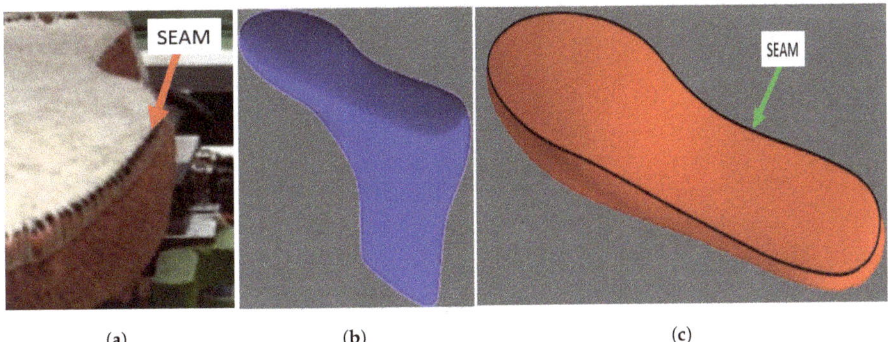

Figure 2. (a) Shoe upper. (b) STL file of shoe upper. (c) Reference glue deposition trace.

The slicing algorithm uses the position of the seam as reference contour for the evaluation of the trajectories where the glue has to be deposited, the seam position being directly computed from the STL file in the slicing software.

As it can be observed in the upper shape reported in Figure 2, the seam correspond to an abrupt variation in the vector normal to the surface. Its points can be therefore found by identifying the mesh triangles where a sudden variation of normal vector n_i direction occurs.

For each triangle, the angular distance between the normal vector n_i and the unit vector defining the vertical reference direction n_{ideal} can then be computed using the inner product

$$\theta_i = \arccos(n_i \cdot n_{ideal})$$

Figure 3a represents the trend of the θ angle obtained for the shoe upper considered in the present work. Two groups of angles can be identified, one in the range from 50° to 100° degree, corresponding to the mesh of the side surface of the upper, and a second one in the range from 0 to 20 degree, corresponding to the mesh triangles belonging to the top surface of the shoe upper. A threshold value to discern the two groups of points can be defined (e.g., 35°) to separate the mesh triangles belonging to the top surface (red normal vectors in Figure 3b) from those belonging to the side surface (green normal vectors in Figure 3b).

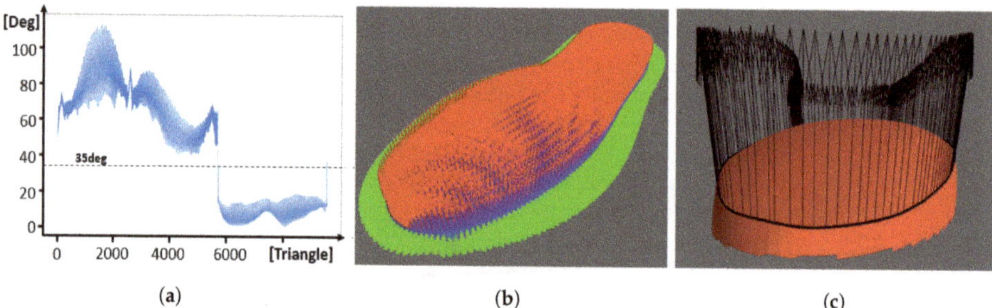

(a) (b) (c)

Figure 3. Seam evaluation algorithm. (**a**) Trend of the θ angle. (**b**) Top and side surfaces. (**c**) Seam line and normal vectors.

The actual points defining the seam contour can then be identified by intersecting the model of the shoe upper with a suitable cutting surface (defined by a list of points and relative connectivity list). The generation of this cutting surface starts with the projection of the triangles belonging to the top surface cluster (characterized by the red arrows in Figure 3b) onto a plane parallel to the X-Y plane (whose normal vector coincides with n_{ideal}). This plane corresponds to the lower base of the cutting surface and it is defined at a height Z_{lower} equal to:

$$Z_{lower} = min(P_z) - \epsilon$$

having indicated with P_z the vector of the Z coordinates of all the points belonging to the shoe upper and being ϵ a safety margin used to guarantee the intersection.

Then an alpha-shapes based algorithm [30] is exploited to get only the boundary points (i.e., the seam) that functionally serves as base for the cutting surface. To characterize the cutting mesh, the base is extruded along the n_{ideal} direction by creating a duplicate of these points on a plane at Z equal to:

$$Z_{top} = max(P_z) + \epsilon$$

The connectivity list is reconstructed using a counter j belonging to [1, Np), where Np is the number of points constituting the lower contour. For each j, two triangles are demarcated:

$$Triangle_1 = [\, Lower(j),\ Lower(j+1),\ Top(j)\,]$$

$$Triangle_2 = [\, Lower(j+1),\ Top(j+1),\ Top(j)\,]$$

where the *Lower* and *Top* labels indicate the points belonging to the lower and upper base respectively.

By intersecting the shoe model with the cutting surface just generated, the actual profile of the seam is computed. The final seam line is represented in Figure 3c, together with the corresponding unit vectors normal to the surface.

Once the seam line has been identified, the desired path for glue deposition is computed in a similar manner, by exploiting the intersection between suitably defined surfaces and the STL mesh of the shoe upper. The seam contour is scaled inward to generate multiple concentric perimeters on the top surface of the shoe upper (using algorithms derived from computer aided machining, such as [31]). Each scaled contour is projected on the X-Y plane, and a new surface is extruded in the n_{ideal} direction to identify the actual intersection with the top surface of the upper. This operation is repeated for the desired number of perimeters. The final path is composed of different lines, perimeters and infills defined as the coordinates of points on the shoe upper surface and the corresponding unit vectors normal to the surface. An example is reported in Figure 4a.

Figure 4a also report the presence of infills, visible in the toe area. Infills are generated by identifying the intersection between additional surfaces, generated by using user-defined primitives with a selected patterns (e.g., lines generating parallel planes), and the shoe upper surface. For instance, parallel lines can be used to define the path for glue deposition in the correspondence of the tip and heal of the top surface, as represented in Figure 4b.

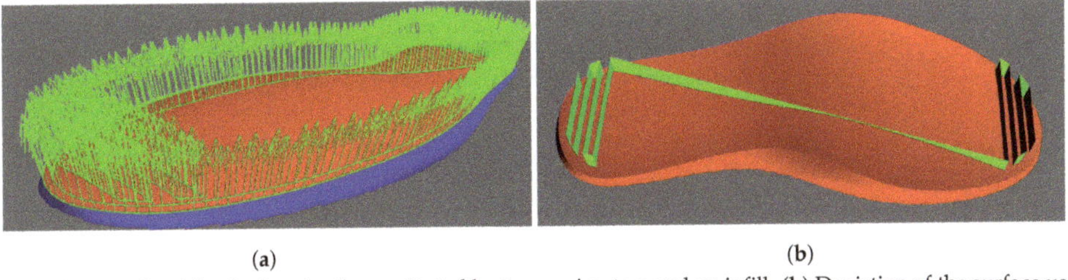

(a)   (b)

Figure 4. (**a**) Example of the final toolpath constituted by two perimeters and an infill. (**b**) Depiction of the surface used to get the infill.

The generation of the path on the lateral surface of the shoe upper still relies on the seam contour, but this time the slicing surface changes. The cutting plane is obtained by lowering the seam along the n_{ideal} direction by the requested quantity and by generating two child curves inward and outward, so that the surface between these two boundaries intersect the portion of the upper.

3.2. Robot Programming

This subsection describes the further mathematical steps needed to program the robot, for both the two considered cell types, with mobile and fixed extruder respectively.

The path generated after the slicing procedure consists of a series of points (position co-ordinates (x, y, z)), to be reached by the tip of the extruder nozzle for glue deposition, and of the direction cosines $(cos(\xi), cos(\psi), cos(\phi))$ of the vector normal to the shoe upper surface in each point. Starting from these path coordinates, the poses of the End Effector (EE) throughout the entire glue deposition process have to be defined. In the cell layout with the mobile extruder the EE consists of the extruder itself, whereas for the cell layout with fixed extruder the EE consists of the gripping system for the shoe upper.

The EE reference frame is conventionally defined with the Z axis exiting the tool (Denavit and Hartenberg's conventions). In the cell layout with mobile extrusion system the EE frame is therefore defined with Z axis parallel to the extruder nozzle and entering the shoe surface. On the other hand, in the layout with mobile shoe and fixed extruder, the EE frame is defined with Z axis perpendicular to the robot wrist flange, oriented outwards of the shoe gripper and exiting the shoe surface.

In both the cases, during the deposition process, the EE has to be oriented so that the vectors normal to the shoe upper surface identified by the slicing procedure lay in the same direction as the extruder, or even inclined at will with respect to it.

The former solution (i.e., mobile extruder) is more straightforward from the programming point of view: when the extruder is mounted on the robot wrist, it can be defined as the robot tool, as its geometry is known. Once the shoe upper is positioned and grounded to the cell frame, the path obtained as an output by the slicing software can be referred to the absolute reference frame of the robot base, and be adopted as the starting point to define the poses of the end effector: based on the geometry of the extruder the position coordinates are set, so that the extruder nozzle gets in contact with the workpiece surface along the deposition path; the vectors of the path normal to the shoe upper surface are then used as a reference direction to orientate the extruder axis.

The second cell layout with fixed extruder and mobile shoe upper requires some further mathematical steps, but also in this second case the sequence of EE poses can be completely generated offline in advance, based on the output of the slicing procedure. The main issue in this case consists of the fact that the path for the glue deposition is traced onto the shoe upper driven by the robot, which makes the relative positioning and orientation of this curve with respect to the glue supply system a bit more tricky. The procedure to define the EE poses requires an intermediate step in which the path generated by the slicing software is referred to a single reference frame belonging to the shoe last, clamped by the gripper mounted on the robot wrist, whose positions and orientation with respect to the reference frame of the robot wrist is therefore known. This intermediate step allows defining the deposition path coordinates with respect to the reference frame of the robot wrist. The vectors normal to the shoe upper surface can then be oriented with respect the axis of the fixed extruder.

For both the considered cell layout, once the orientation of the EE z axis is defined, either the X or Y axis orientation has still to be defined to fully define the EE pose. For a given path generated by the slicing procedure, indeed, an infinite number of EE poses can guarantee the contact between the upper surface and the extruder nozzle with a given orientation of the extruder with respect to the normal direction to the shoe upper surface. Among this infinite number of solutions, the feasible subset is shall accomplish to several constraints, the main ones being:

- **Collision**: no collision are admitted between the bodies in the cell, particularly in this application in which hot surfaces might damage the robot and the electrical wires.
- **Range of motion**: in addition to the need that a waypoint has to be reachable, rotation around the tool axis might be limited in range for a given machines. Therefore, paths that leads to continuous rotation of the wrist joint $J6$ through consecutive points have to be checked.
- **Tool cabling**: tool wires can limit the range of motion of the robot either due to their pulling action (thus increasing the payload and reducing the accuracy) or due to the need of avoiding excessive twist.

Different approaches were adopted in the present work to define a tentative orientation of the EE for each point of the shoe path, later tested for collision or infringement of the range of motion. The techniques exploited to define a full orientation for the EE are: *tangential vector, world direction cosine alignment* and Rodrigues' rotation formula.

The *tangential vector alignment* uses the points in the shoe path to approximate the vector tangent to the curve, and to build a right-handed reference system for each point of the deposition path in which the x axis is oriented as the tangent to the path, and the z axis as the normal to the shoe upper. By imposing the direction of the x unit vector of the extruder reference frame (both in the case of movable or fixed extruder) to be aligned with the x direction of the path local frame, the rotation matrix can be fully defined.

In the *World direction cosine alignment* case, the right-handed local reference frames along the deposition path are defined to have the z axis oriented as the normal to the shoe upper, and the x axis as close as possible to a constant direction defined with respect to the shoe geometry. To this aim, the y unit vector is defined through the vector product between the normal to the upper surface and the reference direction. The x direction of the local reference frame is then determined to complete the right-handed reference frame. The EE poses are then defined by orienting each x unit vector of the path local frame along the x unit vector unit vector of the extruder reference frame (both in the case of movable or fixed extruder).

A third solution uses Rodrigues' rotation formula to compute the rotation matrix to align the upper surface normal vector to the Z axis of the tool. This also maps the entire EE frame to the path as required, even if in this latter cases the actual direction of the X or Y axis of the EE frame is not directly controlled.

In the following, for the sake of conciseness, only the mathematical steps related to the Rodrigues' rotation formula in the case of the cell layout with mobile shoe upper are addressed.

The schematic cell layout is reported in Figure 5: the absolute reference frame adopted is reported in red at the robot base frame; the vector p_k represents the position vector of the tip of the extruder nozzle, p_{ee} the position of the EE reference frame and p_{tool} the position of each point of the shoe path relative to the tool reference frame.

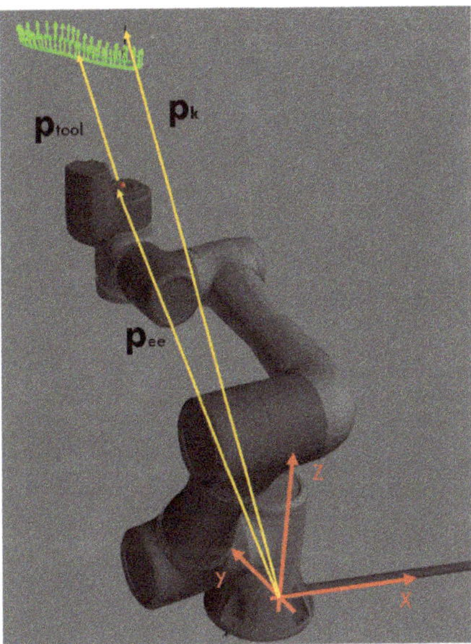

Figure 5. Scheme of the mobile upper layout for defining the EE poses.

The steps adopted for the evaluation of the rotation matrix to define the EE poses can be defined with reference to Figure 6, represented in 2D instead of 3D for ease of reading. The left side Figure 6a represents an initial position of the system, in which the vector v, normal to the shoe surface, is not yet oriented as the extruder axis V_{target}. The robot wrist reference frame is colored in blue. The vector p describes the relative positions of the deposition path points with respect to the robot wrist relative frame.

When considering an initial vector v in $\Re^3$ and a unit vector w defining the rotation axis around which v rotates by an angle θ_r (Figure 6), the rotation matrix can be computed as:

$$\mathbf{R} = \mathbf{I} + (\sin\theta_r)\mathbf{K} + (1 - \cos\theta_r)\mathbf{K}^2 \tag{3}$$

so that the new rotated vector is

$$\mathbf{v_{rot}} = \mathbf{R}\mathbf{v} \tag{4}$$

Both the rotation axis w and the rotation angle θ_r are the unknown to be evaluated in the presented application, v being the initial vector normal to the shoe upper surface and

v_{rot} the final vector oriented as the extruder axis (or at will with respect to it). The term $\sin(\theta_r)\mathbf{K}$ of Equation (3) can be determined as:

$$\sin(\theta_r)\mathbf{K} = \begin{bmatrix} 0 & -k_z & k_y \\ k_z & 0 & -k_x \\ -k_y & k_x & 0 \end{bmatrix} \tag{5}$$

where the elements of the matrix are the components of the k vector, computed as:

$$k = v \times v_{rot} \tag{6}$$

and oriented as the unit vector oriented w.

The remaining term to be computed in Equation (3) is $\cos(\theta_r)$, which can be obtained through the dot product between the normalized vectors v and $vrot$.

The rotation matrix bringing v into $vrot$ is then computed substituting the calculated terms in Equation (3). Please note that a dedicated rotation matrix must be calculated for each one of the points defining the deposition path along the shoe upper surface, but all these step can be defined at the design stage of the cell layout, provided that the CAD model of the shoe upper is available.

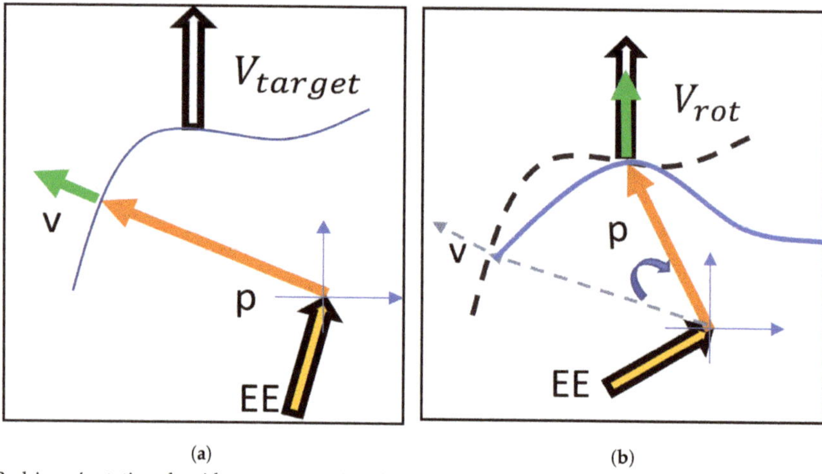

Figure 6. Rodrigues' rotation algorithm representation. (**a**) Exemplary initial position. (**b**) Alignment between the vector normal to the shoe surface and the extruder axis, after rotation.

After the EE poses generating the desired toolpath are fully defined, they can be forwarded to the robot controller (physical or virtual for offline simulation), to compute the joint positions and check their feasibility.

The above-described procedure is rather flexible: in case of any modification to the geometry of shoe upper, to the last, to the holding support or to the nozzle, which might be needed to adapt the robotic cell to the production of a new shoe model, only the algorithm input data (i.e., geometrical data) have to be uploaded, without the need of advanced programming skills, and therefore suitably for the application in craft small and medium size enterprises. The generality of the solution keeps holding for different shoe models, which is a remarkable result to enable flexibility of the proposed automatic cell.

3.3. Collision and Process Checks

The proposed solutionswere developed and tested with a collaborative robot (Techman TM5-700), and with an industrial robot (Mitsubishi RV-2F-Q), for which different approaches were adopted in defining the robot instructions, to better comply with the built-in functionalities.

All the possible control strategies were simulated before being tested in the realized test bench. A dedicated simulator environment for the solution of the inverse kinematic (IK) problem, able to simulate any of the above mentioned robot types, is developed according to references [32,33] and integrated at the end of the slicing software. The solutions obtained as EE poses after the slicing procedure can then be automatically checked and selected on the basis of their feasibility. Among the total number of solutions obtained in the joint-space coordinates (i.e., 8 in case of the 6 d.o.f. manipulators exploited in the present work), some might present indeed singularities or unreachable points (i.e., path coordinates presenting no kinematic solution).

Particular care must also be paid to select a solution in which the orientation of the robot wrist does not generate hazardous twisting of the cables directed to the extruder, to the load cell, and to the glue filament under extrusion. Under this perspective, the cell layout with fixed extruder is definitely preferable.

After discarding the solutions not feasible from a kinematic point of view, the ones remaining need to undergo a collision and self-collision check, before being validated: no contact has to occur between the bodies in the cell, especially in this application in which hot surfaces might damage the robot and electrical wires. Also the software for collision check is self-developed and integrated within the slicing software, since the algorithm for checking the contact between surfaces is the same exploited in the slicing software described in Section 3.1 to identify the intersection between the workpiece model (i.e., stl file) and the auxiliary surfaces. To the aim of collision check, the subroutine of Section 3.1 can be simplified, since there is not the need to evaluate the exact intersection path between the two surfaces, but only to update a Boolean variable that specifies the presence or absence of contact. The procedure developed in the code allows selecting the desired degree of accuracy in modelling the geometries of the involved bodies, which can be even reduced to rectangle-parallelepiped-shaped blocks, to reduce the computational cost to the maximum extent. When a collision between the bodies is identified, the code returns a Boolean value and the indexes of the bodies that collided.

In the case of a single machine cell such as the one analysed in the present paper, collision is most likely to occur between the tool and the robot arms (i.e., self-collision), or between robot parts and other fixed objects in the robotic cell. These types of collisions are not dependent on the simulation time, but only on the joint positions. They can only be avoided by changing the tool-path engineered, so that the first decision the algorithm takes is to discard these configurations. Should all of them be eliminated, the algorithm secondly checks for the presence of any external object interfering with the robot, and report it for removal. If on the other hand the detected collisions are self-collisions, the corresponding toolpath points have to be redesigned. It can be done by joining two different kinematic solutions, or partially redesigning the pose path.

4. Experiments and Results

The objective of the experimental phase is the assessment of the capability of the developed system to guarantee a regular glue deposition over the shoe upper surface, along the designed trajectory. Based on glue producer's specifications and preliminary tests, a target contact force is set at 5 N. In any case, a good quality of glue deposition is expected provided that the contact force is within the range 2–10 N [34], so that the experimental tests are just aimed at verifying the absence of severe contact losses ore force peaks.

According to the manufacturer involved in the present activity, a deposition speed around 100 mm/s could guarantee satisfying cycle times, being a good compromise between the need of getting the lowest as possible cycle time, and the capability of the robot built-in control to actually keep a proper contact force on the shoe 3D surface. The latter gets more challenging as the Tool Contact Point (TCP) speed increases. For these reasons, experimental tests investigated the resulting contact force with TCP linear speed ranging between 50 mm/s and 200 mm/s. The highest limit shall accomplish the maximum speed prescribed by ISO 10218 [35] standard for any robot operating in the proximity of a person (i.e., 250 mm/s), whereas the lower limit is selected to get a low-speed case to be used as a reference for comparison with higher speeds. In the entire speed range considered, the regularity of glue flow (see Equation (1)) shall be guaranteed by the designed hardware.

The control parameters are firstly set up on a linear trajectory going from the shoe heel to the tip, and then exploited for the complete tool-path following the shape of the shoe upper.

For the process of glue deposition in the footwear industry, a positional error up to few millimeters can be accepted [7,36], so that the performance of the robotic cell can only be evaluated by measuring the contact force between the nozzle and the upper surface and by eye inspection. The Industrial robot Mitsubishi RV-2F-Q and the collaborative robot TM5-700 exploited in the tests feature indeed a repeatability equal to $+/-0.02$ mm and $+/-0.05$ respectively. No visual system to check the geometry of the glue strip was developed at this stage of the research activity, even if some methods to inspect the glue dispenser route have already been proposed in the literature for applications requiring higher accuracy in other fields [37,38].

4.1. Mobile Extruder

In the case of the cell layout with mobile extruder, the force measured is affected by the inertia of the extruder (around 0.3 kg), which should therefore be compensated to get the actual contact force between the extruder and the shoe upper surface. The efficacy of the control for different linear speeds was assessed by adopting a low weight feeler (i.e., 0.05 [kg]) built with ABS through FDM process, so that the contact force can be directly assimilated to the force measured by the load cell.

In the case of Mitsubishi industrial robot the final parameters adopted are Gain = 20 [μm/N] and Damping coefficient = 0.1 [N/(mm/s)] [34]. Figure 7 reports the results of the measured contact force for linear speeds of the end effector equal to 50 mm/s (Figure 7a) and 100 mm/s (Figure 7b), and the 3D trajectory followed by the extruder tip on the shoe upper surface (Figure 7c). During the experimental tests the robot is firstly controlled with the position control mode to reach a starting point close to the shoe's surface. The force control mode is then activated (Figure 7c), while the tool keeps on following the predefined trajectory.

As a comparison, Figure 8 reports the contact force results achieved with the mobile extruder in the case of the collaborative TM robot, at the speeds of 100 mm/s (Figure 8a) and 200 mm/s (Figure 8b). When comparing the force results at the speed 100 mm/s in Figures 7b and 8a, no relevant differences are detected for the purposes of the application, so that the suitability of exploiting a collaborative robot in the cell for glue deposition is also confirmed.

The experimental results showed a full capability of the system (no matter which is the kind of robot adopted) to get the linear speed of 200 mm/s, whereas, when dealing with the capability of maintaining the proper contact force, the results show an increase of the standard deviation of the contact force starting from the speed of 100 mm/s. Figure 9 shows, as an example, the trend of the standard deviation of the contact force and the corresponding contact losses obtained with the Mitsubishi robot in the speed range from 50 mm/s to 200 mm/s. The contact loss is defined as the percentage of samples in which the contact force gets lower than 10% of the average target force (i.e., F < 0.5 N), over the total number of samples.

Despite the fact that the standard deviation of the contact force increases for increasing speed, a low percentage of contact loss (0.2%) is detected, only for the speed of 200 mm/s. As recalled in [34], the presence of short detachment does not affect the possibility of depositing the melted glue on the shoe upper surface.

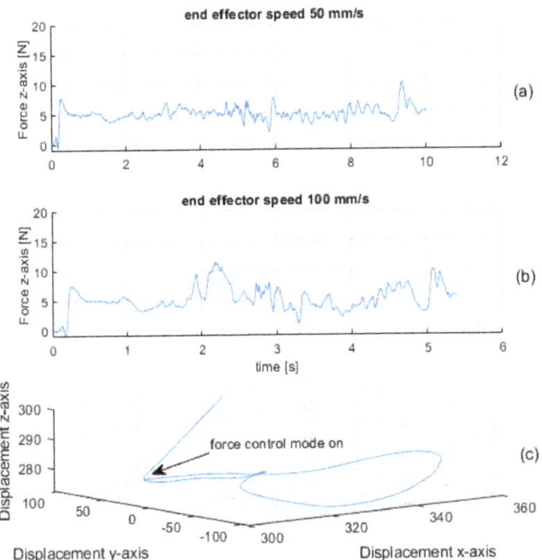

Figure 7. Exemplary contact force results for the cell layout with mobile extruder, Mitsubishi Robot. (**a**) Speed 50 mm/s (**b**) Speed 100 mm/s. (**c**) 3D deposition trajectory.

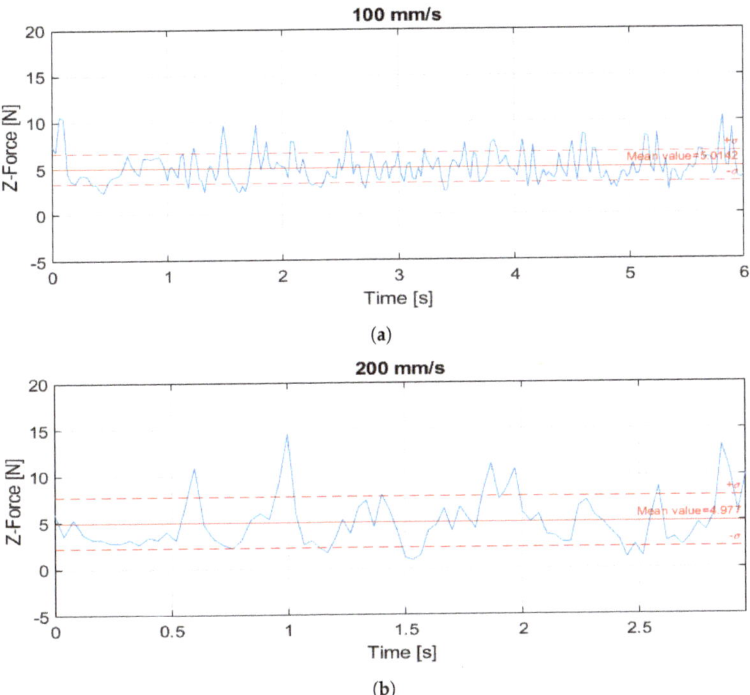

Figure 8. Contact force results for the cell layout with mobile extruder, TM Robot. (**a**) 100 mm/s. (**b**) 200 mm/s.

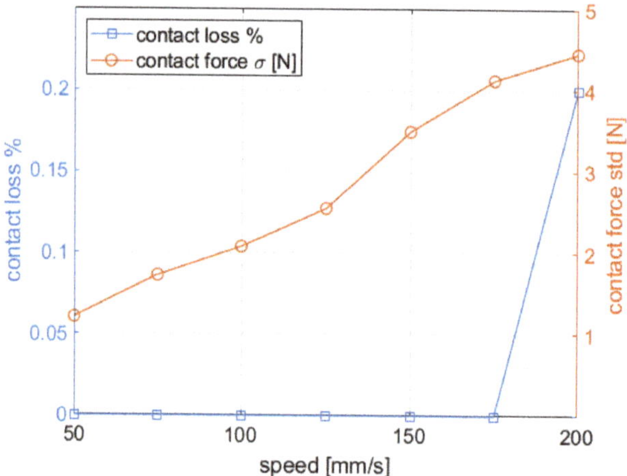

Figure 9. Contact force standard deviation and contact losses as a function of EE speed, exemplary results for Mitsubishi Robot.

4.2. Fixed Extruder and Mobile Shoe Upper

In the case of the test cell with mobile shoe upper and fixed extruder, in which the shoe last is directly attached to the robot wrist, the control performance are expected to be intrinsically worse, due to the higher inertia attached to the robot wrist and to the fact that the moments applied to the robot wrist rapidly vary as a consequence of the relative position between the extruder tip and the shoe upper surface: with reference to the setup represented in Figure 1b, indeed, for a given contact force between the extruder and the shoe upper surface, lower moments are generated when the extruder is in contact on the heel rather than when the contact is on the tip.

Figure 10a,b report as an example the results achieved with TM robot at the speed of 50 mm/s and 100 mm/s respectively. In Figure 10b two detachment are observed around 3 s, corresponding to the zone of the sharp curve of the shoe upper toe.

It can be therefore concluded that the application with the mobile shoe should run at lower speed compared to the case with mobile extruder, which would imply longer glue deposition time for each shoe. However, this drawback would not make the overall cycle time achievable with this cell layout necessarily longer, thanks to other pros this cell configuration leads to. When the shoe upper is directly hold by the robot, indeed, right after the glue deposition phase it could be directly driven to the press for sole fastening, without the need to re-grip the workpiece.

Moreover, when the shoe upper is moved by the robot and the hot extruder is fixed, the robotic cell could more easily be upgraded to become a collaborative robotic cell [39,40], thanks to the fact that this solution shows no hot-mobile parts in the operational space, thus enabling a higher degree of security for the operators.

The use of a collaborative robots would open to relevant advantages in terms of improvement of the organization of manufacturing industries, allowing the introduction of robot in craft enterprises without the need of major changes in the layout of the production floor. This would be particularly relevant in an environment such as the one typical of small and medium-sized shoe manufacturing enterprises, where the shoe production process still involves human labor to a large extent.

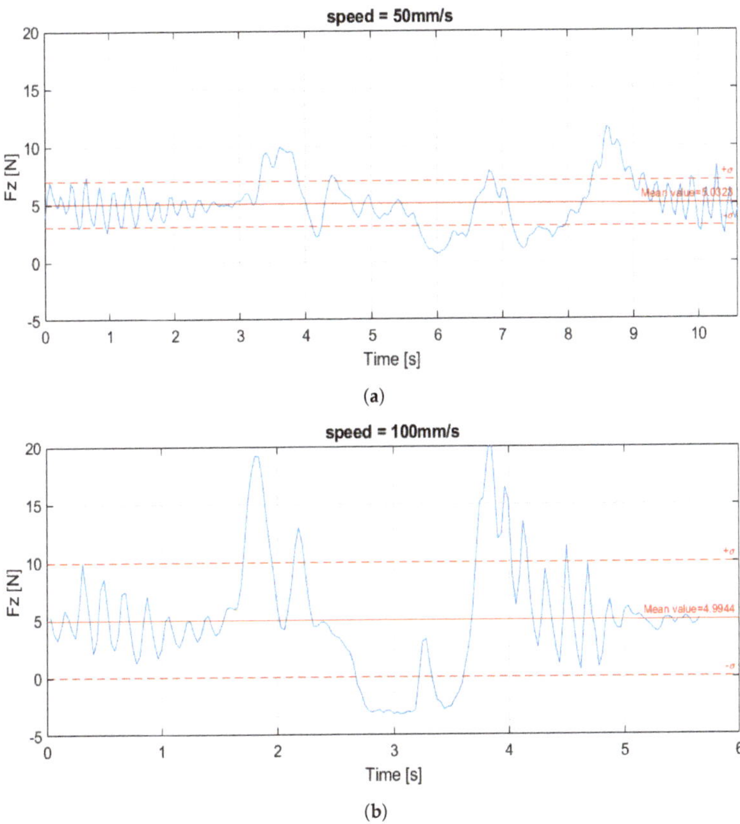

Figure 10. Contact force results for the cell layout with mobile shoe upper and fixed extruder. TM Robot. (**a**) Speed 50 mm/s. (**b**) Speed 100 mm/s.

5. Conclusions

This paper presented all the hardware and software steps developed to study the feasibility of an automatic cell for glue deposition on a shoe upper. A robotic approach is considered, rather than a dedicated automated machine, to increase flexibility in case of relevant variations of the product, and the possibility of the robotic cell to be adapted to a vast variety of processes.

Shoe manufacturing is normally characterized by small and medium-sized craft enterprises, where the shoe production process still involves human labour to a large extent. The possibility of a safe interaction of the operator in the collaborative working area of the machine, and the development of an easy-to-program robotic cell, would therefore pave the way to the introduction of automation in such an environment.

Two robotic cell are developed and compared, the former with the extruder hold by the robot wrist and fixed shoe upper, and the reciprocal with fixed extruder and the shoe upper hold by the robot wrist. A dedicated extruder for the deposition of melted glue on the shoe upper is built and syncronized with the robot movements, and a self-developed slicing software was suitably used to automatically generate the deposition trajectory, starting from the digital model of the shoe upper.

Thanks to the measure of the force exerted between the extrusion tool and the shoe upper, and to the exploitation of the force control mode of the robot, the possibility to maintain a proper contact force between the extruder and the shoe upper is investigated for two different cell layouts, in the speed range from 50 mm/s to 200 mm/s. In the former case it is possible to get to maximum deposition speed of 200 mm/s, with positive force or limited contact loss percentage. On the other hand, the performance of the second cell layout are a bit worse (i.e., relevant detachments at the speed of 100 mm/s), thus requiring lower speeds. However, this second cell layout shows relevant pros, such as the possibility to reduce the number of grips, thus reducing the overall cycle time, and its intrinsically safer applicability in collaborative applications. In the latter case, indeed, the fixed extruder could be placed in a restricted area out of the collaborative zone.

We believe that the proposed system can be used as a safe and flexible tool in small and medium manufacturing enterprises and artisan, for the automation of their production processes and to satisfy market demand.

Author Contributions: Conceptualization, H.G. and M.C.; methodology, H.G., K.C.; software, K.C., Y.D. and A.M.A.Z.; validation, M.C., K.C. and A.M.A.Z.; formal analysis, H.G., M.C.; investigation, M.C., K.C., Y.D., A.M.A.Z.; resources, H.G.; data curation, Y.D., A.M.A.Z.; writing—original draft preparation, M.C.; writing—review and editing, M.C., K.C., A.M.A.Z.; supervision, H.G., M.C.; project administration, H.G.; funding acquisition, H.G. All authors have read and agreed to the published version of the manuscript.

Funding: This research received no external funding.

Institutional Review Board Statement: Not applicable.

Informed Consent Statement: Not applicable.

Data Availability Statement: The data presented in this study are available on request from the corresponding author.

Acknowledgments: The authors would like to thank MITSUBISHI ELECTRIC EUROPE, Viale Colleoni 7, 20041 Agrate B. (Milano) ITALY for providing the Mitsubishi RV-2F-Q robot, and SINTA S.r.l, Via Sant'Uguzzone 5, 20126 Milano, Italy, for providing the Techman Robot TM5- 700.

Conflicts of Interest: The authors declare no conflict of interest.

References

1. Boër, C.R.; Dulio, S. *Mass Customization and Footwear: Myth, Salvation or Reality? A Comprehensive Analysis of the Adoption of the Mass Customization Paradigm in Footwear, from the Perspective of the EUROShoE (Extended User Oriented Shoe Enterprise) Research Project*; Springer Science & Business Media: Berlin/Heidelberg, Germany, 2007.
2. Cocuzza, S.; Fornasiero, R.; Debei, S. Novel automated production system for the footwear industry. In Proceedings of the IFIP International Conference on Advances in Production Management Systems, Rhodos, Greece, 24–26 September 2012; pp. 542–549.
3. Nemec, B.; Žlajpah, L. Automation in shoe assembly. In Proceedings of the 15th International Workshop on Robotics in Alpe-Adria-Danube Region (RAAD 2006), Balatonfüred, Hungary, 15–17 June 2006; pp. 131–136.
4. Nemec, B.; Zlajpah, L. Shoe Grinding Cell using Virtual Mechanism Approach. In Proceedings of the ICINCO-RA, Madeira, Portugal, 11–15 May 2008; pp. 159–164.
5. Hu, Z.; Marshall, C.; Bicker, R.; Taylor, P. Automatic surface roughing with 3D machine vision and cooperative robot control. *Robot. Auton. Syst.* **2007**, *55*, 552–560. [CrossRef]
6. Kim, M.G.; Kim, J.; Shin, D.; Jin, M. Robot-based Shoe Manufacturing System. In Proceedings of the 2018 18th International Conference on Control, Automation and Systems (ICCAS), Daegwallyeong, Korea, 17–20 October 2018; pp. 1491–1494.
7. Pagano, S.; Russo, R.; Savino, S. A vision guided robotic system for flexible gluing process in the footwear industry. *Robot. Comput. Integr. Manuf.* **2020**, *65*, 101965. [CrossRef]
8. Pagano, S.; Russo, R.; Savino, S. A Smart Gluing Process by a Vision Guided Robotic System. *Mech. Mach. Sci.* **2021**, *91*, 414–422. [CrossRef]
9. Lho, T.J.; Park, P.G.; Suh, J.C.; Park, D.J.; Ahn, H.T. A Study on Development of 3D Outsole Profile Scanner for Footwear Bonding Automation. *Control. Robot. Syst. Soc. Conf. Pap.* **2001**, 131–133.
10. Oliver, G.; Gil, P.; Torres, F. Robotic workcell for sole grasping in footwear manufacturing. In Proceedings of the IEEE International Conference on Emerging Technologies and Factory Automation, ETFA, Vienna, Austria, 8–11 September 2020; pp. 704–710. [CrossRef]
11. Choi, H.S.; Hwang, G.D.; You, S.S. Development of a new buffing robot manipulator for shoes. *Robotica* **2008**, *26*, 55–62. [CrossRef]
12. Dulio, S.; Boër, C. Integrated production plant (IPP): An innovative laboratory for research projects in the footwear field. *Int. J. Comput. Integr. Manuf.* **2004**, *17*, 601–611. [CrossRef]

13. Eryilmaz, M.S. Analysis of shoe manufacturing factory by simulation of production processes. *Southeast Eur. J. Soft Comput.* **2012**, *1*. [CrossRef]
14. Staikos, T.; Rahimifard, S. An End-of-Life Decision Support Tool for Product Recovery Considerations in the Footwear Industry. *Int. J. Comput. Integr. Manuf.* **2007**, *20*, 602–615. [CrossRef]
15. Martín-Martínez, J.M. Shoe Industry. In *Handbook of Adhesion Technology*; Springer International Publishing: Cham, Switzerland, 2018; pp. 1483–1532. [CrossRef]
16. Danese, G.; Dulio, S.; Giachero, M.; Leporati, F.; Nazzicari, N. A Novel Standard for Footwear Industrial Machineries. *IEEE Trans. Ind. Inform.* **2011**, *7*, 713–722. [CrossRef]
17. Lee, C.Y.; Kao, T.L.; Wang, K.S. Implementation of a robotic arm with 3D vision for shoes glue spraying system. In Proceedings of the 2018 2nd International Conference on Computer Science and Artificial Intelligence, Shenzhen, CHina, 8–10 December 2018; pp. 562–565. [CrossRef]
18. Lee, J.D.; Hsieh, C.Y.; Jhao, Y.J.; Chang, C.H.; Li, C.Y.; Li, W.C. Implementation of automated gluing and assembly workstation. In Proceedings of the 2018 IEEE International Conference on Advanced Manufacturing, ICAM 2018, Yunlin, Taiwan, 16–18 November 2018; pp. 276–279. [CrossRef]
19. Caruso, L.; Russo, R.; Savino, S. Microsoft Kinect V2 vision system in a manufacturing application. *Robot. Comput. Integr. Manuf.* **2017**, *48*, 174–181. [CrossRef]
20. Bhatt, P.M.; Malhan, R.K.; Shembekar, A.V.; Yoon, Y.J.; Gupta, S.K. Expanding capabilities of additive manufacturing through use of robotics technologies: A survey. *Addit. Manuf.* **2020**, *31*, 100933. [CrossRef]
21. Castelli, K.; Giberti, H. A preliminary 6 Dofs robot based setup for fused deposition modeling. *Mech. Mach. Sci.* **2019**, *68*, 249–257. [CrossRef]
22. Rane, K.; Castelli, K.; Strano, M. Rapid surface quality assessment of green 3D printed metal-binder parts. *J. Manuf. Process.* **2019**, *38*, 290–297. [CrossRef]
23. Flow Math. Available online: https://manual.slic3r.org/advanced/flow-math (accessed on 30 July 2020).
24. Zhao, G.; Ma, G.; Xiao, W.; Tian, Y. Feature-based five-axis path planning method for robotic additive manufacturing. *Proc. Inst. Mech. Eng. Part B J. Eng. Manuf.* **2018**, *233*, 1412–1424. [CrossRef]
25. Kiani, M.A.; Saeed, H.A. Automatic Spot Welding Feature Recognition From STEP Data. In Proceedings of the 2019 International Symposium on Recent Advances in Electrical Engineering (RAEE), Islamabad, Pakistan, 28–29 August 2019; Volume 4, pp. 1–6. [CrossRef]
26. Singamneni, S.; Roychoudhury, A.; Diegel, O.; Huang, B. Modeling and evaluation of curved layer fused deposition. *J. Mater. Process. Technol.* **2012**, *212*, 27–35. [CrossRef]
27. Jee, H.; Kim, M. Spherically curved layer (SCL) model for metal 3-D printing of overhangs. *J. Mech. Sci. Technol.* **2017**, *31*, 5729–5735.' [CrossRef]
28. Ericson, C. *Real-Time Collision Detection*; Morgan Kaufmann Series in Interactive 3D Technology; Taylor & Francis: Abingdon, UK, 2004.
29. Jatta, F.; Zanoni, L.; Fassi, I.; Negri, S. A roughing/cementing robotic cell for custom made shoe manufacture. *Int. J. Comput. Integr. Manuf.* **2004**, *17*, 645–652. [CrossRef]
30. Edelsbrunner, H. Alpha Shapes—A Survey. *Tessellations Sci.* **2010**, *27*, 1–25.
31. Liu, X.Z.; Yong, J.H.; Zheng, G.Q.; Sun, J.G. An offset algorithm for polyline curves. *Comput. Ind.* **2007**, *58*, 240–254. [CrossRef]
32. Tsai, L.W. *Robot Analysis: The Mechanics of Serial and Parallel Manipulators*; Wiley-Interscience: Hoboken, NJ, USA, 1999.
33. Hawkins, K.P. *Analytic Inverse Kinematics for the Universal Robots ur-5/ur-10 Arms*; Technical Report; Georgia Institute of Technology: Atlanta, GA, USA, 2013.
34. Carnevale, M.; Castelli, K.; Zaki, A.M.A.; Giberti, H.; Reina, C. Automation of Glue Deposition on Shoe Uppers by Means of Industrial Robots and Force Control. *Mech. Mach. Sci.* **2021**, *91*, 344–352. [CrossRef]
35. *Robots and Robotic Devices—Safety Requirements for Industrial Robots—Part 1: Robots*; ISO 10218-1:2011; ISO: Geneva, Switzerland, 2011.
36. Yang, Z.; An, Y.; Sun, Y.; Zhang, J. Research on intelligent glue-coating robot based on visual servo. *Phys. Procedia* **2012**, *24*, 2165–2171. [CrossRef]
37. Ting, Y.; Chen, C.H.; Feng, H.Y.; Chen, S.L. Glue dispenser route inspection by using computer vision and neural network. *Int. J. Adv. Manuf. Technol.* **2008**, *39*, 905–918. [CrossRef]
38. Kruesubthaworn, A. Inspection of Defect on Magnetic Disk Surface and Quality of the Glue Dispenser Route. *Vis. Insp. Technol. Hard Disc Drive Ind.* **2015**, 237–263. [CrossRef]
39. Maurtua, I.; Ibarguren, A.; Tellaeche, A. Robotic solutions for footwear industry. In Proceedings of the 2012 IEEE 17th International Conference on Emerging Technologies & Factory Automation (ETFA 2012), Krakow, Poland, 17–21 September 2012; pp. 1–4. [CrossRef]
40. Maurtua, I.; Ibarguren, A.; Tellaeche, A. Robotics for the benefit of footwear industry. In Proceedings of the International Conference on Intelligent Robotics and Applications, Montréal, QC, Canada, 3–5 October 2012; pp. 235–244. [CrossRef]

Author Index

MDPI

St. Alban-Anlage 66

4052 Basel

Switzerland

Tel. +41 61 683 77 34

Fax +41 61 302 89 18

www.mdpi.com

Robotics Editorial Office

E-mail: robotics@mdpi.com

www.mdpi.com/journal/robotics